GROL Plus

General Radiotelephone Operator License
Plus Radar Endorsement

FCC Commercial
Radio License Preparation
Element 1, Element 3
and Element 8
Question Pools

BY

FRED MAIA, W5YI

and

GORDON WEST, WB6NOA

First Edition, 6th Printing, Updated September, 2004

This book was developed and published by:
Master Publishing, Inc.
Lincolnwood, Illinois

Editing by:
Gerald Luecke, KB5TZY
Bob Hurst
Charles Battle

Technical Review:
Roger Boettcher

Printing by:
Arby Graphic Service, Inc.
Lincolnwood, Illinois

Acknowledgements:
All photographs that do not have a source identification are either courtesy
of authors or Master Publishing, Inc. originals.

The authors wish to thank Capt. John L. Thompson, Head, Navigation Section,
Maritime Safety Division, IMO (United Nations), London for his valuable GMDSS
input and documentation, and Matt J. McCullar, KJ5BA, Win Guin, W2GLJ, Harry
Hess, K9MDK, Ron Earl, W6TXK, Brian Borkenhagen, KD6EDM, Frank W. Pfeiffer,
Richard Arland, CE #19329; Nicholas Buchko, KD4SMW; Charles Marcel Klein
of the S.S. United States Preservation Society; Manuel W. Woodall, CET; Patrick
Connally, CET, and Alan Martin, KB3HIP, for their extensive research, text sugges-
tions, contributions and reviews.

The authors wish to acknowledge the fine cooperation of the FCC staff at the
Washington, DC and Gettysburg, PA offices.

A print of the Hoover photograph in Chapter 1 was obtained from the Herbert
Hoover Presidential Library-Museum. Supposedly, Henry Miller News Picture
Service was original copyright owner; however, no such listing is available from the
copyright office.

Trademarks:
Teflon® is a registered trademark of E.I. DuPont DeNemours Co., Inc.
Bakelite® is a registered trademark of Union Carbide Corporation.
MasterCard® is a registered trademark of MasterCard International, Inc.
VISA® is a registered trademark of VISA U.S.A. Inc.
RCA® is a registered trademark of Radio Corporation of America.

First Edition
9   8   7   6

Table of Contents

QUESTION POOL RELEASE AND NOMENCLATURE

The Element 1, Element 3, and Element 8 question pools were released by the FCC on October 1, 1995 and became effective on January 1, 1996. Our interpretation and implementation may be different from other publishers. We are sure that any slight differences will not contribute to improper understanding of the question or its answer. The question pools included in this book are the complete, current pools used to create examinations as of the date of publication of this book.

PUBLISHER'S NOTE:

Since *GROL Plus* was first published in spring, 1996, it has come to our attention that a few questions in the FCC-approved question pools included in this book are incomplete or stated incorrectly. Because the COLEMs that administer the examinations must use the questions as stated in the FCC-approved question pools, we have added an asterisk (*) to mark these questions. In addition, an Editor's Note has been added to explain the error in the way the question is asked or to provide additional information so the reader will have a correct understanding of the concept. —MPI

Preface

Welcome and congratulations on taking a very important step—your decision to prepare for a commercial radio operator's license. Gaining a commercial radio license is official certification by the Federal Communications Commission (FCC) that you have what it takes to operate and/or maintain commercial radio equipment.

Gaining a commercial license is not easy. The examinations are long and difficult. You will need to work hard to learn about operating rules and regulations, electronic devices and their use in electronic circuits, and maintaining electronic equipment.

With your license, you can participate in the rewards—financial security, a wise addition to your resume, increased technical knowledge, personal satisfaction, and, if you're already an amateur, another ticket for the wall of the shack.

GROL Plus contains Element 1, Element 3 and Element 8 question pools. Everything you need to pass the written examinations required for a Marine Radio Operator Permit (MP) or a General Radiotelephone Operator License (PG) is included. Passing the Element 1 and Element 3 examinations not only gains you your MROP or GROL, but puts you in position to add a Ship Radar Endorsement and earn additional commercial licenses. For example, an Element 1 examination is a partial requirement for the Global Maritime Distress and Safety System Radio Operator (DO) and Radio Maintainer (DM) licenses, and the three Radiotelegraph Operator Certificates, T1, T2 and T3. An Element 3 examination is a partial requirement for the DM license. Passing an Element 8 examination earns the examinee a radar endorsement on a PG, DM, a T1, or a T2 license.

GROL Plus begins by providing a brief history of commercial radio license regulation in Chapter 1. Chapter 2 lists the commercial licenses available today, spells out when you do or do not need a commercial radio license, and defines which permit, certificate, license, or endorsement is required. Chapter 3 details which examinations must be passed to earn a particular license. Chapter 4 explains GMDSS. The Element 1 question pool of 170 questions is Chapter 4, the Element 3 question pool of 916 questions is Chapter 5 and the Element 8 question pool of 321 questions is Chapter 8. Chapter 7 provides details on radar systems. The book proper concludes with Chapter 9, which gives details about taking your examination. To aid in your study, especially of Element 1, and to have important FCC Rules and Regulations on hand, the Appendix contains the complete FCC Part 13, and excerpts from Parts 23, 73, 80 and 87.

So step up, study hard, and congratulations will come your way when you pass your examination(s) and have your commercial license in hand.

Fred Maia W5YI
Gordon West WB6NOA

1

History of Radio Regulation

INTRODUCTION

Communications by radio has been called the most regulated business in the land. It is easy to understand why. Progress cannot continue without reliable communication, and this has been understood since biblical times. Many people choose to live without electronic communication, but no one would choose to live with unregulated electronic communication.

IN THE BEGINNING

Table 1-1 lists some significant events and important actions between 1835 and 1910 that furthered electronic communications, which started with the invention of the telegraph. *Figure 1-1* shows a typical system.

Table 1-1. Electronic Communications 1835-1910

Year	Event and Action
1835	Electronic communications began with the invention of the telegraph by Samuel F.B. Morse, a professor at New York University. Morse code is named after him, and is still the international CW code used today.
1849	Two European countries were linked by telegraph. This caused initial international agreements on rules and regulations governing sharing of information.
1850	Other treaties followed.
1865	25 European nations met in Paris and formed the International Telegraphic Convention. Later it became the International Telecommunications Union (ITU).
1865	Guglielmo Marconi, an Italian inventor, proved the feasibility of radio communications for which he received a patent in 1897.
1899	Marconi demonstrated the first transatlantic radio transmission from England to Newfoundland.
1901	Marine radio was born when the U.S. Navy adopted a wireless system.
1903	First international conference on governing radio communications was held in Berlin. Nine nations agreed that public safety took precedence over squabbles between commercial ventures.
1906	International conference in Berlin agreed to require ships to be properly equipped with wireless transmitters and receivers, and to set the first international distress frequency as 500 kHz for ships to use to call for help. Wireless radio was added to ITU's responsibilities.
1909	Steamships *Republic* and *Florida* collided off the coast of New York and 1500 lives were saved by a distress call sent by radio operator Jack Binns. Later in the year, the *S.S. Arapahoe* brought help with "SOS", which was adopted this year as the international radiotelegraph distress call. It is still in use today. "Mayday" was adopted in 1927 as the international distress call for radiotelephony.

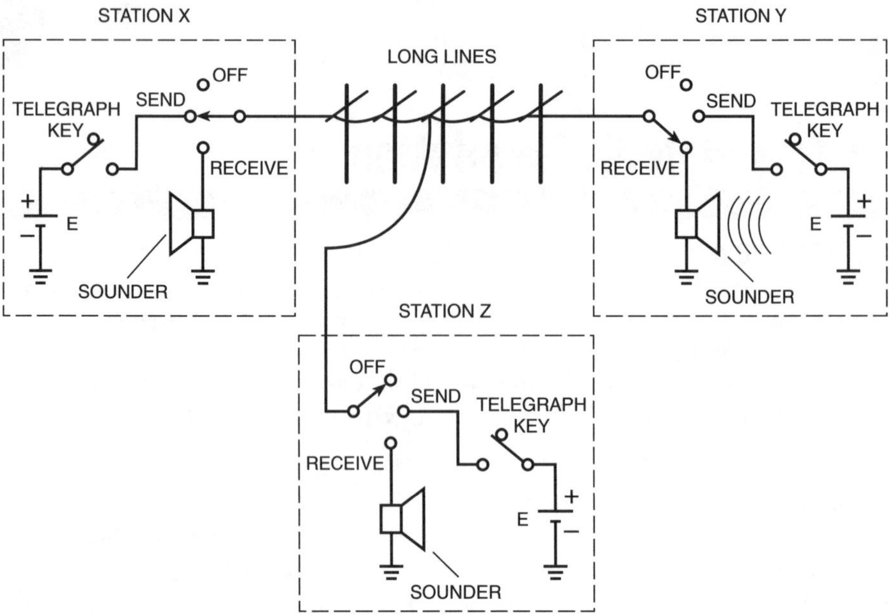

Figure 1-1. Early Long-Line Telegraph (Station X Sending to Station Y) System

AN ACCIDENT BRINGS CHANGE

In the early years of the 20th century, wireless communications remained confined primarily to the sea. Yet not many ships operated radio equipment, and those that did saw no logic in staffing it 24 hours a day. To do so was considered a luxury. That changed after the terrible night in April 1912 when the *Titanic* was ripped open by an iceberg in the North Atlantic and sent to the bottom of the sea three hours later.

The radio operator on board that massive ship frantically called for help over wireless, and the *Carpathia,* 58 miles away, responded and managed to rescue 700 survivors. Some 1,500 passengers perished! Later inquiries learned that another ship, called the *California,* had been only 20 miles away from the *Titanic* at the time of the accident and could have rushed to the scene much faster. Why didn't it? Because the *California's* radio operator had gone to bed, and there was no one to relieve him.

In the United States, *The Wireless Ship Act of 1910* — passed two years before the *Titanic* disaster — should have prevented the loss of so many lives. But not until after 1912 was this Act amended to require at least two operators and a constant watch, with emergency backup power supplies.

THE BEGINNINGS OF RADIO LEGISLATION

Partially as a result of the *Titanic* disaster, the United States Congress enacted the first law for the domestic control of general radio communications — the *Radio Act of 1912*. It regulated the character of emissions, as well as the transmission of distress calls, set aside certain frequencies for government use, and placed the control of all wireless stations under the jurisdiction of the Department of Commerce. The Act was the beginning of government radio licensing. It made access to the electromagnetic spectrum a privilege granted only by government approval.

The *Radio Act of 1912* was praised by many, but not by all. Radio was so new that very few understood its potential. Some barely knew what "wireless" meant. Lee de Forest, the inventor of the Audion three-element vacuum tube amplifier—a very essential ingredient in the advancement of both wired and wireless communications—was prosecuted for using the mails to defraud. The prosecutor accused de Forest of, "...willfully and deliberately misleading the public by stating that soon it will be possible to transmit the human voice across the Atlantic Ocean!"

At the time the *Radio Act of 1912* was passed, the radio spectrum was so spacious and so unoccupied that no one thought that channels would ever have to be assigned, much less share frequencies. If you wanted to operate a transmitter, there was plenty of room in which to do it. All you had to do was apply for a license. The fact that the man in charge of issuing licenses had no legal right to turn down anyone would become an important point years later in court.

WORLD WAR I: CLOSED FOR THE DURATION

When the United States entered the Great War, the Woodrow Wilson administration released two edicts with regard to communications. First, all commercial radio stations would fall under direct government operation for the duration of the war. Second, all amateur radio operators must not only sign off the air until further notice, but also dismantle all of their equipment under penalty of imprisonment. In addition, major radio manufacturers pooled their knowledge and expertise into getting the war won.

THE GREAT RADIO RUSH

World War I ended in November, 1918 and the massive military market for radio receivers came to an end. When the Radio Corporation of America (RCA®) acquired the American Marconi company in 1919, David Sarnoff became its manager and convinced the company to go into the radio business. Sarnoff, a young wireless operator, even before World War I had contemplated a "radio music box" that would receive programs broadcast for public information and entertainment. At about the same time, General Electric and Westinghouse also began making radio receivers. At first, the public demand for receivers was small because there were few stations to listen to. That changed quickly as licensed stations went on the air and began regular broadcasting.

At the end of 1920, only 30 radio stations in the U.S. offered regular broadcasts, and all of them used amplitude modulation (AM). Pittsburgh's 8XK (an amateur radiotelephone station set up in 1916 by Westinghouse engineer Dr. Frank Conrad) became KDKA, the nation's first commercial broadcast station working on a wavelength of 360 meters. It signed on the air on election night, November 2, 1920, and today is the nation's oldest radio station still in operation. Licensing of broadcast stations on a regular basis began in 1921 with WBZ, Springfield, Mass., being the first station licensed. By the end of 1922 there were 382 stations, and by 1927 there were 733 stations! Most were operated by radio manufacturers, dealers and department stores selling receivers. The 1920s brought forth a virtual explosion of growth in the radio industry.

It seems almost impossible to believe that there was once a time when millions of citizens lived their entire lives without ever once hearing the voices of the leaders of their own country. But with a radio, there was so much that could be brought right into the home. Lonely and isolated people eagerly snapped them up to hear what was going on in the world around them. Thousands of musical novices could hear operas on the radio almost every night. Getting the news faster than newspapers could report it was truly exciting, and getting election results in hours instead of days was almost unbelievable.

WHEN BEDLAM REIGNS

The constant flood of radio stations joining the airwaves proved to be overwhelming by the middle of the 1920s. From 1923 to 1926, Secretary of Commerce Herbert Hoover (an engineer in his own right) submitted and resubmitted bills before Congress to straighten out the radio legislative process. Hoover gathered radio conferences each year from 1922 to 1925 in an effort to entice voluntary cooperation from the radio industry. He even listened in on a special receiver at his home as shown in *Figure 1-2,* to better understand the needs of the listening public.

Secretary Hoover (who later became President Hoover) understood that the radio world could no longer be adequately controlled by the *Radio Act of 1912.* As the Federal government's man in charge, he used his power to grant or deny licenses, assign frequencies, and dictate the time of day when a station could operate.

Although radio stations held licenses, they began to bend the rules. Stations lengthened their working hours, changed frequencies, and increased output power without authorization. When one of these station's owners, Zenith, was taken to court, Zenith charged that the Secretary of Commerce had no legal authority to tell them when or

Figure 1-2. Commerce Secretary Herbert Hoover listening to radio receiver installed in his home so he could better understand complaints received by his department.
(Print Received from Herbert Hoover Presidential Library-Museum)

where to transmit a radio signal. On July 8, 1926, the acting Attorney General of the United States decreed that the Secretary of Commerce had no legal authority to assign wavelengths, power levels, or hours of operation, or to restrict the length of a station's license. Officially, no government agency controlled radio for the next six months.

The predictable happened. Without regulations, hundreds of stations had a field day. They cranked up power levels and changed frequencies whenever they desired. New stations, completely unlicensed, went on the air. The result was chaos. There was so much interference that millions of listeners all across America switched off their receivers in disgust, and the sale of new receivers slowed to a trickle.

A NEW AGENCY — THE FRC

It was the lack of income from sales of radio receivers that finally convinced the radio industry that some form of legislation and government control was the answer. This time they were glad to meet with Secretary Hoover. In February, 1927, Congress enacted the *Radio Act of 1927*. The new law created an agency known as the Federal Radio Commission (FRC). Led by a five-man panel, the FRC had power, and used it.

The FRC was granted legal authority to decide how much of the radio spectrum each service would get, and to change it if necessary. License applications could now be legally refused to be granted or renewed. Engineers were placed in charge to keep up with the latest scientific developments, and incorporate them into the rules and regulations as necessary. So sweeping were the changes, and so well were they received, that the *Radio Act of 1927* has been called radio's "Magna Charta."

Also taking place in 1927 was the International Radiotelegraph Convention in Washington. Practically every nation in the world joined in, deciding as a whole who would use what portions of the shortwave bands for different purposes.

THE FINAL REWRITE — THE FCC

By the beginning of the 1930s, it became clear that the FRC, while a good beginning, needed to be expanded to include regulation of other forms of electrical communication. The result was the famous *Communications Act of 1934*; it contained much of the 1927 Act, with several major additions. The FRC was abolished, and in its place the Federal Communications Commission (FCC) was born. Not only would it regulate radio, but also telegraph and telephone services.

Among the duties of the FCC are allocating radio frequency bands along international guidelines, assigning frequencies for various radio services and individual stations, regulating common carriers involved in foreign and interstate commerce, and determining the operational and technical qualifications of radio operators.

The FCC was created from the start to be a powerful agency. Standards were explicitly laid out, and stations that failed to follow them were subject to criminal prosecution. The *Communications Act of 1934* itself is only about 150 pages; however, regulations set up under the law currently occupy several thousand pages. To this day, no radio or television station in the United States can be sold, moved, shut down, or change its operating hours or power level without express permission from the FCC. Because of licensing, the FCC's word is law on whether or not a station can legally exist. In the first 50 years of FCC operation, only 149 stations lost their broadcast licenses. These days the FCC imposes massive fines on stations that break the rules (such as the broadcasting of hoaxes).

FREQUENCY MODULATION

Even though the principle of frequency modulation (FM) was known previously, a patent on FM was not issued until 1902. In addition, its advantages for broadcasting were not developed until much later — shortly before World War II — largely as a result of developmental work by Edwin H. Armstrong. It was not until late 1940 that the FCC granted construction permits for FM stations and provided for authorized commercial FM broadcasting to start January 1, 1941. Several licensed commercial FM stations were operating by the middle of 1941. The first licensed station to secure full commercial status was W47NV[1] at Nashville, TN, which began operation on a regular schedule on March 1, 1941.

[1] Radio and Television Weekly, N.Y., NY, March 19, 1941 issue.

LANDMARK FCC DECISIONS

- Frequency-modulation was first presented to the FCC in 1935.
- Spectrum space was set aside for television in 1939.
- The FCC created the National Television System Committee (NTSC) to develop standards for monochrome television transmission.
- VHF channels 1–13 were assigned. Later, channel 1 was reassigned to land-mobile communication. Remaining channels, 2-13, are in use today.
- Full NTSC monochrome television operation began in 1941.
- 70 new UHF channels were assigned for TV broadcast in 1952.
- First UHF station signed on in 1952.
- Color TV standards were adopted in 1953.
- Stereo FM was approved in 1961.
- It was decreed in 1964 that all TV sets hence forth would be equipped with UHF tuning.
- First maximum-power (5 megawatts) UHF station came on the air in 1974.

KEEPING UP WITH THE TIMES

The President of the United States appoints the five FCC members. No more than three of them may be from the same political party. The five-year terms are staggered so that no one can replace all of them at once. FCC operations are continually monitored by Congress, and, because the FCC is no longer a permanent agency, it must be reauthorized by Congress every two years.

As technology changes, the FCC works steadily to keep up. Cable television, digital data transmission over the telephone lines, satellite broadcasting, high-definition television (HDTV)—all fall under their jurisdiction. Whenever a new format is invented and brought to the FCC's attention, the FCC examines it and makes rules accordingly.

INTERNATIONAL REGULATION

Almost every nation in the world belongs to the International Telecommunication Union (ITU). This is a United Nations agency, headquartered in Geneva, Switzerland. Since radio waves do not respect national boundaries, global telecommunications standards are agreed upon through the ITU. The allocation of radio frequencies consists of setting aside segments or bands of the radio spectrum for the use of particular radio services and assigning specific frequencies within those segments for the operation of individual radio stations.

The ITU divides the world into three general regions. The allocation plan for North America conforms to those for Region 2 shown in *Figure 1-3*. As an aid to enforcement of radio laws, radio station identify themselves using call signs that begin with certain ITU assigned prefixes. All U.S. stations begin with K, N, W and certain A prefixes.

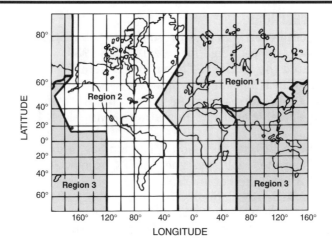

Figure 1-3. ITU Regions

Major World Administrative Radio Conferences (WARCs) used to be held about every ten years to hammer out shares of the radio spectrum. The latest developments in technology help determine who gets to use which frequencies, and for what purposes. As a result, in recent years, the ITU has convened smaller, more specialized World Radio Conferences (WRCs) to deal with the fast changing world of electronic telecommunications. These WRCs have limited agendas and primarily cover specific radio services. WRC-95's major focus was on the mobile satellite services. And agendas to WRC-97 and WRC-99 are already being formed.

The ITU agreements are fine tuned by individual governments. In the United States, the FCC regulates all non-government radio frequencies. Federal government spectrum is under the jurisdiction of the President and managed by the National Telecommunications and Information Administration (NTIA).

IN CONCLUSION

In some respects, spectrum management can be likened to building highways. The planning and development of invisible communication lanes, however, is much more complicated than road building. There is still a wide choice of land highway routes but radio paths are limited in number and many are crowded, and unlike land traffic, radio transmissions cannot be routed by underpasses and overpasses. Neither can they obey traffic signals, for radio waves spread out in all directions, crossing state lines and international borders.

It is for the benefit of the world community that international and national standards be agreed upon, and that engineers and spectrum users follow them. Lives are saved and bettered every day because of this. Anyone studying to further the progress of technical communication should be applauded. For the world to march forward, we must listen to our engineers. And we must listen to each other.

2

Commercial Radio Operator's Licenses – Then and Now

LICENSES AND EXAMINATIONS

There are two basic classes of licensed radio operators, amateur and commercial. Both licensing programs exist for the same reasons: to reduce interference to other radio stations; to insure technician and operator qualifications; and to bring order to use of the radio spectrum. As a general rule, the FCC issues a radio license when an applicant passes an examination demonstrating adequate knowledge of radio rules, operating procedures and electronics. An amateur license is required when the transmissions are for hobby purposes. A commercial license is needed when the radio equipment is used in connection with a business venture.

Besides being able to pass the necessary radio law and technical examinations, applicants for commercial radio licenses must be citizens of the United States (or eligible for employment in the United States) and be able to transmit and receive messages in English. Radiotelephone licenses permit operation, maintenance and repair of any type of transmitting equipment not transmitting Morse code. Before 1985, all radiotelephone and radiotelegraph licenses were issued for a term of five years.

MAJOR CHANGES MADE IN 1984!

On February 24, 1984, the Federal Communications Commission voted to make major changes to its commercial radio operator licensing program. On June 15, 1984, the following amendments were made:
1. The First Class and Second Class Radiotelephone Operator Licenses were discontinued and replaced with a special lifetime General Radiotelephone Operator License. The Third Class Radiotelephone Operator Permit, Aircraft Radiotelegraph Endorsement and the Broadcast Endorsement were eliminated and not replaced.
2. All domestic operating license requirements for the private two-way radio services were abolished. That meant that anyone — commercially licensed or not — could now install, maintain or repair transmitting equipment in any of the private two-way radio services. The licensee of the station is responsible for the proper operation of the station. The radio services affected are
   a. Private Land Mobile Radio Service – Part 90
   b. Private Operational Fixed Microwave Service – Part 94
   c. General Mobile Radio Service (GMRS) – Part 95 – Subpart A

d. Radio Control Services (R/C) – Part 95 – Subpart C

e. Citizens Band Radio Service (CB) – Part 95 – Subpart D

3. Commercial radio licenses still required by international law will be continued indefinitely. These radio services are

a. Aviation Services – Part 87

b. Maritime Services – Part 80

c. International Fixed Service – Part 23

4. A new Marine Radio Operator Permit (MP) was established to be required by operators:

a. aboard certain vessels navigating the Great Lakes, any tidewater or in the open sea,

b. certain aviation radiotelephone stations, and

c. certain maritime coast radiotelephone stations.

WHO NEEDS A COMMERCIAL RADIO OPERATOR LICENSE?

The answer depends on whether you wish to operate only, or to also repair and maintain radio stations.

Radio Operations

You *need* a commercial radio operator license to operate the following:

1. Ship radio stations if:

a. the vessel carries more than six passengers for hire; or

b. the radio operates on medium or high frequencies (300 kHz to 30 MHz); or

c. the ship sails to foreign ports; or

d. the ship is larger than 300 gross tons and is required to carry a radio station for safety purposes.

2. Coast (land) stations which operate on medium or high frequencies, or operate with more than 1500 watts of peak envelope power.

3. Aircraft radio stations, except those that use only VHF frequencies (higher than 30 MHz) on domestic flights;

4. Civil Air Patrol stations on other than VHF frequencies;

5. International fixed public radiotelephone and radiotelegraph stations; and

6. Coast and ship stations transmitting radiotelegraphy.

You *do not need* a commercial radio operator license to operate the following:

1. Coast stations operating on VHF frequencies with 250 watts or less of carrier power.

2. Ship stations operating only on VHF frequencies while sailing on domestic voyages.

3. Aircraft stations which operate only on VHF frequencies and do not make foreign flights.

Radio Maintenance and Repair

You *need* a commercial radio operator license to repair and maintain the following:

1. All ship radio and radar stations,
2. All coast radio stations,
3. All hand-carried units used to communicate with ships and coast stations on marine frequencies.
4. All aircraft stations and aeronautical ground stations (including hand-carried portable units) used to communicate with aircraft.
5. International fixed public radiotelephone and radiotelegraph stations.

You *do not need* a commercial radio operator license to operate, repair or maintain any of the following types of stations:

1. Two-way land mobile radio equipment such as that used by police and fire departments; taxicabs and truckers; businesses and industries; ambulances and rescue squads; local, state and federal government agencies.
2. Personal radio equipment used in the Citizens Band, Radio Control, and General Mobile Radio Services.
3. Auxiliary broadcast stations such as remote pickup stations.
4. Domestic public fixed and mobile radio systems such as mobile telephone systems, cellular systems, rural radio systems, point-to-point microwave systems, multipoint distribution systems, etc.
5. Stations operated in the Cable Television Relay Service.
6. Satellite stations, both uplink and downlink, of all types.
7. AM, FM, TV Broadcast stations. (Eliminated by FCC on 12/01/95. Be aware that, although not required, many licensees of broadcast stations require their engineers and technicians to hold a General Radiotelephone Operator License.)

Important Caution!

These listings only describe when radio *operator* licenses are necessary. *Before you operate any radio* ***station****, make certain that the* ***station*** *is licensed as required.* CB and Radio Control radio stations do not require individual station licenses.

The holder of a commercial radio operator license or permit is not authorized to operate amateur radio stations. Only a person holding an amateur radio operator license may operate an amateur radio station.

TYPES OF LICENSES, PERMITS AND ENDORSEMENTS

The FCC now issues 11 types of commercial radio operator licenses and permits, and two types of endorsements. They are listed in *Table 2-1* through *Table 2-11.* Commercial Radio Operator licenses normally carry a term of five years. The exceptions are the General Radiotelephone Operator License (PG) and the Restricted Radiotelephone Operator Permits (RP), which are valid for the lifetime of the holder.

Licenses may be renewed from anytime in the last year of their term up to five years following expiration without having to retake an examination. After the five year grace period expires, applicants must retake the requisite written and/or telegraphy examination(s). An expired certificate is not valid for any radio operations.

Table 2-1. Restricted Radiotelephone Operator Permit (RP)

Restricted Radiotelephone Operator Permit (RP) holders are authorized to operate most aircraft and aeronautical ground stations. They also can operate marine radiotelephone stations aboard pleasure craft (other than those carrying more than six passengers for hire on the Great Lakes or bays or tidewaters or in the open sea) when operator licensing is required. RPs are issued for the lifetime of the holder.

Residents
An RP is issued without any examination. To qualify for an RP, you *must meet all four* of the following requirements:
1. Be either a legal resident of (or otherwise eligible for employment in) the United States, or hold an aircraft pilot certificate valid in the United States, or hold an FCC radio station license in your own name (see limitation on validity below).
2. Be able to speak or hear.
3. Be able to keep at least a rough written log.
4. Be familiar with provisions of applicable treaties, laws and rules which govern the radio station you will operate.

You *do not need* an RP to operate the following:
1. A voluntarily-equipped ship or aircraft station (including a Civil Air Patrol (CAP) station) which operates only on VHF frequencies and does not make foreign voyages or flights.
2. An aeronautical ground or coast station which operates only on VHF frequencies.
3. On-board stations.
4. A marine utility station, unless it is taken aboard a vessel which makes a foreign voyage.
5. A survival craft station, when using telephony or an emergency position-indicating radio beacon (EPIRB).
6. A ship radar station, if the operating frequency is determined by a fixed-tuned device and if the radar is capable of being operated by only external controls.
7. Shore radar, shore radiolocation, maritime support or shore radio-navigation stations.

Non-Residents
If you are a non-resident alien, you must hold *at least one* of the following three documents to be eligible for an RP:
1. A valid United States pilot certificate issued by the Federal Aviation Administration.
2. A foreign aircraft pilot certificate which is valid in the United States on the basis of reciprocal agreements with foreign governments.
3. A valid radio station license issued by the FCC in your own name. (An RP issued on this basis will authorize you to operate *only* your own station.)

Table 2-2. Marine Radio Operator Permit (MP)

A Marine Radio Operator Permit (MP) is required to operate radiotelephone stations aboard certain vessels that sail the Great Lakes. It also is required to operate radiotelephone stations aboard vessels of more than 300 gross tons and vessels which carry more than six passengers for hire in the open sea or any tidewater area of the United States. It is also required to operate certain aviation radiotelephone stations and certain coast radiotelephone stations.

You *must meet all three* of the following requirements to be eligible for a MP:
1. Be a legal resident of (or eligible for employment in) the United States.
2. Be able to receive and transmit spoken messages in English.
3. Pass a written examination covering basic radio law and operating procedures (Element 1).

Table 2-3. General Radiotelephone Operator License (PG)

A General Radiotelephone Operator License (PG) is required to adjust, maintain or internally repair FCC-licensed radiotelephone transmitters in the aviation, maritime and international fixed public radio services. It conveys all of the operating authority of the Marine Radio Operator Permit (MP). A GROL is issued for the lifetime of the holder.

A GROL is also required to operate the following:
1. Any maritime land radio station or compulsorily equipped ship radiotelephone station operating with more than 1500 watts of peak envelope power.
2. Voluntarily-equipped (pleasure) ship and aeronautical (including aircraft) stations with more than 1000 watts of peak envelope power.

You *must meet all three* of the following requirements to be eligible for a GROL:
1. Be a legal resident of (or otherwise eligible for employment in) the United States.
2. Be able to receive and transmit spoken messages in English.
3. Pass a written examination covering basic radio laws and operating procedures (Element 1), and electronics fundamentals and techniques required to repair and maintain radio transmitters and receivers (Element 3).

Table 2-4. Third Class Radiotelegraph Operator's Certificate (T3)

The Third Class Radiotelegraph Operator's Certificate (T3) authorizes operation of certain coast radiotelegraph stations. It also confers the operating authority of both the Restricted Radiotelephone Operator Permit (RP) and the Marine Radio Operator Permit (MP).

You *must meet all four* of the following requirements to be eligible for a T3:
1. Be a legal resident of (or otherwise eligible for employment in) the United States.
2. Be able to receive and transmit spoken messages in English.
3. Pass Morse code examinations at 16 code groups per minute (16 CG, Telegraphy Element 1) and 20 words-per-minute plain language (20 PL, Telegraphy Element 2) — both receiving and transmitting by hand*.
4. Pass written examinations covering basic radio law and operating procedures for radiotelephony (Element 1), and basic operating procedures for radiotelegraphy (Element 5).

Table 2-5. Second Class Radiotelegraph Operator's Certificate (T2)

The Second Class Radiotelegraph Operator's Certificate (T2) authorizes the holder to operate, repair and maintain ship and coast radiotelegraph stations in the maritime services. It also confers all of the operating authority of the Restricted Radiotelephone Operator's Permit (RP), the Marine Radio Operator's Permit (MP) and the Third Class Radiotelegraph Operator's Certificate (T3).

You *must meet all four* of the following requirements to be eligible for a T2:
1. Be a legal resident of (or otherwise eligible for employment in) the United States.
2. Be able to receive and transmit spoken messages in English.
3. Pass Morse code examinations at 16 code groups per minute (16 CG, Telegraphy Element 1) and 20 words-per-minute plain language (20 PL, Telegraphy Element 2) — both receiving and transmitting by hand.*

4. Pass written examinations covering basic radio law and operating procedures for radiotelephony (Element 1), basic operating procedures for radiotelegraphy (Element 5), and electronics technology applicable to radiotelegraph stations (Element 6).

Current or previous Amateur Extra Class licensees (those holding a current or expired license) receive Telegraphy Element 1 (16 Code Groups per minute) and Telegraphy Element 2 (20 Plain Language words per minute) examination credit towards the Third and Second Class Radiotelegraph Operator's Certificate providing the applicant passed the 20 words-per-minute Amateur Extra Class Morse code telegraphy examination (Element 1C) prior to April 15, 2000. A photo-copy of the Amateur Extra Class license dated prior to April 15, 2000 must be attached to the FCC Form 605 application. See § 13.9(d)(2).

Table 2-6. First Class Radiotelegraph Operator's Certificate (T1)

The First Class Radiotelegraph Operator's Certificate (T1) is required only for those who serve as a chief radio operator on a U.S. passenger ship. The privileges granted are the same as the Second Class Radiotelegraph Operator's Certificate; i.e., it authorizes the holder to operate, repair and maintain ship and coast radiotelegraph stations in the maritime services. It also confers all of the operating authority of the Restricted Radiotelephone Operator's Permit (RP), the Marine Radio Operator's Permit (MP), and the Third and Second Class Radiotelegraph Operator's Certificate (T3 and T2).

You *must meet all six* of the following requirements to be eligible for a T1:
1. Be at least 21 years old.
2. Have at least one year of experience in sending and receiving public correspondence by radiotelegraph at ship stations, public coast stations or both.
3. Be a legal resident of (or otherwise eligible for employment in) the United States.
4. Be able to receive and transmit spoken messages in English.
5. Pass Morse code examinations at 20 code groups per minute (20 CG, Telegraphy Element 3) and 25 words-per-minute plain language (25 PL, Telegraphy Element 4) — both receiving and transmitting by hand*.
6. Pass written examinations covering basic radio law and operating procedures for radiotelephony (Element 1), basic operating procedures for radiotelegraphy (Element 5), and electronics technology applicable to radiotelegraph stations (Element 6).

First, Second and Third Class Radiotelegraph Operator's Certificates are valid for a five-year term. They may be renewed from any-time in the last year of their term up to five years following expiration.

No person may hold more than one unexpired radiotelegraph operator's certificate at the same time — nor may a person hold any class of radiotelegraph operator's certificate and a Marine Radio Operator Permit (MP) or a Restricted Radiotelephone Operator Permit (RP).

* A Morse code hand-sending examination probably will not be required. The FCC has taken the position that applicants who can receive telegraphy by ear can also send code manually.

While the First and Second Class Radiotelegraph Operator's Certificates confer all of the operating authority of a GROL, they *do not* grant Element 3 examination credit toward the GMDSS Radio Maintainer's License.

Table 2-7. GMDSS Radio Operator's License (DO)

The GMDSS Radio Operator's License (DO) qualifies personnel as Global Maritime Distress and Safety System (GMDSS) radio operators for the purposes of operating GMDSS radio installations including some basic equipment adjustments. It also confers the operating authority of the Marine Radio Operator Permit (MP). It does NOT authorize repair and maintenance of GMDSS equipment. The GMDSS Radio Operator's License also confers all of the operating authority of the Restricted Radiotelephone Operator's Permit (RP), and the Marine Radio Operator's Permit (MP).

You *must meet all three* of the following requirements to be eligible for a DO:
1. Be a legal resident of (or otherwise eligible for employment in) the United States.
2. Be able to receive and transmit spoken messages in English.
3. Pass written examinations covering basic radio law and maritime operating procedures (Element 1), and GMDSS radio operating practices (Element 7).

Table 2-8. Restricted GMDSS Radio Operator's License (DO-R)

The Restricted GMDSS Radio Operator's License qualifies personnel as GMDSS radio operators serving aboard compulsory ships that operate exclusively within Sea Area A1. Sea Area A1 is the area within VHF radiotelephone coverage of at least one coast station at which continuous DSC (digital selective calling) is available.

You *must meet all four* of the following requirements to be eligible for this license:
1. Be a legal resident of (or otherwise eligible for employment in) the United States.
2. Be able to receive and transmit spoken messages in English.
3. Be familiar with all of the GMDSS equipment required for vessels sailing within Sea Area A1.
4. Pass written examinations covering basic radio law and maritime operating procedures (Element 1) and GMDSS radio equipment required to be fitted on ships sailing solely within Sea Area A1 (Element 7A).

Table 2-9. GMDSS Radio Maintainer's License (DM)

The DM qualifies personnel as GMDSS radio maintainers to perform at-sea repair and maintenance of GMDSS equipment. It also confers all of the operating authority of the Restricted Radiotelephone Operator's Permit (RP), the Marine Radio Operator's Permit (MP), and the General Radiotelephone Operator License (PG).

You *must meet all three* of the following requirements to be eligible for a DM:
1. Be a legal resident of (or otherwise eligible for employment in) the United States.
2. Be able to receive and transmit spoken messages in English.
3. Pass written examinations covering basic radio law and maritime operating procedures (Element 1), technical examination on electronic fundamentals and techniques (Element 3), and GMDSS radio maintenance practices and procedures (Element 9).

Note: In instances where an applicant qualifies for both a GMDSS Radio Operator's License (DO) and a GMDSS Radio Maintainers License (DM) at one sitting, a GMDSS Radio Operator/Maintainer License (DB) will be issued. A GMDSS Radio Maintainers License (DM) does not authorize *operation* of GMDSS equipment.

Table 2-10. Ship Radar Endorsement

The Ship Radar Endorsement may be placed only on General Radiotelephone Operator Licenses (PG), GMDSS Radio Maintainer's licenses (DM), or on First or Second Class Radiotelegraph Operator's Certificates (T1 or T2). Only persons whose commercial radio operator license bears this endorsement may repair, maintain or internally adjust radar equipment.

You *must meet both* of the following requirements to be eligible for a Ship Radar Endorsement:
1. Hold (or qualify for) a First or Second Class Radiotelegraph Operator's Certificate (T1 or T2), GMDSS Radio Maintainer's license (DM) or a General Radiotelephone Operator License (PG).
2. Pass a written examination (Element 8) covering special rules applicable to ship radar stations and the technical fundamentals of radar and radar maintenance procedures.

Table 2-11. Six-Months Service Endorsement

The Six-Months Service Endorsement is required on the radiotelegraph operator's certificate of anyone who serves as the sole radio operator aboard large U.S. cargo ships sailing the high seas.

You *must meet all five* of the following requirements to be eligible for a Six-Months Service Endorsement:
1. You have been employed as a radio operator on board ships of the United States for a period totaling at least six months.
2. The ships were equipped with radio stations complying with the provisions of Part II of Title III of the Communications Act, or the ships were owned and operated by the U.S. Government.
3. The ships were in service during the applicable six-months period and no portion of any single in-port period included in the qualifying six-months period exceeded seven days.
4. You held a First or Second Class Radiotelegraph Operator's Certificate (T1 or T2) issued by the FCC during this entire six-months qualifying period.
5. You hold a radio officer's license issued by the U.S. Coast Guard at the time the six-months service endorsement is requested.

For those applicants who are qualified by having at least 180 days of creditable service, the following is to be submitted to the Federal Communications Commission, 1270 Fairfield Road, Gettysburg, PA 17325-7245:
1. Certification letter signed by the vessel's master or owner/agent specifying the vessel name, vessel call sign, dates of service (shipment and discharge), total number of days served (minus any in-port periods exceeding seven days); and the name(s) and certificate number of the chief radio officer holding a Six-Months Service Endorsement on the vessel during (shipment) service.
2. Completed FCC Form 605 Commercial Radio Operator License Application and Form 605 Schedule E.
3. Original T1 or T2 certificate.
4. Valid copy of U.S. Coast Guard license.
5. Certificate(s) of Discharge to Merchant Seaman.
6. Two photographs. (See page 19 for details).

PAPERWORK TO OBTAIN A COMMERCIAL RADIO OPERATOR LICENSE

When you take an examination for a new FCC commercial license, you will complete FCC Form 605 and any required schedules, and you will pay an examination fee to the test team. If you pass, you will receive a

Proof of Passing Certificate (PPC) issued on behalf of the COLEM by the test team. The test team will forward the original PPC, your payment, and your exam results to the COLEM. In turn, the COLEM will certify your examination results and electronically file your license application with the FCC. You probably will not have to file any paperwork directly withthe FCC. You should be sure to obtain a copy of your PPC and keep it for your records.

FCC Form 159 — Remittance Advice

Any payment to the FCC must be accompanied by FCC Form 159, Fee Remittance Advice. The current edition (February 2000) collects the Taxpayer Identification Number (TIN) and FCC Registration Number (FRN) for the payer and applicant as required by the *Debt Collection Improvement Act of 1996.*

FCC Form 159 is a Fee Processing Form and not an application. You must always submit application FCC Form 605 accompanied by its Schedule "E" along with your FCC Form 159. The current application fee is $50.00. Make your check or money order payable to FCC or Federal Communications Commission. You also may pay using a VISA or MasterCard credit card.

An FCC Form 159 *is not included* with applications for **new** Marine Radio Operator Permits, GMDSS Radio Operator Licenses, GMDSS Radio Maintainer Licenses, GMDSS Radio Operator/Maintainer Licenses, or First, Second or Third Class Radiotelegraph Operator Certificates since only a testing fee is paid to the COLEM. With the exception of a Restricted Permit, which requires no examination, there are no application fees due to the FCC on newly-issued Commercial Radio Operator licenses.

When an application fee is due, a Payment Type Code (Fee Type Code) must be entered on FCC Form 159. Specific codes have been established according to the type and purpose of your application. It is extremely important to submit your application with the correct Payment Type Code.

- New Restricted Permit: Fee Type Code PARR
- Any request for a Duplicate or Replacement License: Fee Type Code PADM
- Any request for a License Renewal: Fee Type Code PACS

FCC Form 605

Use this form to apply for any Commercial Radio Operator license. This multi-purpose form is entitled *"Quick-Form Application for Authorization in the Ship, Aircraft, Amateur, Restricted and Commercial Operator, and General Mobile Radio Services."*

FCC application Form 605 is used to apply for an authorization to maintain and operate radio stations, amend pending applications, modify existing licenses, and perform a variety of other miscellaneous transactions in the Wireless Telecommunications Bureau (WTB) radio services. For Commercial Radio Operators, FCC Form 605 replaces FCC Forms 753, 755, and 756. A sample of the current FCC Form 605 and Schedules E and F appear at the end of this chapter.

Form 605 consists of two sections. The purpose of the main form is to obtain information sufficient to identify the filer and establish the filer's basic eligibility and qualifications. Schedule E is used to supply additional data needed for the Commercial Radio Operators, Restricted Radiotelephone, and the Restricted Radiotelephone - Limited Use Radio Services. This schedule is used in conjunction with the main form.

Schedule F is a *Temporary Operator Permit* for applicants who have submitted a Restricted Radiotelephone or Restricted Radiotelephone - Limited Use application. It is used as a Temporary Permit for up to 90 days while your application is being processed by the FCC. If you have any questions, you may call the FCC's Consumer Assistance line at 800/225-5322.

FCC Registration Number

As of December 3, 2001, applicants should obtain a unique identifying number called the FCC Registration Number (FRN) and include it on the FCC Form 605. This requirement is to facilitate compliance with the *Debt Collection Improvement Act of 1996* (DCIA). The FRN can be obtained electronically through the FCC web page at www.fcc.gov, then click on Commission Registration System, CORES. (If you don't have a computer, you can register manually using FCC Form 160.)

You also will receive a CORES password which ensures that only you and your authorized representatives are able to update your CORES registration information and make changes to your licenses. If you forget your password or have other password-related questions, contact FCC technical support at 202/414-1250 or on-line at corespassword@fcc.gov.

Applicants applying a new Commercial Radio Operator license who have not obtained an FRN may supply their Social Security Number to the examiners instead and an FRN will automatically be assigned when the license application is submitted to the FCC by the Commercial Radio Operator License Examination Manager (COLEM.)

How to Obtain Forms

All FCC forms are available for downloading on the FCC web site at www.fcc.gov/ and clicking on the Forms link, or by calling the FCC's Forms Distribution Center at 800/418-FORM (3676), or from fax-on-demand by dialing 202/ 418-0177. It is best to always obtain a new copy of any form to assure that you are using the correct, current version.

Required Photographs

If you are applying for a new, renewed or replacement radiotelegraph operator's certificate, or the Six Months Service Endorsement, you must submit two identical, signed photographs of yourself with your completed application.

1. The photographs must be *signed in ink* by the applicant on the front along the left margin. This signature must be clearly visible and match the signature on the FCC Form 605 application.
2. The photographs must not be less than 2 x 2 inches nor more than 2.5 x 3 inches.
3. The photographs must be a front view showing head and shoulders only with clear, full face.
4. The photographs may be color or black and white with a light, plain background.
5. The photographs must have been taken within six months of the application date.
6. Newspaper, magazine or photocopied photographs are not acceptable.
7. Enclose your photos in a plain envelope. Write the word "photos" and your name on the outside. Staple the envelope to the back of your application.

UNIVERSAL LICENSING SYSTEM

The FCC is now using a new, streamlined application processing system called the Universal Licensing System. ULS is an automated process that uses a single, integrated licensee database for all wireless radio services.

The development of ULS represents a fundamental change to the manner in which the FCC receives and processes wireless applications. Previously, wireless applicants and licensees used a myriad of forms for the types of requests to various wireless services, and the information provided on these applications was collected in 11 separate databases, each for a different group of services.

Although in some instances these forms could be filed electronically, many of these older systems did not accommodate electronic filing, instead requiring information to be submitted on paper and then manually keyed into the appropriate database by FCC staff.

The on-line ULS is browser-based for ease of use, provides built-in security, and contains everything needed for electronic filing. You can access your records with a password and immediately make the needed changes. Since strong 128-bit encryption is used, you must use Microsoft Internet Explorer (versions 5.5 or higher) or Netscape Navigator (version 4.5 or higher) with "cookies" enabled.

Since mid-1998, the FCC's Wireless Bureau has been implementing an incremental deployment of ULS. In 2001, the Commercial Radio Service became operational under the Universal Licensing System.

ULS uses just five forms – instead of 41 – and emphasizes electronic filing of all licensee information. The five new forms designed specifically for ULS use are FCC Forms 601 through 605. FCC Form 605 along with its Schedules "E" and "F" are used when applying for or amending Restricted and Commercial Radio Operator permits and licenses. The current edition of these forms carry a July 2002 issue date and are shown on pages 25 and 26 for your information.

The Universal Licensing System also enhances the availability of Commercial Radio Operator licensing information to the public, who for the first time have access to all wireless licensing information on-line. The ULS search engine enables you to look up applications and licenses currently on file with the Wireless Bureau. Just go to: http://wireless.fcc.gov/uls and click on "Search Licenses" to obtain Commercial Radio Operator licensing information.

Currently, applicants seeking Commercial Radio Operator licenses must pass one or more written examinations administered by a Commission-certified *Commercial Operator License Examination Manager* (COLEM) and obtain a *Proof-of-Passing Certificate* (PPC.) In the past, the applicant mailed the original PPC and paper application form to the FCC.

Commercial Radio Operators now have the option of filing renewal or license amendments electronically over the Internet or manually using a paper FCC Form 605. As of 2001, *Commercial Operator License Examination Managers* (COLEMs), may only file their Form 605 applications electronically.

The new Universal Licensing System also requires that applicants in all radio services submit their *Taxpayer Identification Number* (TIN) – which is an applicant's Social Security Number (SSN) – to the FCC.

The collection of the TIN is required by Congress under the requirements of the *Debt Collection Act of 1996*. The TIN information is kept confidential by a registration procedure and issuance of an FCC Registration Number (FRN). Applicants for new licenses have the option of obtaining an FRN by registering their TIN with the FCC or simply supplying their Social Security Number to the examiners on application Form 605. If the SSN is given, the FCC automatically converts this number to an FRN which appears on the license when it is issued by the Commission.

EXAMINATION, REGULATORY AND APPLICATION FEES

There are three separate fees that may be charged to applicants for Commercial Radio Operator licenses, but an applicant only pays a maximum of two of them, depending on the type of license and action required. The three fees are examination fee, regulatory fee, and application fee.

Examination Fee

The examination fee is charged by a COLEM (Commercial Operator License Examination Manager). This fee can vary from $25.00 to $75.00, depending upon which COLEM administers the examination, and the number and type of test elements administered. This fee is paid to the examiners.

Regulatory Fee *(Section 9 of the Communications Act)*

Regulatory fees are determined on a fiscal year basis to assist the FCC in recovering the costs associated with Government regulation, user services and enforcement of the radio service. Beginning with Fiscal Year 1995, the FCC discontinued all regulatory fees associated with commercial radio operator licenses. Regulatory fees continue, however, for certain commercial marine and aviation radio *station* licenses. The FCC has a "Fee Hotline" at (202) 418-0192 to answer questions about fees associated with licenses.

Application Fee *(Section 8 of the Communications Act)*

The application fee (also called a processing fee) currently is $50.00. It is the amount that applicants pay to reimburse the Government for the cost of processing their radio licenses. The application fee is paid for new RP licenses and for duplicate, renewal and replacement licenses. Make check payable to the FCC. An FCC Form 159 Remittance Advice must accompany application form.

Application fees are adjusted upward in $5.00 increments every two years to reflect changes in the Consumer Price Index.

WHERE TO SEND THE APPLICATION AND FEES
New Licenses

Restricted radiotelephone operator permit applications carrying a processing fee must be sent to the FCC's fee collection contractor in Pittsburgh, PA. All other new applications go to Gettysburg, PA.

The FCC also accepts payment by MasterCard® or VISA®. Regardless of the form of payment, an FCC Form 159 (Remittance Advice) must also be completed and attached to the application.

1. RESTRICTED RADIOTELEPHONE OPERATOR PERMIT (RP) (New, Duplicate, or Replacement)

Federal Communications Commission
Wireless Bureau Applications
P.O. Box 358245
Pittsburgh, PA 15251-5130

Fee type code that must be entered on application PARR. Use FCC Form 753* and include FCC Form 159, Remittance Advice.
*Aliens who are not legally eligible for U.S. Employment should use FCC Form 755.)

2. ALL OTHER COMMERCIAL RADIO OPERATOR LICENSES AND ENDORSE-MENTS (New)

Federal Communications Commission
1270 Fairfield Road
Gettysbury, PA 17325-7245

Renewal, Replacement, Duplicate and Modified Licenses

Commercial radio operators applying for a renewed, replacement, duplicate or modified (including name change) license must pay an application processing fee. A check for $50.00, payable to the Federal

Communications Commission, should be attached to the application form. All applications for renewal, replacement, duplicate or modified Commercial Radio Operator licenses must be sent to the appropriate following address (see *Table 2-11* for more detail):

3. ALL LICENSE RENEWALS

Federal Communications Commission
Wireless Bureau Applications
P.O. Box 358245
Pittsburgh, PA 15251-5130

Fee type code that must be entered on application: PACS.
(Use FCC Form 605 and include FCC Form 159, Remittance Advice)

4. ALL REPLACEMENT/DUPLICATE LICENSES

Federal Communications Commission
Wireless Bureau Applications
P.O. Box 358130
Pittsburgh, PA 15251-5130

Fee type code that must be entered on application: PADM.
(Use FCC Form 605 and include FCC Form 159, Remittance Advice)

EXAMINATION AND APPLICATION FEES BY LICENSE TYPE

Table 2-12 details all of the required fees associated with Commercial Radio Operator licenses. The COLEM examination fee is paid to the examiners; the application fee to the FCC.

Table 2-12. Required Fees for Commercial Radio Operator Licenses

LICENSE	Exam	NEW LICENSE Application Section 8	Send To:[1]	RENEWAL LICENSE Application Section 8	Send To:[1]	REPLACEMENT OR DUPLICATE[2] Application Section 8	Send To:[1]
RP	None	$50.00 PARR[3]	1	(Lifetime)		$50.00 PADM[3]	4
MROP [MP]	COLEM[5]	None[6]	2	$50.00 PACS[3]	3	$50.00 PADM[3]	4
GROL [PG]	COLEM[5]	None[6]	2	(Lifetime)		$50.00 PADM[3]	4
T1, T2, T3	COLEM[5]	None[6,4]	2	$50.00[4] PACS[3]	3	$50.00[4] PADM[3]	4
GMDSS/ O/M	COLEM[5]	None[6]	2	$50.00 PACS[3]	3	$50.00 PADM[3]	4
Radar	COLEM[5]	None[6]	2	None		None	
6-Months	None	None[6]	2	None		None	

Important footnote: At present, there are no regulatory fees for Commercial Radio Operator licenses and the applications (processing) fee could be adjusted upward depending upon changes in the Consumer Price Index. Call the FCC's "Fee Hotline" at (202) 418-0192 for the most recent fee information. All submissions involving fees must include Form 159

[1] Refer to item number on pages 19 and 20 for address.
[2] Includes Change of Name.
[3] Fee type code.
[4] Be sure to include the two small signed photographs.
[5] Examination fee charged by COLEM. (See Appendix.)
[6] No regulatory or application fee charged at present.

OPERATING WHILE APPLICATION PENDING

You are immediately authorized to exercise the rights and privileges of the operator license as soon as you pass the required commercial radio operator examinations, provided that:

    1.) your commercial radio operator license has not been revoked or suspended, and

    2.) you are not the subject of an ongoing suspension proceeding, and

    3.) your commercial radio operator license application has been received by the FCC but not yet been acted upon, and

    4.) you hold a Proof of Passing Certificate(s) indicating that you have passed the required examination(s) within the previous 365 days. (Be sure to keep a copy of this certificate!)

    This operating authority is valid for a period of 90 days from the date your application was received by the FCC.

    If you file an application to renew your license before it expires, you may continue to operate under the authority of your license while the FCC processes your renewal application. However, if you fail to renew your license before it expires, you cannot operate equipment that requires that license until it is renewed. You may file to renew your expired license any time during the five year grace period after your license expires. If you fail to renew your license within the grace period, you must apply for a new license and re-take the required examination(s). NOTE: The FCC does not notify you when it is time to renew your license. The expiration date is on your license or permit.

    For purposes of making log entries as required by Part 13.19(c), the rules require an operator awaiting FCC action on an application to enter the PPC serial number and date of issue in place of the FCC-issued serial number and expiration date.

LOST, STOLEN, MUTILATED, OR DESTROYED LICENSES

If your license is lost, stolen, mutilated, or destroyed, you may apply for a duplicate license using the FCC Form 605. There are no provisions for issuing duplicate Restricted Radiotelephone Operator Permits. If you need to replace a lost RP, you must apply for a new one using FCC Form 605. Be sure to include the $50.00 fee and Form 159.

NAME CHANGE

If you change your name and wish a license with your new legal name, you may apply for a new replacement license. It is not required by the FCC, however, that you replace a license when you change your name. If you apply for a replacement license, be sure to indicate the reason for your application and give both your former and new legal names. Restricted Permit holders should apply for a new RP.

    Return your current license to the FCC along with the FCC Form 605 application, FCC Form 159 Remittance Advice, and the $50.00 application fee. Send the package to the FCC address listed in **item 4** under *Renewal, Replacement, Duplicate and Modified Licenses para-*

graph. (See page 22.) Applicants requesting a replacement license will not be granted a new full license term.

POSTING LICENSE OR PERMIT

If you are employed at a station where your commercial operator license must be posted, you may post either your Proof of Passing Certificate or a signed copy of your renewal application instead of your license while your application for license or renewal is being processed.

WHERE TO GET HELP ON COMMERCIAL RADIO OPERATOR APPLICATIONS

If you need assistance with your application or have questions concerning fees, application status or general questions about Commission policies, contact the FCC's Consumer Assistance hotline at (toll free) 1 (888) CALL FCC (or 1 (888) 225-5322).

FEE FILING GUIDE

Fee type codes identify to the FCC the license for which you are paying. Each FCC bureau publishes a Fee Filing Guide that notifies applicants the charges they must pay to the government for their license. Every fee has a three or four letter fee type code with the first letter indicating the appropriate bureau, as shown in *Table 2-12*. Make sure you have used the correct code on your application.

SUMMARY

In this chapter we have summarized when a person does or does not need a commercial radio license; when each permit, license, certificate and endorsement is required; and the forms and paperwork required to complete an application for a license.

An examination fee is paid to the COLEM examiners when an examination is needed. The COLEM will send in your application for a new license, and will include the *original* PPC (Proof-of-Passing Certificate) that was signed by your examiners, and any necessary credit documents. Keep a copy for yourself so that you have a record of passing your examination. No further application fee is due.

There are processing fees required by the FCC for certain new, renewal, replacement and duplicate licenses. *Table 2-12* summarizes these fees which are subject to change. Call the FCC at the telephone number listed in the footnote of *Table 2-12* in case you are in doubt about a fee.

Remember that you need to include a fee type code on your application form. The fee type codes are also included on *Table 2-12* and next to the appropriate FCC address for the application.

Include a check or money order for any appropriate application fee. You may also pay by MasterCard or VISA credit card. In either instance, you must include an FCC Form 159 Remittance Advice along with your application.

Recheck the addresses given in *Where to Send the Applications and Fees* to make sure you have the current address.

FCC Form 605 – Front

FCC 605
Main Form

Quick-Form Application for Authorization in the Ship, Aircraft, Amateur, Restricted and Commercial Operator, and General Mobile Radio Services

Approved by OMB
3060 - 0850
See instructions for public burden estimate

1) Radio Service Code:

2) **Application Purpose** (Select only one) ( )

NE – New
MD – Modification
AM – Amendment

RO – Renewal Only
RM – Renewal / Modification
CA – Cancellation of License

WD – Withdrawal of Application
DU – Duplicate License
AU – Administrative Update

3) If this request if for Developmental License or STA (Special Temporary Authorization) enter the appropriate code and attach the required exhibit as described in the instructions. Otherwise enter 'N' (Not Applicable). ( ) D S N/A

4) If this request is for an Amendment or Withdrawal of Application, enter the file number of the pending application currently on file with the FCC. File Number

5) If this request is for a Modification, Renewal Only, Renewal / Modification, Cancellation of License, Duplicate License, or Administrative Update, enter the call sign (serial number for Commercial Operator) of the existing FCC License. If this is a request for consolidation of DO & DM Operator Licenses, enter serial number of DO. Call Sign/Serial #

6) If this request is for a New, Amendment, Renewal Only, or Renewal Modification, enter the expiration date of the authorization (this item is optional). MM DD

7) Does this filing request a Waiver of the Commissions rules? If "Y", attach the required showing as described in the instructions. ( ) Yes No

8) Are attachments (other than associated schedules) being filed with this application? ( ) Yes No

Applicant/Licensee Information

9) FCC Registration Number (FRN):

10) Applicant /Licensee is a(n): ( ) Individual Corporation Unincorporated Association Limited Liability Corporation Trust Partnership Government Entity Consortium Joint Venture

11) First Name (if individual): MI: Last Name: Suffix:

11a) Date of Birth (required for Commercial Operators (including Restricted Radiotelephone)): (mm)/ (dd)/ (yy)

12) Entity Name (if other than individual):

13) Attention To:

14) P.O. Box: And/Or 15) Street Address:

16) City: 17) State: 18) Zip Code: 19) Country:

20) Telephone Number: 21) FAX Number:

22) E-Mail Address:

FCC 605 – Main Form
October 2002 - Page 1

FCC Form 605 – Back

Ship Applicants/Licensees Only

23) Enter new name of vessel:

Fee Status

24) Is the applicant/licensee exempt from FCC application Fees? ( ) Yes No

25) Is the applicant/licensee exempt from FCC regulatory Fees? ( ) Yes No

General Certification Statements

1) The Applicant/Licensee waives any claim to the use of any particular frequency or of the electromagnetic spectrum as against the regulatory power of the United States because of the previous use of the same, whether by license or otherwise, and requests an authorization in accordance with this application.

2) The applicant/licensee certifies that all statements made in this application and in the exhibits, attachments, or documents incorporated by reference are material, are part of this application, and are true, complete, correct, and made in good faith.

3) Neither the Applicant/Licensee nor any member thereof is a foreign government or a representative thereof.

4) The applicant/licensee nor any other party to the application is subject to a denial of Federal benefits pursuant to Section 5301 of the Anti-Drug Abuse Act of 1988, 21 U.S.C. § 862, because of a conviction for possession or distribution of a controlled substance. **This certification does not apply to applications filed in services exempted under Section 1.2002(b) of the rules, 47 CFR § 1.2002(b).** See Section 1.2002(b) of the rules, 47 CFR § 1.2002(b), for the definition of "party to the application" as used in this certification.

5) Amateur or GMRS Applicant/Licensee certifies that the construction of the station would NOT be an action which is likely to have a significant environmental effect (see the Commission's rules 47 CFR Sections 1.1301-1.1319 and Section 97.13(a) rules (available at web site http://wireless.fcc.gov/rules.html).

6) Amateur Applicant/Licensee certifies that they have READ and WILL COMPLY WITH Section 97.13(c) of the Commission's rules regarding RADIOFREQUENCY (RF) RADIATION SAFETY and the amateur service section of OST/OET Bulletin Number 65 (available at web site http://www.fcc.gov/oet/info/documents/bulletins/).

Certification Statements For GMRS Applicants/Licensees

1) Applicant/Licensee certifies that he or she is claiming eligibility under Rule Section 95.5 of the Commission's rules.

2) Applicant/Licensee certifies that he or she is at least 18 years of age.

3) Applicant/Licensee certifies that he or she will comply with the requirement that use of frequencies 462.650, 467.650, 462.700 and 467.700 MHz is not permitted near the Canadian border North of Line A and East of Line C. These frequencies are used throughout Canada and harmful interference is anticipated.

4) Non-Individual applicants/licensees certify that they have NOT changed frequency or channel pairs, type of emission, antenna height, location of fixed transmitters, number of mobile units, area of mobile operation, or increase in power.

Signature

26) Typed or Printed Name of Party Authorized to Sign

First Name: MI: Last Name: Suffix:

27) Title:

Signature: 28) Date:

Failure to Sign This Application May Result in Dismissal Of The Application And Forfeiture Of Any Fees Paid

WILLFUL FALSE STATEMENTS MADE ON THIS FORM OR ANY ATTACHMENTS ARE PUNISHABLE BY FINE AND/OR IMPRISONMENT (U.S. Code, Title 18, Section 1001) AND / OR REVOCATION OF ANY STATION LICENSE OR CONSTRUCTION PERMIT (U.S. Code, Title 47, Section 312(a)(1)), AND / OR FORFEITURE (U.S. Code, Title 47, Section 503).

FCC 605 – Main Form
October 2002 - Page 2

**FCC 605
Schedule E**

Schedule for Additional Data for the Commercial Radio,
Restricted Radiotelephone, and
Restricted Radiotelephone-Limited Use Radio Services

Approved by OMB
3060 - 0850
See 605 Main Form Instructions
for public burden estimate

LICENSE TYPE / ENDORSEMENTS

1) Operator Class Code: Check only one operator class -Do not apply for more than one kind of license on a single application.

General Radiotelephone Operator License (PG)	First Class Radiotelegraph Operator's Certificate (T1)
GMDSS Radio Operator's License (DO)	Second Class Radiotelegraph Operator's Certificate (T2)
GMDSS Radio Maintainer's License (DM)	Third Class Radiotelegraph Operator's Certificate (T3)
GMDSS Radio Operator/Maintainer License (DB)	Restricted Radiotelephone Operator Permit (RR)
Marine Radio Operator Permit (MP)	Restricted Radiotelephone Operator Permit-Limited Use (RL)

1a) If requesting consolidation of DO & DM Operator licenses, enter serial number of DM: _____

2) Endorsement Type (Check endorsements that apply)

Ship Radar Endorsement (Attach documentation.) (See instructions.)

Six Months Service Endorsement (Attach documentation as required by 47 C.F.R. § 13.9(d). (See instructions).

3) CERTIFICATION CATEGORY - Check only one of the three certification categories below, as appropriate

CERTIFICATION FOR LICENSES AND ENDORSEMENTS OTHER THAN RESTRICTED RADIOTELEPHONE

I certify that:
I am legally eligible for employment in the United States. (All U.S. citizens are considered, for the purposes of this application, to be legally eligible for employment in the U.S.)
I do not have a speech impediment, blindness, acute deafness, or any other disability which will impair or handicap me in properly using the license for which I am applying.

CERTIFICATION FOR RESTRICTED RADIOTELEPHONE OPERATOR PERMIT

I certify that:
I am eligible for employment in the United States. (All U.S. citizens are considered, for the purposes of this application, to be legally eligible for employment in the U.S.)
I can speak and hear.
I can keep at least a rough written log.
I am familiar with the provisions of the applicable laws, treaties, rules, and regulations governing the radio station which will be operated.
I need this permit because of intent to engage in international voyages or flights, international communications, or intent to comply with the requirements of the Vessel Bridge-to-Bridge Radiotelephone Act.

CERTIFICATION FOR RESTRICTED RADIOTELEPHONE OPERATOR PERMIT-LIMITED USE

I certify that:
I am NOT eligible for employment in the United States. (All U.S. citizens are considered, for the purposes of this application, to be legally eligible for employment in the U.S.)
I can speak and hear.
I can keep at least a rough written log.
I am familiar with the provisions of the applicable laws, treaties, rules, and regulations governing the radio station which will be operated.
I need this permit because of intent to engage in international voyages or flights, international communications, or intent to comply with the requirements of the Vessel Bridge-to-Bridge Radiotelephone Act.
I hold an aircraft pilot certificate which is valid in the United States or an FCC Radio Station License issued in my name.

FCC 605 - Schedule E
October 2002 - Page 1

FCC Form 605 – Schedule E

**FCC 605
Schedule F**

Temporary Operator Permit for the Ship,
Aircraft, Restricted Radiotelephone,
Restricted Radiotelephone-Limited Use
and GMRS Radio Services

Approved by OMB
3060 - 0850
See 605 Main Form Instructions
or public burden estimate

DO NOT MAIL THIS SCHEDULE TO THE FCC: KEEP IT FOR YOUR RECORDS

If you need a Temporary Operator Permit for the Ship, Aircraft, Restricted Radiotelephone, Restricted Radiotelephone-Limited Use or the GMRS Radio Services while your application is being processed by the FCC (refer to the instructions for restrictions):
1) Complete the FCC 605 Main Form and the appropriate Additional Data Schedule and submit them to the FCC
2) Complete this FCC 605 Schedule F and keep it for your records.

Certification Statements for a Temporary Operator Permit

All applicants must certify:
I have applied for a Station License by submitting a completed and signed FCC 605 Main Form and appropriate Additional Data Schedule to the FCC, and that the information contained in that application is true and correct.
I have not been denied a license or had my license suspended or revoked by the FCC.
I will obey all applicable laws, treaties, and regulations.

Ship, Aircraft Station, and GMRS Additional Information
Enter your temporary call sign (see instructions):

Signature	
Typed or Printed Name:	
Signature:	
Date of Signature:	Date the FCC 605 Main Form and Additional Data Schedule were submitted to the FCC:

This temporary operator permit is valid for 90 days from the date the FCC 605 Main Form and the appropriate Additional Data Schedule were submitted to the FCC.

Your authority to operate your radio station under this temporary operator permit is subject to all applicable laws, treaties and regulations and is subject to right of control of the Government of the United States.

You must post this temporary operator permit at your radio station location.

FCC 605 - Schedule F
October 2002 - Page 1

FCC Form 605 – Schedule F

Commercial License Examinations – Then and Now

3

FCC CHANGES HAM OPERATOR TESTING

In the early 1980s, in order to conserve resources, a general trend developed at the Federal Communications Commission to privatize many FCC administrative functions. In 1982, President Ronald Reagan signed legislation *(Public Law 97-259)* authorizing Amateur Radio examinations to be prepared and administered by volunteer ham radio organizations.

During 1983, the FCC developed new guidelines for Amateur Radio operator license testing, patterned after a system used by the Federal Aviation Administration (FAA), in which all examination questions are in the public domain and widely published. The FCC, with help from the amateur community, developed and released to the public examination question pools that contained all of the possible verbatim questions that could be asked of amateur operators.

VEC SYSTEM ADOPTED

The FCC then identified testing system administrators who would recruit and manage teams of volunteer examiners (VEs) who would administer Amateur Radio operator examinations. The result was the development of the VEC System, a network of Volunteer Examiner Coordinators that act as the link between the VEs that administer the examinations and the FCC, which issues Amateur Radio licenses. VECs provide their VE teams with license testing materials; collect, screen, approve, and forward amateur radio license applications to the FCC; and maintain records of all test sessions. The FCC acts only in an oversight capacity and provides guidance to the VECs. The VEC System is funded by license applicants who pay small testing fees to cover the out-of-pocket costs of the testing VEs.

In 1986, maintenance of the various Amateur Radio question pools was turned over to a committee of VECs that is responsible for keeping all ham examination question pools up-to-date. Since its inception, the VEC System has proven to be highly successful, very efficient, and networks of VE teams have developed throughout the country making it easy for applicants to find a nearby testing site and convenient examination date.

COMMERCIAL RADIO TESTING DILEMMA

All through the 1980s, applicants wishing to take a commercial radio examination had to travel to their nearest of FCC field office, located in

Table 3-1. Commercial Radio Licenses — Examination Requirements

License Class	Test Element	Type of Examination	Pass[3]
Marine Radio Operator Permit (MP)	Written Element 1	24-question written examination	18
General Radiotelephone Operator License (PG)	Written Element 1 Written Element 3	24-question written examination 76-question written examination	18 57
First Class Radiotelegraph Operator Certificate (T1) (Must be 21 years old and have 1 year of experience)	Telegraphy Element 3[2] Telegraphy Element 4[2] Written Element 1 Written Element 5 Written Element 6	20 code groups (CG) per minute 25 plain language (PL) words per minute 24-question written examination 50-question written examination 100-question written examination	18 38 75
Second Class Radiotelegraph Operator Certificate (T2)	Telegraphy Element 1[2] Telegraphy Element 2[2] Written Element 1 Written Element 5 Written Element 6	16 code groups (CG) per minute[4] 20 plain language (PL) words per minute[4] 24-question written examination 50-question written examination 100-question written examination	18 38 75
Third Class Radiotelegraph Operator Certificate (T3)	Telegraphy Element 1[2] Telegraphy Element 2[2] Written Element 1 Written Element 5	16 code groups (CG) per minute[4] 20 plain language (PL) words per minute[4] 24-question written examination 50-question written examination	18 38
GMDSS Radio Operator[1] (DO[5])	Written Element 1 Written Element 7	24-question written examination 100-question written examination	18 75
Restricted GMDSS Operator (DO-R)	Written Element 1 Written Element 7A	24-question written examination 50-question written examination	18 38
GMDSS Radio Maintainer[1] (DM[5])	Written Element 1 Written Element 3 Written Element 9	24-question written examination 76-question written examination 50-question written examination	18 57 38
Ship Radar Endorsement (RADAR)	Written Element 8	50-question written examination	38
Six Months Endorsement (Must hold First or Second Class Radiotelegraph Operator's Certificate)	No examination requirement, but requires a minimum of six months service as a radio operator aboard a U.S. ship.		
Restricted Radiotelephone Operator Permit (RP[6])	No examination requirement, but holder must be eligible for employment in the U.S., be familiar with radio law, be able to keep a log and be able to speak and understand English.		

Notes:
1. The Global Maritime Distress and Safety System (GMDSS) is an automated ship-to-shore distress alerting system using satellite and advanced terrestrial communications systems.
2. Passing telegraphy examination requires copying CW (Morse code) for one minute without error.
3. Number of correct answers to pass examination.
4. Amateur Extra Class operators who hold licenses dated on or before April 14, 2000, automatically receive credit for this element examination without testing.
5. A special combined GMDSS Operator/Maintainer (DB) license is available to those who pass both requirements at one sitting.
6. A "Limited Use" RP is available to non-resident aliens.

25 cities throughout the nation, where commercial radio examinations were only administered quarterly. In spite of this inconvenience and lack of available testing, demand for commercial radio operator licenses continued to grow. The demand for testing and the need to update examination questions far outstripped the FCC's manpower and fiscal capability. Public comments indicated that a more efficient commercial radio examination system was needed.

Recognizing the success of the VEC System, the FCC came to believe that a similar mechanism could be implemented in the Commercial Radio Services as well. In 1990, the FCC received legislative authority from Congress to delegate the examination of commercial radio operators to private groups.

COMMERCIAL RADIO OPERATOR TESTING PRIVATIZED

In October, 1992, the FCC transferred the responsibility for its commercial radio operator license testing program to the Private Radio Bureau (later renamed the Wireless Telecommunications Bureau — WTB), the same department that handles Amateur Radio operator examinations. The final *Report and Order* to private commercial radio license testing was released on February 12, 1993.

Since then, commercial radio license testing operates very much like the amateur service's successful VEC System. In the process of reinventing commercial radio license testing administration, the FCC also completely rewrote Part 13 of its Rules, which covers commercial radio operator examination requirements (see Appendix for the complete text of Part 13).

There are now 13 types of commercial radio operator licenses, certificates, permits or endorsements that are either required by international radio law or private industry. They are listed in *Table 3-1* and *Table 3-2* along with the examinations required. They are granted after passing examinations selected from eight written and four telegraphy examinations on the subjects shown in *Table 3-3*.

COMMERCIAL OPERATOR LICENSE EXAMINATION MANAGERS

The FCC's new Commercial Radio Operator testing program is directed by nine private groups known as Commercial Operator License Examination Managers (COLEMs). The FCC said it believed "...a system with multiple entities managing operator examinations will encourage competition between the entities and result in good service, responsiveness, and lower prices to the applicants." The COLEMs are chosen by the FCC's Wireless Telecommunications Bureau.

One of those chosen was The W5YI Group, Inc., founded by one of your authors, Fred Maia. The W5YI Group was not only the first organization to provide amateur radio operator examinations on a national basis, but, through its National Radio Examiners Division, nationwide commercial radio operator testing as well.

Table 3-2. Summary of Elements Required for Commercial Licenses

License	Written Elements								Telegraphy Elements			
	1*	3	5	6	7	7A	8	9	1	2	3	4
MROP (MP)	X											
GROL (PG)	X	X										
3rd R/T (T3)	X		X						X	X		
2nd R/T (T2)	X		X	X					X	X		
1st R/T (T1)	X		X	X							X	X
GMDSS/O (DO)	X				X							
GMDSS/RO (DO/R)	X					X						
GMDSS/M (DM)	X	X						X				
RADAR							X**					

*Formerly Elements 1 and 2   **Endorsement on PG, T1, T2, DM or DB

A complete listing of all COLEMs is provided in the Appendix on pages 427-428.

COLEMs are responsible for:
- Announcing examination sessions
- Verifying the identity of each examinee
- Preparing, administering and grading examinations
- Notifying examinees of examination results (pass/fail)
- Certifying that an applicant has passed the test elements required to qualify for a commercial operator license
- Electronically filing all new commercial radio operator licenses using the FCC's new Universal Licensing System.
- Issuing to the examinee a Proof-of-Passing Certificate (PPC) within 10 days of the examination
- Ensuring that no activity takes place that would compromise the examination and that no unauthorized material is permitted in the examination room
- Handling post-examination questions and problems
- Treating all applicants equally regarding fees or services rendered.

Examiners are prohibited from administering an examination to an employee, relative, or relative of an employee.

COMMERCIAL QUESTION POOLS

All commercial radio operator examinations are prepared from question pools developed by the FCC in 1993 and later updated in 1996. It is expected that these questions will remain current until further notice. Each question pool contains a minimum of five times the number of questions required for a single examination. There are no secret questions. All possible written examination questions, multiple choices and answers for all written examinations have been released to the public. The Element 1, 3 and 8 questions contained in this book are the exact ones that will appear in your Marine Radio Operator Permit,

Table 3-3. Commercial Radio Operator Question Pool Element Subjects

Element	Element Subject
Telegraphy Element 1 (Formerly Elements 1 and 2) 2nd R/T (T2) 3rd R/T (T3)	16 Code Groups (CG) per minute. (A code group is random groups of three to six numerals, punctuation and prosigns.)
Telegraphy Element 2 2nd R/T (T2) 3rd R/T (T3)	20 Plain Language (PL) words per minute text.
Telegraphy Element 3 1st R/T (T1)	20 Code Groups (CG) per minute.
Telegraphy Element 4 1st R/T (T1)	25 Plain Language (PL) words per minute text.
Written Element 1 GROL (PG) MROP (MP) 1st R/T (T1) 2nd R/T (T2) 3rd R/T (T3) GMDSS/O (DO) GMDSS/O-R (DO-R) GMDSS/M (DM)	Basic radio law and radiotelephone operating practice with which every maritime operator should be familiar.
Written Element 3 GROL (PG) GMDSS/M (DM)	Electronic fundamentals and techniques required to adjust, repair and maintain radio equipment in the aviation, maritime and international fixed public radio services. The exam consists of questions from the following categories: operating procedures, radio wave propagation, radio practice, electrical principles, circuit components, practical circuits, signals and emissions, and antennas and feed lines.
Written Element 5 1st R/T (T1) 2nd R/T (T2) 3rd R/T (T3)	Radiotelegraph operating procedures and practices generally followed or required in communicating by means of radiotelegraph stations.
Written Element 6 1st R/T (T1) 2nd R/T (T2)	Advanced Radiotelegraph. Technical, legal, and other matters applicable to the operation of all classes of radiotelegraph stations.
Written Element 7 GMDSS/O (DO)	GMDSS Radio Operating Practices. GMDSS radio operating procedures and practices sufficient to show detailed practical knowledge of the operation of all GMDSS sub-systems and equipment. The exam consists of questions from the following categories: general information, narrow band direct printing, INMARSAT, NAVTEX, digital selective calling, and survival craft.
Written Element 7A GMDSS/RO (DO-R)	GMDSS Radio Operating Practices/Restricted. GMDSS radio operating procedures and practices used in Sea Area 1-A. Exam questions on VHF communications used within ±30 miles of shore.
Written Element 8 RADAR	Ship Radar Techniques. Specialized theory and practice applicable to the proper installation, servicing, and maintenance of ship radar equipment in general use for marine navigation purposes.
Written Element 9 GMDSS/M (DM)	GMDSS Radio Maintenance Practices and Procedures. Requirements set forth in IMO assembly on Training for Radio Personnel (GMDSS), Annex 5 and IMO Assembly on Radio Maintenance Guidelines for the Global Maritime Distress and Safety System related to Sea Areas A3 and A4. The exam consists of questions from the following categories: radio system theory, amplifiers, power sources, troubleshooting, digital theory, and GMDSS equipment and regulations.

GROL, and Ship Radar Endorsement examinations. Contact National Radio Examiners at (800) 669-9594 if you need the question pools for Elements 5 and 6 (required for the commercial radiotelegraph certificates) and Elements 7 and 9 (required for the GMDSS licenses.)

Table 3-4. General Radiotelephone Operator License Question Pools

Examination	Subelement Topic	Page	Total Questions	Examination Questions
Element 1	Maritime Radio Law and Operating Practices	46	170	24
Element 1	Totals		170	24
Element 3	A Operating Procedures	92	40	3
	B Radio Wave Propagation	102	42	3
	C Radio Practice	114	69	6
	D Electrical Principles	133	202	17
	E Circuit Components	195	150	10
	F Practical Circuits	231	139	17
	G Signals and Emissions	265	131	10
	H Antennas and Feed Lines	298	143	10
Element 3	Totals		916	76
Element 8	Ship Radar Techniques	340	321	50
Element 8	Totals		321	50
Total Questions in GROL Plus			1407	

Table 3-5. Balance of Commercial Radio Operator Question Pools

Examination	Subelement Topic	Total Questions	Examination Questions
Element 5	Radiotelegraph Procedures	286	50
Element 6	Advanced Radiotelegraph	616	100
Element 7	GMDSS Operating Practices	600	100
Element 7A	Restricted GMDSS Operating Practices	300	50
Element 9	GMDSS Radio Maintenance	250	50

WRITTEN EXAMINATIONS — CONTENT AND PASS RATE

The pass rate for all written examinations is 75 percent. All examination questions must be taken from the respective question pools. *Table 3-4* shows the number of questions in each question pool in this book, and how many examination questions will be selected from the question pool topics. *Table 3-5* covers the remaining question pools. *Table 3-1* identifies the number of correct answers needed to pass the written examinations.

Handicapped applicants must be accommodated and special examination procedures employed if necessary. A doctor's certification indicating the nature of the disability may be required by the examination manager. Applicants with uncorrected disabilities which adversely affect their performance as a commercial radio operator will be issued a license with a restrictive endorsement. (See § 13.7(c) (6).)

Use of reference materials in FCC examinations is not permitted. Do not bring any books, papers, notes, study guides, or other unauthorized aids to an examination. You may use a non-programmable calculating device during any examination, so long as the memory is erased when the device is turned off.

PASSING THE COMMERCIAL TELEGRAPHY CODE TESTS

Applicants are required to copy a telegraphy test message by ear for a period of one minute. As in the amateur service, the sending examination need not be administered since the FCC has taken the position that: "Passing a telegraphy-receiving examination is adequate proof of an examinee's ability to send and receive telegraphy." Each Morse code test message must contain all letters of the alphabet, numerals 0-9, the period, comma, question mark, slant mark and prosigns $\overline{AR}$, $\overline{BT}$, and $\overline{SK}$. All numerals, punctuation and prosigns count as two characters. Examinees may use their own typewriter to copy the 25 words-per-minute receiving test, but must copy tests at lower speeds by hand.

CREDIT FOR AMATEUR OPERATORS

Of particular interest to amateur radio operators is that examination credit toward commercial telegraphy Elements 1 and 2 will be routinely allowed to Amateur Extra Class operators who have already passed the amateur Element 1(C) 20 words-per-minute code test. This means that Extra Class amateurs who hold licenses dated on or before April 14, 2000 can qualify for the Second or Third Class Radiotelegraph Operator's Certificate by passing only the associated written examinations. Examinees who merely pass amateur examination Element 1(C) without fulfilling all requirements for the Extra Class license are not granted credit for commercial telegraphy Elements 1 and 2. An additional advantage is that several of the Element 3 question pool questions are similar to those found in the amateur radio Element 4 question pool for Extra Class examinations.

EXAMINATION QUESTION POOLS

The complete Element 1 question pool of 170 questions is contained in Chapter 5, the complete Element 3 question pool of 916 questions is contained in Chapter 6 and the complete Element 8 Question pool is contained in Chapter 8. These question pools were released into the public domain in 1995 and must be used in all Element 1, 3 and 8 examinations effective January 1, 1996. Again, these question pools will remain current until further notice.

QUESTION CODING

In *Figure 3-1,* a number from the Element 3 pool is used as an example to explain the coding used in the question pools. 3A2 means Element 3; A means subelement A, Operating Procedures; and 2 means it is the second question in subelement A.

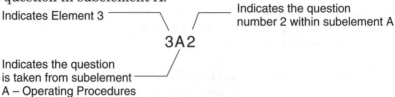

Figure 3-1. Examination Question Coding

HINTS FOR STUDYING FOR THE EXAMINATIONS

The Element 1 Radio Law questions are based on the *Code of Federal Regulations,* Title 47 *(Telecommunication)* Part 80 *(Maritime Service Rules.)* Part 80 is more than 200 pages long, but answers to the Element 1 questions can be found by reading only the important Part 80 pages found in the Appendix — about 24 pages. All Element 1 questions have been cross indexed to the exact Part 80 section that applies.

Go through all the questions and study the easy ones first. Mark the ones to which you already know the answers. Then go back and concentrate on the harder questions (or topics). Memorize the formulas and learn how to use them. Obtain reference texts from your library, or ask your teacher, amateur friends, or work companions for recommendations. Study with the confidence you are only going to miss two or three questions. You can miss up to 6 questions on Element 1, 19 on Element 3, and 12 on Element 8 and still pass! *NEVER* leave an answer blank. You have a 25% chance of getting it right just by taking a wild guess!

ADDITIONAL STUDY MATERIALS

The W5YI Group has Windows®-based software programs and printed study material for all commercial radio operator exams. The software allows you to take sample tests right at your keyboard and lets you know when you have attained the skill level needed to pass any of the examinations. Call 1-800-669-9594 for more information.

CREDIT FOR PASSING ELEMENTS

If you are striving for a GROL, you get a credit document (called a PPC —for *Proof-of-Passing Certificate*) if you pass only one element and not the other. You have a whole year to complete the failed element before you lose examination credit. Also, if you pass Element 1 and fail Element 3, you still qualify for the Marine Radio Operator Permit. While you do not have to accept the MROP, you can obtain additional time credit for Element 1 if you do because the MROP grants Element 1 credit as long as the license is valid. (MROPs are issued for a 5-year term.)

4

GMDSS – Global Maritime Distress and Safety System

GMDSS — GLOBAL MARITIME DISTRESS AND SAFETY SYSTEM
Morse Code

Since the invention of radio at the end of the 19th Century, ships at sea have relied on Morse code. In 1906 the ITU, then the *International Telegraph Union,* agreed at a Berlin conference that ships of every nation should be equipped with wireless radio, and 500 kHz was set aside as the first international distress frequency. The need for ship and coast radio stations to have and use radiotelegraph equipment, and to constantly monitor a common radio frequency for Morse encoded distress calls, was heightened after the sinking of the *Titanic* in the North Atlantic in 1912.

Morse encoded distress calling has saved thousands of lives since its inception almost a century ago, but its use requires skilled radio operators spending many hours listening to the radio distress frequency. Its range on the medium frequency (MF) distress band (500 kHz) is limited, as is the amount of traffic Morse signals can carry.

IMO

Over fifteen years ago the *International Maritime Organization (IMO),* a United Nations agency specializing in safety of shipping and preventing ships from polluting the seas, began looking at ways to improve maritime distress and safety communications. Established in 1959, the IMO is the international governing body for the maritime service. Its member nations account for more than 97 percent of the world's ocean shipping. Among its duties is the specification of equipment to be carried aboard certain classes of ships.

In 1979, a group of experts drafted the *International Convention on Maritime Search and Rescue,* which called for development of a global search and rescue plan. This group also passed a resolution calling for development by IMO of a *Global Maritime Distress and Safety System (GMDSS)* to provide the communication support needed to implement the search and rescue plan.

THE GMDSS

The Global Maritime Distress and Safety System — a global plan for emergency alerting and search and rescue on the high seas — is known primarily by its acronym GMDSS. It represents the biggest improvement in marine safety since the introduction of radio. GMDSS was officially adopted in November, 1988 when world shipping leaders met in London for a two-week conference of the International Maritime Organization. An overview of the system is shown in *Figure 4-1.*

Under the GMDSS, licensed radio operators on board GMDSS-equipped ships use modern equipment to send distress alerts over long distances with assurance that they will be received on shore. The GMDSS represents more than a decade of work by the IMO and the *International Telecommunication Union (ITU),* which is headquartered in Geneva.

Nearly 200 nations strong, the ITU meets regularly to agree on radio operating procedures and on the allocation of radio frequencies. At the 1983 and 1987 *World Administrative Radio Conference for Mobile Services* (WARC-MOB-83 and 87) and WARC-92, the ITU adopted GMDSS-associated amendments to the *International Radio Regulations,* which established the frequencies, operational procedures and radio personnel requirements for GMDSS.

A year later, world shipping leaders gave the go ahead for the introduction of new automatic communications that would mean the end of Morse code for ships at sea. The IMO amended the 1974 *SOLAS (Safety of Life at Sea) Convention* to implement the Global Maritime Distress and Safety System internationally. The amendments were initially effective on February 1, 1992, and the GMDSS was fully implemented by February 1, 1999. In fact, the satellite-based GMDSS has completely replaced shipboard radiotelegraphers. The implementation schedule is shown in *Table 4-1.*

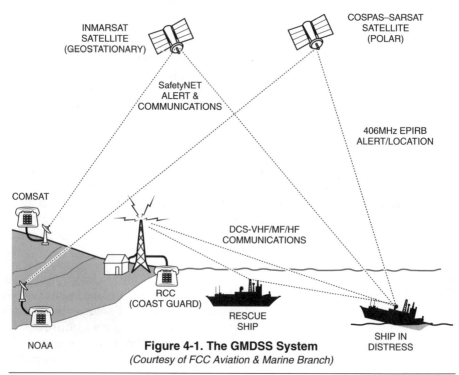

Figure 4-1. The GMDSS System
(Courtesy of FCC Aviation & Marine Branch)

The New System

The Global Maritime Distress and Safety System is an automated ship-to-shore distress alerting system that relies on satellite and advanced land-based communications. The system allows the crew to send a distress signal by simply pushing a button. Such action should prevent ships from disappearing without a trace when messages cannot be sent in time. In addition, ships must carry a radio beacon which will give the ship's position, and this device must be able to float free if the vessel sinks suddenly. By incorporating these innovative techniques into the safety system and using ship-to-shore communications links, GMDSS significantly improves the safety of life and property at sea throughout the world.

The GMDSS consists of several systems, some of which are new, but many of which have been in operation for many years. The system is able to reliably perform the following functions: alerting (including position determination of the unit in distress), search and rescue coordination, locating (homing), maritime safety information broadcasts, general communications, and bridge-to-bridge communications. Specific radio carriage requirements depend upon the ship's area of operation, rather than its tonnage. The system also provides redundant means of distress alerting, and emergency sources of power.

This system, which the world's maritime nations, including the United States, have implemented has changed international distress communications from being primarily ship-to-ship based to ship-to-shore. It spelled the end of Morse code communications for all but a few users, such as Amateur Radio. The GMDSS provides for automatic distress alerting and locating in cases where a radio operator doesn't have time to send an SOS or MAYDAY call, and, for the first time, requires ships to receive broadcasts of maritime safety information which could prevent a distress from happening in the first place.

Table 4-1. GMDSS Implementation Schedule

Date:	Compliance Schedule:
Feb. 1, 1992	Voluntary compliance, any ship may be GMDSS equipped
Aug. 1, 1993	All compulsory ships must have 406-MHz EPIRB and carry a NAVTEX receiver. Manual telegraphy and watch keeping on the 500 kHz distress frequency discontinued by U.S. Coast Guard.
Feb. 1, 1995	Newly constructed compulsory ships must be GMDSS equipped. All ships must be fitted with one radar capable of operating in the 9-GHz band.
Feb. 1, 1999	All compulsory ships must be GMDSS equipped. The Coast Guard will continue to monitor all distress frequencies.
Feb. 1, 2005	The date ships subject to the *Safety of Life Convention* can suspend their watch on VHF Channel 16.

U.S. Support

The United States has been a strong advocate of the GMDSS internationally. One of the biggest advantages of the GMDSS over the former system is that it eliminates reliance on a single person for communications. It requires at least two licensed GMDSS operators and typically two maintenance methods to ensure distress communications capability at all times. It also eliminates the need for manual watchkeeping. The Channel 16 VHF and 2182 kHz MF networks have been upgraded to include the GMDSS Digital Selective Calling (DSC) on Channel 70 VHF and 2187.5 kHz MF. The Coast Guard plans to maintain the shore watch on Channel 16 VHF, the 2182 kHz MF and HF distress frequencies for the foreseeable future.

Former System

The former distress and safety plan at sea was primarily a manual, ship-to-ship system that relied on Morse code radiotelegraphy on 500 kHz and voice telephony on 2182 kHz (MF) and 156.8 MHz (VHF Channel 16). It frequently was found to be unreliable since its effectiveness depended on the location of the nearest vessel, radio-wave propagation conditions, and the technical proficiency of the radio officer. The range of MF (medium frequency) is approximately 150 miles, so for ships beyond this distance from the nearest coast station, the distress alerting depended on the availability of a nearby ship.

The Morse code, invented by American Samuel Morse and first used in 1844, was the backbone of the former system. It was the foundation of ship distress and safety communications since the turn of the last century. Radiotelegraphy was used to inform the world in 1912 that the supposedly unsinkable ocean liner *Titanic* was going down in the North Atlantic. The Titanic sank swiftly and 1,500 people perished. Later that year, the first maritime regulations were enacted.

Morse code enjoys a romantic history of gallant radio operators sending off SOS distress calls as the ship sinks. But radiotelegraphy is signing off, and Morse code communications are going away.

ARE RADIO OFFICERS STILL REQUIRED?

The former system was based on a radio officer, when at sea, maintaining a watch on specific distress frequencies. The change from manual Morse code to GMDSS was fiercely opposed by members of the various Maritime Radio Officers Unions worldwide. The primary reason for the opposition was job security. Under the new GMDSS almost anyone without Morse code knowledge could now participate in distress communications.

In an October 1990 proceeding, the Federal Communications Commission said it would adopt the international GMDSS provisions for U.S. "compulsory" vessels. A compulsory ship is defined as a large cargo ship of 300 tons gross tonnage and over and all passenger vessels that carry more than twelve passengers (regardless of size) on international voyages or in the open sea. Compulsory ships are required to carry certain radio equipment and personnel for safety purposes.

The FCC telecommunications regulations are contained in Title 47 of the Communications Act. Part 13 of Title 47 covers commercial radio operator qualifications, licenses and examinations. The Part 80 Maritime Service rules specify the radio operator, practices and equipment requirements aboard U.S. vessels. These regulations are based on the international and domestic requirements of the 1974 SOLAS Convention and the U.S. Communications Act.

Radio officers (trained in manual Morse code) are not part of the GMDSS regulations or system. Instead of a single radio officer, the GMDSS regulations require at least two GMDSS radio operators – one of whom must be dedicated to communications during a distress situation, with the second operator available as backup.

Each operator must hold a GMDSS Radio Operator's License. The GMDSS radio operator is an individual licensed to handle radio communications aboard ships in compliance with the GMDSS regulations, including basic equipment and antenna adjustments. One of the GMDSS licensed radio operators can be the current radio officer or any other qualified member of the crew holding the appropriate FCC license.

Those ships that elect at-sea repair as one of its maintenance options must also have on-board a person holding a GMDSS Radio Maintainer's License. Previously, holders of the First or Second Class Radiotelegraph Operator's Certificate or the General Radiotelephone Operator License were authorized to maintain shipboard GMDSS equipment, but this authorization expired on February 1, 1999.

HOW DOES THE GMDSS WORK?

Basically, GMDSS is a sophisticated ship-to-shore alerting system with ship-to-ship capability. Its purpose is to automate and improve distress communications and search and rescue (SAR) for the ocean shipping industry on a global basis. GMDSS regulations do not apply to fishing, recreational, or small passenger ships operated on domestic voyages.

Using Existing Systems

GMDSS is made up of several communications systems, some of which have been in operation for many years.

First, there is COSPAS-SARSAT, a joint international satellite-based search and rescue system established in Canada, France, the former USSR and the United States to locate emergency radio beacons transmitting on 121.5 and 406 MHz. The U.S. satellites in this system not only receive on 121.5 and 406 MHz, but also receive on 243 MHz. The COSPAS-SARSAT satellite system, which has been in operation since 1982, provides distress alerting using Emergency Position Indicating Radio Beacons (EPIRBs). The radio beacons located aboard ship automatically give the ship's position and the EPIRB must be able to float free if the ship sinks. There are at present three types of satellite beacons, namely airborne emergency locator transmitters (ELTs), EPIRBs (maritime) and land-based personal locator beacons (PLBs.)

Second, there is the International Maritime Satellite Organization's maritime mobile satellite system (INMARSAT), in operation since 1982, that forms a major component for distress alerting and communications. The four INMARSAT satellites operate in the 1.5 and 1.6 GHz "L-band." They are in geostationary orbits 22,300 miles above the equator and cover most of the world's ocean regions.

Third, automated terrestrial data systems and existing communications systems have been combined into one overall communications system which, together with the satellites, make up the Global Maritime Distress and Safety System.

New Systems

The GMDSS provides for new digital selective calling (DSC) services on the high-frequency (HF), medium-frequency (MF) or very-high-frequency (VHF) bands, depending upon the location of ship in distress. These new DSC services are used for ship-to-ship, ship-to-shore, and shore-to-ship automatic alerting, while existing terrestrial HF, MF and VHF radiotelephony equipment provides distress, urgent and safety-related communications.

The GMDSS enhances search and rescue operations at sea through the use of the 9-GHz search and rescue radar transponder (SART). And it creates a global network for the dissemination of maritime safety information (MSI) using three systems: NAVTEX, INMARSAT's enhanced group calling (EGC), and HF narrow-band direct-printing (NBDP) radiotelegraphy.

NAVTEX is an English language direct printing meteorological, warning and urgent safety information dissemination service that operates on 518 kHz. A selective message-rejection feature allows only certain communications to be received.

The enhanced group call (EGC) system was developed by INMARSAT to enable pre-determined groups of ships (or all vessels) to receive warning and distress messages. All ships must carry equipment for receiving MSI broadcasts. Manual Morse code is not part of GMDSS at all.

MOST NOTABLE FEATURES

The two most notable features of the GMDSS are that it is based on sea areas of operation and that it offers multiple communications options. The first of these notable features, sea area basing, divides the seas into four communications areas.

Sea Areas

GMDSS sea areas serve two purposes: to describe areas where GMDSS services are available, and to define what GMDSS equipment ships must carry. Prior to the GMDSS, the number and type of radio safety equipment ships had to carry depended upon its tonnage. With GMDSS, the number and type of radio safety equipment ships must carry depends upon the areas in which they travel.

The sea areas described in *Table 4-2* are established by individual countries, which equip their shore stations with appropriate VHF, MF, HF or satellite facilities to cover the particular segments of ocean.

Table 4-2. Ocean Communications Areas

Sea Area	Description
Sea Area A1	Sea area A1 is the area within VHF radiotelephone coverage of at least one coast station at which continuous DSC (digital selective calling) is available (approximately 20-30 miles). A1 ships must carry VHF equipment and either a satellite EPIRB or a VHF EPIRB.
Sea Area A2	Sea area A2 is the area within MF (medium frequency) radiotelephone coverage of at least one coast station at which continuous DSC is available (approximately 75 to 150 miles), excluding Sea Area A1. A2 ships must carry VHF and MF equipment and a satellite EPIRB.
Sea Area A3	Sea area A3 is the area within the coverage of an INMARSAT geostationary satellite in which continuous alerting is available (approximately 70° North to 70° South), excluding Sea Areas A1 and A2. A3 ships must carry VHF and MF equipment, a satellite EPIRB and either HF or satellite communications. (Most of the world's ocean area is in sea area A3.)
Sea Area A4	Sea area A4 is the remainder of the seas of the world (essentially the polar regions) and relies primarily on HF communications. A4 ships must carry VHF, MF and HF equipment and a satellite EPIRB.

Multiple Communications Options

The second notable feature of the GMDSS, multiple communications options, ensures that each ship using the GMDSS has at least two options of distress alerting appropriate to its sea area. This redundancy minimizes the chance that a ship in distress will be unable to communicate because of weather, radio propagation difficulties, equipment failure, or other circumstances. The multiple communications options represent a significant improvement over the former system.

OVERALL SEARCH AND RESCUE CONCEPT

The basic concept of the GMDSS is that the search and rescue (SAR) authorities on shore, as well as shipping in the immediate vicinity of the ship in distress, can be rapidly alerted to a distress incident. The shore-based authorities designated as a Rescue Coordination Center (RCC) can then direct rescue operations with minimal delay.

In the United States, the Coast Guard is the designated maritime SAR organization and it operates the necessary RCCs. The particular GMDSS equipment used to communicate varies by sea area and may have several alternatives. *Table 4-3* is a simplified chart of GMDSS equipment and its primary functions.

Table 4-3. GMDSS Shipboard Radio Equipment

Equipment:	Function:
406-MHz EPIRB	Ship-to-shore alerts via COSPAS-SARSAT satellite
VHF radio (DSC and voice)	SAR Communications (Search and Rescue)
MF radio (DSC and voice)	Ship-to-shore alerts and communications
HF radio – (required for sea area A4)	Ship-to-shore alerts and communications
INMARSAT ship earth station plus EGC capability	Ship-to-shore alerts, communications and MSI (SafetyNET)
NAVTEX receiver	MSI (SafetyNET) 518 kHz
9-GHz SART	SAR locating beacon
2-way VHF portable radios	SAR communications
2182-kHz watch receiver/auto alarm	Receipt of 2182-kHz alerts

EGC=Enhanced group calling, MSI=Marine safety information
SAR=Search and rescue, SART=Search and rescue transponder

GMDSS LICENSES

The IMO conference prescribes two levels of GMDSS operators: GMDSS radio operators (DO) and GMDSS radio maintainers (DM). These two GMDSS commercial radio licenses were introduced by the FCC in late 1993 to accommodate GMDSS and are now two of the most popular commercial licenses – second only to the General Radiotelephone Operator License. There also is a combined GMDSS Radio Operator/Maintainer license (DB) for examinees who pass all examination requirements at one sitting.

In late 2002, the FCC added the Restricted GMDSS radio operator license (DO-R), for ships that operate exclusively in Sea Area A1.

While anyone can obtain a GMDSS Operator or Restricted Operator license by passing the required FCC examinations, *only those persons who take and pass the U.S. Coast Guard 70-hour GMDSS Certification Course can serve on board a ship as a GMDSS operator.* For more information about the U.S. Coast Guard requirements and courses, call the Coast Guard at (202) 493-1015, or visit their Merchant Mariner Information Center website at: www.uscg.mil/STCW/m-pers.htm and then click on Approved Courses, and finally click on GMDSS for a list of organizations offering USCG Approved GMDSS Courses.

GMDSS Radio Operators

Each GMDSS ship must carry two GMDSS-qualified radio operators for distress and safety radio communications purposes. Each must hold the GMDSS Radio Operator's License (DO). One operator is designated as having primary responsibility for radio communications during distress incidents. The second operator provides back-up.

The GMDSS Radio Operator is an individual qualified to handle radio communications aboard ships in compliance wit the GMDSS regulations, including basic equipment and antenna adjustments.

Examination Requirements

The GMDSS Radio Operator (DO) license is obtained by passing commercial radio license examinations on Element 1 (basic marine radio law) and Element 7 (GMDSS radio operating practices). Holders of the Marine Radio Operator Permit receive examination credit for Element 1. The GMDSS DO license term is for 5 years, and can be renewed.

Note that persons who take the 70-hour U.S. Coast Guard GMDSS radio operators certification course and pass that examination receive examination credit for Element 7. They can obtain their GMDSS radio operators license by taking the FCC examination for Element 1, and presenting their proof of passing certificate to the COLEM for application processing. Only a COLEM can process and submit an application for a new GMDSS license. Individuals are allowed to renew their license directly with the FCC.

Restricted GMDSS Radio Operators

The FCC now issues a Restricted GMDSS Radio Operator license (DO-R), in keeping with the International Telecommunications Union (ITU) regulations. The ITU regulations provide for two types of GMDSS radio operators certificates: a General Operators Certificate (equivalent to the GMDSS Radio Operators License), and a Restricted Operators Certificate for compulsory ships that operate exclusively within Sea Area A1 (see Table 4-2).

As of late 2002, FCC rules permit compulsory vessels that are required to carry only the GMDSS equipment described for Sea Area A1 to be staffed with GMDSS operators who hold only the Restricted GMDSS Radio Operators License (DO-R). A Restricted Operator's License requires familiarity with all of the GMDSS equipment required for vessels sailing within Sea Area A1, including VHF DSC procedures, and basic radio law and operating practice with which every maritime radio operator should be familiar.

Examination Requirements

The Restricted GMDSS Radio Operator (DO-R) license is obtained by passing commercial radio license examinations on Element 1 (basic marine radio law) and Element 7A (GMDSS radio operating practices). Element 7A consists of questions taken from Element 7 that pertain to the use of equipment required on ships sailing solely within Sea Area A1, or, if appropriate, Sea Area A2. Holders of the Marine Radio

Operator Permit receive examination credit for Element 1. The GMDSS DO-R license term is for 5 years, and can be renewed.

GMDSS Radio Maintainers

International GMDSS regulations provide three methods to ensure that a ship's radio equipment is functionally capable of providing communications in distress situations: 1. duplication of equipment; 2. shore-based maintenance; 3. at-sea maintenance, or a combination of these methods. Two of these three methods are required for ships operating in Sea Areas A3 and A4.

Duplication of equipment does not mean that GMDSS ships must carry "two sets of everything." *It does mean that certain equipment critical to radio communications during a distress situation is required to be duplicated.*

Ships that use duplication and/or shore-based maintainance options, are not required to carry an on-board, licensed GMDSS Radio Maintainer. Ships electing at-sea maintenance are required to carry a licensed GMDSS Radio Maintainer. The GMDSS Radio Maintainer (DM) my be one of the GMDSS Radio Operators or a different member of the crew as long as he/she holds the license.

Examination Requirements

The GMDSS Radio Maintainer license (DM) is obtained by passing commercial radio license examinations on Element 1 (basic marine radio law), Element 3 (general radiotelephone electronics), and Element 9 (GMDSS radio maintenance practice and procedures). Holders of the Marine Radio Operator Permit receive examination credit for Element 1. Holders of the General Radiotelephone Operators License receive examination credit for Element 1 and Element 3. The GMDSS Radio Maintainer License term is for 5 years, and may be renewed.

As explained in Chapter 3, Commercial Operator License Examination Managers (COLEMs) administer the testing program for all 3 of the GMDSS licenses. See the Appendix for a complete listing of COLEM organizations. Each of them has testing organizations, and some (e.g. National Radio Examiners) provide copies of the GMDSS Elements 7, 7A, and 9 question pools, as well as the Elements 5 and 6 radiotelegraph examination question pools.

WHAT ABOUT COMMERCIAL RADIOTELEGRAPH?

For all practical purposes, radio telegraphers when the way of the horse and buggy on February 1, 1999. They are no longer needed. Element 5 and 6 are the examinations needed for the 1st, 2nd, and 3rd Class Radiotelegraph licenses. Extra Class amateur operators who hold licenses dated on or before April 14, 2000, and who passed the amateur Element 1-C 20 wpm Morse code examination, receive credit for the 2nd Class commercial telegraphy examination without testing.

5

Question Pool – Element 1

The questions asked in each Element 1 examination must be taken from the most recent question pool. The release date of this pool is October 1, 1995 and the effective date is January 1, 1996; therefore, all Element 1 examinations administered after January 1, 1996 must use these questions.

An Element 1 examination is used to prove that the examinee possesses the qualifications to operate licensed radio stations required of a person holding a Marine Radio Operator Permit (MP). An examination on Element 1 is also necessary to properly perform the duties required of a person holding a General Radiotelephone Operator License (GROL) [PG]. Element 1 is also a partial requirement for the Global Maritime Distress and Safety System Maintainer's License and the three Radio-telegraph Operator Certificates, T1, T2, and T3.

An MP must be held by a person required to operate radiotelephone stations (1) aboard certain vessels that sail the Great Lakes; (2) aboard vessels of more than 300 gross tons; (3) aboard vessels that carry more than six passengers for hire; and (4) in certain aviation and coast stations.

Each Element 1 examination is administered by a Commercial Operator License Examination Manager (COLEM). The COLEM must construct the examination by selecting 24 questions from this 170-question pool. You must have 18 or more answers correct from the 24 in order to pass the examination.

In this book, each question in the pool has the question, its four multiple-choice answers, the right answer identified, and a detailed explanation of why the right answer is correct. The COLEM may change the order of the answer and distracters (incorrect choices) of questions from the order that appears in this pool.

Suggestions concerning improvement to the questions in the pool or submittal of new questions for consideration in updated pools may be sent to technical author Gordon West, 2414 College Drive, Costa Mesa, CA 92626.

1A1 What is the Global Maritime Distress and Safety System (GMDSS)?
- A. An automated ship-to-shore distress alerting system using satellite and advanced terrestrial communications systems
- B. An emergency radio service employing analog and manual safety apparatus
- C. An association of radio officers trained in emergency procedures
- D. The international organization charged with the safety of ocean-going vessels

ANSWER A: This system, coordinated worldwide by the International Maritime Organization (IMO), provides rapid transfer of a ship's distress message to units best suited for providing assistance. With the GMDSS, certain frequency bands are set aside and each station is assigned a unique call sign. See FCC Rule § 80.5.

1A2 What authority does the Marine Radio Operator Permit confer?
- A. Grants authority to operate commercial broadcast stations and repair associated equipment
- B. Allows the radio operator to maintain equipment in the Business Radio Service
- C. Confers authority to operate licensed radio stations in the Aviation, Marine and International Fixed Public Radio Services
- D. The non-transferable right to install, operate and maintain any type-accepted radio transmitter

ANSWER C: The MROP confers authority to operate licensed radio stations. It is earned by passing an exam containing 24 questions from this question pool. 18 correct answers pass. Element 1 question pool concentrates on operating techniques and practices in the maritime service. It is not concerned with maintenance and repair. See *Table 2-2* of Chapter 2.

1A3 Which of the following persons are ineligible to be issued a commercial radio operator license?
- A. Individuals who are unable to send and receive correctly by telephone spoken messages in English
- B. Handicapped persons with uncorrected disabilities which affect their ability to perform all duties required of commercial radio operators
- C. Foreign maritime radio operators unless they are certified by the International Maritime Organization (IMO)
- D. U.S. Military radio operators who are still on active duty

ANSWER A: By FCC Rule § 13.9, persons who cannot transmit or receive correctly voice messages in English are not eligible for a commercial operator license. This includes the mute and deaf. Persons afflicted with other physical handicaps may be issued a commercial radio license, if found qualified, with certain restrictive endorsements.

1A4 Who is required to make entries on a required service or maintenance log?
- A. The licensed operator or a person whom he or she designates
- B. The operator responsible for the station operation or maintenance

C. Any commercial radio operator holding at least a Restricted Radio-telephone Operator Permit

D. The technician who actually makes the adjustments to the equipment

ANSWER B: Since the person servicing the radio equipment knows better than anyone else what was required to fix a problem, that person is best qualified to state the problem and the steps taken to repair it in the maintenance log. See FCC Rule § 80.409.

1A5 What is a requirement of every commercial operator on duty and in charge of a transmitting system?

A. A copy of the Proof-of-Passing Certificate (PPC) must be on display at the transmitter location.

B. The original license or a photocopy must be posted or in the operator's personal possession and available for inspection.

C. The FCC Form 756 certifying the operator's qualifications must be readily available at the transmitting system site.

D. A copy of the operator's license must be supplied to the radio station's supervisor as evidence of technical qualification.

ANSWER B: The person in charge of maintaining correct operation of a commercial radio station must be able to provide proof of technical expertise and competence by prominently displaying his or her FCC commercial radio license while on duty. See FCC Rule § 13.19(c).

1A6 What is distress traffic?

A. In radiotelegraphy, SOS sent as a single character; in radiotelephony, the speaking of the word, "Mayday"

B. Health and welfare messages concerning the immediate protection of property and the safety of human life

C. Internationally recognized communications relating to emergency situations

D. All messages relative to the immediate assistance required by a ship, aircraft or other vehicle in imminent danger

ANSWER D: While choice A is technically correct, choice D covers the entire spectrum of possible situations in which distress traffic may occur. See definitions in § 80.5 and § 80.314.

1A7 What is a maritime mobile repeater station?

A. A fixed land station used to extend the communications range of ship and coast stations

B. An automatic on-board radio station which facilitates the transmissions of safety communications aboard ship

C. A mobile radio station which links two or more public coast stations

D. A one-way, low-power communications system used in the maneuvering of vessels

ANSWER A: A repeater automatically retransmits everything it hears, adding strength to the original signal, and extending its range. Public correspondence telephone service is a repeater station. See definitions in § 80.5.

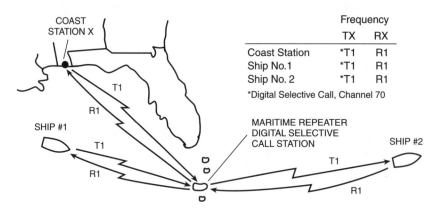

| | Frequency | |
	TX	RX
Coast Station	*T1	R1
Ship No.1	*T1	R1
Ship No. 2	*T1	R1

*Digital Selective Call, Channel 70

Maritime Repeater DSC Station

1A8 What is an urgency transmission?
A. A radio distress transmission affecting the security of humans or property
B. Health and welfare traffic which impacts the protection of on-board personnel
C. A communications alert that important personal messages must be transmitted
D. A communications transmission concerning the safety of a ship, aircraft or other vehicle, or of some person on board or within sight

ANSWER D: Urgency traffic covers situations which potentially could worsen into distress traffic. A person on board who is ill or near death could require an urgency message. See definitions in § 80.5, also § 80.327.

1A9 What is a ship earth station?
A. A maritime mobile-satellite station located at a coast station
B. A mobile satellite location located on board a vessel
C. A communications system which provides line-of-sight communications between vessels at sea and coast stations
D. An automated ship-to-shore distress alerting system

ANSWER B: A ship earth station allows messages to be exchanged between a ship and a satellite orbiting the earth. Many navigation systems use such equipment. While an earth station may be located anywhere on the planet, a ship earth station resides only on a ship. See definitions in § 80.5.

1A10 What is the internationally recognized urgency signal?
A. The letters "TTT" transmitted three times by radiotelegraphy
B. Three oral repetitions of the word "safety" sent before the call
C. The word "PAN PAN" spoken three times before the urgent call
D. The pronouncement of the word "Mayday"

ANSWER C: Remember that "Mayday" is reserved for distress traffic only; "PAN PAN PAN" is for urgency. See § 80.327.

1A11 What is a safety transmission?
A. A radiotelephony warning preceded by the words "PAN"
B. Health and welfare traffic concerning the protection of human life

C. A communications transmission which indicates that a station is preparing to transmit an important navigation or weather warning

D. A radiotelegraphy alert preceded by the letters "XXX" sent three times

ANSWER C: Choices A and B are incorrect because they pertain to urgency traffic. Safety traffic concerns only navigation and weather information. See definitions in § 80.5, also § 80.329.

1A12 What is a requirement of all marine transmitting apparatus used aboard United States vessels?

A. Only equipment that has been type-accepted by the FCC for Part 80 operations is authorized.

B. Equipment must be approved by the U.S. Coast Guard for maritime mobile use.

C. Certification is required by the International Maritime Organization (IMO).

D. Programming of all maritime channels must be performed by a licensed Marine Radio Operator.

ANSWER A: All the rules and regulations covered in this question pool come directly from Part 80 — Stations in the Maritime Services. Technical standards for radio equipment are covered in Subpart E. See § 80.43 and § 80.203(a).

1A13 Where do you submit an application for inspection of a ship radio station?

A. To a Commercial Operator Licensing Examination Manager (COLE Manager)

B. To the Federal Communications Commission, Washington, DC 20554

C. To the Engineer-in-Charge of the FCC District Office nearest the proposed place of inspection

D. To the nearest International Maritime Organization (IMO) review facility

ANSWER C: Per FCC Rule § 80.59, an application for inspection and certification must be submitted to the engineer in charge of the FCC District Office at least three days before the proposed inspection date.

1A14 What are the antenna requirements of a VHF telephony coast, maritime utility or ship station?

A. The shore or on-board antenna must be vertically polarized.

B. The antenna array must be type-accepted for 30-200 MHz operation by the FCC

C. The horizontally-polarized antenna must be positioned so as not to cause excessive interference to other stations.

D. The antenna must be capable of being energized by an output in excess of 100 watts.

ANSWER A: A vertically-polarized antenna takes up less space on board a ship and avoids the potential problem of cross-polarization. By requiring all antennas to be polarized the same way, the chance of receiving radio signals properly is greatly increased. Also, a non-directional antenna, such as an omni-directional one, reduces the chance that a distress call could be missed simply because the ship's antenna is pointing the wrong way. See § 80.72 and § 80.81.

1A15 What regulations govern the use and operation of FCC-licensed ship stations in international waters?

A. The regulations of the International Maritime Organization (IMO) and Radio Officers Union

B. Part 80 of the FCC Rules plus the international Radio Regulations and agreements to which the United States is a party

C. The Maritime Mobile Directives of the International Telecommunication Union

D. Those of the FCC's Aviation and Marine Branch, PRB, Washington, DC 20554

ANSWER B: Part 80 specifies rules for operating procedures, technical standards, safety watch requirements, emission classes, transmitter power, transmitter licensing, frequencies and tolerances, and more.

1A16 Which of the following transmissions are not authorized in the Maritime Service?

A. Communications from vessels in dry dock undergoing repairs

B. Message handling on behalf of third parties for which a charge is rendered

C. Needless or superfluous radiocommunications

D. Transmissions to test the operating performance of on- board station equipment

ANSWER C: With so many vessels on the water, spending time on the air exchanging trivial information could prevent a station from getting through with an important message. Keep all messages short and to the point. See § 80.89.

1A17 What are the highest-priority communications from ships at sea?

A. All critical message traffic authorized by the ship's master

B. Navigation and meteorological warnings

C. Distress calls, and communications preceded by the international urgency and safety signals

D. Authorized government communications for which priority right has been claimed

ANSWER C: The ultimate priority in radio traffic goes to those messages whose reception could mean the difference between life and death. In such cases, rules and regulations pertaining to the legality of frequency use and duration are temporarily suspended. See § 80.312 and § 80.91.

1A18 What is the best way for a radio operator to minimize or prevent interference to other stations?

A. By using an omni-directional antenna pointed away from other stations

B. Reducing power to a level that will not affect other on-frequency communications

C. By changing frequency when notified that a radiocommunication causes interference

D. Determine that a frequency is not in use by monitoring the frequency before transmitting

ANSWER D: Per Answer D, even a few watts of power used on board a ship can be heard quite clearly thousands of miles away. This can disrupt other radio services, perhaps even masking a distress call from another ship. You may only be able to hear one of two stations, so listen for a few minutes before transmitting—it's not only polite, but it's also the law. See § 80.92 and § 80.87. Answer A is not correct because an omni-directional antenna transmits equally well in all directions; you can't point it away from other stations. Answers B and C try to solve a problem after causing it.

1A19 Under what circumstances may a ship or aircraft station interfere with a public coast station?
A. Under no circumstances during on going radiocommunications
B. During periods of government priority traffic handling
C. When it is necessary to transmit a message concerning the safety of navigation or important meteorological warnings
D. In cases of distress

ANSWER D: Since distress messages have the ultimate priority, operators may break in on any station at any time. If the operator of the distress station feels that the best way to attract attention is by disrupting another station, then (and only then!) it is legal to do so. All other messages must be sent in such a manner as to not disrupt other stations. See § 80.312.

1A20 Who determines when a ship station may transmit routine traffic destined for a coast or government station in the maritime mobile service?
A. Shipboard radio officers may transmit traffic when it will not interfere with ongoing radiocommunications.
B. The order and time of transmission and permissible type of message traffic is decided by the licensed on-duty operator.
C. Ship stations must comply with instructions given by the coast or government station.
D. The precedence of conventional radiocommunications is determined by FCC and international regulation.

ANSWER C: Ensure the smooth flow of information by following a coast or government station's requests to the letter. Making sudden demands on the coast station could upset their methods of handling paperwork, with the possible result that your message gets damaged or even lost. See § 80.116(g).

1A21 Who is responsible for payment of all charges accruing to other facilities for the handling or forwarding of messages?
A. The licensee of the ship station transmitting the messages
B. The third party for whom the message traffic was originated
C. The master of the ship, jointly with the station licensee
D. The licensed commercial radio operator transmitting the radiocommunication

ANSWER A: The person who owns the station license is held responsible for paying any charges that another station might present for handling message traffic. Most communications between a ship and a land station involve scheduling of vessel movements, obtaining supplies, and arranging for repairs. No charge is made for a distress, urgency, or safety message. See § 80.95.

1A22 Ordinarily, how often would a station using a telephony emission identify?

A. At least every 10 minutes
B. At 15-minute intervals, unless public correspondence is in progress
C. At the beginning and end of each transmission and at 15-minute intervals
D. At 20-minute intervals

ANSWER C: Stations engaged in conversation must identify their stations periodically. This not only lets listeners know who is communicating, but it also satisfies the FCC's rule that only licensed operators are allowed to use the airwaves. Operators must give their station call sign in English. See § 80.102.

1A23 When does a maritime radar transmitter identify its station?

A. By radiotelegraphy at the onset and termination of operation
B. At 20-minute intervals, using an automatic transmitter identification system
C. Radar transmitters must not transmit station identification.
D. By a transmitter identification label (TIL) secured to the transmitter

ANSWER C: Per FCC Rule § 80.104, a radar station is a self-contained transmitter and receiver that is designed to respond only to its own emissions and is not meant to be received by any other stations.

1A24 What is the general obligation of a coast or marine-utility station?

A. To accept and dispatch messages without charge, which are necessary for the business and operational needs of ships
B. To acknowledge and receive all calls directed to it by ship or aircraft stations
C. To transmit lists of call signs of all fixed and mobile stations for which they have traffic
D. To broadcast warnings and other information for the general benefit of all mariners

ANSWER B: The purpose of coast stations and marine-utility stations is to exchange messages. They cannot legally ignore any call. A coast station is simply a maritime radio station on land. A marine-utility station is simply a handheld radiotelephone unit, usually used while aboard a vessel. See § 80.105.

1A25 How does a coast station notify a ship that it has a message for the ship?

A. By making a directed transmission on 2182 kHz or 156.800 MHz
B. The coast station changes to the vessel's known working frequency
C. By establishing communications using the eight-digit maritime mobile service identification
D. The coast station may transmit, at intervals, lists of call signs in alphabetical order for which they have traffic.

ANSWER D: This "bulletin-board" method of traffic notification works well. It allows ships to find out if they have messages without querying the coast station. You should listen several times during the day, as the message queue constantly changes. See § 80.108.

1A26 Under what circumstances may a coast station using telephony transmit a general call to a group of vessels?
A. Under no circumstances
B. When announcing or preceding the transmission of distress, urgency, safety or other important messages
C. When the vessels are located in international waters beyond 12 miles
D. When identical traffic is destined for multiple mobile stations within range

ANSWER B: Messages concerning health and welfare are of importance to everyone. This timely information flows much more smoothly when offered to all stations simultaneously, rather than one ship at a time. See § 80.111(a).

1A27 Who has ultimate control of service at a ship's radio station?
A. The master of the ship
B. A holder of a First Class Radiotelegraph Certificate with a six months' service endorsement
C. The Radio Officer-in-Charge authorized by the captain of the vessel
D. An appointed licensed radio operator who agrees to comply with all Radio Regulations in force

ANSWER A: The master of the ship (usually the owner or "captain" of the vessel) owns the station license and enjoys ultimate authority on the air. He is permitted to designate a qualified operator to handle radio messages. See § 80.114.

1A28 What is the power limitation of associated ship stations operating under the authority of a ship station license?
A. The power level authorized to the parent ship station
B. Associated vessels are prohibited from operating under the authority granted to another station licensee.
C. The minimum power necessary to complete the radiocommunications
D. Power is limited to one watt.

ANSWER D: An associated ship station is a VHF transmitter, usually a hand-held transceiver. It may be used only in the vicinity of the ship station with which it is associated. It must not communicate with any other stations, so its power is limited. It may not be used from shore. See definitions in § 80.5 and § 80.115(a).

1A29 How is an associated vessel operating under the authority of another ship station license identified?
A. All vessels are required to have a unique call sign issued by the Federal Communications Commission.
B. With any station call sign self-assigned by the operator of the associated vessel
C. By the call sign of the station with which it is connected and an appropriate unit designator
D. Client vessels use the call sign of their parent plus the appropriate ITU regional indicator.

ANSWER C: More than one associated ship station may be in use at one time. A unique suffix added to the ship station's call sign identifies individual stations. See § 80.115(a).

1A30 On what frequency should a ship station normally call a coast station when using a radiotelephony emission?
- A. On a vacant radio channel determined by the licensed radio officer
- B. Calls should be initiated on the appropriate ship-to-shore working frequency of the coast station.
- C. On any calling frequency internationally approved for use within ITU Region 2
- D. On 2182 kHz or 156.800 MHz at any time

ANSWER B: A few VHF channel frequencies are set aside strictly for this type of message traffic. Commercial pilots use them for the movement and docking of vessels near ports, locks and waterways. See § 80.116(a).

1A31 On what frequency would a vessel normally call another ship station when using a radiotelephony emission?
- A. Only on 2182 kHz in ITU Region 2
- B. On the appropriate calling channel of the ship station at 15 minutes past the hour
- C. On 2182 kHz or 156.800 MHz, unless the station knows the called vessel maintains a simultaneous watch on another intership working frequency
- D. On the vessel's unique working radio channel assigned by the Federal Communications Commission

ANSWER C: Not only are 2182 kHz and 156.800 MHz distress frequencies, but they are also calling frequencies. If you do not know the intership working frequency in advance, make a call on the calling frequency. When you get a response, ask about the intership frequency and move to it. A radio "watch" is the act of listening on a designated frequency for any possible distress messages. See definitions in § 80.5, and § 80.116(b).

1A32 What is required of a ship station which has established initial contact with another station on 2182 kHz or 156.800 MHz?
- A. The stations must check the radio channel for distress, urgency and safety calls at least once every ten minutes.
- B. The stations must change to an authorized working frequency for the transmission of messages.
- C. Radiated power must be minimized so as not to interfere with other stations needing to use the channel
- D. To expedite safety communications, the vessels must observe radio silence for two out of every fifteen minutes.

ANSWER B: After making contact on the calling frequency, move to another frequency to keep the distress channel clear. See § 80.116(c).

1A33 What type of communications may be exchanged by radioprinter between authorized private coast stations and ships of less than 1600 gross tons?
- A. Public correspondence service may be provided on voyages of more than 24 hours.
- B. All communications, providing they do not exceed 3 minutes after the stations have established contact

C. Only those communications which concern the business and operational needs of vessels

D. There are no restrictions.

ANSWER C: Per FCC Rule § 80.1155, radioprinter and facsimile communications
are also allowed between ships to accomplish the business and operation of the
ships.

1A34 What are the service requirements of all ship stations?

A. Each ship station must receive and acknowledge all communications
with any station in the maritime mobile service.

B. Public correspondence services must be offered for any person during
the hours the radio operator is normally on duty.

C. All ship stations must maintain watch on 500 kHz, 2182 kHz and
156.800 MHz.

D. Reserve antennas, emergency power sources and alternate communications installations must be available.

ANSWER A: All maritime radio operators must exchange information without
delay or concealment. See § 80.141(b).

**1A35 When may the operator of a ship radio station allow an unlicensed
person to speak over the transmitter?**

A. At no time. Only commercially-licensed radio operators may modulate the transmitting apparatus.

B. When the station power does not exceed 200 watts peak envelope
power

C. When under the supervision of the licensed operator

D. During the hours that the radio officer is normally off duty

ANSWER C: In some cases a telephone call must be made from ship to shore;
e.g., a passenger may want to call home. The passenger does not own the
station license, therefore, the station operator supervises the communication to
keep it legal. A marine operator is contacted over a VHF frequency and the call
is placed just like any other phone call. The holder of the ship station license is
charged for the call. See § 80.156.

**1A36 What are the radio operator requirements of a cargo ship equipped
with a 1000 watt peak-envelope-power radiotelephone station?**

A. The operator must hold a General Radiotelephone Operator License
or higher-class license.

B. The operator must hold a Restricted Radiotelephone Operator
Permit or higher-class license.

C. The operator must hold a Marine Radio Operator Permit or higher-
class license.

D. The operator must hold a GMDSS Radio Maintainer's License.

ANSWER C: Per FCC Rule § 80.159, operating a radiotelephone station aboard a
cargo ship requires an MROP as long as the transmitter's output power is less
than 1500 watts peak envelope power. If it exceeds 1500 watts, then a General
Radiotelephone Operator License is required.

1A37 What are the radio operator requirements of a small passenger ship carrying more than six passengers equipped with a 1000-watt carrier power radiotelephone station?
A. The operator must hold a General Radiotelephone Operator or higher-class license.
B. The operator must hold a Marine Radio Operator Permit or higher-class license.
C. The operator must hold a Restricted Radiotelephone Operator Permit or higher-class license.
D. The operator must hold a GMDSS Radio Operator's License.
ANSWER A: FCC Rule § 80.159 also covers passenger ships. As long as the carrier power of the transmitter is less than 250 watts, an MROP is all that is required. In this case, however, the carrier power is over 250 watts, so a General Radiotelephone Operator License is needed. The key word is "carrier" power. A Marine Radio Operator Permit is all that is required when the station power is 1500 watts or less "peak envelope power."

1A38 Which commercial radio operator license is required to operate a fixed tuned ship radar station with external controls?
A. A radio operator certificate containing a Ship Radar Endorsement
B. A Marine Radio Operator Permit or higher
C. Either a First or Second Class Radiotelegraph certificate or a General Radiotelephone Operator License
D. No radio operator authorization is required.
ANSWER D: FCC Rule § 80.177 authorizes operation without a radio operator license.

1A39 Which commercial radio operator license is required to install a VHF transmitter in a voluntarily-equipped ship station?
A. A Marine Radio Operator Permit or higher class of license
B. None, if installed by, or under the supervision of, the licensee of the ship station and no modifications are made to any circuits
C. A Restricted Radiotelephone Operator Permit or higher class of license
D. A General Radiotelephone Operator License
ANSWER B: FCC Rule § 80.5 says that a "voluntary ship" is "any ship which is not required by treaty or statute to be equipped with radiotelecommunication equipment." No license is required to own or operate its radio equipment; however, the ship must have a station license. See § 80.177(b).

1A40 What transmitting equipment is authorized for use by a station in the maritime services?
A. Transmitters that have been certified by the manufacturer for maritime use
B. Unless specifically excepted, only transmitters type-accepted by the Federal Communications Commission for Part 80 operations
C. Equipment that has been inspected and approved by the U.S. Coast Guard
D. Transceivers and transmitters that meet all ITU specifications for use in maritime mobile service
ANSWER B: If any equipment does not meet Part 80 requirements, it could cause interference to other radio services. Refer to questions 1A12 and 1A15.

1A41 What is the Communication Act's definition of a "passenger ship"?

A. Any ship which is used primarily in commerce for transporting persons to and from harbors or ports.

B. A vessel that carries or is licensed or certificated to carry more than 12 passengers.

C. Any ship transporting more than six passengers for hire.

D. A vessel of any nation that has been inspected and approved as a passenger carrying vessel.

ANSWER B: According to Part II of Title III of the Communications Act, a ship is a *"passenger ship"* if it carries or is licensed or certificated to carry more than twelve passengers. Do not confuse this with a *"passenger carrying vessel"* or small passenger vessel which means any ship transporting more than six passengers for hire See §80.5, *Categories of ships*.

1A42 What is a distress communication?

A. An internationally recognized communication indicating that the sender is threatened by grave and imminent danger and requests immediate assistance

B. Communications indicating that the calling station has a very urgent message concerning safety

C. Radiocommunications which, if delayed, will adversely affect the safety of life or property

D. An official radiocommunications notification of approaching navigational or meteorological hazards

ANSWER A: *Grave and imminent danger* defines a distress communication; it has top priority over messages of urgency and safety. See § 80.5 and § 80.314.

1A43 Who may be granted a ship station license in the maritime service?

A. Anyone, including foreign governments

B. Only FCC-licensed operators holding a First or Second Class Radiotelegraph Operator's Certificate or the General Radiotelephone Operator License

C. Vessels that have been inspected and approved by the U.S. Coast Guard and Federal Communications Commission

D. The owner or operator of a vessel, or their subsidiaries

ANSWER D: Per FCC Rule § 80.15, "A ship station license may only be granted to the owner or operator of the vessel, a subsidiary communications corporation of the owner or operator of the vessel, a state or local government subdivision, or any agency of the U.S. Government" as the FCC sees fit.

The same rule also specifies who may not receive a station license. "A station license cannot be granted to or held by a foreign government or its representative." An alien must request a station license from the communications bureau of his or her own country.

1A44 Who is responsible for the proper maintenance of station logs?

A. The station licensee and the radio operator in charge of the station

B. The station licensee

C. The commercially-licensed radio operator in charge of the station

D. The ship's master and the station licensee

ANSWER A: FCC Rule § 80.409. "The station licensee and the radio operator in charge of the station are responsible for the maintenance of station logs. ...

These persons must keep the log in an orderly manner. Key letters or abbreviations may be used if their proper meaning or explanation is contained elsewhere in the same log." Erasures are not allowed. See question 1A71.

1A45 How long should station logs be retained when there are entries relating to distress or disaster situations?
A. Until authorized by the Commission in writing to destroy them
B. Indefinitely, or until destruction is specifically authorized by the U.S. Coast Guard
C. For a period of three years from the date of entry, unless notified by the FCC
D. For a period of one year from the date of entry
ANSWER C: Logs are always required to be retained by the station licensee for at least one year, but for at least three years if a distress message was received or transmitted. See § 80.409(b).

1A46 Where must ship station logs be kept during a voyage?
A. At the principal radiotelephone operating position
B. They must be secured in the vessel's strongbox for safekeeping.
C. In the personal custody of the licensed commercial radio operator
D. All logs are turned over to the ship's master when the radio operator goes off duty.
ANSWER A: Per FCC Rule § 80.409, station logs (both radiotelephone and radiotelegraph) must be kept at the principal radio operating room during the entire voyage.

1A47 What is the antenna requirement of a radiotelephone installation aboard a passenger vessel?
A. The antenna must be located a minimum of 15 meters from the radiotelegraph antenna.
B. An emergency reserve antenna system must be provided for communications on 156.8 MHz.
C. The antenna must be vertically polarized and as non-directional and efficient as is practicable for the transmission and reception of ground waves over seawater.
D. All antennas must be tested and the operational results logged at least once during each voyage.
ANSWER C: See explanation at question 1A14.

1A48 Where must the principal radiotelephone operating position be installed in a ship station?
A. At the principal radio operating position of the vessel
B. In the room or an adjoining room from which the ship is normally steered while at sea
C. In the chart room, master's quarters or wheel house
D. At the level of the main wheel house or at least one deck above the ship's main deck
ANSWER B: To get the person in command correct and timely information, it makes sense to keep the radio close to the steering room at all times, particularly in a busy shipping channel or at sea during a rough storm. See § 80.853(d).

1A49 What are the technical requirements of a VHF antenna system aboard a vessel?

A. The antenna must provide an amplification factor of at least 2.1 dbi.

B. The antenna must be vertically polarized and non-directional.

C. The antenna must be capable of radiating a signal a minimum of 150 nautical miles on 156.8 MHz.

D. The antenna must be constructed of corrosion-proof aluminum and capable of proper operation during an emergency.

ANSWER B: See explanation at question 1A14.

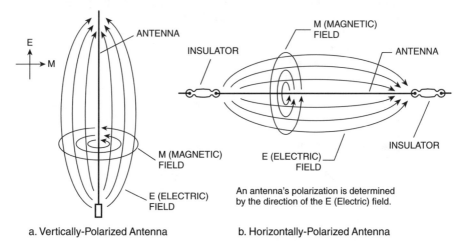

a. Vertically-Polarized Antenna   b. Horizontally-Polarized Antenna

Polarized Antennas

1A50 How often must the radiotelephone installation aboard a small passenger boat be inspected?

A. Equipment inspections are required at least once every 12 months.

B. When the vessel is first placed in service and every 2 years thereafter

C. At least once every five years

D. A minimum of every 3 years, and when the ship is within 75 statute miles of an FCC field office

ANSWER C: Radio systems in large ships must be inspected once a year. Because small passenger boats have much smaller radio systems with fewer parts, inspection is required only every five years. Also, the station licensee is in a better position to detect equipment failure in a small passenger boat. See § 80.903.

1A51 How far from land may a small passenger vessel operate when equipped only with a VHF radiotelephone installation?

A. No more than 20 nautical miles from the nearest land if within the range of a VHF public coast or U.S. Coast Guard station

B. No more than 100 nautical miles from the nearest land

C. No more than 20 nautical miles unless equipped with a reserve power supply

D. The vessel must remain within the communications range of the nearest coast station at all times.

ANSWER A: VHF signals travel by line-of-sight, and, therefore, cannot be heard reliably by any station more than 20 miles away. Without radio equipment that operates at lower frequencies (which offer longer propagation), it makes sense to stay close to shore. See § 80.905(a).

1A52 What is the minimum transmitter power level required by the FCC for a medium-frequency transmitter aboard a compulsorily fitted vessel?

A. At least 100 watts, single-sideband, suppressed-carrier power
B. At least 60 watts PEP
C. The power predictably needed to communicate with the nearest public coast station operating on 2182 kHz
D. At least 25 watts delivered into 50 ohms effective resistance when operated with a primary voltage of 13.6 volts DC

ANSWER B: An international radiotelephone distress frequency, 2182 kHz, is classified in the medium-frequency range. Since all compulsorily fitted vessels must carry radio equipment capable of transmitting on this frequency, 60 watts peak envelope power (PEP) is considered the minimum power level necessary to ensure contact with another station. See § 80.807(c)(2) and § 80.855(d)(2).

1A53 What is a Class "A" EPIRB?

A. An alerting device notifying mariners of imminent danger
B. A satellite-based maritime distress and safety alerting system
C. An automatic, battery-operated emergency position-indicating radiobeacon that floats free of a sinking ship
D. A high-efficiency audio amplifier

ANSWER C: Per FCC Rule § 80.1053, "EPIRB" means Emergency Position Indicating Radio Beacon. It is designed to attract attention by broadcasting a distress signal. A Class "A" EPIRB must be able to ballast itself into upright position in less than one second; its antenna must deploy automatically; it must contain a visual indicator that shows when it is operating; and it must be waterproof.

1A54 What are the radio watch requirements of a voluntary ship?

A. While licensees are not required to operate the ship radio station, general-purpose watches must be maintained if they do.
B. Radio watches must be maintained on 500 kHz, 2182 kHz and 156.800 MHz, but no station logs are required.
C. Radio watches are optional but logs must be maintained of all medium-, high-frequency and VHF radio operation.
D. Radio watches must be maintained on the 156-158 MHz, 1600-4000 kHz and 4000-23,000 kHz bands

ANSWER A: Per Rule § 80.310, "Voluntary vessels must maintain a watch on 156.800 MHz whenever the radio is operating and is not being used to communicate."

1A55 What is the Automated Mutual-Assistance Vessel Rescue System?

A. A voluntary organization of mariners who maintain radio watch on 500 kHz, 2182 kHz and 156.800 MHz
B. An international system operated by the Coast Guard, providing coordination of search and rescue efforts

C. A coordinated radio direction-finding effort between the Federal
Communications Commission and U.S. Coast Guard to assist ships
in distress

D. A satellite-based distress and safety-alerting program operated by
the U.S. Coast Guard

ANSWER B: The Automated Mutual-Assistance Vessel Rescue System, known
as AMVER for short, has saved many lives over the years. Information is
available to ships or search-and-rescue agencies of any country to increase
marine safety. It is truly an international effort. See definitions in § 80.5.

1A56 What is a bridge-to-bridge station?

A. An internal communications system linking the wheel house with
the ship's primary radio operating position and other integral ship
control points

B. An inland waterways and coastal radio station serving ship stations
operating within the United States

C. A portable ship station necessary to eliminate frequent application to
operate a ship station on board different vessels

D. A VHF radio station located on a ship's navigational bridge or main
control station that is used only for navigational communications

ANSWER D: A bridge-to-bridge station operates on designated frequency
Channel 13 in the 156-162 MHz band. These operations take place strictly
between the bridges of two ships, as a means of letting each other know exactly
who they are and where they are going. See definitions in § 80.5.

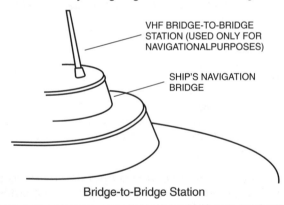

VHF BRIDGE-TO-BRIDGE
STATION (USED ONLY FOR
NAVIGATIONALPURPOSES)

SHIP'S NAVIGATION
BRIDGE

Bridge-to-Bridge Station

1A57 Which of the following statements is true as to ships subject to the Safety Convention?

A. A cargo ship participates in international commerce by transporting
goods between harbors.

B. Passenger ships carry six or more passengers for hire as opposed to
transporting merchandise.

C. A cargo ship is any ship that is not licensed or certificated to carry
more than 12 passengers.

D. Cargo ships are FCC-inspected on an annual basis, while passenger
ships undergo U.S. Coast Guard inspections every six months.

ANSWER C: Per FCC Rule § 80.5, a passenger ship carries, or is licensed or
certificated to carry, more than twelve passengers. A cargo ship is any ship that
is not a passenger ship.

1A58 What is a "passenger-carrying vessel" when used in reference to the Great Lakes Radio Agreement?

A. A vessel that is licensed or certificated to carry more than 12 passengers

B. Any ship carrying more than six passengers for hire

C. Any ship, the principal purpose of which is to ferry persons on the Great Lakes and other inland waterways

D. A ship which is used primarily for transporting persons and goods to and from domestic harbors or ports

ANSWER B: FCC Rule § 80.5 states that a "...'passenger-carrying vessel', when used in reference to the Great Lakes Radio Agreement, means any ship transporting more than six passengers for hire." Note the difference in "passenger-carrying vessel" here and "passenger ship" in question 1A57.

1A59 How do the FCC's rules define a power-driven vessel?

A. A ship that is not manually propelled or under sail

B. Any ship propelled by machinery

C. A watercraft containing a motor with a power rating of at least 3 HP

D. A vessel moved by mechanical equipment at a rate of 5 knots or more

ANSWER B: Rule § 80.5 mentions this definition specifically. If a ship is propelled by engines, jets, paddlewheels, or any other machinery, it is considered a power-driven vessel.

1A60 How do the rules define "navigational communications"?

A. Safety communications pertaining to the maneuvering or directing of vessels' movements

B. Important communications concerning the routing of vessels during periods of meteorological crisis

C. Telecommunications pertaining to the guidance of maritime vessels in hazardous waters

D. Radio signals consisting of weather, sea conditions, notices to mariners and potential dangers

ANSWER A: At first glance, the other three answers appear to be legitimate, but they cover dangerous situations. Answer A covers the movements of ships in all kinds of weather. See definitions in § 80.5.

1A61 What traffic management service is operated by the U.S. Coast Guard in certain designated water areas to prevent ship collisions, groundings and environmental harm?

A. Water Safety Management Bureau (WSMB)

B. Vessel Traffic Service (VTS)

C. Ship Movement and Safety Agency (SMSA)

D. Interdepartmental Harbor and Port Patrol (IHPP)

ANSWER B: The VTS operates in the 156-MHz range and protects, among other regions, the Seattle (Puget Sound), New York, New Orleans and Houston shipping channels. Other regions use VTS when there is no interference to the regions just described. See definitions in § 80.5.

1A62 What action must be taken by the owner of a vessel who changes its name?

A. A Request for Ship License Modification (RSLM) must be submitted to the FCC's licensing facility.

B. The Engineer-in-Charge of the nearest FCC field office must be informed.

C. The Federal Communications Commission in Gettysburg, PA, must be notified in writing.

D. Written confirmation must be obtained from the U.S. Coast Guard.

ANSWER C: Per FCC Rule § 80.29, a station licensee must notify the FCC *in writing* whenever any of the following is changed: mailing address, licensee name, or vessel name. Be as complete and specific as possible with all information.

1A63 When may a shipboard radio operator make a transmission in the maritime services not addressed to a particular station or stations?

A. General CQ calls may only be made when the operator is off duty and another operator is on watch.

B. Only during the transmission of distress, urgency or safety signals or messages, or to test the station

C. Only when specifically authorized by the master of the ship

D. When the radio officer is more than 12 miles from shore and the nearest ship or coast station is unknown

ANSWER B: A station may attempt to reach any other station by making a general call during distress, when any contact is desired because immediate assistance is needed, or when testing equipment. When testing equipment, stating over the air that you are doing so avoids confusion, and all stations listening understand what you are doing and do not respond unless you ask for a radio check. See § 80.89.

1A64 What is the order of priority of radiotelephone communications in the maritime services?

A. Distress calls and signals, followed by communications preceded by urgency and safety signals

B. Alarm, radio direction-finding, and health and welfare communications

C. Navigation hazards, meteorological warnings, priority traffic

D. Government precedence, messages concerning safety of life and protection of property, and traffic concerning grave and imminent danger

ANSWER A: Distress calls always take precedence over all other types of communication. Distress means lives are in danger and immediate assistance is required. Urgency calls concern the safety of a ship or person. Safety traffic concerns navigation problems and weather warnings. See § 80.91.

1A65 What should a station operator do before making a transmission?

A. Transmit a general notification that the operator wishes to utilize the channel

B. Except for the transmission of distress calls, determine that the frequency is not in use by monitoring the frequency before transmitting

C. Check transmitting equipment to be certain it is properly calibrated

D. Ask if the frequency is in use

ANSWER B: Always listen before transmitting; however, it is legal to transmit a distress call immediately. See question 1A18 for consequences of interference on the airwaves.

1A66 What is the proper procedure for testing a radiotelephone installation?

A. Transmit the station's call sign, followed by the word "test" on the radio channel being used for the test.

B. A dummy antenna must be used to insure the test will not interfere with ongoing communications.

C. Permission for the voice test must be requested and received from the nearest public coast station.

D. Short tests must be confined to a single frequency and must never be conducted in port.

ANSWER A: While answers B and C sound plausible, it is perfectly legal to test a radiotelephone installation — provided that it is done legally. Using the word "test" informs all stations listening that you are simply testing the radio equipment and do not require any emergency assistance. Using the ship's antenna during this test (instead of a dummy load) also verifies the antenna system. See § 80.101(a)(2).

1A67 What is the minimum radio operator requirement for ships subject to the Great Lakes Radio Agreement?

A. Third Class Radiotelegraph Operator's Certificate

B. General Radiotelephone Operator License

C. Marine Radio Operator Permit

D. Restricted Radiotelephone Operator Permit

ANSWER C: Interestingly, the FCC gives this information in a rule (§ 80.161) by itself: "Each ship subject to the Great Lakes Radio Agreement must have on board an officer or member of the crew who holds a Marine Radio Operator Permit or higher class license." The Great Lakes carry heavy marine traffic, and effective radio communications help prevent navigation mishaps.

1A68 What FCC authorization is required to operate a VHF transmitter on board a vessel voluntarily equipped with radio and sailing on a domestic voyage?

A. No radio operator license or permit is required.

B. Marine Radio Operator Permit

C. Restricted Radiotelephone Operator Permit

D. General Radiotelephone Operator License

ANSWER A: The key word is *voluntarily*. Voluntarily equipped vessels do not require operator permits for radio operation, but a station license is still needed. See definitions in § 80.177(a).

1A69 On what frequencies does the Communications Act require radio watches by compulsory radiotelephone stations?

A. Watches are required on 500 kHz and 2182 kHz.

B. Continuous watch is required on 2182 kHz only.

C. On all frequencies between 405-535 kHz, 1605-3500 kHz and 156-162 MHz
D. Watches are required on 2182 kHz and 156.800 MHz.

ANSWER D: The *watch* frequencies for *radiotelephone* stations are 2182 kHz and 156.800 MHz. Watches must be on these frequencies at all times. This is a common question on the examination, so be ready for it. See § 80.147 and § 80.148.

1A70 What is the purpose of the international radiotelephone alarm signal?
A. To notify nearby ships of the loss of a person or persons overboard
B. To call attention to the upcoming transmission of an important meteorological warning
C. To alert radio officers monitoring watch frequencies of a forthcoming distress, urgency or safety message
D. To actuate automatic devices giving an aural alarm to attract the attention of the operator where there is no listening watch on the distress frequency

ANSWER D: While a constant watch is required on distress frequencies, it may not always be possible to keep an operator on duty around the clock. In these cases, an automatic alarm signal, triggered when a message with an encoded alarm signal is received on a distress frequency, alerts the operator. See § 80.305(a).

1A71 What is the proper procedure for making a correction in the station log?
A. The ship's master must be notified, approve and initial all changes to the station log.
B. The mistake may be erased and the correction made and initialized only by the radio operator making the original error.
C. The original person making the entry must strike out the error, initial the correction and indicate the date of the correction.
D. Rewrite the new entry in its entirety directly below the incorrect notation and initial the change.

ANSWER C: Per FCC Rule § 80.409, "Erasures, obliterations or willful destruction within the retention period are prohibited. Corrections may be made only by the person originating the entry by striking out the error, initialing the correction and indicating the date of correction." Though not required, it is a good idea to keep station logs in ink.

1A72 What authorization is required to operate a 350-watt, PEP maritime voice station on frequencies below 30 MHz aboard a small, non-commercial pleasure vessel?
A. Third Class Radiotelegraph Operator's Certificate
B. General Radiotelephone Operator License
C. Restricted Radiotelephone Operator Permit
D. Marine Radio Operator Permit

ANSWER C: The Restricted Radiotelephone Operator Permit is often abbreviated as RP. A chart showing what licenses are required to operate various types of transmitters is in FCC Rule § 80.165, which is specifically for voluntary stations.

1A73 What is selective calling?

A. A coded transmission directed to a particular ship station
B. A radiotelephony communication directed at a particular ship station
C. An electronic device which uses a discriminator circuit to filter out unwanted signals
D. A telegraphy transmission directed only to another specific radiotelegraph station

ANSWER A: Selective calling operates very much like dialing a telephone. Without it, a station operator must continually monitor the airwaves, listening for a specific call sign. A digital coding system permits the radio to be squelched until a message containing the proper digital code is received. The selection code activates the radio and normal communication continues. See definitions in § 80.5.

1A74 In the International Phonetic Alphabet, the letters D, N, and O are represented by the words:

A. Delta, November, Oscar.
B. Denmark, Neptune, Oscar.
C. December, Nebraska, Olive.
D. Delta, Neptune, Olive.

ANSWER A: The following is the International Phonetic Alphabet. All radio operators around the world use it:

ITU/ICAO	Alternates	ITU/ICAO	Alternates
A - Alpha	(Able, Adam)	N - November	(Norway, Nancy)
B - Bravo	(Baker)	O - Oscar	(Ocean, Otto)
C - Charlie	(Cocoa)	P - Papa	(Peter, Pacific)
D - Delta	(Dog, David)	Q - Quebec	(Queen)
E - Echo	(Easy, Edward)	R - Romeo	(Roger, Radio)
F - Foxtrot	(Fox, Frank)	S - Sierra	(Sugar, Susan)
G - Golf	(George, Gulf)	T - Tango	(Thomas, Tokyo)
H - Hotel	(Henry)	U - Uniform	(United, Union)
I - India	(Item, Ida, Italy)	V - Victor	(Victory, Victoria)
J - Juliette	(Japan, Jig)	W - Whiskey	(Willie, William)
K - Kilo	(King, Kilowatt)	X - X-Ray	(Xylophone)
L - Lima	(Love, Lewis)	Y - Yankee	(Yoke, Young)
M - Mike	(Mexico, Mary)	Z - Zulu	(Zebra, Zed, Zanzibar)

Phonetics in parentheses are sometimes used on the ham bands, especially by older, long-term amateurs, but the ITU/ICAO phonetics are preferable.

ITU/ICAO Phonetic Alphabet
Adopted by the International Telecommunication Union and International Civil Aviation Organization
Source: *Amateur Radio Logbook,* Contributions by F. Maia, ©1993 Master Publishing, Inc.

English is the closest we have to a universal language. Even so, it can be troublesome for some radio operators from other countries. That is why the International Telecommunication Union (ITU) created this internationally recognized standard phonetic alphabet. It helps all radio operators to identify their stations and receive information when interference makes communication difficult. Whenever you cannot understand the other operator, ask him or her to spell the word phonetically. Station call signs, upon initial contact, are always spelled out this way to lessen confusion. This is especially important during an emergency.

Learning this phonetic alphabet is easier than it looks. Try to spell your own name with it. Practice spelling ordinary words with it. Very soon it will become second nature to you.

1A75 When is it legal to transmit high power on channel 13?
A. Failure of vessel being called to respond
B. In a blind situation such as rounding a bend in a river
C. During an emergency
D. All of these

ANSWER D: Channel 13 messages must be about navigation; for example, passing or meeting other vessels. Your power must not be more than one watt unless you are declaring an emergency, the other vessel fails to respond, or if an obstruction prevents low-power communication. In such cases, higher output power is allowed. See § 80.331(c).

1A76 What must be in operation when no operator is standing watch on a compulsory radio-equipped vessel while out at sea?
A. An auto alarm
B. Indicating Radio Beacon signals
C. Distress-Alert signal device
D. Radiotelegraph transceiver set to 2182 kHz

ANSWER A: An auto alarm activates the receiver's speaker only when a special alarm signal is received. Such a signal is transmitted by a station in distress to specifically activate any stations within range which are equipped with devices to react to this alarm signal. See § 80.305 and also question 1A70.

1A77 When may a bridge-to-bridge transmission be more than 1 watt?
A. When broadcasting a distress message
B. When rounding a bend in a river or traveling in a blind spot
C. When calling the Coast Guard
D. When broadcasting a distress message and rounding a bend in a river or traveling in a blind spot

ANSWER D: Bridge-to-bridge communications are usually on Channel 13 using only 1 watt. See also question 1A75.

1A78 When are EPIRB batteries changed?
A. After emergency use; after battery life expires
B. After emergency use; as per manufacturer's instructions marked on outside of transmitter within month and year replacement date
C. After emergency use; every 12 months when not used
D. Whenever voltage drops to less than 50% of full charge

ANSWER B: All companies that make EPIRB batteries must stamp on the outside of the battery case the month and year of the battery's manufacture and the month and year when 50% of its useful life is due to expire. When the expiration date is reached, the battery must be replaced; but if the EPIRB is actually used before that date, the battery must be replaced immediately afterward. Refer to question 1A53 for EPIRB definition, and see also § 80.1053(e).

1A79 The radiotelephone distress message consists of:
- A. MAYDAY spoken three times, call sign and name of vessel in distress.
- B. Particulars of its position, latitude and longitude, and other information which might facilitate rescue, such as length, color and type of vessel, and number of persons on board.
- C. Nature of distress and kind of assistance required.
- D. All of these.

ANSWER D: Many people become so frightened during an emergency that they forget how to call for help. It is important to stay calm and help the authorities help you. Provide them with as much information as possible to make their job easier. "MAYDAY MAYDAY MAYDAY" captures their attention; your call sign and vessel name tell them who you are; your latitude and longitude tell them where you are. State the nature of your distress so the authorities can bring along the necessary equipment. Give the number of persons aboard and conditions of any injured. Briefly describe your boat's length and color of the hull. If a ship is swamped, count heads immediately. If someone is missing, tell the authorities. See § 80.315(b).

1A80 If a ship sinks, what device is designed to float free of the mother ship, is turned on automatically and transmits a distress signal?
- A. EPIRB on 121.5 MHz/243 MHz or 406.025 MHz
- B. EPIRB on 2182 kHz and 405.025 kHz
- C. Bridge-to-bridge transmitter on 2182 kHz
- D. Auto alarm keyer on any frequency

ANSWER A: An EPIRB automatically transmits a distress call on a specially assigned frequency when a vessel sinks. A water-activated battery is the central power source. Note that these frequencies are in the VHF and UHF ranges, and that an EPIRB is meant to be tracked by orbiting satellites, part of the SARSAT COSPAS system. Direction-finding techniques help home in on the signal. (Refer to question 1A53.)

1A81 International laws and regulations require a silent period on 2182 kHz:
- A. For three minutes immediately after the hour.
- B. For three minutes immediately after the half-hour.
- C. For the first minute of every quarter-hour.
- D. For three minutes immediately after the hour and the half-hour.

ANSWER D: Each ship with a radiotelephone station must listen on 2182 kHz for three minutes immediately after the hour, and for three minutes immediately after the half-hour, Coordinated Universal Time (UTC). UTC is a worldwide standard of timekeeping. Since all ship clocks are synchronized to one universal time, all ships will listen during the same silent period. Silent periods give low-power stations and those with emergencies a chance to be heard above the din of other stations.

There is a different silent period for radiotelegraphy stations. For 500 kHz, the silent periods run for three minutes immediately after the 15-minute mark, and for three minutes immediately after the 45-minute mark. Don't get them mixed up. See § 80.304(b).

1A82 How should the 2182-kHz auto-alarm be tested?
A. On a different frequency into antenna
B. On a different frequency into dummy load
C. On 2182 kHz into antenna
D. Only under U.S. Coast Guard authorization

ANSWER B: While it is important to test emergency communications equipment to make sure it functions properly, doing so must not allow a false signal to be received by any other station and possibly trigger a needless rescue mission. Testing the auto-alarm on a different frequency ensures that the signal won't activate any other equipment, and the dummy load ensures that the radio waves emitted travel no farther than the interior of the ship's radio room. See § 80.855(h).

1A83 What is the average range of VHF marine transmissions?
A. 150 miles
B. 50 miles
C. 20 miles
D. 10 miles

ANSWER C: VHF radio waves cannot travel much farther than you can see. Beyond 20 to 30 miles, VHF radio waves cannot bring help if needed. That is why vessels equipped only with VHF radio equipment may not travel more than 20 miles from shore. See also question 1A51.

1A84 A ship station using VHF bridge-to-bridge Channel 13:
A. May be identified by call sign and country of origin.
B. Must be identified by call sign and name of vessel.
C. May be identified by the name of the ship in lieu of call sign.
D. Does not need to identify itself within 100 miles from shore.

ANSWER C: Calling another ship station by that ship's name is perfectly legal because it is often impossible to know another ship's call sign in advance. When you make contact on Channel 13, 156.65 MHz, the other ship station will identify itself with its own call sign. See § 80.102(c).

1A85 When using a SSB station on 2182 kHz or VHF-FM on Channel 16:
A. Preliminary call must not exceed 30 seconds.
B. If contact is not made, you must wait at least 2 minutes before repeating the call.
C. Once contact is established, you must switch to a working frequency.
D. All of these.

ANSWER D: The idea is to take up as little time on the distress or calling frequency as possible. Keep in mind that some vessel in distress may need the frequency at any time. See § 80.116.

1A86 By international agreement, which ships must carry radio equipment for the safety of life at sea?
A. Cargo ships of more than 300 gross tons and vessels carrying more than 12 passengers
B. All ships traveling more than 100 miles out to sea
C. Cargo ships of more than 100 gross tons and passenger vessels on international deep-sea voyages
D. All cargo ships of more than 100 gross tons

ANSWER A: Cargo ships travel all over the world, most of the time far away from shore, and may require immediate assistance. Passenger ships can be in danger at a moment's notice. For these reasons, these ships must carry radio equipment. See § 80.851 and § 80.901.

1A87 What is the most important practice that a radio operator must learn?
- A. Monitor the channel before transmitting
- B. Operate with the lowest power necessary
- C. Test a radiotelephone transmitter daily
- D. Always listen to 121.5 MHz

ANSWER A: Always listen first! See explanation at question 1A18.

1A88 Portable ship radio transceivers operated as associated ship units:
- A. Must be operated on the safety and calling frequency 156.8 MHz (Channel 16) or a VHF intership frequency.
- B. May not be used from shore without a separate license.
- C. Must only communicate with the ship station with which it is associated or with associated portable ship units.
- D. All of these.

ANSWER D: Associated ship stations operating under the authority of the ship station license are restricted to only one watt of output power. The portable station must be identified by the call sign of the ship station with which it is associated, followed by an appropriate unit designator. See questions 1A28 and 1A29 for clarification.

1A89 Which is a radiotelephony calling and distress frequency?
- A. 500 kHz
- B. 2182 kHz
- C. 156.3 MHz
- D. 3113 kHz

ANSWER B: This is an important frequency to remember. The FCC requires all ship stations to maintain a watch on 2182 kHz. See also questions 1A69 and 1A81.

1A90 What is the priority of communications?
- A. Distress, urgency, safety and radio direction-finding
- B. Safety, distress, urgency and radio direction-finding
- C. Distress, safety, radio direction-finding, search and rescue
- D. Radio direction-finding, distress and safety

ANSWER A: Distress calls always take precedence over all other messages. An urgency message warns other stations of dangerous situations. Safety messages warn stations of navigation warnings or meteorological warnings. See also questions 1A17 and 1A64.

1A91 Cargo ships of 300 to 1600 gross tons should be able to transmit a minimum range of:
  A. 75 miles.
  B. 150 miles.
  C. 200 miles.
  D. 300 miles.
ANSWER B: Per FCC Rule § 80.855, the minimum range is 150 *nautical* miles. Since cargo ships must travel throughout international waters, they must carry radio equipment capable of reaching across great distances.

1A92 Radiotelephone stations required to keep logs of their transmissions must include:
  A. Station, date and time.
  B. Name of operator on duty.
  C. Station call signs with which communication took place.
  D. All of these.
ANSWER D: And that's not all. Logs must contain the following and the time of their occurrence: a summary of all distress, urgency and safety traffic; the position of the ship at least once a day; the time the watch is discontinued, the reason for it, and the time the watch resumes; the times when storage batteries are placed on charge and taken off charge; results of required equipment tests; and a daily statement of the condition of the required radiotelephone equipment. See § 80.409.

1A93 Each cargo ship of the United States which is equipped with a radiotelephone station for compliance with Part II of Title III of the Communications Act shall, while being navigated outside of a harbor or port, keep a continuous and efficient watch on:
  A. 2182 kHz.
  B. 156.8 MHz.
  C. 2182 kHz and 156.8 MHz.
  D. Monitor all frequencies within the 2000 kHz to 27500 kHz band used
     for communications.
ANSWER C: Per FCC Rule § 80.305, each cargo ship equipped with a radiotelephone station must maintain a continuous watch on 2182 kHz and 156.800 MHz. You will almost certainly have a question about this topic on your examination.

1A94 What call should you transmit on Channel 16 if your ship is sinking?
  A. SOS three times
  B. MAYDAY three times
  C. PAN three times
  D. URGENCY three times
ANSWER B: "SOS" is reserved for telegraphy only; it can be sent in a very short time. "MAYDAY" is easily and universally understood when spoken, so it's used in telephony. See § 80.315.

Message Type	Telephony	Telegraphy
Distress	MAYDAY MAYDAY MAYDAY	SOS SOS SOS
Urgency	PAN PAN PAN PAN PAN PAN	XXX XXX XXX
Safety	SECURITY SECURITY SECURITY	TTT TTT TTT

International Distress, Urgency, and Safety Signals

1A95 Under normal circumstances, what do you do if the transmitter aboard your ship is operating off-frequency, overmodulating or distorting?
A. Reduce to low power
B. Stop transmitting
C. Reduce audio volume level
D. Make a notation in station operating log

ANSWER B: Operating a transmitter out of its specifications is not only illegal, it is also dangerous. It could interfere with other radio services. The operator must constantly be aware of the radio's behavior and repair any problems as soon as possible. See § 80.90.

1A96 The urgency signal has lower priority than:
A. Direction-finding.
B. Distress.
C. Safety.
D. Security.

ANSWER B: See explanations at questions 1A8, 1A10, 1A17 and 1A64.

1A97 The primary purpose of bridge-to-bridge communications is:
A. Search and rescue emergency calls only.
B. All short-range transmission aboard ship.
C. Transmission of Captain's orders from the bridge.
D. Navigational communications.

ANSWER D: Bridge-to-bridge communications are only for navigation. VHF Channel 13 is known as the bridge-to-bridge channel. It is available to all ships. See also question 1A56.

1A98 What is the international VHF digital selective calling channel?
A. 2182 kHz
B. 156.35 MHz
C. 156.525 MHz
D. 500 kHz

ANSWER C: There are several international digital selective calling (DSC) frequencies, but the only one in the VHF band is 156.525 MHz. This is Channel 70. See questions 1A73 and 1A105.

During normal communications, a ship and a coast station talk to each other over digital selective calling on two different frequencies: one to transmit and one to receive. When calling another station or making a distress call, however, the same frequency is used for both transmitting and receiving.

1A99 When your transmission is ended and you expect no response, say:
   A. BREAK.
   B. OVER.
   C. ROGER.
   D. CLEAR.
ANSWER D: Radio operators say "over and out" only in the movies. "CLEAR" means "I am ceasing transmission." The station operator may continue to listen for other stations, however.

1A100 When attempting to contact other vessels on Channel 16:
   A. Limit calling to 30 seconds.
   B. If no answer is received, wait 2 minutes before calling vessel again.
   C. Channel 16 is used for emergency calls only.
   D. Limit calling to 30 seconds and if no answer is received, wait 2 minutes before calling vessel again.
ANSWER D: Answer C is incorrect because Channel 16 is used not only for distress calls, but also for a calling frequency. See also question 1A85, and remember that Channel 16 is VHF-FM.

1A101 When a message has been received and will be complied with, say:
   A. MAYDAY.
   B. OVER.
   C. ROGER.
   D. WILCO.
ANSWER D: This one is easy to remember because "WILCO" is short for "will comply." Remember never to say "MAYDAY" unless you are in a distress situation!

CLEAR	I am ceasing transmission.
WILCO	I have received your message and will comply.
BREAK	Do you acknowledge receipt?
OVER	I am awaiting your response.
ROGER	I understand your entire message.

Common Over-the-Air Protocol

1A102 The FCC may suspend an operator license upon proof that the operator:
   A. Has assisted another to obtain a license by fraudulent means.
   B. Has willfully damaged transmitter equipment.
   C. Has transmitted obscene language.
   D. Any of these.
ANSWER D: If the FCC can prove that an operator is involved in any of the above, that person is in serious trouble. Radio communication, particularly at sea, is serious business. Sabotaging radio equipment or helping an unqualified

operator obtain a license can bring forth the wrath of the FCC, and the consequences can range from a severe fine to jail time. Good station operators take
pride in their work. See § 13.9(e).

**1A103 What channel must compulsorily-equipped vessels monitor at all
times in the open sea?**
- A. Channel 8, 156.4 MHz
- B. Channel 16, 156.8 MHz
- C. Channel 22A, 157.1 MHz
- D. Channel 6, 156.3 MHz

ANSWER B: Per FCC Rule § 80.148, Channel 16, 156.800 MHz is an international radiotelephone distress frequency in the VHF range. All vessels using
radio equipment are required by law to maintain a watch on this frequency. VHF
Channel 16 is also a calling channel.

**1A104 When testing is conducted on 2182 kHz or 156.8 MHz, testing
should not continue for more than _____ in any 5-minute period.**
- A. 10 seconds
- B. 1 minute
- C. 2 minutes
- D. None of the above

ANSWER A: Per FCC Rule § 80.101, "Test signals must not exceed ten seconds,
and must not be repeated until at least one minute has elapsed. On these
distress frequencies, the time between tests must be a minimum of five minutes."

Testing radio equipment in this manner requires transmitting the station's
call sign, followed by the word "test." If another station responds with "wait,"
then you must cease transmitting for at least 30 seconds.

1A105 Which VHF channel is used only for digital selective calling?
- A. Channel 70
- B. Channel 16
- C. Channel 22A
- D. Channel 6

ANSWER A: The international VHF digital selective calling frequency is 156.525
MHz, which is Channel 70. With this method of communication, digital codes
trigger circuits on specially designed receivers. Ordinarily, once you make
contact with another station set up for DSC, both of you can switch to another
frequency. DSC keeps the conversation private, since no one else can hear it.
See also questions 1A73 and 1A98.

A distress message on DSC is different because it triggers all Class A DSC
radios. Ordinary messages trigger only those radios set up to receive messages
that include a specific code.

**1A106 VHF ship station transmitters must have the capability of reducing
carrier power to:**
- A. 1 watt.
- B. 10 watts.
- C. 25 watts.
- D. 50 watts.

ANSWER A: FCC rules require radio operators to use as little output power as possible to maintain reliable communications. Restricting VHF transmissions to only one watt is common when calling for another ship, because it is usually within eyeshot and one watt is more than powerful enough. If the station does not respond, then try calling with higher power. See § 80.873(c).

1A107 The system of substituting words for corresponding letters is called:

A. International code system.

B. Phonetic system.

C. Mnemonic system.

D. 10 codes.

ANSWER B: The International Phonetic Alphabet allows messages to be transmitted and received through noisy conditions. Do you remember them? Refer to question 1A74.

1A108 How long should station logs be retained when there are no entries relating to distress or disaster situations?

A. For a period of three years from the date of entry, unless notified by the FCC

B. Until authorized by the Commission in writing to destroy them

C. Indefinitely, or until destruction is specifically authorized by the U.S. Coast Guard

D. For a period of one year from the date of entry

ANSWER D: Logs in which no distress traffic was handled must be retained for at least one year. If distress traffic is handled, however, the logs must be retained for at least three years. See also question 1A45.

1A109 The auto alarm device for generating signals shall be:

A. Tested monthly, using a dummy load.

B. Tested every three months, using a dummy load.

C. Tested weekly, using a dummy load.

D. None of these.

ANSWER C: It is imperative that a dummy load be used during this test. Refer to question 1A82.

1A110 Licensed radiotelephone operators are not required on board ships for:

A. Voluntarily equipped ship stations on domestic voyages operating on VHF channels.

B. Ship radar, provided the equipment is non-tunable, pulse-type magnetron and can be operated by means of exclusively-external controls.

C. Installation of a VHF transmitter in a ship station where the work is performed by or under the immediate supervision of the licensee of the ship station.

D. Any of these.

ANSWER D: FCC Rule § 80.177 also states that no license is required to operate a survival craft station or an emergency position indicating radio beacon.

1A111 Under what license are hand-held transceivers covered when used on board a ship at sea?
- A. The ship station license
- B. Under the authority of the licensed operator
- C. Walkie-talkie radios are illegal to use at sea.
- D. No license is needed

ANSWER A: Per FCC Rule § 80.13, "One ship station license will be granted for operation of all maritime services transmitting equipment on board a vessel." Refer to question 1A88.

1A112 What should an operator do to prevent interference?
- A. Turn off transmitter when not in use
- B. Monitor channel before transmitting
- C. Transmissions should be as brief as possible
- D. Monitor channel before transmitting and make transmissions as brief as possible

ANSWER D: A good radio operator would never dream of turning on the power with the transmit button pushed! Always listen before transmitting. See also explanation at question 1A18.

1A113 Identify a ship station's radiotelephone transmissions by:
- A. Country of registration.
- B. Call sign.
- C. Port of registry.
- D. Name of vessel operator.

ANSWER B: Per FCC Rule § 80.102, stations must give their call signs in English at the beginning and end of each communication with any other station, and at least every 15 minutes in between. See also question 1A22.

1A114 Maritime emergency radios should be tested:
- A. Before each voyage.
- B. Weekly while the ship is at sea.
- C. Every 24 hours.
- D. Before each voyage and weekly while the ship is at sea.

ANSWER D: Emergency equipment must be inspected and tested, before a voyage and weekly while underway, to make sure it will be operational when you need it. The sea is a hazardous environment for electrical and electronic circuits. See § 80.832(a).

1A115 The URGENCY signal concerning the safety of a ship, aircraft or person shall be sent only on the authority of:
- A. Master of ship.
- B. Person responsible for mobile station.
- C. Either Master of ship or person responsible for mobile station.
- D. An FCC-licensed operator.

ANSWER C: The FCC rules state this clearly in § 80.327 a): "The urgency signal must be sent only on the authority of the master or person responsible for the mobile station." The person sending urgency traffic has a "very urgent message to transmit concerning the safety of a ship, aircraft, or other vehicle, or the safety of a person." It is important that such a message be made by a dependable person. If the master of the ship is busy (as often happens in urgency situations), then the responsibility falls upon the person so designated.

1A116 Survival craft emergency transmitter tests may NOT be made:
A. For more than 10 seconds.
B. Without using station call sign, followed by the word "test".
C. Within 5 minutes of a previous test.
D. All of these.

ANSWER D: Follow these rules to the letter. If anyone monitoring the airwaves picks up a signal that does not conform to the rules of testing, they will assume that it is a genuine distress call. Make survival radio equipment tests as quickly as possible, and take pains to restrict their RF waves to within the testing area. Remember to use the word "test." See § 80.101(a)(3).

1A117 International laws and regulations require a silent period on 2182 kHz:
A. For three minutes immediately after the hour.
B. For three minutes immediately after the half-hour.
C. For the first minute of every quarter-hour.
D. For three minutes immediately after the hour and half-hour.

ANSWER D: See the explanation at question 1A81.

1A118 How should the 2182-kHz auto-alarm be tested?
A. On a different frequency into antenna
B. On a different frequency into dummy load
C. On 2182 kHz into dummy load
D. On 2182 kHz into antenna

ANSWER B: Never test auto-alarm equipment on the international distress frequency! Not even when using a dummy load. Clarify by referring to question 1A82.

1A119 Each cargo ship of the United States which is equipped with a radiotelephone station for compliance with the Safety Convention shall, while at sea:
A. Not transmit on 2182 kHz during emergency conditions.
B. Keep the radiotelephone transmitter operating at full 100% carrier power for maximum reception on 2182 kHz.
C. Reduce peak envelope power on 156.8 MHz during emergencies.
D. Keep continuous watch on 2182 kHz using a watch receiver having a loudspeaker and auto-alarm distress-frequency watch receiver.

ANSWER D: Auto alarms allow a watch to be carried out without actually stationing someone to monitor it 24 hours a day. See also questions 1A70 and 1A76.

1A120 What is the procedure for testing a 2182-kHz ship radiotelephone transmitter with full carrier power while out at sea?
A. Reduce to low power, then transmit test tone
B. Switch transmitter to another frequency before testing
C. Simply say: "This is (call letters) testing." If all meters indicate normal values, it is assumed transmitter is operating properly.
D. It is not permitted to test on the air.

ANSWER C: Remember to use the word "test" and your call letters. If your test equipment reveals any erroneous readings, shut down the transmitter immediately and correct the problem. Take as little time with this kind of test as possible. See § 80.101(a).

1A121 If your transmitter is producing spurious harmonics or is operating at a deviation from the technical requirements of the station authorization:
A. Continue operating until returning to port.
B. Repair problem within 24 hours.
C. Cease transmission.
D. Reduce power immediately.
ANSWER C: Any malfunctions of the ship station transmitter must be noted by the station operator, and the transmitter must be taken off the air immediately. Faulty equipment can be very dangerous. See also question 1A95.

1A122 As an alternative to keeping watch on a working frequency in the band 1600-4000 kHz, an operator must tune station receiver to monitor 2182 kHz:
A. At all times.
B. During distress calls only.
C. During daytime hours of service.
D. During the silence periods each hour.
ANSWER A: This is law. The international maritime radiotelephone distress frequency is 2182 kHz. It must be monitored constantly for any possible distress calls. Radiotelegraph stations must maintain a watch on 500 kHz. Ships equipped with VHF radio equipment must listen on 156.800 MHz for any possible distress calls. See § 80.301(b).

1A123 An operator or maintainer must hold a General Radiotelephone Operator License to:
A. Adjust or repair FCC-licensed transmitters in the aviation, maritime and international fixed public radio services.
B. Operate voluntarily-equipped ship maritime mobile or aircraft transmitters with more than 1,000 watts of peak envelope power.
C. Operate radiotelephone equipment with more than 1,500 watts of peak envelope power on cargo ships over 300 gross tons.
D. All of these.
ANSWER D: Chapter 2 details when a GROL is needed. To earn a General Radiotelephone Operator License, you must pass two written exams which are taken from the question pools for Element 1 and Element 3, both of which are contained in this book.

1A124 What is the radiotelephony calling and distress frequency?
A. 500 kHz
B. 500R122JA
C. 2182 kHz
D. 2182R2647
ANSWER C: Since frequencies are always given in kHz or MHz, answers B and D are obviously wrong. 500 kHz is the radiotelegraphy calling and distress frequency; since this is telephony, answer C is correct. See § 80.313.

1A125 If a ship radio transmitter signal becomes distorted:
  A. Cease operations.
  B. Reduce transmitter power.
  C. Use minimum modulation.
  D. Reduce audio amplitude.
ANSWER A: Defective radio equipment must be taken off the air immediately. This topic will almost certainly appear on your examination. Faulty transmitters can really mess up the airwaves. See § 80.90.

1A126 Tests of survival craft radio equipment, EXCEPT EPIRBs and two-way radiotelephone equipment, must be conducted:
  A. At weekly intervals while the ship is at sea.
  B. Within 24 hours prior to departure when a test has not been conducted within a week of departure.
  C. At weekly intervals while the ship is at sea and within 24 hours prior to departure when a test has not been conducted within a week of departure.
  D. When required by the Commission.
ANSWER C: Better to be sure, than to think, you're safe. Confirm that the survival craft transmitters work properly before you leave. During the voyage, make certain that the batteries maintain a healthy charge. They could save your life. See § 80.832(a).

1A127 Each cargo ship of the United States which is equipped with a radiotelephone station for compliance with Part II of Title III of the Communications Act shall, while being navigated outside of a harbor or port, keep a continuous watch on:
  A. 2182 kHz.
  B. 156.8 MHz.
  C. 2182 kHz and 156.8 MHz.
  D. Cargo ships are exempt from radio watch regulations.
ANSWER C: Keep radio watches on both frequencies at all times. Question 1A93 is almost identical!

1A128 When may you test a radiotelephone transmitter on the air?
  A. Between midnight and 6:00 AM local time
  B. Only when authorized by the Commission
  C. At any time as necessary to assure proper operation
  D. After reducing transmitter power to 1 watt
ANSWER C: Rather than risk operating with broken equipment, it is better to test it on the air and find out for sure. Don't take any more time than is necessary. See § 80.101.

1A129 What is the required daytime range of a radiotelephone station aboard a 900-ton, ocean-going cargo vessel?
  A. 25 miles
  B. 50 miles
  C. 150 miles
  D. 500 miles
ANSWER C: Heavy cargo ships must be able to communicate across great distances, since they travel throughout the world's oceans. The minimum range is 150 *nautical* miles. See § 80.855(c).

1A130 What do you do if the transmitter aboard your ship is operating off-frequency, overmodulating or distorting?
A. Reduce to low power
B. Stop transmitting
C. Reduce audio volume level
D. Make a notation in the station operating log

ANSWER B: Defective radio equipment is dangerous. Remember to always monitor your transmissions and equipment. Chances are good that you will be tested on this. Refer to questions 1A95, 1A121, and 1A125.

1A131 What is the authorized frequency for an on-board ship repeater for use with a mobile transmitter operating at 467.750 MHz?
A. 457.525 MHz
B. 467.775 MHz
C. 467.800 MHz
D. 467.825 MHz

ANSWER A: A repeater receives on one frequency and transmits on another. The separation between TX and RX is 10.225 MHz. To avoid unnecessary propagation, an on-board ship repeater's antenna must not be more than 10 feet above the deck. In addition, an on-board ship repeater must be able to shut off the transmitter if it receives a constant signal for more than three minutes.

Channel	Mobile	Repeater
1	467.750 MHz	457.525 MHz
2	467.775 MHz	457.550 MHz
3	467.800 MHz	457.575 MHz
4	467.825 MHz	457.600 MHz

1A132 Survival craft EPIRBs are tested:
A. With a manually-activated test switch.
B. With a dummy load having the equivalent impedance of the antenna affixed to the EPIRB.
C. With radiation reduced to a level not to exceed 25 microvolts per meter.
D. All of these.

ANSWER D: An EPIRB contains test switches that are covered with switch guards so they cannot be activated accidentally. When the switch is placed in the test position, the outside antenna is disabled and an equivalent dummy load absorbs the attenuated RF energy from the transmitter. Full power occurs only when in actual use. An indicator light shows that the unit is working properly. When the switch is released, the EPIRB shuts off and the outside antenna is automatically reconnected. See § 80.1053.

1A133 What safety signal call word is spoken three times, followed by the station call letters spoken three times, to announce a storm warning, danger to navigation, or special aid to navigation?
A. PAN
B. MAYDAY
C. SECURITY
D. SAFETY

ANSWER C: Distress messages, urgency messages and safety messages each have their own special call word. The word for safety messages is SECURITY.

Be careful not to choose answer D. When you hear "SECURITY SECURITY SECURITY," you know that a message concerning a storm warning or danger to navigation is about to be sent. See definitions in § 80.5 and § 80.329.

1A134 When should both the call sign and the name of the ship be mentioned during radiotelephone transmissions?
A. At all times
B. During an emergency
C. When transmitting on 2182 kHz
D. Within 100 miles of any shore

ANSWER B: During ordinary radio communications, the ship station's call sign is enough for identification purposes. During an emergency, however, stating the name of the ship in addition to the call sign lessens confusion. The more information presented to those assisting, the more efficiently they can provide help. See § 80.316.

1A135 How often is the auto alarm tested?
A. During the 5-minute silent period
B. Monthly on 121.5 MHz, using a dummy load
C. Weekly on frequencies other than the 2182-kHz distress frequency, using a dummy antenna
D. Each day on 2182 kHz, using a dummy antenna

ANSWER C: Never test emergency alarm equipment on international distress frequencies! Ship station equipment must be tested weekly due to the highly corrosive nature of seawater. See also question 1A82, and § 80.855(h).

1A136 One nautical mile is approximately equal to how many statute miles?
A. 1.61 statute miles
B. 1.83 statute miles
C. 1.15 statute miles
D. 1.47 statute miles

ANSWER C: The nautical mile is recognized throughout the world as the fundamental unit of distance for navigation. It is defined as the length of 1 minute of arc on a great circle drawn on the surface of a sphere that has the same area as the earth. The nautical mile is slightly longer (6076.1 feet) than the statute mile (5,280 feet). To convert nautical miles to statute miles, multiply by 1.15.

Miles		
Nautical	**Statute**	
1.0	1.152	
5.0	5.758	
10.0	11.516	
20.0	23.030	

Statute mile = 5280 ft.
*Nautical mile = 6076.1 ft.
Nautical mile = 1.1515 statue miles
*A Nautical mile is equal to $\frac{1}{21,600}$ of the great circle of the earth.

Converting Nautical Miles to Statute Miles

1A137 A reserve power source must be able to power all radio equipment, plus an emergency light system, for how long?
A. 24 hours
B. 12 hours
C. 8 hours
D. 6 hours

ANSWER D: The rules governing a ship station's reserve power supply are quite strict. The reserve power supply is usually made up of batteries. The owner of the station license must be able to prove, upon request of an FCC inspector, that the reserve power supply can last for at least six hours. The reserve power source must be located as close to the backup transmitter and receiver as practicable, and illuminated if it is not located within the radio room. Batteries must remain fully charged every day while the ship is at sea. All of this equipment, plus emergency lighting, must be ready to activate within one minute at all times. Take shortcuts with reserve equipment at your peril. See § 80.808(a)(9).

1A138 Frequencies used for portable communications on board ship:
A. 9300-9500 MHz.
B. 1636.5-1644 MHz.
C. 2900-3100 MHz.
D. 457.525-467.825 MHz.

ANSWER D: These are UHF frequencies authorized for stations on board ship and may also be used for portable ship stations. On-board repeater stations also use these frequencies. See also question 1A131.

1A139 In the FCC rules, the frequency band from 30 to 300 MHz is also known as:
A. Very-High Frequency (VHF).
B. Ultra-High Frequency (UHF).
C. Medium Frequency (MF).
D. High Frequency (HF).

ANSWER A: All ships with an FCC station license must carry VHF equipment. Bridge-to-bridge transmissions and the international distress frequency fall within this frequency range. See spectrum at question 1A140.

1A140 What channel must VHF-FM-equipped vessels monitor at all times the station is operated?
A. Channel 8; 156.4 MHz
B. Channel 16; 156.8 MHz
C. Channel 5A; 156.25 MHz
D. Channel 1A; 156.07 MHz

ANSWER B: Per FCC Rule § 80.148, each VHF ship station during its hours of operation must maintain a watch on 156.800 MHz (Channel 16) whenever such station is not being used for exchanging communications. See also question 1A103.

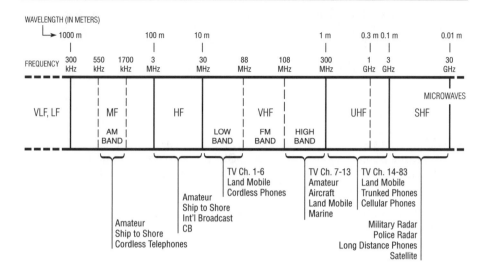

Radio Frequency Spectrum
Source: *Mobile 2-Way Radio Communications,* G. West, ©1993 Master Publishing, Inc.

1A141 When testing is conducted within the 2170-2194-kHz and 156.75-156.85-MHz bands, transmissions should not continue for more than _____ in any 15-minute period.
  A. 30 seconds
  B. 1 minute
  C. 5 minutes
  D. No limitation

ANSWER A: Section §80.101(a)(3) states that "Test signals must not exceed 10 seconds, and must not be repeated until at least one minute has elapsed. On the frequency 2182 kHz or 156.800 MHz, the time between tests must be a minumum of five minutes. Since there are three 5-minute periods in 15 minutes, three times 10 seconds is 30 seconds.

1A142 What emergency radio testing is required for cargo ships?
  A. Tests must be conducted weekly while ship is at sea.
  B. Full-power carrier tests into dummy load
  C. Specific-gravity check in lead-acid batteries, or voltage under load for dry-cell batteries
  D. All of these

ANSWER D: Take care of your radio and it will take care of you. Be especially careful about the batteries. Check all parameters constantly. Check for broken cables, broken antennas, loose connectors, etc. See § 80.855 (h) and § 80.811(a)(1).

1A143 The master or owner of a vessel must apply how many days in advance for an FCC ship inspection?
  A. 60 days
  B. 30 days
  C. 3 days
  D. 24 hours

ANSWER C: Three days is minimum; however, waiting until the last minute may result in some stiff charges for the ship's master or owner if FCC engineers have to work overtime to check your radio equipment. See § 80.59(a).

1A144 Marine transmitters should be modulated between:
A. 75% - 100%.
B. 70% - 105%.
C. 85% - 100%.
D. 75% - 120%.

ANSWER A: There are several different ways to modulate a radio signal, but all face the same percentage of modulation. Station radio equipment must include a modulation limiter to prevent modulation over 100% to prevent distortion. Always monitor your signal. See § 80.213(a).

1A145 What is a good practice when speaking into a microphone in a noisy location?
A. Overmodulation
B. Change phase in audio circuits
C. Increase monitor audio gain
D. Shield microphone with hands

ANSWER D: If noise enters a transmitter, it must be filtered out electronically, and no filter is perfect. It is best to prevent extraneous noise from entering the transmitter in the first place, and the best way to do so is to shield the microphone. Shielding the microphone with your hands narrows the pickup area to your voice only.

1A146 When pausing briefly for station copying message to acknowledge, say:
A. BREAK.
B. OVER.
C. WILCO.
D. STOP.

ANSWER A: Due to the unpredictability of radio communication, it is not unusual for a station operator to miss part of a message during copying. This is especially true when a proper name is spoken, which is why spelling words with the phonetic alphabet is required. Using "BREAK" periodically gives the other station a chance to either confirm reception or ask for a repeat.

1A147 Overmodulation is often caused by:
A. Turning down audio gain control.
B. Station frequency drift.
C. Weather conditions.
D. Shouting into microphone.

ANSWER D: Overmodulation occurs when the amplitude of an incoming audio signal extends beyond a given level. It is "clipped," which reduces signal quality. Shouting or talking too closely into the microphone can cause overmodulation.

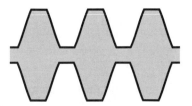

Overmodulation of audio in transmitter causes distortion of modulation envelope.

Oscilloscope Waveform Showing "Flattopping"

1A148 To indicate a response is expected, say:
A. WILCO.
B. ROGER.
C. OVER.
D. BREAK.

ANSWER C: "OVER" informs the other station operator that it is now his or her turn to speak. This is a universally understood phrase and eliminates confusion, particularly during heavy interference.

1A149 When all of a transmission has been received, say:
A. ATTENTION.
B. ROGER.
C. RECEIVED.
D. WILCO.

ANSWER B: Note that "ROGER" is spoken only when *ALL* of a transmission is understood. If you hear only part of a message, ask for a repeat.

1A150 What information must be included in a *DISTRESS* message?
A. Name of vessel
B. Location
C. Type of distress and specifics of help requested
D. All of these

ANSWER D: People can become so scared during an emergency that they may panic and forget to include even the most basic information needed to receive assistance. On land, a 911 operator may be able to learn the location from which a phone call is received, but no such luxury exists at sea. The more information you provide over the air, the better your chances of receiving help. See § 80.316.

1A151 The maritime MF radiotelephone silence periods begin at _____ and _____ minutes past the UTC hour.
A. :15, :45
B. :00, :30
C. :20, :40
D. :05, :35

ANSWER B: Don't transmit on the distress frequency during silent periods unless you are declaring an emergency. See also question 1A81, and § 80.304(b).

1A152 A marine public coast station operator may not charge a fee for what type of communication?
A. Port Authority transmissions
B. Storm updates

C. Distress

D. All of these

ANSWER C: Sec. § 80.95(a)(3) states that "Charges must not be made for distress calls and related traffic." *Navigation hazard warnings* must not be charged if they are preceded by the SAFETY signal.

1A153 Which of the following represent the first three letters of the phonetic alphabet?

A. Alpha Bravo Charlie

B. Adam Baker Charlie

C. Alpha Baker Crystal

D. Adam Brown Chuck

ANSWER A: If you cannot remember the phonetic alphabet, refer to question 1A74 and keep practicing.

1A154 Two-way communications with both stations operating on the same frequency is:

A. Radiotelephone.

B. Duplex.

C. Simplex.

D. Multiplex.

ANSWER C: A simplex system allows both stations to use the same frequency for transmitting and receiving, but only one station can transmit at a time. A duplex system allows both stations to transmit and receive simultaneously, but not on the same frequency.

1A155 When a ship is sold:

A. New owner must apply for a new license.

B. FCC inspection of equipment is required.

C. Old license is valid until it expires.

D. Continue to operate; license automatically transfers with ownership.

ANSWER A: FCC Rule § 80.56 states this plainly: "Whenever the vessel ownership is transferred, the previous authorization must be forwarded to the Commission for cancellation. The new owner must file for a new authorization." In other words, when you purchase a boat, the radio license does not come with it automatically.

What happens if the boat is sold for scrap, to a collector, or for some other reason is not meant to be used any more? Then the radio equipment aboard permanently discontinues operation, and the new owner of the boat must return the station license to the Commission for cancellation. This is covered in FCC Rule § 80.31.

1A156 What is the second in order of priority?

A. URGENT

B. DISTRESS

C. SAFETY

D. MAYDAY

ANSWER A: An urgency message is second in importance only to distress traffic. An urgent situation could evolve into a distress situation unless assistance is given. See § 80.327(d).

1A157 Portable ship units, hand-helds or walkie-talkies used as an associated ship unit:
A. Must operate with 1 watt and be able to transmit on Channel 16
B. May communicate only with the mother ship and other portable units and small boats belonging to mother ship
C. Must not transmit from shore or to other vessels
D. All of these

ANSWER D: Portable ship stations are meant to be used for low-power communications, using the same ship station license. See also questions 1A88, 1A111 and § 80.115.

1A158 The HF (High-Frequency) band is:
A. 3 - 30 MHz.
B. 3 - 30 GHz.
C. 30 - 300 MHz.
D. 300 - 3000 MHz.

ANSWER A: The High-Frequency band is much lower in the RF spectrum than VHF or UHF. HF signals propagate much farther than VHF signals. See the figure at question 1A139.

1A159 Omega operates in what frequency band?
A. Below 3 kHz
B. 3 - 30 kHz
C. 30 - 300 kHz
D. 300 - 3000 kHz

ANSWER B: Omega is a form of radio navigation using very low frequencies. It currently resides in the 10-14 kHz band. What makes Omega special is the way in which radio navigation is determined. Rather than measure the differences in time between signals, Omega measures the difference in phase between signals.

Fewer than 10 Omega transmitting stations are required for the entire world. While all Omega stations use the same frequency, only one is on at any time — they are turned on and off in sequence, approximately one second at a time. All Omega transmitters send out the same signal and they are phase-synchronized with atomic clocks. The Omega receiver on board a ship measures the difference in phase between one received signal and another. From these comparisons, a precise location of the ship can be extrapolated to within 10 nautical miles.

1A160 Shipboard transmitters using F3E emission (FM voice) may not exceed what carrier power?
A. 500 watts
B. 250 watts
C. 100 watts
D. 25 watts

ANSWER D: FM radio equipment aboard a vessel cannot exceed 25 watts carrier power. It also must be able to reduce output power to one watt (refer to question 1A106). Such a wide range ensures that ship-to-ship communications will not interfere with other stations, and that a distress message has a good chance of being received elsewhere. See § 80.215(g).

1A161 Loran C operates in what frequency band?

A. VHF; 30 - 300 MHz
B. HF; 3 - 30 MHz
C. MF; 300 - 3000 kHz
D. LF; 30 - 300 kHz

ANSWER D: Loran was invented before World War II, and has been undergoing refinement ever since. Loran-C has been around for over 20 years. It is being used by the Federal Aviation Administration, the U.S. Coast Guard, and the marine community for precise navigation assistance. It uses a low frequency band, 90 kHz to 110 kHz.

Loran-C consists of transmitters arranged in chains. At least three transmitters make up a chain. All transmitters emit bursts of digital information, with each station sending certain unique bits of data at different intervals. A Loran-C receiver detects all these bits at different times, depending on the receiver's location. Loran-C extrapolates the receiver's location this way with atomic-clock timing, allowing one-quarter mile accuracy.

Loran-C chains extend around the world, including over 20 million square miles around the United States and 600,000 users.

1A162 What has most priority:

A. URGENT
B. DISTRESS
C. SAFETY
D. SECURITY

ANSWER B: Distress traffic always has the ultimate priority. All other traffic must stand by while distress traffic is acknowledged. See § 80.312.

1A163 When and how may Class A and B EPIRBs be tested?

A. Within the first 5 minutes of the hour; tests not to exceed 3 audible sweeps or one second, whichever is longer
B. Within first 3 minutes of hour; tests not to exceed 30 seconds
C. Within first 1 minute of hour; test not to exceed 1 minute
D. At any time ship is at sea

ANSWER A: A Class A EPIRB floats free when a vessel sinks, while a Class B EPIRB must be manually activated. If at all possible, testing any EPIRB must be coordinated with the Coast Guard to prepare them for receiving such a signal. Otherwise, brief operational tests must be conducted very quickly and within certain time windows. If tests of these devices are to be expected within specified time frames, then the "blips" these tests make over the air are known by other stations to be tests. Testing an EPIRB at any other time, even for only an instant, could trigger a false rescue mission. See also explanation at question 1A53.

1A164 When is the Silent Period on 2182 kHz, when only emergency communications may occur?

A. One minute at the beginning of every hour and half hour
B. At all times
C. No designated period; silence is maintained only when a distress call is received
D. Three minutes at the beginning of every hour and half hour

ANSWER D: 2182 kHz is the radiotelephone distress frequency, so the silent periods are at the top of the hour and at every half-hour. See also questions 1A81, 1A151, and § 80.304(b).

1A165 What is the frequency range of UHF?

    A. 0.3 to 3 GHz
    B. 0.3 to 3 MHz
    C. 3 to 30 kHz
    D. 30 to 300 MHz

ANSWER A: UHF means <u>U</u>ltra-<u>H</u>igh <u>F</u>requency. GHz means gigahertz; 1 GHz equals 1000 MHz. UHF ranges from 300 MHz to 3000 MHz; therefore, 0.3 to 3 GHz. UHF frequencies permit extremely high rates of data transmission, but the range on earth is limited, usually line-of-sight. However, these frequencies are often used in satellite communications.

1A166 A room temperature of +30.0 degrees Celsius is equivalent to how many degrees Fahrenheit?

    A. 104
    B. 83
    C. 95
    D. 86

ANSWER D: The Celsius temperature scale is much more widely accepted throughout the world than Fahrenheit. The mathematical formulas for converting between the two scales are:

$$°F = 9/5°C + 32$$
$$°C = 5/9 \, (°F - 32)$$

The figure below shows the derivation of the formulas and a conversion chart for selected values.

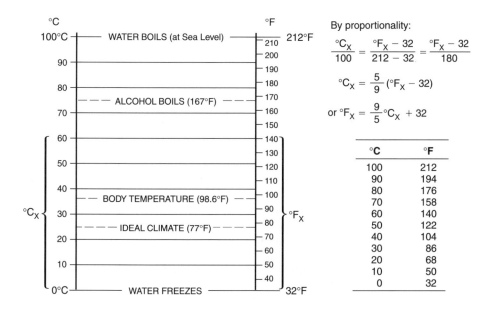

Relationship of Celsius (°C) to Fahrenheit (°F)

1A167 Atmospheric noise or static is not a great problem:
A. At frequencies below 20 MHz
B. At frequencies below 5 MHz
C. At frequencies above 1 MHz
D. At frequencies above 30 MHz

ANSWER D: Most atmospheric noise and static takes place at rather low frequencies in the RF spectrum. Lightning, while rather broad-band, centers near 7 MHz, but its harmonics disrupt many other frequencies. The earth itself has its own geographic resonant frequency of only 7.6 Hz! Geomagnetic storms, sponsored by the sun, can wipe out radio communications all over the world in this low band.

1A168 Frequencies which have substantially straight-line propagation characteristics similar to that of light waves are:
A. Frequencies below 500 kHz
B. Frequencies between 500 kHz and 1000 kHz
C. Frequencies between 1000 kHz and 3000 kHz
D. Frequencies above 50,000 kHz

ANSWER D: Above 50,000 kHz, radio waves behave very much like light waves and do not propagate much farther than the visual horizon. Television transmitters operate in this range. They use extremely tall masts and towers for their transmitting antennas to extend their line of sight. Even so, most VHF television signals fade out after about 50 miles. Conversely, extremely low-frequency (below 500 kHz) radio waves can travel around the earth unimpeded several times, no matter what the height of the transmitting antenna.

1A169 In the International Phonetic Alphabet, the letters E, M, and S are represented by the words:
A. Echo, Michigan, Sonar
B. Equator, Mike, Sonar
C. Echo, Mike, Sierra
D. Element, Mister, Scooter

ANSWER C: Another phonetic alphabet question! If you have forgotten the International Phonetic Alphabet, refer to question 1A74. You will almost certainly receive one of these questions on your examination.

1A170 What is the international radiotelephone distress call?
A. "SOS, SOS, SOS; THIS IS" followed by the call sign of the station (repeated 3 times)
B. "MAYDAY, MAYDAY, MAYDAY; THIS IS" followed by the call sign (or name, if no call sign assigned) of the mobile station in distress, spoken three times
C. For radiotelephone use, any words or message which will attract attention may be used
D. The alternating two-tone signal produced by the radiotelephone alarm signal generator

ANSWER B: The key words are "radiotelephone" and "distress". "SOS" is meant to be sent only in radiotelegraphy distress traffic. "MAYDAY" is meant to be spoken in radiotelephony distress traffic. See § 80.315.

6

Question Pool – Element 3

The questions asked in each Element 3 examination must be taken from the most recent pool. The release date of this pool is *October 1, 1995,* and the effective date is January 1, 1996; therefore, all Element 3 examinations administered after January 1, 1996 must use these questions.

An Element 3 examination is used to prove that the examinee possesses the operational and technical qualifications necessary to properly perform the duties required of a person holding a General Radiotelephone Operator License (GROL) [PG]. To pass, an examinee must answer correctly at least 57 out of 76 questions from this pool. Element 3 is also a partial requirement for the Global Maritime Distress and Safety System Maintainer's License (DM).

A GROL must be held by any person who adjusts, maintains, or internally repairs a radiotelephone transmitter at a station licensed by the FCC in the aviation, maritime, or international fixed public radio services. A GROL must also be held by the operator of (1) certain aviation and maritime land radio stations; (2) compulsory equipped ship radiotelephone stations transmitting with more than 1,500 watts of peak envelope power output; and (3) voluntarily equipped ship stations transmitting with more than 1,000 watts peak envelope power.

Each Element 3 examination is administered by a Commercial Operator License Examination Manager (COLEM). The COLEM must construct the examination by selecting 76 questions from this 916-question pool using the following algorithm:

Subelement 3A — Operating Procedures (3 questions)
Subelement 3B — Radio Wave Propagation (3 questions)
Subelement 3C — Radio Practice (6 questions)
Subelement 3D — Electrical Principles (17 questions)
Subelement 3E — Circuit Components (10 questions)
Subelement 3F — Practical Circuits (17 questions)
Subelement 3G — Signals and Emissions (10 questions)
Subelement 3H — Antennas and Feed Lines (10 questions)

The COLEM may change the order of the answer and distracters (incorrect choices) of questions from the order that appears in this pool.

Suggestions concerning improvement to the questions in the pool or submittal of new questions for consideration in updated pools may be sent to technical author Gordon West, 2414 College Drive, Costa Mesa, CA 92626.

Subelement 3A – Operating Procedures   (3 questions)

3A1 What is facsimile?
A. The transmission of characters by radioteletype that form a picture when printed
B. The transmission of still pictures by slow-scan television
C. The transmission of video by television
D. The transmission of printed pictures for permanent display on paper

ANSWER D: In the marine and aviation radio services, the most common use of radio facsimile is for the printed pictures of weather charts. This is called "WEFAX", and is the term used to refer to the broadcast of facsimile images containing weather charts and satellite imagery. In the United States, these broadcasts are typically prepared by the National Weather Service (NWS), a branch of the National Oceanographic and Atmospheric Administration (NOAA).

Facsimile System
Courtesy of ALDEN ELECTRONICS

3A2 What is the standard scan rate for VHF 137 MHz polar orbiting weather facsimile reception?
A. 240 lines per minute
B. 50 lines per minute
C. 150 lines per second
D. 60 lines per second

ANSWER A: Low earth orbit (LEO) VHF polar orbiting satellites offer incredible resolution of 240 lines per minute. Both infrared and high-resolution imagery may be received during a pass. Commercial fishermen rely heavily on polar weather facsimile reception because water temperatures are graphically presented in color. Polar orbiting weather facsimile transmissions at 240 lines per minute will continue into the next decade without interruption or budget cuts.

3A3 What is the standard scan rate for high-frequency 3 MHz-23 MHz weather facsimile reception from associated shore stations?
A. 240 lines per minute
B. 120 lines per minute
C. 150 lines per second
D. 60 lines per second

ANSWER B: High frequency short-wave weather facsimile transmissions are sent at 120 lines per minute, roughly half the resolution of polar orbiting satellites. HF weather facsimile reception takes 10 minutes for each chart product, and weather forecasters give you all forms of synopsis and forecasts with the appropriate weather symbols. High frequency weather facsimile may go off the air at anytime due to budget constraints.

3A4 The Distance Measuring Equipment (DME) measures the distance from the aircraft to the DME ground station. This is referred to as:

A. The slant range.
B. DME bearing.
C. Glide Slope angle of approach.
D. Localizer course width.

ANSWER A: The term "range," found in the correct answer, refers to *distance to* or *distance from*. The question asks the "distance from," and there is only one answer with the word "range" in it. Because the aeronautical station is well above the ground, the direct distance to the DME station is called the "slant range."

3A5 What is an ascending pass for a low-earth-orbit communications satellite?

A. A pass from west to east
B. A pass from east to west
C. A pass from south to north
D. A pass from north to south

ANSWER C: Constellations of low-earth-orbit satellites are just being launched, and they will be the perfect answer for worldwide data and digital telecommunications from portable pocket phones and data sender/receivers. When a constellation of low-earth-orbit satellites is completed, there will always be satellites in view, and you will be able to make a call from just about anywhere! Satellites moving across North America from south to north are considered "ascending."

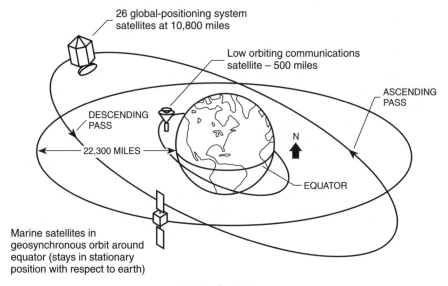

Orbiting Satellites

3A6 What is a descending pass for a low-earth-orbit communications satellite?

A. A pass from north to south
B. A pass from west to east
C. A pass from east to west
D. A pass from south to north

ANSWER A: Low-earth-orbit satellites moving across North America from the north and to the south are considered "descending."

3A7 What is the period of a satellite?

A. An orbital arc that extends from 60 degrees west longitude to 145 degrees west longitude
B. The point on an orbit where satellite height is minimum
C. The amount of time it takes for a satellite to complete one orbit
D. The time it takes a satellite to travel from perigee to apogee

ANSWER C: Polar orbiting weather satellites complete two orbits, or two periods, every 24 hours. Four U.S. and two Russian polar orbiting satellites operate on the 137-MHz band and give us visible light imagery for the daytime pass, and infrared imagery for the nighttime pass. At 137 MHz, an inexpensive computer program in a laptop PC, a wide-band scanner, and a cross dipole is all that's necessary to easily pull in weather facsimile imagery when any one of these satellites is within view on its ascending or descending orbit.

3A8 What is the accuracy of a global positioning system (GPS) fix with selective availability (SA) turned on?

A. 100 meters 95% of the time
B. 100 feet 50% of the time
C. 150 meters 95% of the time
D. 10 feet 95% of the time

ANSWER A: The Department of Defense runs the global positioning satellite system. An *on-purpose* "clock dither" prevents exact, spot-on accuracy to protect our national defense system. This *on-purpose* dithering causes our fix to fluctuate within a 100-meter circle. The fix is accurate within 100 meters 95 percent of the time.

3A9 What is the probable cause for a docked vessel to occasionally register speed and direction on a global positioning system (GPS) receiver?

A. Interference entering the antenna cable
B. A misconnection to the associated electronic chart display
C. A loose coaxial cable connector in the antenna unit
D. A normal condition caused by Department of Defense (DOD) selective availability

ANSWER D: A global positioning system receiver always shows an approximate one knot speed in various directions. This is caused by an *on-purpose* dithering of the transmitted satellite clock time. If you plot this *phantom* course, it would remain within a 100-meter radius 95 percent of the time.

3A10 What voice communications system will lead to a slight delay and possible echo in a telephone conversation?
 A. Propagational delay in communications through a geostationary satellite
 B. Acoustic delay and echo over cellular phone frequency
 C. Propagational delays and echoes on low-earth-orbit satellites
 D. Propagational delays on marine VHF 157/162 MHz bands

ANSWER A: Voice communications through 22,300-mile-high geostationary satellites always contain a slight delay because of the distance the signal travels (22,300 × 2 = 44,000 miles). Modern satellite equipment suppresses the echo, but fast-paced conversations between two satellite telephone users still can be a problem because of the delay. Sometimes the word "over" can help in a satellite conversation!

3A11 Why does the received signal from a satellite, stabilized by a computer-pulsed electromagnet, exhibit a fairly rapid pulsed fading effect?
 A. Because the satellite is rotating
 B. Because of ionospheric absorption
 C. Because of the satellite's low orbital altitude
 D. Because of the Doppler effect

ANSWER A: Some satellites are continuously rotated to keep all sides of the "bird" equally bathed in sunlight. This distributes the heat buildup on the solar panels, and also allows the panels to receive energy from the sun throughout the rotation. A circularly-polarized antenna will help minimize the pulsed fading.

3A12 The majority of airborne Distance Measuring Equipment systems automatically tune their transmitter and receiver frequencies to the paired _____ channel.
 A. VOR/marker beacon
 B. VOR/LOC
 C. Marker beacon/glideslope
 D. LOC/Glideslope

ANSWER B: The distance measuring equipment (DME) receives the VHF omnidirectional range (VOR) NAVAID and localizer (LOC) at the same time. The LOC is the left-right information portion of an instrument landing system.

3A13 The Distance Measuring Equipment (DME) ground station has a built-in delay between reception of an interrogation and transmission of the reply to allow:
 A. Someone to ANSWER the call.
 B. The VOR to make a mechanical hook-up.
 C. Clear other traffic for a reply.
 D. Operation at close range.

ANSWER D: Distance measuring equipment computes ground speed by timing the interval between distant changes. When operating at close range, the DME uses a built-in delay to calculate the frequent updating necessary to sense changes in wind direction or velocity from one area to another.

3A14 During soldering operations, "wetting" is:

A. The action of the flux on the surface of the connection.
B. The blending of solder and copper to form a new alloy.
C. The dipping of the connection in water to cool the connection.
D. The operation of cooling the completed connection with flux.

ANSWER B: The common solder for repair work is 60 percent tin and 40 percent lead, abbreviated "60-40 solder." This has the best wetting qualities. It spreads rapidly and conforms to materials' shapes at a relatively low melting point.

3A15 One nautical mile is equal to how many statute miles?

A. 1.5
B. 8.3
C. 1.73
D. 1.15

ANSWER D: One nautical mile equals 1.15 statute miles.

3A16 Solder most commonly used for hand soldering in electronics is:

A. 50% tin 50% lead
B. 60% tin 40% lead
C. 40% tin 60% lead
D. 30% tin 30% lead

ANSWER B: "60-40" is the common solder; it contains 60% tin and 40% lead.

3A17 Which of the following is true about the soldering tip?

A. One size soldering tip is right for all jobs, only the heat is changed.
B. If the temperature of the soldering tip is the same in each case, the larger work mass will heat up faster than the smaller one will.
C. Iron-plated soldering tips require no preparation before using.
D. The soldering tip should be tinned during the initial heating up of the soldering iron.

ANSWER D: Always melt a little solder on the tip of your soldering gun or iron to equally distribute the heat. You can clean the tip by putting the hot tip on a wet sponge, or by applying a mixture of ammonium chloride and then using the wet sponge to clean the tip. Before you continue soldering, be sure the tip is absolutely free of the ammonium chloride mixture; this prevents contamination of your soldering project.

3A18 3:00 PM Central Standard Time is:

A. 1000 UTC.
B. 2100 UTC.
C. 1800 UTC.
D. 0300 UTC.

ANSWER B: UTC stands for universal time coordinated, formerly called "GMT", and sometimes abbreviated as Zulu time, Z for the zero meridian.

From table on next page we see that:
        Eastern Standard Time plus 5 = UTC.
        Central Standard Time plus 6 = UTC.
        Mountain Standard Time plus 7 = UTC.
        Pacific Standard Time plus 8 = UTC.

In this question, 3:00 PM is 1500 hours in 24-hour time, and 1500 plus 600 (6 hours) gives 2100 hours UTC.

UTC	AST/EDT	EST/CDT	CST/MDT	MST/PDT	PST
0000	2000	1900	1800	1700	1600
0100	2100	2000	1900	1800	1700
0200	2200	2100	2000	1900	1800
0300	2300	2200	2100	2000	1900
0400	0000	2300	2200	2100	2000
0500	0100	0000	2300	2200	2100
0600	0200	0100	0000	2300	2200
0700	0300	0200	0100	0000	2300
0800	0400	0300	0200	0100	0000
0900	0500	0400	0300	0200	0100
1000	0600	0500	0400	0300	0200
1100	0700	0600	0500	0400	0300
1200	0800	0700	0600	0500	0400
1300	0900	0800	0700	0600	0500
1400	1000	0900	0800	0700	0600
1500	1100	1000	0900	0800	0700
1600	1200	1100	1000	0900	0800
1700	1300	1200	1100	1000	0900
1800	1400	1300	1200	1100	1000
1900	1500	1400	1300	1200	1100
2000	1600	1500	1400	1300	1200
2100	1700	1600	1500	1400	1300
2200	1800	1700	1600	1500	1400
2300	1900	1800	1700	1600	1500

Remember that the day changes at 0000 UTC. [6:00 p.m. Pacific Standard Time on Monday evening is actually 2:00 a.m. UTC the following day, Tuesday!]
To convert local time to UTC: For AST/EDT, add 4 hours. For EST/CDT, add 5 hours. For CST/MDT, add 6 hours. For MST/PDT, add 7 hours. For PST, add 8 hours.

Coordinated Universal Time (UTC) to Local Time Conversion Chart

Source: *Amateur Radio Logbook,* ©1993, Master Publishing, Inc.

3A19 Which of the following is an acceptable method of solder removal from holes in a printed board?
A. Compressed air
B. Toothpick
C. Soldering iron and a suction device
D. Power drill

ANSWER C: Solder flux will clean the surfaces to be joined and prevent oxides from forming during soldering. Using acid flux will cause corrosion.

3A20 6:00 PM PST is equal to what time in UTC?
A. 0200
B. 1800
C. 2300
D. 1300

ANSWER A: Since PST is plus 8 and 6:00 PM is 1800 hours, 800 + 1800 = 2600, and 2600 − 2400 = 0200 UTC. See question 3A26.

3A21 The ideal method of removing insulation from wire is:
A. The thermal stripper
B. The pocket knife

C. A mechanical wire stripper.

D. The scissor action stripping tool.

ANSWER A: The best way to remove insulation from old or new wire is with a thermal stripper, which literally melts the insulation for easy removal. Removing insulation with a simple pocket knife, *diagonals*, or scissors can sometimes cut into the wire, making it weaker, or can remove an anti-corrosion coating from the wire. The thermal stripper is your best tool.

3A22 2300 UTC time is:
   A. 2 PM CST.
   B. 3 PM PST.
   C. 10 AM EST.
   D. 6 AM EST.

ANSWER B: Refer to question 3A18 for conversion factors. All the answers using the conversion factors are calculated. B is correct answer.
   A. 2 PM CST = 1400 + 600 = 2000 UTC
   B. 3 PM PST = 1500 + 800 = 2300 UTC
   C. 10 AM EST = 1000 + 500 = 1500 UTC
   D. 6 AM EST = 0600 + 500 = 1100 UTC

3A23 One statute mile equals how many nautical miles?
   A. 3.8
   B. 1.5
   C. 0.87
   D. 0.7

ANSWER C: Did you spot that "ST" as a "87"? One statute mile is less than one nautical mile — 0.87 nautical mile.

3A24 When soldering electronic circuits be sure to:
   A. Wet wires with flux after soldering.
   B. Use maximum heat.
   C. Heat wires until wetting action begins.
   D. Use maximum solder to insure a good bond.

ANSWER C: Heat the wires to a temperature at which the solder begins to flow on the hot wire or circuit. This brings about the wetting action that insures a uniform coating of the hot solder.

3A25 What is the purpose of flux?
   A. Removes oxides from surfaces and prevents formation of oxides during soldering
   B. To act as a lubricant to aid the flow of the solder
   C. Acid cleans printed circuit connections
   D. To cool the soldered connection

ANSWER A: The purpose of solder flux is to remove oxides from circuit surfaces; flux also helps prevent undesirable oxides from forming during the soldering process.

3A26 Which of these will be useful for insulation at UHF?
   A. Rubber
   B. Mica
   C. Wax impregnated paper
   D. Lead

ANSWER B: At UHF frequencies, Teflon® and mica are good insulators. Lead would be considered a conductor, and rubber is not appropriate as an insulator at UHF or microwave frequencies. Wax paper was used as insulation in capacitors back in the old days, but not at UHF frequencies!

3A27 The condition of a lead-acid storage battery is determined with a(n):
A. Hygrometer.
B. Manometer.
C. FET.
D. Hydrometer.

ANSWER D: When you take your GROL examination, pay particular attention to answers that look similar in spelling. Answer A is incorrect, but it's only one letter different from the correct answer! It is the hydrometer that is used to check the condition of a lead-acid storage battery.

3A28 Lines drawn from the VOR station in a particular magnetic direction are:
A. Radials.
B. Quadrants.
C. Bearings.
D. Headings.

ANSWER A: The lines from a VOR station are called radials, and all VOR stations transmit 360 distinct radial signals. The signals are magnetically created by phase shifts between the VOR sub-carriers.

3A29 The horizontal dipole VOR transmit antenna rotates at _____ revolutions per second.
A. 60
B. 2400
C. 30
D. 1800

ANSWER C: A VOR station transmits two signals simultaneously. One signal is omnidirectional, and the other signal is rotated to create a phase shift. The horizontal dipole VOR transmitting antenna rotates 30 times per second, creating an in-phase condition at the 360 degree radial, and an out-of-phase condition at the 180 degree radial. This signal is interpreted by the airborne receiver, and the pilot knows which radial the aircraft is flying on.

3A30 The direction from the aircraft's nose to the VOR station is referred to as the:
A. Heading.
B. Bearing.
C. Deflection.
D. Inclination.

ANSWER B: When referring to a VOR signal, the term "bearing" means the aircraft is located at some point along a particular radial. The VOR system cannot measure distance, only bearing to or from the VOR.

3A31 All directions associated with a VOR station are related to:
A. North pole.
B. North star.

C. Magnetic north.

D. None of these.

ANSWER C: All VOR bearings are related to magnetic north, not true north.

3A32 If a shunt motor, running with a load, has its shunt field opened, how would this affect the speed of the motor?

A. Slow down

B. Stop suddenly

C. Speed up

D. Unaffected

ANSWER C: A shunt motor running under load immediately speeds up if the field winding develops an open circuit. The field acts as a governor, and without a governor to create a counter EMF, the motor speeds up.

3A33 On runway approach, an ILS Localizer shows:

A. Deviation left or right of runway center line.

B. Deviation up and down from ground speed.

C. Deviation percentage from authorized ground speed.

D. Wind speed along runway.

ANSWER A: The localizer in an instrument landing system shows deviation to the left or right of the runway center line. The localizer approach is a non-precision approach.

3A34 What radio navigation aid determines the distance from an aircraft to a selected VORTAC station by measuring the length of time the radio signal takes to travel to and from the station?

A. Radar

B. Loran C

C. Distance Marking (DM)

D. Distance Measuring Equipment (DME)

ANSWER D: It is the distance measuring equipment, abbreviated DME, that measures the distance from an aircraft to a selected VORTAC station. The DME determines the distance based on the time delay of the radio signal.

3A35 Which of the following is a feature of an Instrument Landing System (ILS)?

A. Localizer: shows aircraft deviation horizontally from center of runway

B. Altimeter: shows aircraft height above sea-level

C. VHF Communications: provide communications to aircraft

D. Distance Measuring Equipment: shows aircraft distance to VORTAC station

ANSWER A: The localizer is part of the instrument landing system (ILS). The altimeter, VHF COM radio, and DME equipment are not part of the instrument landing system, although they have a vital roll in air safety.

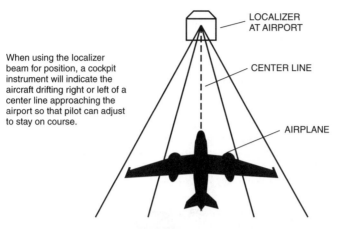

When using the localizer beam for position, a cockpit instrument will indicate the aircraft drifting right or left of a center line approaching the airport so that pilot can adjust to stay on course.

LOCALIZER AT AIRPORT

CENTER LINE

AIRPLANE

Flight Control with Localizer

3A36 Why do we tin component leads?
A. It helps to oxidize the wires
B. Prevents rusting of circuit board
C. Decreases heating time and aids in connection
D. Provides ample wetting for good connection

ANSWER C: Tinning the component leads decreases the amount of time you must apply heat to the component to get the solder *wet*. Too much heat during the soldering process could cause the component to fail.

3A37 What is the purpose of using a small amount of solder on the tip of a soldering iron just prior to making a connection?
A. Removes oxidation
B. Burns up flux
C. Increases solder temperature
D. Aids in wetting the wires

ANSWER D: Tinning the tip of the soldering iron dramatically aids in the transfer of solder from the iron to the wires; this transfer is called *wetting*.

3A38 What occurs if the load is removed from an operating series DC motor?
A. It will stop running.
B. Speed will increase slightly.
C. No change occurs.
D. It will accelerate until it flies apart.

ANSWER D: A series DC motor must be matched to the load; the load keeps the motor spinning at a constant rate. If the load is removed from a series DC motor, the motor speeds up and could possibly fly apart.

3A39 Why are concentric transmission lines sometimes filled with nitrogen?
A. Reduces resistance at high frequencies
B. Prevent water damage underground

C. Keep moisture out and prevent oxidation

D. Reduce microwave line losses

ANSWER C: Concentric transmission lines are sometimes *back-filled* with nitrogen to minimize moisture settling in the lines, which can cause undesirable oxidation.

3A40 Nitrogen is placed in transmission lines to:

A. Improve the "skin-effect" of microwaves.

B. Reduce arcing in the line.

C. Reduce the standing wave ratio of the line.

D. Prevent moisture from entering the line.

ANSWER D: You can sometimes see large nitrogen tanks linked to radio tower transmission lines. The lines are filled with nitrogen to prevent moisture from getting in that an causing unwanted corrosion.

Subelement 3B – Radio Wave Propagation   (3 questions)

3B1 What is a selective fading effect?

A. A fading effect caused by small changes in beam heading at the receiving station

B. A fading effect caused by phase differences between radio wave components of the same transmission, as experienced at the receiving station

C. A fading effect caused by large changes in the height of the ionosphere, as experienced at the receiving station

D. A fading effect caused by time differences between the receiving and transmitting stations

ANSWER B: If you ever tried to watch television with a pair of rabbit ears, or an outside antenna, chances are you noticed what happened when a jet plane flew overhead. Your picture went in and out. This is caused by selective fading. As two signals arrive at your antenna, the reflection off the jet plane is slightly delayed from the direct signal, and this creates a phase difference. When two signals are out of phase, they tend to cancel each other. You can also hear selective fading at its best (at its worst?) on the 10-meter FM band at 29.6 MHz. While the signals are strong, they are many times distorted by their sidebands being out of phase. Worldwide AM broadcast stations are best received in upper or lower sideband, rather than AM double sideband, because AM double-sideband (DSB) reception brings in phase differences on the incoming DSB shortwave signal.

3B2 What is the propagation effect called when phase differences between radio wave components of the same transmission are experienced at the recovery station?

A. Faraday rotation

B. Diversity reception

C. Selective fading

D. Phase shift

ANSWER C: When received signals from the same original radio transmission come in out of phase, this is called selective fading. Watch out for answer D.

The question reads "What is the effect of a phase difference," and the **effect** is "selective fading."

3B3 What is the major cause of selective fading?
A. Small changes in beam heading at the receiving station
B. Large changes in the height of the ionosphere, as experienced at the receiving station
C. Time differences between the receiving and transmitting stations
D. Phase differences between radio wave components of the same transmission, as experienced at the receiving station

ANSWER D: The key words *phase differences* depict an incoming signal that will sound distorted due to selective fading of the sidebands.

3B4 Which emission modes suffer the most from selective fading?
A. CW and SSB
B. FM and double sideband AM
C. SSB and image
D. SSTV and CW

ANSWER B: Selective fading affects most wideband signals. Since FM is about 10 kHz wide, and AM double-sideband is 6 kHz wide, these two emissions are most influenced by selective fading.

3B5 How does the bandwidth of the transmitted signal affect selective fading?
A. It is more pronounced at wide bandwidths.
B. It is more pronounced at narrow bandwidths.
C. It is equally pronounced at both narrow and wide bandwidths.
D. The receiver bandwidth determines the selective fading effect.

ANSWER A: The wider the bandwidth, the more severely the signal will suffer from selective fading. Aeronautical and marine VHF stations may encounter selective fading when operating in the proximity of other nearby aircraft or metal-hull ship stations. Marine and aeronautical high-frequency SSB is a much narrower bandwidth signal, and is less susceptible to selective fadi ng than older AM double-sideband equipment used years ago.

3B6 What phenomenon causes the radio-path horizon distance to exceed the geometric horizon?
A. E-layer skip
B. D-layer skip
C. Auroral skip
D. Radio waves may be bent

ANSWER D: During periods of temperature inversion caused by tropospheric ducting, radio wave refraction causes signals on VHF and UHF frequencies to extend past the geometric horizon. The radio waves are bent by the temperature inversion.

3B7 How much farther does the radio-path horizon distance exceed the geometric horizon?
A. By approximately 15% of the distance
B. By approximately twice the distance
C. By approximately one-half the distance
D. By approximately four times the distance

ANSWER A: This 15 percent increase in distance over the geometric horizon is enhanced during periods of stable high-pressure systems overriding a cool water mass. This could allow ship stations on marine VHF to intercommunicate in excess of 100 miles. It's the same phenomenon—tropospheric bending—that may allow aeronautical stations to communicate with grou nd stations hundreds, and sometimes, thousands of miles away.

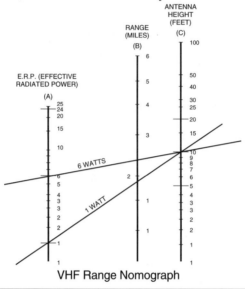

VHF Range Nomograph

3B8 What propagation condition is usually indicated when a VHF signal is received from a station over 500 miles away?
  A. D-layer absorption
  B. Faraday rotation
  C. Tropospheric ducting
  D. Moonbounce

ANSWER C: Some incredible extra-long-range VHF and UHF propagation may take place in the presence of a high-pressure system. Sinking warm air, called a subsidence, overlays cool air just above the surface of the earth and over water. A well-defined inversion layer occurs when ther e is a sharp boundary between the sinking warm air and the cool air beneath. Over big cities, we call it smog. To the VHF and UHF DX enthusiast, we call it a for-sure key for VHF *tropospheric ducting* (key words) super range. Your author has communicated many times from his station in California to Hawaii on the 2-meter band during summer months when tropospheric ducting is at its best!

3B9 What happens to a radio wave as it travels in space and collides with other particles?
  A. Kinetic energy is given up by the radio wave.
  B. Kinetic energy is gained by the radio wave.
  C. Aurora is created.
  D. Nothing happens since radio waves have no physical substance.

ANSWER A: It doesn't take much power for radio waves to travel over millions of miles in space. Look at the signals coming in from deep-space probes! But the further the radio wave travels, the more kinetic energy is given up.

3B10 When the earth's atmosphere is struck by a meteor, a cylindrical region of free electrons is formed at what layer of the ionosphere?
A. The F1 layer
B. The E layer
C. The F2 layer
D. The D layer
ANSWER B: High power meteor burst systems for remote data transmissions normally operate between 30 MHz and 50 MHz. These transmitting stations may be used to monitor climatic conditions in remote areas of the world, and it's surprising how predictable the incoming packets of digitized information come in from the meteor burst transmitter.

3B11 What is transequatorial propagation?
A. Propagation between two points at approximately the same distance north and south of the magnetic equator
B. Propagation between two points on the magnetic equator
C. Propagation between two continents by way of ducts along the magnetic equator
D. Propagation between any two stations at the same latitude
ANSWER A: A TE (transequatorial) signal is propagated between stations north and south of the equator. Some of the best contacts have been between Florida and Venezuela. If you look at a map, you will see that each station is about the same distance away from the equator—ideal conditions for a TE contact. Both CW and SSB are the normal modes for TE on VHF and UHF bands. Ship stations and aeronautical stations operating near the equator may be influenced by the ionosphere's strong effect in these regions. See figure at 3B12.

3B12 What is the maximum range for signals using *transequatorial propagation*?
A. About 1,000 miles
B. About 2,500 miles
C. About 5,000 miles
D. About 7,500 miles
ANSWER C: The total distance for a TE (transequatorial) contact is about 5,000 miles.

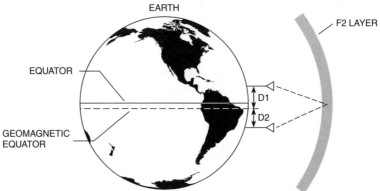

For transequatorial propagation to occur stations at D1 and D2 must be equidistance from geomagnetic equator. Popular theory is that reflections occur in and off the F2 layer during peak sunspot activity.

Transequatorial Propagation

3B13 What is the best time of day for *transequatorial propagation?*
A. Morning
B. Noon
C. Afternoon or early evening
D. Transequatorial propagation only works at night

ANSWER C: Similar to sporadic E, the best time of day for TE propagation is mid-afternoon or early evening. The time of day has allowed ultraviolet radiation from the sun to "warm up" the possible path.

3B14 What is knife-edge diffraction?
A. Allows normally line-of-sight signals to bend around sharp edges, mountain ridges, buildings and other obstructions
B. Arcing in sharp bends of conductors
C. Phase angle image rejection
D. Line-of-sight signals causing distortion to other signals

ANSWER A: Vertically polarized VHF and UHF signals will literally drag the bottom edge of the wave front over tall mountains that would normally block line-of-sight communications. Directional antennas can help fill in dead areas that you might think could never be reached by a VHF or UHF base station.

3B15 The bending of radio waves passing over the top of a mountain range that disperses a weak portion of the signal behind the mountain is:
A. Eddy-current phase effect.
B. Knife-edge diffraction.
C. Shadowing.
D. Mirror refraction effect.

ANSWER B: Knife-edge diffraction and knife-edge refraction may fill in VHF and UHF signals into shadow areas of supposedly no coverage.

3B16 Knife-edge diffraction:
A. Is the bending of UHF frequency radio waves around a building, mountain or obstruction.
B. Causes the velocity of wave propagation to be different from original wave.
C. Is the bending of UHF frequency radio waves around a building, mountain or obstruction and causes the velocity of wave propagation to be different from original wave.
D. Attenuates UHF signals.

ANSWER C: The "knife-edge" effect causes VHF and UHF signals to bend around buildings, mountains, and obstructions. This is caused by a change in velocity at the lower portion of the wave front that makes its velocity different from that at the top part of the wave front.

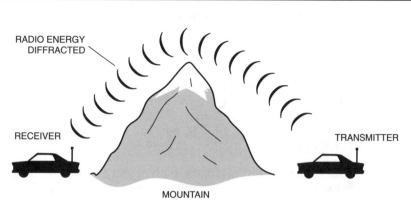

RADIO ENERGY DIFFRACTED

RECEIVER

TRANSMITTER

MOUNTAIN

Knife-Edge Diffraction

3B17 If the elapsed time for a radar echo is 62 microseconds what is the distance in nautical miles to the object?
   A. 5 nautical miles
   B. 87 nautical miles
   C. 37 nautical miles
   D. 11.5 nautical miles

ANSWER A: It takes a radar signal 6.2 microseconds to travel 1 nautical mile. In 62 microseconds, the radar signal will travel 10 total miles — 5 miles to the target and 5 miles back. $62 \div (6.2 \times 2) = 5$

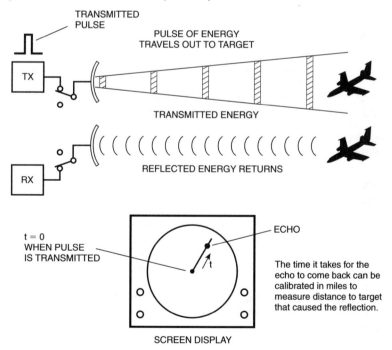

TRANSMITTED PULSE

PULSE OF ENERGY TRAVELS OUT TO TARGET

TX

TRANSMITTED ENERGY

RX

REFLECTED ENERGY RETURNS

t = 0 WHEN PULSE IS TRANSMITTED

ECHO

The time it takes for the echo to come back can be calibrated in miles to measure distance to target that caused the reflection.

SCREEN DISPLAY

Radar Echo

Source: *Technology Dictionary,* ©1987, Master Publishing, Inc.

3B18 What is the wavelength of a signal at 500 MHz?

A. 0.062 cm
B. 6 meters
C. 60 cm
D. 60 meters

ANSWER C: To calculate the wavelength in meters of a signal if you know the frequency in MHz, or to calculate MHz if you know the wavelength in meters, simply remember the number 300. Then divide 300 by the frequency in MHz. In this problem, 300 ÷ 500 = 0.6 meter = 60 cm. You have to convert meters to centimeters (multiply by 100) to get the right answer.

3B19 The radar range in nautical miles to an object can be found by measuring the elapsed time during a radar pulse and dividing this quantity by:

A. 0.87 seconds.
B. 1.15 microseconds.
C. 12.36 microseconds.
D. 1.73 microseconds.

ANSWER C: We know that it takes 6.173 (previously rounded to 6.2) microseconds for a radar wave to travel 1 nautical mile. Therefore, it will take twice that, or 12.35 microseconds, to make the trip from the radar antenna to the target 1 nautical mile away and then back to the antenna.

3B20 The band of frequencies least susceptible to atmospheric noise and interference is:

A. 30 − 300 kHz.
B. 300 − 3000 kHz.
C. 3 − 30 MHz.
D. 300 − 3000 MHz.

ANSWER D: Atmospheric noise and onboard electrical noise from ship and aircraft stations will dominate the low-frequency, medium-frequency, high-frequency, and even VHF-frequency bands up to 300 MHz. Signals between 300 MHz and 3 GHz are least susceptible to atmospheric noise.

3B21 What is the relationship in degrees of the electrostatic and electromagnetic fields of an antenna?

A. 0 degrees
B. 45 degrees
C. 90 degrees
D. 180 degrees

ANSWER C: The relationship between the electric field and the magnetic field of an antenna is 90 degrees. A horizontal antenna radiates an electric field parallel to the earth, and a magnetic field perpendicular to the earth. We base the polarization of an antenna on the electric field it radiates.

3B22 For a space wave transmission, the radio horizon distance of a transmitting antenna with a height of 100 meters is approximately:

A. 10 km.
B. 40 km.
C. 100 km.
D. 400 km.

ANSWER B: Radio waves—especially VHF, UHF, and radar radio waves traveling over the water—travel approximately 1.1 times further than the optical horizon. The formula to calculate the distance of the radio horizon in kilometers is d = √17h , where d is equal to the distance to the radio horizon in kilometers, and h is equal to the height of the antenna above the average terrain in meters. The calculation of the problem is

E▶ $d = \sqrt{17 \times 100} = \sqrt{1700} = 41.23$ km   Where:

d = Distance to radio horizon in **kilometers**

h = Height of antenna in **meters**

40 is your closest answer.

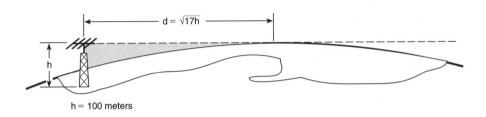

h = 100 meters

Radio Wave Transmission Distance

3B23 For a space wave transmission, the radio horizon distance of a receiving antenna with a height of 64 meters is approximately:

A. 8 km.
B. 32 km.
C. 64 km.
D. 256 km.

ANSWER B: To solve this problem, multiply $17 \times 64 = \sqrt{1088} = 32.98$ km.

3B24 If a transmitting antenna is 100 meters high and a separate receiving antenna is 64 meters high, what is the maximum space wave communication distance possible between them?

A. 18 km
B. 72 km
C. 164 km
D. 656 km

ANSWER B: In this problem, you separately calculate each antenna's communication distance. Then, you combine the distances to get the answer.

Coincidentally, the double calculations are the same as the preceding two questions. A 100-meter antenna *sees* the radio horizon at 40 kilometers, and the 64-meter antenna *sees* the radio horizon at 32 kilometers. Simply add the two distances to get the correct answer of 72 kilometers. Be careful! If you combine the two antenna heights and then apply the formula, you end up with a wrong answer. Calculate each antenna's communication distance; then add the two answers.

 This symbol identifies formulas that are used for question solutions. A list of the most used formulas is located in the Appendix for easy reference and for study before examinations.

3B25 A receiver is located 64 km from a space wave transmitting antenna that is 100 meters high. Find the required height of the receiving antenna.
A. 36 meters high
B. 64 meters high
C. 100 meters high
D. 182.25 meters high

ANSWER A: To solve this problem, first calculate the communication range for the 100-meter transmitter antenna; as we know from previous questions, this is 40 kilometers. We know the total distance (64 km). We need the receiving antenna distance, which is 24 (64 − 40) more kilometers. Now we can solve for the height by modifying the formula $d = \sqrt{17h}$ slightly:

$$d^2 = 17h \text{ or } d^2 \div 17 = h = 576/17 \text{ which is 33.9 meters.}$$

The question really should read "minimum required height of the receiving antenna." In any case, the closest answer is 36.

3B26 Which of the following is *not* one of the natural ways a radio wave may travel from transmitter to receiver?
A. Ground wave
B. Microwave
C. Sky wave
D. Space wave

ANSWER B: This question asks what type of radio wave propagation can travel from a transmitter to a receiver. The correct names for propagation are ground wave, sky wave, and space wave. But this question asks "which is not" a propagation phenomena, and the term "micro wave" refers more to the wave length of radar waves, than to the propagation type.

3B27 Which of the following terrain types permits a ground wave to travel the farthest?
A. Salt water
B. Fresh water
C. Sandy
D. Rocky

ANSWER A: Salt water has the highest conductivity of the four *answers* by a long shot! A highly conductive ground plane—in this case, sea water—leads to an extension of ground waves. It also causes the second hop of a sky wave to get a good bounce back up to the ionosphere.

3B28 Which of the following frequency bands is best suited for ground wave propagation?
A. 30 kHz to 300 kHz
B. 300 kHz to 3 MHz
C. 3 MHz to 30 MHz
D. 30 MHz to 300 MHz

ANSWER A: Ground waves are best propagated over land and sea when wave lengths are extremely long. 30 kHz to 300 kHz have traditionally been used for long range Omega navigation, Loran A and C navigation, and some radio beacons, too. The biggest problem with these long wave length signals is their susceptibility to on-board ship radio frequency interference from the ship's electrical system.

3B29 Which of the following frequency bands is best suited for sky-wave propagation?
   A. 30 kHz to 300 kHz
   B. 3 MHz to 30 MHz
   C. 30 MHz to 300 MHz
   D. 3 GHz to 30 GHz
ANSWER B: The region from 3 MHz to 30 MHz is called high frequency, and this is where sky wave SSB and digital communications are best bounced off the ionosphere. Marine bands at 4 MHz, 6 MHz, and 8 MHz offer communications out to approximately 3,000 miles via sky waves. Marine bands at 12 MHz, 16 MHz, 22 MHz, and 26 MHz offer worldwide communication capabilities for long sky wave communications.

3B30 Which of the following layers of the Ionosphere has no effect on sky-wave propagation during the hours of darkness?
   A. D
   B. E
   C. F
   D. None of these
ANSWER A: During daylight hours, the ionosphere is composed of layers of free electrons: the D layer, the E layer, the F1 layer, and the F2 layer. During the hours of darkness, the D layer disappears, which means it has no effect on sky wave propagation at night.

3B31 Which of the following least affects refraction of sky waves?
   A. Frequency of the radio wave.
   B. Density of the ionized layer.
   C. Angle at which the radio wave enters the ionosphere.
   D. Geographical variations.
ANSWER D: Here is another one of those questions where they are looking for a "negative" answer. Geographical location has the least effect on sky wave refraction. All the other answers have a major effect on sky wave refraction, including frequency, density of the ionosphere, and critical angle.

3B32 The area that lies between the outer limit of the ground-wave range and the inner edge of energy returned from the Ionosphere is called:
   A. The critical angle.
   B. The skip zone.
   C. The skip distance.
   D. The shadow.
ANSWER B: The skip zone is the area that is too far away for ground wave reception, yet too close for sky wave reception. Marine radio bands at 8, 12, 16, and 22 MHz all exhibit a skip zone, or no-reception zone, from approximately 100 miles to 400 miles. The skip zone varies from day to night, but after a few weeks at sea in the radio room, you can begin to estimate when you'll be able to reliably communicate near the skip zone, and when you won't.

3B33 Skip Distance can be maximized by using the ____ radiation angle possible and the ____ frequency that will be refracted at that angle.
   A. Lowest, lowest
   B. Lowest, highest

C. Highest, lowest

D. Highest, highest

ANSWER B: To establish the furthest communications range on marine SSB and digital channels on high frequency, choose the highest frequency that leads to the lowest radiation angle possible. But watch out for those incorrect answers—the question is looking for the *lowest* radiation angle possible and the *highest* frequency. Don't get your answers reversed when you read over your four possible choices.

3B34 To obtain the most reliable skywave propagation the ____ should be used.

A. Lowest useable frequency (LUF)

B. Maximum useable frequency (MUF)

C. Optimum useable frequency (OUF)

D. Critical frequency

ANSWER C: "Optimum useable frequency" is the frequency that leads to the most reliable sky wave to the communication station you want to contact. This frequency can vary depending on the time of day.

3B35 Tropospheric scatter is a method of skywave propagation for which of the following frequency bands?

A. 300 kHz to 3 MHz

B. 3 MHz to 30 MHz

C. 30 MHz to 300 MHz

D. 300 MHz to 3 GHz

ANSWER D: Tropospheric scatter is most pronounced at frequencies from 300 MHz to 3 GHz. 1000-watt transmitters and highly directive antenna arrays *forward scatter* and *back scatter* signals off the troposphere—the weather layer atmosphere—establishing communications dramatically further than might be anticipated by calculating the 4/3 radio horizon. But it takes high power levels and large antennas to do the job.

3B36 Which of the following methods are used for diversity reception to overcome the effects of tropospheric scattering of a sky wave?

A. Frequency diversity

B. Phase diversity

C. Amplitude diversity

D. Critical diversity

ANSWER A: Tropospheric scatter communications can change dramatically in seconds because of the ever-changing weather conditions in the troposphere. Frequency diversity optimizes the reception capabilities of tropo scatter.

3B37 Which of the following will not significantly reduce the effects of fading?

A. Use an antenna with a good front to back ratio

B. Use an antenna with a sharp frontal lobe

C. Use an antenna with a minimum number of spurious side and back lobes

D. Use an antenna with good omni directional pattern

ANSWER D: Fading of an incoming high-frequency signal is caused by an antenna system receiving incoming energy in slightly different time domains, creating a phase difference of incoming signals that tends to cancel out the signals. To minimize the reception of any delayed signal that is passing through a different ionospheric layer, use an antenna with a good front-to-back ratio, a sharp main lobe, and a minimum number of side lobes. The question asks which of the following *will not* reduce selective fading, and an omnidirectional antenna won't help reduce selective fading at all. It might make it worse!

3B38 Which of the following terms is not used to define ionospheric variations?
A. Seasonal variations
B. Geographical variations
C. Cyclical variations
D. Tropospheric scatter variations

ANSWER D: The ionosphere gets its "charge" from solar radiation, charged particles, and ultraviolet radiation. Seasonal, geographical, and 11-year cyclical variations influence the ionosphere. The question asks, "Which ... is not ..." The answer is local weather conditions like tropospheric scatter variations. This has nothing to do with the ionosphere, so this is the correct answer.

3B39 The polarization of a radio wave:
A. Is perpendicular to the electrostatic field of the antenna.
B. Is the same direction as the electrostatic field of the antenna.
C. Is the same direction as the magnetic field of the antenna.
D. Is perpendicular to both the electrostatic and magnetic fields of the antenna.

ANSWER B: The polarization of a radio wave is determined by the direction of the electric (electrostatic) field of the antenna. The director of the electric field is at 90° (90 degrees) to the direction of the magnetic field.

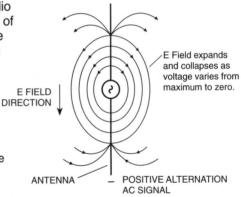

E Field expands and collapses as voltage varies from maximum to zero.

E FIELD DIRECTION

Polarization is Vertical Because Electric Field is Vertical

ANTENNA — POSITIVE ALTERNATION AC SIGNAL

3B40 The direction of propagation of a radio wave is ____ to the electro-static field of the antenna and ____ to the magnetic field of the antenna.
A. Parallel, parallel
B. Parallel, perpendicular
C. Perpendicular, parallel
D. Perpendicular, perpendicular

ANSWER D: Electromagnetic (radio) waves consist of magnetic and electric fields that are at right angles (90°) to each other. The two fields, E for electric and M for magnetic, can be represented by two vectors at 90° to each other (perpendicular to each other). The direction of travel of the wave is at 90° to the E and M vectors and is shown as another vector W. The direction of travel of W, the wave, is 90° (perpendicular) to both the E and M vectors, which are 90° (perpendicular) to each other. (See illustration at 3H143, page 329).

3B41 Most AM broadcasts employ ____ polarization while most FM broadcasts employ ____ polarization of the radio wave.

A. Vertical, vertical
B. Vertical, horizontal
C. Horizontal, vertical
D. Horizontal, horizontal

ANSWER B: AM broadcast stations usually employ vertical polarization for best ground wave propagation over rough terrain and populated cities. VHF broadcast services like FM stereo and television normally employ horizontal polarization.

3B42 Ducts often form over:

A. Dry and arid deserts.
B. Cold arctic regions.
C. Highly industrialized regions.
D. Water.

ANSWER D: Tropospheric ducting usually occurs over large oceans and lakes. The cool air from the water keeps the surface temperature uniformly colder than the warm air just above the local cloud layer. If the warm air inversion is 10 degrees warmer than the cool air above and below it, it triggers a tropospheric duct that could dramatically extend VHF and UHF communications range. The greater the temperature inversion, the more pronounced the duct, and the greater the signal strength at the distant receiving station. The duct usually forms as high pressure air subsides down over the water and warms up because of the subsidence pressure heating effect. The most common VHF/UHF tropospheric duct occurs every August between Hawaii and California.

Subelement 3C – Radio Practice   (6 questions)

3C1 What is a frequency standard?

A. A well known (standard) frequency used for transmitting certain messages
B. A device used to produce a highly accurate reference frequency
C. A device for accurately measuring frequency to within 1 Hz
D. A device used to generate wide-band random frequencies

ANSWER B: You may already have a frequency standard in your test equipment pool, but the world's most expensive frequency standards are stations WWV and WWVH. You can pick up time signals from them on 5 MHz, 10 MHz, 15 MHz, and 20 MHz. These signals are absolutely on frequency, 24 hours a day, for your equipment and test apparatus calibration.

3C2 What is a frequency-marker generator?

A. A device used to produce a highly accurate reference frequency
B. A sweep generator
C. A broadband white noise generator
D. A device used to generate wide-band random frequencies

ANSWER A: Older radio equipment might have a "CAL" band switch position. This is the calibrate position which energizes an internal marker generator. Every 100 kHz there would be a relatively accurate signal. But guess how you calibrate the internal marker generator? That's right, with the on-the-air WWV time signals referred to in question 3CI.

3C3 How is a frequency-marker generator used?
A. In conjunction with a grid-dip meter
B. To provide reference points on a receiver dial
C. As the basic frequency element of a transmitter
D. To directly measure wavelength

ANSWER B: The older marker generators came in handy when your digital readout went haywire on your older transceiver. You could generally find where you were on the dial by listening to the signals occurring every 100 kHz.

3C4 How is a frequency counter used?
A. To provide reference points on an analog receiver dial thereby aiding in the alignment of the receiver
B. To heterodyne the frequency being measured with a known variable frequency oscillator until zero beat is achieved, thereby indicating what the unknown frequency is
C. To measure the deviation in an FM transmitter in order to determine the percentage of modulation
D. To measure the time between events, or the frequency which is the reciprocal of the time

ANSWER D: You must never directly couple a frequency counter to your transmitter output. Most portable counters have a small rubber antenna that will pick up the signal from a nearby antenna quite nicely. Most counters can read out a signal within 50 feet. The more expensive the counter, the further away it can read a transmitting signal. If you plan to use the counter outside, make sure you purchase a counter with a LCD display. Stay away from LED displays for outside use because you can't see their readout in the bright sunlight!

Frequency Counter
Courtesy of OPTOELECTRONICS

3C5 What is the most the actual transmitter frequency could differ from a reading of 156,520,000 hertz on a frequency counter with a time base accuracy of ± 1.0 ppm?
A. 165.2 Hz
B. 15.652 kHz
C. 156.52 Hz
D. 1.4652 MHz

ANSWER C: You will probably have 1 out of the following 4 questions on your test regarding the error readout on a frequency counter with a specific part per million time base. It's easy to solve these problems with a simple calculator.

3C6 What is the most the actual transmitter frequency could differ from a reading of 156,520,000 Hertz on a frequency counter with a time base accuracy of ± 10 ppm?
  A. 146.52 Hz
  B. 10 Hz
  C. 156.52 kHz
  D. 1565.20 Hz

ANSWER D: Keystroke the following on your calculator: Clear 156.52 × 10 = 1565.2 Hz. You are multiplying the frequency in MHz by the time base accuracy of 10 ppm.

     Readout error = f × a   Where: f = Frequency in **MHz**
                                         a = Counter accuracy in **parts per million**

3C7 What is the most the actual transmitter frequency could differ from a reading of 462,100,000 hertz on a frequency counter with a time base accuracy of ± 1.0 ppm?
  A. 46.21 MHz
  B. 10 Hz
  C. 1.0 MHz
  D. 462.1 Hz

ANSWER D: How refreshing, a new frequency. From Hz to MHz, we end up with 462.1 MHz, which multiplied by 1.0 ppm gives you the answer. The following are the keystrokes: Clear 462.1 × 1 = 462.1 Hz.

3C8 What is the most the actual transmit frequency could differ from a reading of 462,100,000 Hertz on a frequency counter with a time base accuracy of ± 0.1 ppm?
  A. 46.21 Hz
  B. 0.1 MHz
  C. 462.1 Hz
  D. 0.2 MHz

ANSWER A: Remember, multiply the frequency in MHz times the time base accuracy. Try this on your calculator: Clear 462.1 × 0.1 = 46.21.

3C9 What is a dip-meter?
  A. A meter used as a nonradiating load (dummy load) to measure transmitter output power
  B. A meter used to measure the reflection coefficient of an RF transmission path
  C. A variable LC oscillator with metered feedback current.
  D. A fixed tuned LC oscillator used to troubleshoot RF tank circuits

ANSWER C: A dip-meter is useful for determining the resonant frequency of tuned circuits, especially for a box full of unmarked antenna traps. It could also be used as an absorption meter with the meter indicating a peak. Older dip-meters have an assortment of coils to preselect a general group of frequencies to be tuned in. For antenna SWR matching, a more convenient all-in-one test instrument might be the SWR analyzer, rather than the dip meter.

SWR Analyzer

Courtesy of MFJ Enterprises, Inc.

3C10 How does a dip-meter function?

A. Reflected waves at a specific frequency desensitize the detector coil
B. Power absorbed by a resonant circuit causes a decrease in dip-meter current
C. Power from a transmitter cancels feedback current
D. Harmonics of the oscillator cause an increase in resonant circuit Q resulting in an increase in transmitter output power

ANSWER B: Your little portable dip-meter couples power from its built-in oscillator to your antenna. At resonance, the antenna grabs all of the current, and this causes a decrease in metered current, seen as a dip on the indicating meter. The meter dips because all of the current is getting soaked up at resonance within the antenna system. If ever you are performing maintenance on a high-frequency antenna system aboard an aircraft, a boat, or an authorized base station, the dip meter is an invaluable tool to carry in your fix-it bag.

3C11 What two ways could a dip-meter be used in a radio station?

A. To measure resonant frequency of antenna traps and to measure percentage of modulation
B. To measure antenna resonance and to measure percentage of modulation
C. To measure antenna resonance and to measure antenna impedance
D. To measure resonant frequency of antenna traps and to measure a tuned circuit resonant frequency

ANSWER D: The dip-meter is a handy way to determine the resonant frequency of a box full of unmarked antenna traps. You can also measure a tuned circuit's resonant frequency, such as an antenna coupler. The dip-meter is a great one to measure a *tuned circuit* (key words).

3C12 How tight should the dip-meter be coupled with the tuned circuit being checked?

A. As loosely as possible, for best accuracy
B. As tightly as possible, for best accuracy

C. First loose, then tight, for best accuracy
D. With a soldered jumper wire between the meter and the circuit to be checked, for best accuracy

ANSWER A: If you get too close to the circuit you are testing, you could de-tune it with your dip-meter coil. For best accuracy, you want to keep the coupling as loose as possible.

3C13 What factors limit the accuracy, frequency response, and stability of an oscilloscope?
A. Sweep oscillator quality and deflection amplifier bandwidth
B. Tube face voltage increments and deflection amplifier voltage
C. Sweep oscillator quality and tube face voltage increments
D. Deflection amplifier output impedance and tube face frequency increments

ANSWER A: Every service shop should have a full-featured oscilloscope ("scope" for short). Many new marine and aeronautical transceivers may be employing digital signal processing (DSP). This demands a stable, *high-quality sweep oscillator* (key words) circuit inside the scope, and the scope must also have enough *bandwidth* (key word) to handle high-frequency applications, as well as VHF and UHF.

3C14 What factors limit the accuracy, frequency response, and stability of a D'Arsonval movement type meter?
A. Calibration, coil impedance and meter size
B. Calibration, series resistance and electromagnet current
C. Coil impedance, electromagnet voltage and movement mass
D. Calibration, mechanical tolerance and coil impedance

ANSWER D: Marine and aviation radios are now produced without the traditional needle movement meter. However, you may be working on older equipment that uses an electromechanical meter movement. Mechanical abuse and overload are the primary causes of inaccuracy and failure in these meters.

3C15 What factors limit the accuracy and stability of a frequency counter?
A. Number of digits in the readout, speed of the logic and time base stability
B. Time base accuracy, speed of the logic and time base stability
C. Time base accuracy, temperature coefficient of the logic and time base stability
D. Number of digits in the readout, external frequency reference and temperature coefficient of the logic

ANSWER B: Here we are back to counters again! Everyone should have one. Be sure to get one that has *accuracy, speed* and *stability* (key words).

3C16 How can the frequency response of an oscilloscope be improved?
A. By using a triggered sweep and a crystal oscillator as the time base
B. By using a crystal oscillator as the time base and increasing the vertical sweep rate
C. By increasing the vertical sweep rate and the horizontal amplifier frequency response
D. By decreasing the minimum rise time of the vertical amplifier

ANSWER D: By decreasing the minimum rise time of the vertical amplifier of your oscilloscope, you may dramatically increase the frequency response of that scope.

3C17 How can the accuracy of a frequency counter be improved?

A. By using slower gating circuitry and increasing the number of digits used for display

B. By increasing the amount of time the control gate is held open, more pulses can be counted

C. By using a crystal controlled oscillator mounted in a thermal oven for the time base

D. By using faster gating circuitry and decreasing the number of digits used for display

ANSWER C: The professional frequency counter used by land mobile and radio marine radio technicians has a thermal crystal oven that keeps the crystal at a specific temperature for a more accurate time base. On portable frequency counters, you must let the crystal oven heat up for approximately 20 minutes before taking a reading. Also keep in mind that the crystal oven draws considerably more current than the counter itself, so you better have a spare set of counter batteries handy at a remote site!

3C18 What is the name of the condition that occurs when the signals of two transmitters in close proximity mix together in one or both of their final amplifiers, and unwanted signals at the sum and difference frequencies of the original transmissions are generated?

A. Amplifier desensitization

B. Neutralization

C. Adjacent channel interference

D. Intermodulation interference

ANSWER D: Land mobile repeater stations and remote bases require regular maintenance. A problem called *intermodulation* may be caused by other close proximity transmitters. When you listen to the repeater's local speaker, you may hear two signals coming in at once. This is the effect commonly called intermodulation.

3C19 How does intermodulation interference between two transmitters usually occur?

A. When the signals from the transmitters are reflected out of phase from airplanes passing overhead

B. When they are in close proximity and the signals mix in one or both of their final amplifiers

C. When they are in close proximity and the signals cause. feedback in one or both of their final amplifiers

D. When the signals from the transmitters are reflected in phase from airplanes passing overhead

ANSWER B: "Intermod," in its true sense, occurs within the repeater final amplifier. When signals *mix* in one or both repeater amplifiers, sum and difference frequencies on many different frequencies can be heard.

3C20 How can intermodulation interference between two transmitters in close proximity often be reduced or eliminated?
  A. By using a Class C final amplifier with high driving power
  B. By installing a terminated circulator or ferrite isolator in the feed line to the transmitter and duplexer
  C. By installing a band-pass filter in the antenna feed line
  D. By installing a low-pass filter in the antenna feed line
ANSWER B: At repeater and base station installations atop buildings and mountains, you may need to install *circulators* and *isolators* in the feed line to the transmitter and duplexer to minimize other nearby base stations getting into your system. This is an ongoing problem because most mountain tops will have different radios systems that will be coming and going, causing your repeater or remote base station to continually hang up on specific signals from a nearby antenna setup. A spectrum analyzer is a good piece of test equipment to use when you suspect the problem is intermodulation interference.

3C21 What will occur when a non-linear amplifier is used with a single-sideband phone transmitter?
  A. Reduced amplifier efficiency
  B. Increased intelligibility
  C. Sideband inversion
  D. Distortion
ANSWER D: Maritime mobile stations operating on marine SSB may legally employ single-sideband amplifiers for additional power output. The added power amplifier must be linear to prevent the signal from becoming *distorted* (key word). Older tube-type amplifiers may not be rated for good linearity on an SSB emission.

3C22 How can even-order harmonics be reduced or prevented in transmitter amplifier design?
  A. By using a push-push amplifier
  B. By using a push-pull amplifier
  C. By operating class C
  D. By operating class AB
ANSWER B: Most worldwide amplifiers run a pair of output tubes. The tubes are in push-pull, and this reduces even-order harmonics. Watch out for answer A— if you don't read it carefully, you could mistake it for "push-pull."

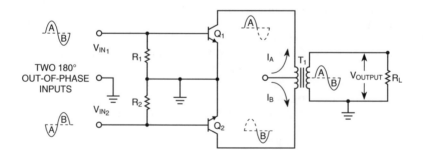

Push-Pull Amplifier

3C23 What is receiver desensitizing?

A. A burst of noise when the squelch is set too low

B. A burst of noise when the squelch is set too high

C. A reduction in receiver sensitivity because of a strong signal on a nearby frequency

D. A reduction in receiver sensitivity when the AF gain control is turned down

ANSWER C: You are up servicing a base station on a building top, and all of a sudden the incoming signal abruptly drops, and then pops up again at its normal level. Chances are the *receiver* of the equipment you are working on is losing *sensitivity* because of the presence of a *strong signal* (key words) on an adjacent channel. Time to consult your spectrum analyzer, or a sensitive portable frequency counter, and determine what steps you are going to take to minimize receiver desensitization.

3C24 What is the term used to refer to a reduction in receiver sensitivity caused by unwanted high-level adjacent channel signals?

A. Intermodulation distortion

B. Quieting

C. Desensitizing

D. Overloading

ANSWER C: If signals mysteriously come in strong, then abruptly get weak, and then come back strong again, chances are there is someone nearby transmitting on an adjacent frequency, desensitizing your receiver. The best way to check for adjacent signals is through the use of a frequency counter.

3C25 What is cross-modulation interference?

A. Interference between two transmitters of different modulation type

B. Interference caused by audio rectification in the receiver preamp

C. Harmonic distortion of the transmitted signal

D. Modulation from an unwanted signal is heard in addition to the desired signal

ANSWER D: If you have driven your little handheld transceiver down to a big city, chances are you have heard the effects of cross-modulation interference. "Cross-mod" brings in modulation from an *unwanted signal* (key words) that rides on top of the signal you are presently listening to. This mix occurs inside your receiver with handheld transceivers. Do not confuse cross-modulation with intermodulation. "Intermod" occurs in the transmitters, and cross-modulation occurs in the receivers.

3C26 What is the term used to refer to the condition where the signals from a very strong station are superimposed on other signals being received?

A. Intermodulation distortion

B. Cross-modulation interference

C. Receiver quieting

D. Capture effect

ANSWER B: If in your receiver's signal you hear more than one transmitted frequency, chances are it is cross-modulation interference. Remember, if they say the word *received,* it means "cross-mod."

3C27 What is the capture effect?
A. All signals on a frequency are demodulated by an FM receiver.
B. All signals on a frequency are demodulated by an AM receiver.
C. The strongest signal received is the only demodulated signal.
D. The weakest signal received is the only demodulated signal.

ANSWER C: A unique property of an FM receiver is the ability to hear only the loudest incoming FM signal. If a distant station is transmitting, and a strong local station comes on the air, you will only hear the local station. The stronger signal is "capturing" your receiver.

3C28 If a strong FM-phone signal and a weak FM-phone signal, both using the same carrier frequency, are received simultaneously, why is only one demodulated?
A. Strong signal desensitizes the receiver circuitry to block the weak signal
B. Strong signal causes cross-modulation interference to prevent the weaker signal from being received
C. Strong signal captures the local oscillator preventing the weak signal from being detected
D. Strong signal forces the receiver RF amps to discriminate against the weaker signal

ANSWER C: When an incoming FM signal is blocked by a stronger FM signal, it is called "capture effect." Sometimes you hear the two signals coming in at once, sometimes you hear nothing more than a garbled message, and if there is a big difference between the two signal strengths, all you will hear is the stronger signal completely. In the AM aviation radio service, two signals coming in at once will usually lead to a tell-tale heterodyne tone. When the tower hears a heterodyne tone, the personnel know that two stations are attempting to transmit at once. In AM radio, you do not normally experience "capture effect."

3C29 How does a spectrum analyzer differ from a conventional oscilloscope?
A. The oscilloscope is used to display electrical signals while the spectrum analyzer is used to measure ionospheric reflection.
B. The oscilloscope is used to display electrical signals in the frequency domain while the spectrum analyzer is used to display electrical signals in the time domain.
C. The oscilloscope is used to display electrical signals in the time domain while the spectrum analyzer is used to display electrical signals in the frequency domain.
D. The oscilloscope is used for displaying audio frequencies and the spectrum analyzer is used for displaying radio frequencies.

ANSWER C: The spectrum analyzer displays the strength of signals at particular frequencies according to frequency (in a frequency domain) along a horizontal axis. A signal can be locked in and centered in the middle of the screen with 25 kHz to the left and 25 kHz to the right for a close examination of any off-frequency spurs it might have. A simple oscilloscope *cannot* display the signal in the frequency domain—it displays the signals on a time axis (in a time domain) going from left to right. With your output signal displayed on a spectrum analyzer, you may adjust your transmitter output network for minimum spurious signals. See figure at 3C30.

3C30 What does the horizontal axis of a spectrum analyzer display?

A. Amplitude
B. Voltage
C. Resonance
D. Frequency

ANSWER D: The spectrum analyzer analyzes the frequency of signals by displaying the amplitude of signals at particular frequencies on a frequency axis.

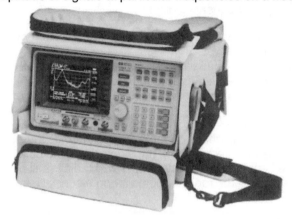

Spectrum Analyzer
Courtesy of Hewlett-Packard Company

3C31 What does the vertical axis of a spectrum analyzer display?

A. Amplitude
B. Duration
C. Frequency
D. Time

ANSWER A: The vertical axis of the spectrum analyzer displays the amplitude of the signal.

3C32 What test instrument can be used to display spurious signals in the output of a radio transmitter?

A. A spectrum analyzer
B. A wattmeter
C. A logic analyzer
D. A time-domain reflectometer

ANSWER A: You can check for spurious signals with a spectrum analyzer.

3C33 What test instrument is used to display intermodulation distortion products from an SSB transmitter?

A. A wattmeter
B. An audio distortion analyzer
C. A logic analyzer
D. A time-domain reflectometer

ANSWER B: An audio distortion analyzer is the best instrument to display intermodulation distortion products from an SSB transmitter. Displaying intermod distortion from a transmitter is relatively difficult unless you have the right type of equipment on hand.

3C34 How can ferrite beads be used to suppress ignition noise?
A. Install them in the resistive high-voltage cable every 2 years.
B. Install them between the starter solenoid and the starter motor.
C. Install them in the primary and secondary ignition leads.
D. Install them in the antenna lead.

ANSWER C: It is the primary and secondary ignition leads which radiate strong spark pulses that create ignition noise. Ferrite beads strung on the leads may help reduce the higher-frequency noises.

3C35 How can you determine if a line-noise interference problem is being generated within a building?
A. Check the power-line voltage with a time-domain reflectometer.
B. Observe the AC wave form on an oscilloscope.
C. Turn off the main circuit breaker and listen on a battery-operated radio.
D. Observe the power-line voltage on a spectrum analyzer.

ANSWER C: A small transistorized portable radio that tunes the AM broadcast band is a good noise "sniffer". Shut off the electricity to the building, and then listen to see whether or not your battery-operated radio is still picking up the interference. Use that same radio to "home in" on electrical interference within a building.

3C36 An electrical relay:
A. Is a current limiting device.
B. Is a device used for supplying 3 or more voltages to a circuit.
C. Is concerned mainly with HF audio amplifiers.
D. Is a remotely controlled switching device.

ANSWER D: An electrical relay is usually used as a switch. Its open contacts provide a near infinite resistance open circuit, and its closed contacts provide a near zero resistance closed circuit.

3C37 A high standing wave ratio on a transmission line can be caused by:
A. Excessive modulation.
B. An increase in output power.
C. Detuned antenna coupling.
D. Poor B+ regulation.

ANSWER C: Water inside an antenna mount will many times detune the antenna circuit. This causes high SWR, which can be detected with an SWR meter or analyzer.

3C38 What may be the results of adding a GaAsFET pre-amplifier circuit to a marine VHF transceiver?
A. A slight increase in sensitivity, but susceptibility to overload and intermodulation
B. Increased selectivity
C. Increase in sensitivity, and a decrease in the susceptibility to overload and intermodulation
D. The GaAsFET pre-amplifier will minimize out-of-band unwanted reception

ANSWER A: Marine radio manufacturers might advertise the GaAsFET pre-amp as a major improvement in marine VHF reception. While it might boost the gain compared to a traditional field-effect transistor, the GaAsFET also increases the radio's susceptibility to overload and intermodulation. We normally see the GaAsFET in extremely selective VHF equipment or in *weak-signal* UHF equipment.

3C39 Which of the following contains a multirange AF voltmeter calibrated in dB and a sharp, internal 1000-Hz bandstop filter, both used in conjunction with each other to perform quieting tests?
    A. Reflectometer
    B. Dip-meter
    C. SINAD meter
    D. Vector-impedance meter
ANSWER C: The term "SINAD" stands for Signal Plus Noise And Distortion. The SINAD meter is most commonly used to measure the sensitivity of a VHF FM receiver.

3C40 A(n) ____ and ____ can be combined to measure the characteristics of transmission lines and such an arrangement is known as a time-domain reflectometer (TDR).
    A. Frequency spectrum analyzer, RF generator
    B. Oscilloscope, pulse generator
    C. AC millivolt meter, AF generator
    D. Frequency counter, linear detector
ANSWER B: You might be sent out to troubleshoot a marine communications system aboard a big ship with hundreds of feet of coaxial cable feedline. An oscilloscope and a pulse generator could help determine how far up the line there is a pinch, crack, or break. This device is known as a *time-domain reflectometer* and is very handy when troubleshooting long runs of coaxial cable.

3C41 Which of the following can not be measured with an admittance meter?
    A. Conductance and susceptance
    B. Voltage standing wave ratio (VSWR)
    C. Reflection coefficient
    D. Capacitance and inductance
ANSWER D: The admittance meter measures the reciprocal of impedance as the siemen. Admittance is directly proportional to current. This is because resistance, reactance, and impedance are inversely proportional to current. The meter is a good way to test parallel circuit impedances.

3C42 One piece of equipment to indicate neutralization is:
    A. A neon bulb.
    B. A tachometer.
    C. Linear detector.
    D. Ceramic filter.
ANSWER A: When you work on large shipboard or shore-side SSB power amplifiers that have power output tubes instead of transistors, it might be necessary to ensure you do not have undesirable feedback from the amplifier section. Runaway oscillation could actually destroy the tubes within minutes. A

neon bulb is a simple device to hold next to the chassis of the equipment to see whether or not there is any RF present on the outside of the cabinet. You can also use the neon bulb to help test for the presence of RF aboard a ship station's associated electronics.

3C43 Pulse type interference to radio receivers in automobiles can be reduced by:
  A. Connecting resistances in series with the spark plugs.
  B. Using heavy conductors between the starting battery and the starting motor.
  C. Connecting resistances in series with the starting battery.
  D. Grounding the negative side of the starting battery.

ANSWER A: Factory-recommended resistance spark plugs might be necessary to reduce the interference in certain mobile two-way radio installations. There are also resistance spark-plug wires that help *knock down* the noise pulses.

3C44 Magnetron oscillators are used for:
  A. Generating SHF signals.
  B. Multiplexing.
  C. Generating rich harmonics.
  D. FM demodulation.

ANSWER A: The magnetron is found in high-power radar and microwave installations; it generates super-high frequency (SHF) signals. Because of the lethal high voltages present, observe extreme caution when working near a magnetron.

3C45 Which of the following pieces of test equipment will permit you to measure the percentage of modulation of a transmitter signal on an oscilloscope when the frequency of the RF exceeds the frequency response of the oscilloscope's vertical amplifiers?
  A. Wow and flutter meter
  B. Curve tracer
  C. Linear detector
  D. Vector-impedance meter

ANSWER C: Sometimes the frequency of the input signal is greater than the frequency response of the vertical amplifier of the oscilloscope. When this happens, use a linear detector to measure the percentage of modulation of the transmitted signal. Within a linear detector, a stable oscillator works as a LO (local oscillator) to mix with higher frequency RF to produce an intermediate frequency that is easily displayed by the oscilloscope at a lower frequency within the bandwidth of the vertical amplifier.

3C46 Typical airborne HF transmitters usually provide a nominal RF power output to the antenna of _____ watts, compared with _____ watts RF output from a typical VHF transmitter.
  A. 10, 50
  B. 50, 10
  C. 20, 100
  D. 100, 20
ANSWER D: Most airborne and marine single-sideband, high-frequency, 3-30 MHz transmitters have an output of 100 watts. All marine VHF shipboard transceivers have a typical output between 20 watts and 25 watts. Twenty-five watts is the FCC legal limit for marine ship-station VHF transmitters. Fifty watts is permitted for an on-shore marine station with a *private coast station* or *public coast station* license.

3C47 Which of the following devices is a key element in modern day ADF systems?
  A. Goniometer
  B. Deflection yoke
  C. UART integrated circuit
  D. Swinging choke
ANSWER A: High-frequency automatic direction finders with an oscilloscope-type readout tube use a goniometer to trace the incoming radio bearing. A propeller-shaped pattern on the scope represents the actual incoming signal and its direction.

3C48 An absorption wave meter is useful in measuring:
  A. Field strength
  B. Output frequencies to conform with FCC tolerance
  C. Standing wave frequencies
  D. The resonant frequency of a powered LC tank circuit
ANSWER D: The absorption wave meter is used to measure the resonant frequency of a powered LC tank circuit in a high-frequency marine transceiver.

3C49 Which of the following navigational methods utilizes the lowest frequency for a carrier?
  A. LORAN
  B. OMEGA
  C. ADF
  D. VOR
ANSWER B: The OMEGA signal coming from eight worldwide stations is soon going off the air. OMEGA operates at 10 kHz and enjoyed popularity when ships and aircraft were well beyond typical Loran A and Loran C reception areas. Now that we have the global positioning system, OMEGA has outlived its usefulness and will soon be shut down.

3C50 Neglecting line losses, the RMS voltage along an RF transmission line having no standing waves:
  A. Is equal to the impedance.
  B. Is one-half of the surge impedance.
  C. Is the product of the surge impedance times the line current.
  D. Varies sinusodially along the line.

ANSWER C: You can calculate the root-mean-square (RMS) voltages along an RF transmission line by multiplying the surge impedance by the line current. These measurements are important to ensure your transmission line is capable of handling the power going through it.

3C51 Waveguides are:
A. A hollow conductor that carries RF.
B. Solid copper conductor for RF.
C. Special type of coaxial cables.
D. Special alloy wire able to conduct high frequencies.

ANSWER A: The wave guide is a transmission "line" that is hollow in the center; it carries the microwave RF energy from the transmitter to the antenna. In a radar receiver, a wave guide can also carry microwave echoes from the antenna back to the radar receiver section. Now that radar TR units are located in the antenna, long runs of waveguide are not found on modern vessels.

3C52 When a vacuum tube operates at VHF or higher frequencies as compared to lower frequencies:
A. Transit time of electrons becomes important.
B. It is necessary to make larger components.
C. It is necessary to increase grid spacing.
D. Only a pentode is satisfactory.

ANSWER A: Vacuum tubes operating in VHF or UHF power amplifiers are extremely expensive, and internal component size and make-up is small and critical. One important consideration is the transit time of electrons within the tube at VHF/UHF and microwave frequencies.

3C53 The principle of OMEGA navigation depends on measuring the _____ between received pulses.
A. Frequency difference
B. Phase angle
C. Time interval
D. Amplitude difference

ANSWER B: Although OMEGA navigation is just about extinct, it's important to know your history and to remember that OMEGA was the measuring of phase angles between received pulses that was then decoded as invisible lanes on major air and ocean routes.

3C54 A circulator:
A. Cools DC motors during heavy loads by turning on a cooling fan for the motors' windings.
B. Allows two or more antennas to feed one transmitter.
C. Allows one antenna to feed two separate microwave transmitters and receivers at the same time.
D. Insulates UHF frequencies on transmission lines and helps to reduce skin effect at high frequencies.

ANSWER C: The circulator allows a single repeater antenna to feed a microwave transmitter and a receiver at the same time. With more and more satellite systems going aboard big ships, circulators have become an important part of the marine electronics industry.

3C55 Coaxial transmission line shielding is grounded:
A. At the input only.
B. At both input and output.
C. At the output only.
D. Only when a balanced transmission line is needed.

ANSWER B: Coaxial cable transmission line shielding should be grounded at the transceiver and at the antenna. Leaving the antenna shielding "open" at either end creates high capacitance in the line at the antenna end, and sometimes causes the line to radiate most of your radio signal.

3C56 Motorboating (low frequency oscillations) in an amplifier can be stopped by:
A. Grounding the screen grid.
B. Bypassing the screen grid resistor with a .1 mfd capacitor.
C. Connect a capacitor between the B+ lead and ground.
D. Grounding the plate.

ANSWER C: Occasionally certain older RF amplifiers may go into a low frequency oscillation at specific frequencies. Connecting a capacitor between the B+ lead and chassis ground, keeping the leads extremely short, will help minimize this problem.

3C57 What effect could transmitting on marine SSB have on a running autopilot?
A. Seldom is there an effect.
B. The autopilot may sharply turn left or right.
C. The autopilot will make a slow clockwise circle.
D. The autopilot will make a slow counterclockwise circle.

ANSWER B: After installing a marine SSB transceiver, it's important to sea-trial the entire system to ensure your transmitter does not effect the operation of an automatic pilot. Some automatic pilots do not have adequate shielding on the rudder feedback control, and this could cause the pilot to turn sharply port or starboard (left or right) when the operator modulates the SSB transceiver.

3C58 An antenna radiates a primary signal of 500 watts output. If there is a 2nd harmonic output of 0.5 watt, what attenuation of the 2nd harmonic has occurred?
A. 3 dB
B. 10 dB
C. 20 dB
D. 30 dB

ANSWER D: It's relatively simple to calculate a dB power change. Remember this table:

DB	Power Change	
3 dB	2X	Power change
6 dB	4X	Power change
9 dB	8X	Power change
10 dB	10X	Power change
20 dB	100X	Power change
30 dB	1000X	Power change
40 dB	10,000X	Power change

In this question, a signal of 500 watts output has a second harmonic at 0.5 watt. This is a 1000X power change, so the second harmonic is down 30 dB.

Derivation:

If $dB = 10 \log_{10} \dfrac{P_1}{P_2}$

then what power ratio is 20 dB?

$20 = 10 \log_{10} \dfrac{P_1}{P_2}$

$\dfrac{20}{10} = \log_{10} \dfrac{P_1}{P_2}$

$2 = \log_{10} \dfrac{P_1}{P_2}$

Remember: logarithm of a number is the exponent to which the base must be raised to get the number.

$\therefore 10^2 = \dfrac{P_1}{P_2}$

$100 = \dfrac{P_1}{P_2}$

Or $P_1 = 100 \, P_2$

20 dB means P_1 is 100 times P_2

Decibels

3C59 An oscilloscope can be used to accomplish all of the following *except:*

A. Measure electron flow with the aide of a resistor.
B. Measure phase difference between two signals.
C. Measure electrical pressure.
D. Measure velocity of light with the aide of a light emitting diode.

ANSWER D: There is plenty that you can do with an oscilloscope, but one thing you **cannot** do is measure the velocity of light with the aid of a light emitting diode. You can measure voltage, phase, and electron flow with an oscilloscope, but not the velocity of light.

3C60 SSB transmitters require a(n) _____ wattmeter and AM or FM transmitters require a(n) _____ wattmeter.

A. RMS-reading, peak-reading
B. Peak-reading, rms-reading
C. Peak-reading, PEP
D. PEP, Peak-reading

ANSWER B: Use a peak-reading wattmeter to measure single-sideband output. This is necessary because of the different characteristics of our voice. On FM, you can use an average reading (RMS) wattmeter.

3C61 What type of antenna system allows you to receive and transmit at the same time in both directions?

A. Simplex
B. Duplex
C. Multiplex
D. Digital diplex

ANSWER B: Duplex allows you to transmit and receive at the same time in both directions. We would use a "duplexer" on the antenna system to keep the transmit signal from coming back down and getting into the receiver that may be several hundred kilohertz or a couple of megahertz away. You normally use duplexers in VHF and UHF systems.

3C62 A microwave device that allows RF energy to pass through in one direction with very little loss but absorbs RF power in the opposite direction:
  A. Circulator
  B. Wave trap
  C. Multiplexer
  D. Isolator

ANSWER D: In microwave systems, the isolator keeps RF energy going in one direction, but opposes it from coming back in the opposite direction.

3C63 Which of the following is found in an improved MOPA but not in a basic MOPA?
  A. Oscillator
  B. RF amplifier
  C. Buffer amplifier
  D. Key

ANSWER C: The term MOPA stands for Master Oscillator Power Amplifier. The oscillator uses either a crystal-controlled or a high-stability VFO, a buffer amp, and frequency multipliers, to arrive at the output frequency and power levels. The older MOPA CW transmitters did not use a buffer amplifier, but the new MOPA amplifiers do.

3C64 Which of the following is the only service using self-excited, or variable-frequency, oscillators in transmitters?
  A. Aviation
  B. Amateur
  C. Maritime
  D. Government

ANSWER B: It is the amateur radio service that uses variable frequency oscillators (with large tuning knobs) in their transmitters. Aviation, marine, and government radio systems use phase-locked loop devices that have individual channel steps.

3C65 Aviation services use predominantly _____ microphones.
  A. Carbon
  B. Dynamic
  C. Condenser
  D. Piezoelectric crystal

ANSWER B: The aviation service regularly uses dynamic microphones because of their rugged characteristics and their ability to withstand vibration and heat. Dynamic microphones require impedance-matching amplifiers to match their very low output impedance to the high-impedance audio input stage on the avaiation radio set.

3C66 What geometric patterns will be displayed on an oscilloscope if the modulating audio is fed to the horizontal plates while the modulated RF is fed to the vertical plates, assume 100 % modulation?
  A. Circle
  B. Line
  C. Square
  D. Triangle

ANSWER D: An oscilloscope shows a perfect triangle when the modulated RF is at 100 percent and is fed to the vertical plates while the audio is fed to the horizontal plates.

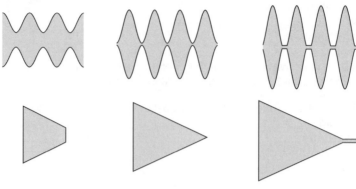

a. Undermodulated   b. 100% Modulated   c. Overmodulated

Measuring Percent Modulation on a Oscilliscope

3C67 To produce a single-sideband suppressed carrier transmission it is necessary to ____ the carrier and to ____ the unwanted sideband.
   A. Filter, filter
   B. Filter, cancel
   C. Cancel, filter
   D. Cancel, cancel
ANSWER C: Marine and aviation SSB transceivers cancel the carrier for A3J operation and filter the unwanted sideband.

3C68 Which of the following *is only* required for CW or SSB demodulation?
   A. BFO
   B. BPF
   C. VFO
   D. VOX
ANSWER A: To demodulate and receive both SSB and CW, the receiver requires a beat frequency oscillator to produce the desired signal in the audio frequency stage.

3C69 Which of the following FM detectors does not require IF limiter stage to precede it?
   A. Foster-Seeley
   B. Ratio
   C. Quadrature
   D. Stagger-tuned
ANSWER B: The ratio detector in an FM receiver does not require an intermediate frequency limiter stage ahead of it. However, the limiter stage is still preferred to recover full-fidelity speech.

Subelement 3D – Electrical Principles   (17 questions)

3D1 What is reactive power?
A. Wattless, non-productive power
B. Power consumed in wire resistance in an inductor
C. Power lost because of capacitor leakage
D. Power consumed in circuit Q

ANSWER A: The letters AC are found in the word reactive (meaning out-of-phase). It indicates *non-productive power* (key words) in circuits containing inductors and capacitors. Since out-of-phase power in inductors will tend to cancel out-of-phase power in capacitors, reactive power is wattless. It is not converted to heat and dissipated. It is energy being stored temporarily in a field and then returned to the circuit. It circulates back and forth in a coil's magnetic field and/or a capacitor's electric field.

3D2 What is the term for an out-of-phase, non-productive power associated with inductors and capacitors?
A. Effective power
B. True power
C. Peak envelope power
D. Reactive power

ANSWER D: If it is non-productive, it is reactive power in an ac circuit. It is energy stored temporarily in a coil's magnetic field or a capacitor's electric field and then returned to the circuit. It is power not converted to heat and dissipated.

3D3 What is the term for energy that is stored in an electromagnetic or electrostatic field?
A. Potential energy
B. Amperes-joules
C. Joules-coulombs
D. Kinetic energy

ANSWER A: When you go into the examination room to take your Commercial General Radiotelephone Element 3 test, you will have a great deal of information energy stored as a result of the studying you have done. You will have potential energy (key words) all stored up, ready to react to your test.

3D4 What is responsible for the phenomenon when voltages across reactances in series can often be larger than the voltages applied to them?
A. Capacitance
B. Resonance
C. Conductance
D. Resistance

ANSWER B: Coils (having inductance) and capacitors (having capacitance) will be at resonance when the capacitive reactance equals the inductive reactance. At resonance, large voltages greater than the applied voltage can be present across these components in the circuit. You can see this with a mobile whip antenna. If something were to touch the whip in transmit, you may see quite a spark!

X_L = Inductive Reactance

$X_L = 2\pi fL$

X_C = Capacitive Reactance

$X_C = \dfrac{1}{2\pi fC}$

At Resonance $X_L = X_C$

$$2\pi f_r L = \dfrac{1}{2\pi f_r C}$$

Solving for f_r

$$f_r^2 = \dfrac{1}{(2\pi)^2 LC}$$

The Resonant Frequency, f_r, is:

E $\quad f_r = \dfrac{1}{2\pi\sqrt{LC}}$

Equation for Resonant Frequency

3D5 What is resonance in an electrical circuit?
A. The highest frequency that will pass current.
B. The lowest frequency that will pass current.
C. The frequency at which capacitive reactance equals inductive reactance.
D. The frequency at which power factor is at a minimum.

ANSWER C: At resonance, XL (inductive reactance) equals XC (capacitive reactance). In a mobile series resonant whip antenna, there will be maximum current through the loading coil when the whip is tuned to resonance. On your base station vertical trap antenna or trap beam antenna, parallel resonant circuits offer high impedance, trapping out a specific length of the antenna for a certain frequency. At resonance, the resonant parallel circuit looks like a high impedance to any of your power getting through to the longer length of the antenna system.

3D6 Under what conditions does resonance occur in an electrical circuit?
A. When the power factor is at a minimum.
B. When inductive and capacitive reactances are equal.
C. When the square root of the sum of the capacitive and inductive reactances is equal to the resonant frequency.
D. When the square root of the product of the capacitive and inductive reactances is equal to the resonant frequency.

ANSWER B: When a coil's inductive reactance equals a capacitor's capacitive reactance, resonance occurs. On antenna systems, you fine-tune your elements for resonance at a certain frequency.

3D7 What is the term for the phenomena that occurs in an electrical circuit when the inductive reactance equals the capacitive reactance?
A. Reactive quiescence
B. High Q
C. Reactive equilibrium
D. Resonance

ANSWER D: Resonance in a circuit occurs at the frequency where inductive reactance equals capacitive reactance.

3D8 What is the approximate magnitude of the impedance of a series R-L-C circuit at resonance?
A. High, as compared to the circuit resistance
B. Approximately equal to the circuit resistance
C. Approximately equal to XL
D. Approximately equal to XC

ANSWER B: On the 2-MHz marine band whips, you will see that manufacturers use an extremely large loading coil for this series resonant antenna. The bigger the loading coil, the less the coil resistance. At resonance, the magnitude of the impedance is equal to the circuit resistance (key words).

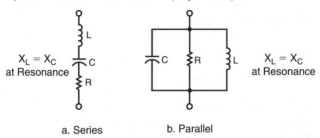

a. Series   b. Parallel

Series and Parallel Resonant Circuits

3D9 What is the approximate magnitude of the impedance of a parallel R-L-C circuit at resonance?
A. Approximately equal to the circuit resistance
B. Approximately equal to XL
C. Low, as compared to the circuit resistance
D. Approximately equal to XC

ANSWER A: Whether the system is parallel or series R-L-C resonant, the impedance will always be equal to the circuit resistance (key words).

3D10 What is the characteristic of the current flow in a series R-L-C circuit at resonance?
A. It is at a minimum.
B. It is at a maximum.
C. It is DC.
D. It is zero.

ANSWER B: A mobile whip antenna will have maximum current in a series R-L-C circuit at resonance. One way you can tell whether or not a mobile whip is working properly is to feel the coil after the transmitter has been shut down. If it's warm, there has been current through the coil and up to the whip tip stinger.

3D11 What is the characteristic of the current flow in a parallel R-L-C circuit at resonance?
A. The current circulating in the parallel elements is at a minimum.
B. The current circulating in the parallel elements is at a maximum.
C. The current circulating in the parallel elements is DC.
D. The current circulating in the parallel elements is zero.

ANSWER B: A parallel circuit at resonance is similar to a tuned trap in a multi-band trap antenna. At resonance, the trap keeps power from going any further to the longer antenna elements. But be assured there is maximum circulating current within the parallel R-L-C circuit. Now don't get confused with this question—a parallel circuit has maximum current within it, and minimum current going to the next stage. That's why they call those parallel resonant circuits "traps."

3D12 What is the skin effect?
A. The phenomenon where RF current flows in a thinner layer of the conductor, close to the surface, as frequency increases
B. The phenomenon where RF current flows in a thinner layer of the conductor, close to the surface, as frequency decreases
C. The phenomenon where thermal effects on the surface of the conductor increase the impedance
D. The phenomenon where thermal effects on the surface of the conductor decrease the impedance

ANSWER A: If you ever had a chance to inspect an old wireless station, you would have seen that the antenna "plumbing" leading out of the transmitter was usually constructed of hollow copper tubing. Radio-frequency current always travels along the thinner outside layer of a conductor, and as frequency increases, there is almost no current in the center of the conductor. The higher the frequency, the greater the skin effect. This is why we use wide copper ground foil to minimize the resistance to AC current that we need to pass to ground.

3D13 What is the term for the phenomenon where most of an RF current flows along the surface of the conductor?
A. Layer effect
B. Seeburg effect
C. Skin effect
D. Resonance

ANSWER C: Marine single-sideband whip antennas may use hollow tubing as their radiators. This is because most of the RF current travels along the surface of the conductor. This phenomenon is called skin effect.

3D14 Where does practically all of the RF current flow in a conductor?
A. Along the surface
B. In the center of the conductor
C. In the magnetic field around the conductor
D. In the electromagnetic field in the conductor center

ANSWER A: As frequency increases, RF currents travel in the thin surface layers of the conductors, not within the body. On large high-power shipboard SSB transmitters, small copper tubing is an excellent way to pass the energy out of the ship's radio room, and up to the antenna system.

3D15 Why does practically all of an RF current flow within a few thousandths-of-an-inch of the conductor's surface?
A. Because of skin effect
B. Because the RF resistance of the conductor is much less than the DC resistance
C. Because of heating of the metal at the conductor's interior
D. Because of the AC-resistance of the conductor's self inductance

ANSWER A: It is the skin effect that keeps the RF current traveling within a few thousandths of an inch of a conductor's surface.

3D16 Why is the resistance of a conductor different for RF current than for DC?

A. Because the insulation conducts current at radio frequencies
B. Because of the Heisenburg effect
C. Because of skin effect
D. Because conductors are non-linear devices

ANSWER C: The higher the frequency of an alternating current (AC) signal, the greater the skin effect. For DC, the internal cross-sectional area of a conductor is very important, but for radio frequency signals, it's not important.

3D17 What is a magnetic field?

A. Current flow through space around a permanent magnet
B. A force set up when current flows through a conductor
C. The force between the plates of a charged capacitor
D. The force that drives current through a resistor

ANSWER B: In marine and aeronautical installations, you would always route your power cable well away from the magnetic or fluxgate compass assembly. As current passes through a conductor, the current produces a magnetic field as a force around the conductor. Stronger current will produce a stronger magnetic field, which could affect the reading of a magnetic or fluxgate compass.

3D18 In what direction is the magnetic field about a conductor when current is flowing?

A. In the same direction as the current
B. In a direction opposite to the current flow
C. In all directions; omnidirectional
D. In a direction determined by the left hand rule

ANSWER D: If you grab a big straight cable with your left hand, with your thumb pointing in the direction the electrons are flowing in the conductor (opposite from conventional current), your fingers will naturally point in the direction of the magnetic lines of force. This is known as the left-hand rule. Have you ever been to a junk yard and watched the electromagnet coil magically lift a car off the ground and drop it into a nearby boxcar? When the coil is energized, it becomes a magnet. As quickly as the energy is cut off, magnetism stops, and the car drops off.

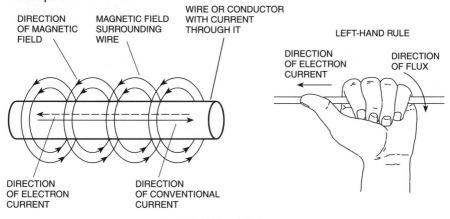

Left-Hand Rule

3D19 What device is used to store electrical energy in an electrostatic field?

A. A battery
B. A transformer
C. A capacitor
D. An inductor

ANSWER C: A capacitor is made up of parallel plates separated by a dielectric (non-conductor). When a voltage is placed across a capacitor, energy is stored in the electrostatic field developed between the capacitor plates. As a reminder, notice the letters "AC" in the word capacitor, and the letters "ATIC" in the word electrostatic. It should lead you to the correct answer.

3D20 What is the term used to express the amount of electrical energy stored in an electrostatic field?

A. Coulombs
B. Joules
C. Watts
D. Volts

ANSWER B: The joule is a measure of energy, or the capacity to do work. The amount of electrical energy stored in an electrostatic field associated with capacitors is expressed in joules. If a capacitor is charged with one coulomb of electrons at a voltage of one volt, one joule of energy is stored in the capacitor's electrostatic field.

3D21 What factors determine the capacitance of a capacitor?

A. Area of the plates, voltage on the plates and distance between the plates
B. Area of the plates, distance between the plates and the dielectric constant of the material between the plates
C. Area of the plates, voltage on the plates and the dielectric constant of the material between the plates
D. Area of the plates, amount of charge on the plates and the dielectric constant of the material between the plates

ANSWER B: Some of the SSB equipment you may work on will still use variable capacitors. The amount of capacitance is determined by the area of the plates as you turn the variable capacitor, the distance between the plates, and the dielectric constant of the material between the plates, usually air in variable capacitors. As you engage more of the plates next to each other, capacitance increases.

Variable Capacitor
Courtesy of AEA, Inc.

3D22 What is the dielectric constant for air?

A. Approximately 1
B. Approximately 2
C. Approximately 4
D. Approximately 0

ANSWER A: The dielectric constant for air is approximately 1. For atmospheric propagation studies at VHF frequencies, the dielectric constant of air at 1 is normally abbreviated "n", for normal. Glass has a dielectric constant of 8; Bakelite®, 5; and Teflon®, 2. Teflon is used extensively at VHF and UHF frequencies because it can firmly hold components in place, and has approximately the dielectric constant of air.

3D23 What determines the strength of the magnetic field around a conductor?
    A. The resistance divided by the current
    B. The ratio of the current to the resistance
    C. The diameter of the conductor
    D. The amount of current
ANSWER D: Back to the electromagnetic field that develops around conductors— the more current passing through a conductor, the greater the magnetic field strength around the conductor.

3D24 Why would the rate at which electrical energy is used in a circuit be less than the product of the magnitudes of the AC voltage and current?
    A. Because there is a phase angle that is greater than zero between the current and voltage
    B. Because there are only resistances in the circuit
    C. Because there are no reactances in the circuit
    D. Because there is a phase angle that is equal to zero between the current and voltage
ANSWER A: When the *phase angle is greater than zero* (key words) between voltage and current, the rate at which electrical energy is used in a circuit is going to be less than the product of the magnitudes of the AC voltage and current. Since the voltage and current aren't in phase, the true power will be less because there is some reactive power present. See question 3D25 and 26.

3D25 In a circuit where the AC voltage and current are out of phase, how can the true power be determined?
    A. By multiplying the apparent power times the power factor
    B. By subtracting the apparent power from the power factor
    C. By dividing the apparent power by the power factor
    D. By multiplying the RMS voltage times the RMS current
ANSWER A: If we multiply apparent power times the power factor of an AC voltage and current out of phase, we can determine true power. (See figure at 3D26.)

3D26 What does the power factor equal in an R-L circuit having a 60 degree phase angle between the voltage and the current?
    A. 1.414
    B. 0.866
    C. 0.5
    D. 1.73
ANSWER C: If you are using a scientific calculator, you will find the cosine of 60 degrees, given in the problem, comes out 0.5, answer C. The power factor is $\cos \phi$. If you don't have a scientific calculator, you might memorize this table:

ϕ	$\cos \phi$
30°	0.866
45°	0.707
60°	0.5

True power is power dissipated as heat, while reactive power of a circuit is stored in inductances or capacitances and then returned to the circuit.

Apparent power is voltage times current without taking into account the phase angle between them. According to right angle trigonometry,

$$\text{cosine } \phi = \frac{\text{side adjacent}}{\text{hypotenuse}}$$

therefore,

$$\text{cosine } \phi = \frac{\text{True Power}}{\text{Apparent Power}}$$

therefore,

True Power = Apparent Power $\times$ Cosine ϕ

Cosine ϕ is called the *power factor* (PF) of a circuit.

Since apparent power is E $\times$ I,

True Power = E $\times$ I $\times$ Cos ϕ

or

True Power = E $\times$ I $\times$ PF

Apparent and True Power

3D27 What does the power factor equal in an R-L circuit having a 45 degree phase angle between the voltage and the current?
 A. 0.866
 B. 1.0
 C. 0.5
 D. 0.707
ANSWER D: Using your scientific calculator, or from memory, remember that the cosine of 45 degrees is 0.707. Think of an airplane, the very popular 707 jet aircraft, taking off at an angle of 45 degrees. See table at 3D26.

3D28 What does the power factor equal in an R-L circuit having a 30 degree phase angle between the voltage and the current?
 A. 1.73
 B. 0.5
 C. 0.866
 D. 0.577
ANSWER C: At 30 degrees, the power factor is 0.866. If you don't pass your Commercial exam, you could be 86'd out of the room! See table at 3D26.

3D29 How many watts are being consumed in a circuit having a power factor of 0.2 when the input is 100-V AC and 4-amperes is being drawn?
 A. 400 watts
 B. 80 watts
 C. 2000 watts
 D. 50 watts

ANSWER B: Here come two simple problems that are based on the formula:

 True power = E (volts) × I (amps) × PF (power factor).

True power works out to be 100 volts × 4 amps × 0.2, giving us 80 watts. Easy enough, right?

3D30 How many watts are being consumed in a circuit having a power factor of 0.6 when the input is 200-V AC and 5-amperes is being drawn?
 A. 200 watts
 B. 1000 watts
 C. 1600 watts
 D. 600 watts

ANSWER D: True power = E × I × PF; therefore, it is 200 volts × 5 = amps × 0.6 = 600 watts. Easy!

3D31 What is the effective radiated power of a repeater with 50 watts transmitter power output, 4 dB feedline loss, 3 dB duplexer and circulator loss, and 6 dB antenna gain?
 A. 158 watts, assuming the antenna gain is referenced to a half-wave dipole
 B. 39.7 watts, assuming the antenna gain is referenced to a half-wave dipole
 C. 251 watts, assuming the antenna gain is referenced to a half-wave dipole
 D. 69.9 watts, assuming the antenna gain is referenced to a half-wave dipole

ANSWER B: You will have one question out of the following 10 questions based on simple dB calculations at a repeater site. Don't worry about logarithms, you can do these in your head, and all you must remember is: 3 dB = a 2X increase; to twice the value; and −3 dB = a 2X loss to one half the value.

To solve the problems, just add and subtract all the dB's, and if it ends up about +3 dB, then double your transmitter power output to end up with effective radiated power. If your answer ends up around −3-dB gain, cut your transmitter power in half for effective radiated power. In this question, we have −7-dB loss, offset by 6-dB gain for 1-dB net loss. If we started out with 50 watts, our effective radiated power (ERP) will be slightly less, so 39.7 is the only answer in the ballpark.

3D32 What is the effective radiated power of a repeater with 50 watts transmitter power output, 5 dB feedline loss, 4 dB duplexer and circulator loss, and 7 dB antenna gain?
 A. 300 watts, assuming the antenna gain is referenced to a half-wave dipole
 B. 315 watts, assuming the antenna gain is referenced to a half-wave dipole
 C. 31.5 watts, assuming the antenna gain is referenced to a half-wave dipole
 D. 69.9 watts, assuming the antenna gain is referenced to a half-wave dipole

ANSWER C: Here we have 9-dB loss, and 7-dB gain. That's just about a half-power loss (−3 dB), and the only answer that is reasonably close is answer C, 31.5 watts.

3D33 What is the effective radiated power of a repeater with 75 watts transmitter power output, 4 dB feedline loss, 3 dB duplexer and circulator loss, and 10 dB antenna gain?

A. 600 watts, assuming the antenna gain is referenced to a half-wave dipole

B. 75 watts, assuming the antenna gain is referenced to a half-wave dipole

C. 18.75 watts, assuming the antenna gain is referenced to a half-wave dipole

D. 150 watts, assuming the antenna gain is referenced to a half-wave dipole

ANSWER D: We have 7-dB loss, and a great antenna system giving us 10-dB gain. This gives us a +3 dB, which is 2 times our power output. There it is, only one answer at 150 watts!

3D34 What is the effective radiated power of a repeater with 75 watts transmitter power output, 5 dB feedline loss, 4 dB duplexer and circulator loss, and 6 dB antenna gain?

A. 37.6 watts, assuming the antenna gain is referenced to a half-wave dipole

B. 237 watts, assuming the antenna gain is referenced to a half-wave dipole

C. 150 watts, assuming the antenna gain is referenced to a half-wave dipole

D. 23.7 watts, assuming the antenna gain is referenced to a half-wave dipole

ANSWER A: 9-dB loss, offset by 6-dB antenna gain gives a 3-dB net loss, or half power. Half of 75 watts is answer A, 37.6 watts.

3D35 What is the effective radiated power of a repeater with 100 watts transmitter power output, 4 dB feedline loss, 3 dB duplexer and circulator loss, and 7 dB antenna gain?

A. 631 watts, assuming the antenna gain is referenced to a half-wave dipole

B. 400 watts, assuming the antenna gain is referenced to a half-wave dipole

C. 25 watts, assuming the antenna gain is referenced to a half-wave dipole

D. 100 watts, assuming the antenna gain is referenced to a half-wave dipole

ANSWER D: 7 dB down the tubes, offset by 7-dB gain. No change on power output, giving us still 100 watts ERP.

3D36 What is the effective radiated power of a repeater with 100 watts transmitter power output, 5 dB feedline loss, 4 dB duplexer and circulator loss, and 10 dB antenna gain?

A. 800 watts, assuming the antenna gain is referenced to a half-wave dipole

B. 126 watts, assuming the antenna gain is referenced to a half-wave dipole

C. 12.5 watts, assuming the antenna gain is referenced to a half-wave dipole

D. 1260 watts, assuming the antenna gain is referenced to a half-wave dipole

ANSWER B: A total of 9-dB loss, offset by 10-dB gain, net +1 dB. Since we had 100 watts to start with and the net gain is 1 dB, the only close answer is answer B, 126 watts.

3D37 What is the effective radiated power of a repeater with 120 watts transmitter power output, 5 dB feedline loss, 4 dB duplexer and circulator loss, and 6 dB antenna gain?

A. 601 watts, assuming the antenna gain is referenced to a half-wave dipole

B. 240 watts, assuming the antenna gain is referenced to a half-wave dipole

C. 60 watts, assuming the antenna gain is referenced to a half-wave dipole

D. 379 watts, assuming the antenna gain is referenced to a half-wave dipole

ANSWER C: Here we have 9-dB loss with 6-dB gain. That's a net loss of 3 dB. It cuts our 120-watt transmitter to an effective radiated power of only 60 watts. Quit buying junk coax!

3D38 What is the effective radiated power of a repeater with 150 watts transmitter power output, 4 dB feedline loss, 3 dB duplexer and circulator loss, and 7 dB antenna gain?

A. 946 watts, assuming the antenna gain is referenced to a half-wave dipole

B. 37.5 watts, assuming the antenna gain is referenced to a half-wave dipole

C. 600 watts, assuming the antenna gain is referenced to a half-wave dipole

D. 150 watts, assuming the antenna gain is referenced to a half-wave dipole

ANSWER D: This one is a wash. You have −7 dB offset by +7 dB. This would make that 150-watt transmitter have an effective radiated power of 150 watts.

3D39 What is the effective radiated power of a repeater with 200 watts transmitter power output, 4 dB feedline loss, 4 dB duplexer and circulator loss, and 10 dB antenna gain?

A. 317 watts, assuming the antenna gain is referenced to a half-wave dipole

B. 2000 watts, assuming the antenna gain is referenced to a half-wave dipole

C. 126 watts, assuming the antenna gain is referenced to a half-wave dipole

D. 260 watts, assuming the antenna gain is referenced to a half-wave dipole

ANSWER A: An 8-dB loss, offset by 10-dB gain for a net of +2 dB is not quite a double in ERP. Your only close answer is a 200-watt transmitter with an effective radiated power of 317 watts. It's almost 3 dB, which would give you an answer of 400 watts. 317 watts is closer than an incorrect answer of 126 watts (actually a loss), and a slight gain of only 260 watts.

3D40 What is the effective radiated power of a repeater with 200 watts transmitter power output, 4 dB feedline loss, 3 dB duplexer and circulator loss, and 6 dB antenna gain?

    A. 252 watts, assuming the antenna gain is referenced to a half-wave dipole

    B. 63.2 watts, assuming the antenna gain is referenced to a half-wave dipole

    C. 632 watts, assuming the antenna gain is referenced to a half-wave dipole

    D. 159 watts, assuming the antenna gain is referenced to a half-wave dipole

ANSWER D: You have a total loss of 7 dB, and a puny antenna gain of only +6 dB. This means that you are down 1 dB, causing your 200-watt transmitter to radiate more like a 159-watt transmitter, the only logical answer below 200 watts.

3D41 What is the photoconductive effect?

    A. The conversion of photon energy to electromotive energy

    B. The increased conductivity of an illuminated semiconductor junction

    C. The conversion of electromotive energy to photon energy

    D. The decreased conductivity of an illuminated semiconductor junction

ANSWER B: The photo cell has high resistance when no light shines on it, and a varying resistance when light shines on it. It can be used in an ON or OFF state as a simple beam alarm across a doorway, or may be found in almost all modern SLR cameras as a sensitive light meter to judge the amount of light present.

3D42 What happens to photoconductive material when light shines on it?

    A. The conductivity of the material increases

    B. The conductivity of the material decreases

    C. The conductivity of the material stays the same

    D. The conductivity of the material becomes temperature dependent

ANSWER A: In the dark, a photo cell has high resistance. When you shine light on it, the conductivity increases as the resistance decreases.

3D43 What happens to the resistance of a photoconductive material when light shines on it?

    A. It increases.

    B. It becomes temperature dependent.

    C. It stays the same.

    D. It decreases.

ANSWER D: More light, less resistance. The resistance varies inversely with the light.

Light Incidence	Current I	Resistance	Test Conditions
Dark	5nA	2000MΩ	$V_R = 10V$ *$E_e = 0$
Light	15µA	666.7kΩ	$V_R = 10V$ *$E_e = 250µW/cm^2$ at 940nm

*Irradiance (E_e) is the radiant power per unit area incident on surface

Resistance Variation of a Photo Diode

3D44 What happens to the conductivity of a semiconductor junction when it is illuminated?
A. It stays the same.
B. It becomes temperature dependent.
C. It increases.
D. It decreases.

ANSWER C: Conductivity increases when light shines on a semiconductor junction—just like the photo cell. More light; less resistance.

3D45 What is an *optocoupler?*
A. A resistor and a capacitor
B. A frequency modulated helium-neon laser
C. An amplitude modulated helium-neon laser
D. An LED and a phototransistor

ANSWER D: Optocouplers are found in many modern marine and aviation transceivers. When you tune the dial for a new channel, you are no longer tuning a big variable capacitor. Rather, you are interrupting a light beam on a phototransistor.

3D46 What is an *optoisolator?*
A. An LED and a phototransistor
B. A P-N junction that develops an excess positive charge when exposed to light.
C. An LED and a capacitor
D. An LED and a solar cell

ANSWER A: Another name for an optocoupler is an optoisolator. A LED shines through a window at a phototransistor in an optoisolator.

3D47 What is an *optical shaft encoder?*
A. An array of optocouplers chopped by a stationary wheel
B. An array of optocouplers whose light transmission path is controlled by a rotating wheel
C. An array of optocouplers whose propagation velocity is controlled by a stationary wheel
D. An array of optocouplers whose propagation velocity is controlled by a rotating wheel

ANSWER B: When you spin the big knob of your new high-frequency transceiver it is connected to an optical shaft encoder that determines the tuned frequency.

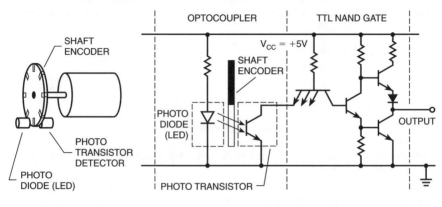

| OPTOCOUPLER | TTL NAND GATE |

a. Mechanical Setup   b. Schematic

Optocoupler Used for Shaft Encoder

3D48 What does the photoconductive effect in crystalline solids produce a noticeable change in?
  A. The capacitance of the solid
  B. The inductance of the solid
  C. The specific gravity of the solid
  D. The resistance of the solid
ANSWER D: The resistance of the crystalline solid varies when light shines on it because of the photoconductive effect.

3D49 What is the meaning of the term time constant of an RC circuit?
  A. The time required to charge the capacitor in the circuit to 36.8% of the supply voltage
  B. The time required to charge the capacitor in the circuit to 36.8% of the supply current
  C. The time required to charge the capacitor in the circuit to 63.2% of the supply current
  D. The time required to charge the capacitor in the circuit to 63.2% of the supply voltage
ANSWER D: If you have a separate power supply feeding your equipment, when you turn off your power supply with the equipment still turned on, you will notice that the power supply output voltage decreases slowly. The slow decay is because of the voltage still across the large electrolytic filter capacitors on the output of the power supply. This is a visual example that capacitors charge up, and discharge, on a curve. The time required to charge (or discharge in this case) the capacitor in an RC (Resistance-Capacitance) circuit to 63.2 percent of the supply voltage is called a time constant, τ, where:

    $\tau = RC$ τ = Greek letter tau, the time constant in **seconds**
                               R = total resistance in **ohms**
                                C = capacitance in **farads**

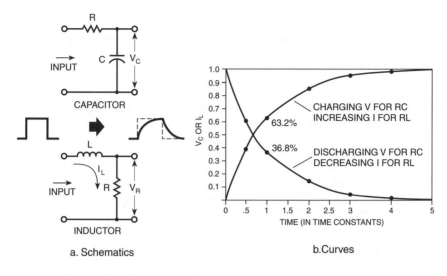

a. Schematics

b.Curves

RC (RxC) and RL (L/R) Time Constant

3D50 What is the meaning of the term time constant of an RL circuit?

A. The time required for the current in the circuit to build up to 36.8% of the maximum value

B. The time required for the voltage in the circuit to build up to 63.2% of the maximum value

C. The time required for the current in the circuit to build up to 63.2% of the maximum value

D. The time required for the voltage in the circuit to build up to 36.8% of the maximum value

ANSWER C: In an RL circuit, the coil will develop a back EMF which will oppose the flow of current. In the RL circuit, the time required for the current to build up to 63.2 percent of the maximum value is also called a time constant, τ, where:

$$\tau = \frac{L}{R}$$

τ = Greek letter tau, the time constant in **seconds**

L = total inductance in **henries**

R = total resistance in **ohms**

3D51 What is the term for the time required for the capacitor in an RC circuit to be charged to 63.2% of the supply voltage?

A. An exponential rate of one

B. One time constant

C. One exponential period

D. A time factor of one

ANSWER B: In an RC circuit, assuming there is no initial charge on the capacitor, it takes one time constant to charge a capacitor to 63.2 percent of its final value.

3D52 What is the term for the time required for the current in an RL circuit to build up to 63.2% of the maximum value?

A. One time constant

B. An exponential period of one

C. A time factor of one

D. One exponential rate

ANSWER A: In an RL circuit, assuming there is no initial current through the inductance, it takes one time constant for the current in the circuit to build up to 63.2 percent of its final value. Measuring V_R tracks I_L.

3D53 What is the term for the time it takes for a charged capacitor in an RC circuit to discharge to 36.8% of its initial value of stored charge?

A. One discharge period

B. An exponential discharge rate of one

C. A discharge factor of one

D. One time constant

ANSWER D: Remember the example of the filter capacitor in the power supply, where the capacitors discharged when the input power was removed? The time it takes a capacitor in an RC circuit to discharge to 36.8 percent of its initial value of stored charge is one time constant.

3D54 What is meant by back EMF?

A. A current equal to the applied EMF

B. An opposing EMF equal to R times C (RC) percent of the applied EMF

C. A current that opposes the applied EMF

D. A voltage that opposes the applied EMF

ANSWER D: A back EMF which opposes the applied EMF is developed in an RL circuit by the inductor. This reduces the current flow in the inductor and the circuit.

3D55 After two time constants, the capacitor in an RC circuit is charged to what percentage of the supply voltage?

A. 36.8%

B. 63.2%

C. 86.5%

D. 95%

ANSWER C: To calculate the percentage of charge after two time constants in an RC circuit, write down the percent after a single time constant as 63.2 percent. The remaining percent of charge is 36.8 percent. It is found by deducting 63.2 percent from 100 percent. In the next time constant the capacitor will charge to 63.2 percent of the remaining 36.8 percent. In other words, another amount of charge is added to the capacitor in the second time constant equal to:

$$36.8\% \times 63.2\% = 23.26\% \text{ (rounded to 23.3\%)}$$

This really needs to be calculated in decimal format as:

$$0.368 \times 0.632 = 0.2326$$

Percent values are converted to decimals by moving the decimal point two places to the left. Decimal values are converted to percent values by multiplying by 100 (moving decimal point two places to the right.)

Since the final percent of charge in two time constants is desired, it is only necessary to add the two percent charges together:

	Percent	Decimal
Charge in first time constant =	63.2%	0.632
Charge in second time constant =	23.3%	0.233
Total charge after two time constants	86.5%	0.865

The process can be continued for the third time constant. Since the capacitor has charged to 86.5 percent in two time constants, if you start from that point, the final value of charge would be another 13.5 percent. Therefore, in the next time constant, the third, the capacitor would charge another:

13.5% $\times$ 63.2% = 8.53% (rounded to 8.5%)

After the third time constant, the total charge is:

86.5% + 8.5% = 95%

If the process is continued through five time constants, the capacitor charge is 99.3 percent. Therefore, in electronic calculations, the capacitor is considered to be fully charged (or discharged) after five time constants.

3D56 After two time constants, the capacitor in an RC circuit is discharged to what percentage of the starting voltage?
  A. 86.5%
  B. 63.2%
  C. 36.8%
  D. 13.5%
ANSWER D: This question is actually the reverse of the previous question. It is asking to what percent the capacitor has discharged. The capacitor discharges the same percent in a time constant as it charges (for the same RC circuit values, of course.) However, the question is asking for the amount of charge remaining. At one time constant, the capacitor discharges 63.2 percent (in the previous question, it charged 63.2 percent in one time constant); therefore, the percent charge remaining on the capacitor is 36.8 percent (100%-63.2%). From the previous question, the capacitor charges another 23.3 percent in the second time constant. Thus, the capacitor will discharge another 23.3 percent in the second time constant. After two time constants the capacitor has discharged by 86.5 percent; therefore, the remaining charge is 13.5 percent (100%-86.5%). In five time constants the capacitor would have discharged 99.3 percent of its charge, so 0.7 percent of charge remains—it is considered to be fully discharged.

3D57 What is the time constant of a circuit having a 100-microfarad capacitor and a 470-kilohm resistor in series?
  A. 4700 seconds
  B. 470 seconds
  C. 47 seconds
  D. 0.47 seconds
ANSWER C: First of all, this is an RC circuit, so remember the formula:

$\tau = RC$

Time constant (**seconds**) = Resistance (**ohms**) $\times$ Capacitance (**farads**)

Remember that "micro" means $\times 10^{-6}$ (the multiplication is accomplished by moving the decimal point 6 places to the left), and "kilo" means $\times 10^{+3}$ (the multiplication is accomplished by moving the decimal point 3 places to the right). The problem solution is as follows:

$$\tau = 470 \times 10^{+3} \times 100 \times 10^{-6}$$
$$\tau = 4.7 \times 10^{+2} \times 10^{+3} \times 1 \times 10^{+2} \times 10^{-6}$$
$$\tau = 4.7 \times 10^{+1} = 47 \text{ seconds}$$

Although this is fairly easy, you can make it easier by using a calculator. Turn on your calculator, press the clear button (several times for good luck), and then execute the following keystrokes:

470000 $\times$ .0001 = and presto, 47 appears on the display.

This is the correct answer. Be sure you understand that 470 kilohms equals 470000 and that 100 microfarads equals .0001.

3D58 What is the time constant of a circuit having a 220-microfarad capacitor and a 1-megohm resistor in parallel?
  A. 220 seconds
  B. 22 seconds
  C. 2.2 seconds
  D. 0.22 seconds
ANSWER A: Since it is an RC circuit, use $\tau = RC$.

Using powers of 10:

$$\tau = 1 \times 10^{+6} \times 220 \times 10^{-6} \qquad \tau = 220 \text{ seconds}$$

Using only decimal values:

$$\tau = 1,000,000 \times .000220 \qquad \tau = 220 \text{ seconds}$$

Remember that when using powers of ten: to multiply, you add the exponents; to divide, you subtract the exponent of the denominator (lower one) from the exponent of the numerator (upper one). Since "meg" = 10^{+6} and "micro" = 10^{-6}, multiplying "megs" times "micros" means you add $+6$ and -6, and the result is 0 or 10^0, which is 1. In this case, $220 \times 1 = 220$. Actually, this stuff is pretty simple once you get the hang of it.

   To calculate using decimal values on your calculator, the calculator keystrokes are: Clear, 1000000 $\times$ .000220 = 220.

3D59 What is the time constant of a circuit having two 100-microfarad capacitors and two 470-kilohm resistors all in series?
  A. 470 seconds
  B. 47 seconds
  C. 4.7 seconds
  D. 0.47 seconds.
ANSWER B: For an RC circuit $\tau = RC$, but first we must combine resistor and capacitor values to come up with a single value for each. Since this is a series circuit, the resistor values add ($R_T = R_1 + R_2$) to arrive at a total resistance, but for the capacitance total you have to solve the formula $C_T = (C_1 \times C_2) \div (C_1 + C_2)$. Therefore,

Powers of 10:

$$C_T = \frac{(1 \times 10^{+2} \times 10^{-6} \times 1 \times 10^{+2} \times 10^{-6})}{(100 \times 10^{-6} + 100 \times 10^{-6})}$$

$$C_T = \frac{(1 \times 10^{-8})}{(2 \times 10^{+2} \times 10^{-6})}$$

$$C_T = \frac{(1 \times 10^{-8})}{2 \times 10^{-4}} = 0.5 \times 10^{-4} = 50 \times 10^{-6}$$

$$C_T = 0.00005 \text{ farads or 50 microfarads}$$

For the resistance:

$$R_T = 470,000 + 470,000 = 940,000$$
$$R_T = 4.7 \times 10^{+5} + 4.7 \times 10^{+5} = 9.4 \times 10^{+5}$$

Now that we have calculated the total series capacitance and the total resistance, we simply multiply $0.00005 \times 940,000 = 47$ seconds. In powers of ten this is $\tau = 0.5 \times 10^{-4} \times 9.4 \times 10^{+5} = 4.7 \times 10^{+1} = 47$.

Of course, the old timers that recognize that two equal capacitors in series equals a single capacitor at half the value would immediately jump to $\tau = 50$ microfarads $\times$ 940 kilohms, pull out the calculator and press the following: Clear, $.00005 \times 940000 = 47$. The time constant is in seconds. Notice if you miss a single decimal point, they have an incorrect answer for you. Watch out!

Remember:

(For Two Cs)	(For Two Rs)
Capacitors in Series:	Resistors in Series:
$C_T = \dfrac{C_1 \times C_2}{C_1 + C_2}$	$R_T = R_1 + R_2$
Capacitors in Parallel:	Resistors in Parallel:
$C_T = C_1 + C_2$	$R_T = \dfrac{R_1 \times R_2}{R_1 + R_2}$
C = Capacitance in **farads**	R = Resistance in **ohms**

Capacitors, Resistors in Series and Parallel

3D60 What is the time constant of a circuit having two 100-microfarad capacitors and two 470-kilohm resistors all in parallel?
- A. 470 seconds
- B. 47 seconds
- C. 4.7 seconds
- D. 0.47 seconds

ANSWER B: Again a RC circuit, only this time everything is in parallel. Since the capacitors are in parallel, the capacitance values add ($C_T = C_1 + C_2$), but for the total resistance you have to solve the formula $R_T = (R_1 \times R_2) \div (R_1 + R_2)$. Therefore,

Powers of 10:

$$R_T = \frac{(4.7 \times 10^{+5} \times 4.7 \times 10^{+5})}{(4.7 \times 10^{+5} + 4.7 \times 10^{+5})}$$

$$R_T = \frac{(22.09 \times 10^{+10})}{9.4 \times 10^{+5}} = 2.35 \times 10^{+5}$$

$$R_T = 235,000 \text{ ohms or } 235 \text{ kilohms}$$

for the capacitance:

$$C_T = 0.000100 + 0.000100 = 0.000200$$
$$C_T = 100 \times 10^{-6} + 100 \times 10^{-6} = 200 \times 10^{-6}$$
$$C_T = 200 \text{ microfarads or } 0.0002 \ (2 \times 10^{-4}) \text{ farads}$$

Since the total resistance and the total capacitance are known, just multiply 235,000 × .000200 to end up with 47 seconds. Now that wasn't too hard, was it?

This is really easy for old timers. They know that the capacitor values add for parallel capacitors—so they have 200 microfarads. For the resistance, they know that two equal value resistors in parallel result in a single resistor at half the value or 235 kilohms. They would go immediately to $\tau = 200$ microfarads × 235 kilohms. On their calculator they would press the following: Clear, .0002 × 235000 = 47. The time constant is in seconds, and keep exact track of the decimal points.

3D61 What is the time constant of a circuit having two 220-microfarad capacitors and two 1-megohm resistors all in series?
   A. 55 seconds
   B. 110 seconds
   C. 220 seconds
   D. 440 seconds

ANSWER C: This is another series RC circuit, however, now there are two capacitors in series and two resistors in series. Review the calculations in question 3D59 The procedure is the same. Just remember that two equal capacitors in series have a total capacitance of half of one of the capacitors, and two equal resistors in series have a total resistance that is twice the value of one of the resistors. Now you can do this one in your head. Two 220 micro-farad capacitors in series is really one big 110 microfarad capacitor. Write down 0.000110 as you recall that 110 microfarads is 110×10^{-6} — "micro" means move the decimal point six places to the left. Two one-megohm resistors in series would be two megohms. Write down 2,000,000 as you recall that 2 megohms is $2 \times 10^{+6}$ — "meg" means move the decimal point six places to the right. Since $\tau = 2 \times 10^{+6} \times 110 \times 10^{-6}$, you can see by the powers of ten that "megs" times "micros" cancel each other and you end up with $2 \times 110 = 220$ seconds. Anytime you have both "megs" and micros" multiplied together, all those zeroes to the left or right of the stated numerical value cancel.

3D62 What is the time constant of a circuit having two 220-microfarad capacitors and two 1-megohm resistors all in parallel?
   A. 22 seconds
   B. 44 seconds

C. 220 seconds

D. 440 seconds

ANSWER C: Another RC circuit with all components in parallel. Refer to question 3D60. It will be easy if you remember that two equal capacitors in parallel have a total capacitance of twice the value of one of the capacitors, and that two equal value resistors in parallel have a total resistance equal to one half the value of one of the resistors. And more good news—"megs" are being multiplied by "micros" (refer to question 3D61) so the powers of ten after the numerical value cancel. Two 220 microfarad capacitors in parallel is a larger 440 micro-farad capacitor. Two one-megohm resistors in parallel is a resistance of 0.5 megohms. Solve for the time constant by multiplying .5 × 440 ("megs" times "micros") and you end up with 220 seconds as the correct answer. The calcula-tor keystrokes are: Clear, .5 × 440 = 220.

3D63 What is the time constant of a circuit having one 100-microfarad capacitor, one 220-microfarad capacitor, one 470-kilohm resistor and one 1-megohm resistor all in series?

A. 68.8 seconds

B. 101.1 seconds

C. 220.0 seconds

D. 470.0 seconds

ANSWER B: This time another RC circuit with all components in series, but now the series capacitors are not equal and neither are the two resistors. You can't do this one in your head. You need to calculate the total capacitance and the total resistance. First resolve the two capacitors in series:

$$C_T = \frac{(100 \times 10^{-6} \times 220 \times 10^{-6})}{(100 \times 10^{-6} + 220 \times 10^{-6})}$$

$$C_T = \frac{(22,000 \times 10^{-12})}{320 \times 10^{-6}} = 68.75 \times 10^{-6}$$

$$C_T = 68.75 \text{ microfarads}$$

Now find the total resistance. Since the resistors are in series, the resistors add (470 k is 0.47 meg), therefore

$$R_T = (0.47 + 1) \text{ megohms}$$

Multiplying R times C to calculate the time constant (remembering "meg"s times "micros" cancel), gives $\tau = 1.47 \times 68.75 = 101.1$ seconds, the correct answer.

3D64 What is the time constant of a circuit having a 470-microfarad capacitor and a 1-megohm resistor in parallel?

A. 0.47 seconds

B. 47 seconds

C. 220 seconds

D. 470 seconds

ANSWER D: Easy one here—do it in your head. It's an RC circuit—one capacitor and one resistor in parallel—so t = RC. The time constant calculations work the same for series or parallel circuits. It's made easy by using "megs" times "micros"—the powers of ten cancel. Only one simple step: multiply 1 "megs" times 470 "micros" to get 470 seconds. There's your answer.

3D65 What is the time constant of a circuit having a 470-microfarad capacitor and a 470-kilohm resistor in series?

A. 221 seconds
B. 221,000 seconds
C. 470 seconds
D. 470,000 seconds

ANSWER A: Same circuit as previous question except the resistor and capacitor values are changed. Use "megs" times "micros" to make it easy, but you have one additional step—you must convert the 470 kilohm resistor value to megohms. 470,000 ohms is equal to 0.47 megohm ($0.47 \times 10^{+6}$). Therefore, τ = $0.47 \times 470 = 220.9$. Round this off to 221 seconds, which is answer A.

3D66 What is the time constant of a circuit having a 220-microfarad capacitor and a 470-kilohm resistor in series?

A. 103 seconds
B. 220 seconds
C. 470 seconds
D. 470,000 seconds

ANSWER A: Same circuit as in previous question, so use "megs" and "micros" and convert the resistance value to megohms. 470,000 ohms is 0.47 megohm. The time constant $\tau = 0.47 \times 220 = 103.4$ seconds, rounded off to 103 seconds.

Do you have RC time constants down pat now? You'll probably have one examination question that will require you to calculate an RC time constant.

3D67 How long does it take for an initial charge of 20 V DC to decrease to 7.36 V DC in a 0.01-microfarad capacitor when a 2-megohm resistor is connected across it?

A. 12.64 seconds
B. 0.02 seconds
C. 1 second
D. 7.98 seconds

ANSWER B: The questions now change emphasis a bit. The next five questions are still on RC circuits, but now a 0.01 microfarad capacitor is charged to 20 volts DC and then discharged through a 2-megohm resistor across it. The three values stay the same for the five problems, making it easy to calculate the progression of correct answers.

Note on the chart below that a capacitor discharges (or charges) 63.2 percent in one time constant. Thus, after discharging for one time constant, 36.8 percent of the charge remains on the capacitor. Therefore, in one time constant, the 20 volts will drop down to 36.8 percent or 7.36 volts (.368 $\times$ 20 = 7.36). Write down one time constant = 7.36 volts for later reference. Since 7.36 volts matches the voltage referred to in the question, the time asked for is one time constant. Use the "megs" times "micros" again to calculate the time constant as 2 $\times$ .01 = .02 second. That's the correct answer.

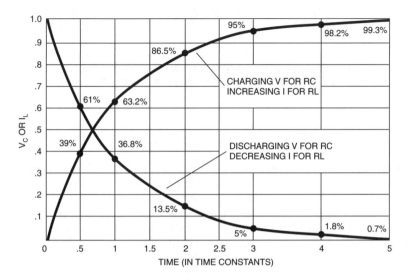

Time Constants Showing V and I Percentages

3D68 How long does it take for an initial charge of 20 V DC to decrease to 2.71 V DC in a 0.01-microfarad capacitor when a 2-megohm resistor is connected across it?
    A. 0.04 seconds
    B. 0.02 seconds
    C. 7.36 seconds
    D. 12.64 seconds
ANSWER A: Refer to the chart at question 3D67. From the previous question, the voltage across the capacitor after discharging for one time constant is 7.36 volts. If the capacitor continues to discharge for another time constant, it again will discharge 63.2 percent of its charge, and have 36.8 percent of the 7.36 volts left. Thus, after two time constants of discharge, the voltage on the capacitor will be 2.71 volts (7.36 × .368 = 2.71). Write down 2.71 = two time constants. It is the time the question is asking for. If one time constant is 0.02 second (as you calculated in question 3D67), two time constants will be 0.04 second, the correct answer.

3D69 How long does it take for an initial charge of 20 V DC to decrease to 1 V DC in a 0.01-microfarad capacitor when a 2-megohm resistor is connected across it?
    A. 0.01 seconds
    B. 0.02 seconds
    C. 0.04 seconds
    D. 0.06 seconds
ANSWER D: The discharge of the capacitor continues from the previous question. Refer to chart in question 3D67. The voltage across the capacitor is down to 2.71 volts at the start of this question. Repeating the same process as the last question, in one more time constant—the third—the capacitor will discharge another 63.2 percent. The voltage after three time constants will be 36.8 percent of 2.71 volts or 1.0 volt (2.71 × .368 = 1). Write down 1.0 V = three

time constants. 1.0 V DC matches the voltage asked for, so the time is three time constants. Three time constants = 3 × 0.02 = 0.06 second, your correct answer.

3D70 How long does it take for an initial charge of 20 V DC to decrease to 0.37 V DC in a 0.01-microfarad capacitor when a 2-megohm resistor is connected across it?
- A. 0.08 seconds
- B. 0.6 seconds
- C. 0.4 seconds
- D. 0.2 seconds

ANSWER A: The discharge continues. Refer to chart in question 3D67. Now the voltage asked for is down to 0.37 volt. Continuing the same thought process as the last three questions, a fourth time constant discharge reduces the voltage across the capacitor to 1 V × 0.368 = 0.368, rounded to 0.37 volt. Write down 0.37 volt = four time constants. Four time constants is 4 × 0.02 = 0.08 second, the correct answer. Remember the time constant for this circuit is 2 × 0.01 = 0.02 second.

3D71 How long does it take for an initial charge of 20 V DC to decrease to 0.13 V DC in a 0.01-microfarad capacitor when a 2-megohm resistor is connected across it?
- A. 0.06 seconds
- B. 0.08 seconds
- C. 0.1 seconds
- D. 1.2 seconds

ANSWER C: Now the discharge voltage asked for is all the way down to 0.13 volt DC. As in previous questions, this voltage may be calculated as the discharge voltage after one additional time constant from the previous question by multiplying 0.37 × 0.368 volt = 0.1362 volt. The question calls it 0.13 V DC. Write down 0.13 volt = five time constants. Five time constants = 5 × .02 = 0.1 second. If you plot the voltages against the time constants, you will have the same curve as the chart in question 3D67, except the chart's vertical scale is normalized, so the curve comes out in percentage.

3D72 How long does it take for an initial charge of 800 V DC to decrease to 294 V DC in a 450-microfarad capacitor when a 1-megohm resistor is connected across it?
- A. 80 seconds
- B. 294 seconds
- C. 368 seconds
- D. 450 seconds

ANSWER D: Since we're having so much fun, let's go for some new numbers, such as 800 volts DC and a 450 microfarad capacitor with a 1 megohm resistor connected across it. These values apply to the next five questions.

Let's continue the same steps as in the previous questions. Since the voltage across a capacitor drops to 36.8 percent of its initial value, what will the voltage be across the capacitor in one time constant? 800 × .368 = 294.4 volts. Write down 294 volts = one time constant. That's the voltage for this question, so we know the time wanted is one time constant. How long is the time constant? Using "megs" × "micros", the time constant = 1 × 450 = 450 seconds.

3D73 How long does it take for an initial charge of 800 V DC to decrease to 108 V DC in a 450-microfarad capacitor when a 1-megohm resistor is connected across it?
   A. 225 seconds
   B. 294 seconds
   C. 450 seconds
   D. 900 seconds
ANSWER D: The capacitor voltage is down to 294 volts from the previous question. What will the voltage be after discharging for another time constant— the second one? 294 volts $\times$ 0.368 = 108.3 volts, the voltage asked for. Write down 108 volts = two time constants. Two time constants = 2 $\times$ 450 = 900 seconds, the correct answer.

3D74 How long does it take for an initial charge of 800 V DC to decrease to 39.9 V DC in a 450-microfarad capacitor when a 1-megohm resistor is connected across it?
   A. 1,350 seconds
   B. 900 seconds
   C. 450 seconds
   D. 225 seconds
ANSWER A: You guessed it, the discharge continues for another time constant— the third. Let's check to see if this is the voltage asked for. The voltage from the last question is 108 volts times 36.8 percent is 39.9 volts. Sure enough, the voltage checks. Write down 39.9 volts = three time constants. $\times$ 3 "megs" $\times$ 450 $\times$ "micros" = 3 $\times$ 450 = 1350 seconds.

3D75 How long does it take for an initial charge of 800 V DC to decrease to 40.2 V DC in a 450-microfarad capacitor when a 1-megohm resistor is connected across it?
   A. Approximately 225 seconds
   B. Approximately 450 seconds
   C. Approximately 900 seconds
   D. Approximately 1,350 seconds
ANSWER D: This voltage is almost the same as 39.9 volts in the problem above. We'll still consider this the third time constant, so 3(1 $\times$ 450) = 1350 seconds is close enough, since "approximately" is used in the answer.

3D76 How long does it take for an initial charge of 800 V DC to decrease to 14.8 V DC in a 450-microfarad capacitor when a 1-megohm resistor is connected across it?
   A. Approximately 900 seconds
   B. Approximately 1,350 seconds
   C. Approximately 1,804 seconds
   D. Approximately 2,000 seconds
ANSWER C: The discharge of the capacitor is now into the fourth time constant. The initial voltage is 39.9 volts at the end of the third time constant. At the end of the fourth time constant the voltage is 39.9 volts $\times$ 0.368 = 14.67 volts, very close to the voltage required. Write down 14.7 volts = four time constants. 4 $\times$ 450 = 1800 seconds. Answer C with an approximate time of 1804 is the correct answer. Take a well-deserved break. You have finished the discussion of time constants!

3D77 What is the impedance of a network composed of a 0.1-microhenry inductor in series with a 20-ohm resistor, at 30 MHz? (Specify your answer in rectangular coordinates.)

    A. 20 + j19
    B. 20 − j19
    C. 19 + j20
    D. 19 − j20

ANSWER A: Don't panic, we are not going to send you back to high school for algebra, trigonometry, and calculus to solve these problems. Here we have a circuit containing both reactance and resistance, giving us *impedance*, represented by the letter Z. Impedance has both a resistive (R) and a reactive part (X_C or X_L). The parts are represented by vectors that are at right angles (perpendicular) to each other. As a result, $Z = \sqrt{R^2 + X^2}$. Each part forms a leg of a right triangle and contributes to the magnitude of the impedance, which is the hypotenuse of the triangle. You determine the impedance's magnitude by calculating each part, R and X, and then Z. Let's first calculate the inductive reactance X_L of this question's circuit with the formula for inductive reactance:

E▶ $X_L = 2\pi fL$ Where: L is inductance in **henries**
                                f is frequency in **hertz**
                                $\pi = 3.14$

At 30 MHz,

    $X_L = 2 \times 3.14 \times 30 \times 10^{+6} \times 0.1 \times 10^{-6}$
    $X_L = 18.84$ rounded to 19 ohms

The resistance, R is given as 20 ohms

Therefore,

    $Z = \sqrt{20^2 + 19^2} = \sqrt{400 + 361} = \sqrt{761}$
    $Z = 27.59$ rounded to 27.6 ohms

But the examination question reads, "Specify your answer in rectangular coordinates." This is good news because it allows us to simplify the problem solving—even to the point of doing them in our heads! Rectangular coordinates consist of the two parts of the impedance discussed above—the first part is the resistance, and the second part is the reactance. The reactance is preceded by a (+j) or a (−j) which indicates whether the circuit's reactance is inductive (+j) or capacitive (−j). The resistance is placed first followed by the reactance; therefore, the impedance of an inductive circuit is $Z = R + jX_L$, and for a capacitive circuit it is $Z = R - jX_C$.

    This first problem has an inductor in series with a 20-ohm resistor. Since it's an inductor, its reactance is inductive and a +j precedes it. Therefore, for this problem, the impedance in rectangular coordinates is 20 + j19. That's not so hard, right?

3D78 What is the impedance of a network composed of a 0.1-microhenry inductor in series with a 30-ohm resistor, at 5 MHz? (Specify your answer in rectangular coordinates.)

    A. 30 − j3
    B. 30 + j3
    C. 3 + j30
    D. 3 − j30

ANSWER B: Here is another inductive circuit; therefore, in the rectangular coordinates required, the reactance is inductive and will have a +j in front of it. Thus, two of the possible answers are already eliminated (A and D because they have a −j reactance). Normally, you would calculate the inductance from $X_L = 2\pi fL$ and place a +j in front of it to determine the correct answer, but in this case, it's much easier. The question gives the resistance as 30 ohms, so you know the answer is 30 + jsomething. Answer B is the only one that has "30 +j". It is the correct answer.

Rectangular Coordinates

A quantity can be represented by points that are measured on an X-Y plane with an X and Y axis perpendicular to each other. The point where the axes cross is the origin, the points on the X and Y axis are called coordinates, and the magnitude is represented by the length of a vector from the origin to the unique point defined by the *rectangular coordinates*.

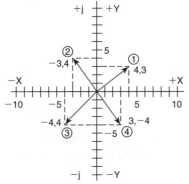

To be able to identify ac quantities and their phase angle, the X axis is called the real axis and the Y axis is called the imaginary axis. There are positive and negative real axis coordinates and positive and negative imaginary coordinates. The imaginary axis coordinates have a **j** operator to identify they are imaginary coordinates.

In rectangular coordinates:

Vector 1 is 4 + j3
Vector 2 is −3 + j4
Vector 3 is −4 − j4
Vector 4 is 3 − j4

▶ Z = R±JX

Polar Coordinates

A quantity can be represented by a vector starting at an origin, the length of which is the magnitude of the quantity, and rotated from a zero axis through 360°.

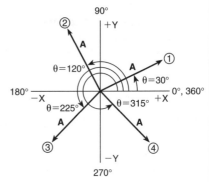

The position of the Vector A can be defined by the angle through which it is rotated. When we do so we are defining the ac quantity in *polar coordinates*.

In polar coordinates:

Vector 1 is A /30°

Vector 2 is A /120°

Vector 3 is A /225°

Vector 4 is A /315°

▶ Z = Z /±θ

Representing an AC Quantity

3D79 What is the impedance of a network composed of a 10-microhenry inductor in series with a 40-ohm resistor, at 500 MHz? (Specify your answer in rectangular coordinates.)
  A. 40 + j31400
  B. 40 − j31400
  C. 31400 + j40
  D. 31400 − j40

ANSWER A: This will also be a (+j) answer because the network is inductive. The resistance is given as 40 ohms, so you already have the resistor value for the rectangular coordinates. The answer will be of the form "40 +j", so look for the answer that begins "40 +j". Only correct answer A has that form. Look, you've picked the correct answer without calculating a thing! You could doublecheck your work by computing the inductive reactance in ohms, $X_L = 2\pi f L$, to agree with the j operator. Remember, f = 500 MHz. $X_L = 2 \times 3.14 \times 500 \times 10^{+6} \times 10 \times 10^{-6}$.

3D80 What is the impedance of a network composed of a 0.001-micro-farad capacitor in series with a 400-ohm resistor, at 500 kHz? (Specify your answer in rectangular coordinates.)
  A. 400 − j318
  B. 318 − j400
  C. 400 + j318
  D. 318 + j400

ANSWER A: This is still rectangular coordinates. Easy one here—do it in your head. It's a 400 ohm resistor in series with a capacitor. You know that gives you a −j, thus, answer A because none of others have "400 −j". If you want to check X_C use the formula in question 3D81.

3D81 What is the impedance of a network composed of a 100-picofarad capacitor in parallel with a 4000-ohm resistor, at 500 kHz? (Specify your answer in polar coordinates.)
  A. 2490 ohms, $\angle 51.5$ degrees
  B. 4000 ohms, $\angle 38.5$ degrees
  C. 5112 ohms, $\angle -38.5$ degrees
  D. 2490 ohms, $\angle -51.5$ degrees

ANSWER D: Shift gears now and be alert because this test question requires your answer in polar coordinates. Refer to the figure at question 3D78. Disconnect yourself from the thought process used for rectangular coordinates in the previous question. Because the resistance is given as 4000, *DO NOT* look at the 4000-ohm resistor as your best choice for the correct answer. Polar coordinate values are calculated differently. In polar coordinates, the ac quantity—let's say impedance Z in this case—is described as a vector at a particular phase angle. You solve for the vector magnitude Z separately and for the phase angle separately. The vector Z (which is the hypotenuse of a right triangle) is defined by a resistance vector (R) and a reactance vector (X) at right angles to each other. If the reactance vector is inductive, it is a $+X_L$; if it is capacitive the reactance is $-X_C$. Positive inductive reactances produce positive angles and negative capacitive reactances produce negative angles. Let's begin the necessary calculations by finding the value of the capacitive reactance with the formula:

$$X_c = \frac{1}{2\pi fC}$$

Where: C is capacitance in **farads**
f is frequency in **hertz**
$\pi = 3.14$

$$X_c = \frac{1}{2 \times 3.14 \times 0.5 \times 10^{+6} \times 100 \times 10^{-12}}$$

Clearing the powers of ten:

$$X_c = \frac{10^{+6}}{6.28 \times 0.5 \times 100}$$

This equation can be used directly when:
f is frequency in **MHz**
C is in **picofarads**

$$X_c = \frac{1,000,000}{314}$$

$$X_c = 3184.7 \text{ ohms}$$

The resistance is already given at 4000 ohms, so R = 4000.

You now need to find the magnitude of the impedance, Z in ohms, of the *parallel* circuit. This must be done with vector (phasor) algebra, as follows:

$$Z = \frac{Z_1 \times Z_2}{Z_1 + Z_2}$$

Special Case:
When $Z_1 = R \underline{/0°}$ and $Z_2 = X_c \underline{/-90°}$

$$Z = \frac{R \underline{/0°} \times X_c \underline{/-90°}}{(R + j0) + (0 - jX_c)}$$

$$Z = \frac{RX_c \underline{/-90°}}{\sqrt{R^2 + X_c^2} \underline{/\arctan \frac{-X_c}{R}}}$$

$$Z = \frac{4000 \underline{/0°} \times 3185 \underline{/-90°}}{(4000 - j3185)} = \frac{4000 \times 3185 \underline{/-90°}}{\sqrt{4000^2 + 3185^2} \underline{/\arctan -3185/4000}}$$

$$Z = \frac{12.74 \times 10^{+6} \underline{/-90°}}{\sqrt{16 \times 10^{+6} + 10.144 \times 10^{+6}} \underline{/\arctan -0.7963}}$$

$$Z = \frac{12.74 \times 10^{+6} \underline{/-90°}}{5.113 \times 10^3 \underline{/-38.5°}}$$

$$Z = 2.493 \times 10^3 \underline{/-90° + 38.5°} = 2493 \underline{/-51.5°} \text{ ohms}$$

Answer D is the correct answer. You will need a scientific calculator to look up the tangent of the phase angles. The solution is not easy. You must keep your wits about you because you cannot add and subtract vectors directly. You must deal with the magnitude and the angle in polar coordinates and the real (resistance) and imaginary (reactance) parts in rectangular coordinates. See the discussion at questions 3D87. You can further doublecheck your answer by recognizing that there must be a minus sign in front of the phase angle because the circuit has capacitive reactance.

3D82 What is the impedance of a network composed of a 100-ohm-reactance inductor in series with a 100-ohm resistor? (Specify your answer in polar coordinates.)
    A. 121 ohms, $\underline{/35 \text{ degrees}}$
    B. 141 ohms, $\underline{/45 \text{ degrees}}$
    C. 161 ohms, $\underline{/55 \text{ degrees}}$
    D. 181 ohms, $\underline{/65 \text{ degrees}}$

ANSWER B: This one is relatively easy—impedance equals the square root of the sum of the resistance squared and the inductive reactance squared. Z in polar coordinates is:

$$Z = \sqrt{100^2 + 100^2} \: / \arctan 100/100$$
$$Z = \sqrt{20000} \: / \arctan 1$$
$$Z = 141.4 \: / +45°$$

You might recognize the square root easier if you work in powers of 10. The magnitude of $Z = \sqrt{(1 \times 10^2)^2 + (1 \times 10^2)^2} = \sqrt{2 \times 10^4} = \sqrt{2} \times 10^2 = 1.414 \times 10^2$. You may recognize the $\sqrt{2}$ as 1.414, so the answer is 141.4. From just the magnitude, you could have selected the correct answer, or you might have used the phase angle. A right triangle with equal R and X has a phase angle of 45° because, as you probably recall, the tangent of 45° is 1.

3D83 A nickel-cadmium cell has an operating voltage of about:
   A. 1.25 volts
   B. 1.4 volts
   C. 1.5 volts
   D. 2.1 volts

ANSWER A: The popular nickel-cadmium cell has an operating voltage of 1.25 volts; a similar-sized alkaline cell has a slightly higher voltage. It's important that all nickel-cadmium batteries be properly charged and *exercised*. Monitor both temperature and voltage during charging to ensure the batteries last as long as possible.

3D84 What is the impedance of a network composed of a 400-ohm-reactance capacitor in series with a 300-ohm resistor? (Specify your answer in polar coordinates.)
   A. 240 ohms, $/ 36.9$ degrees
   B. 240 ohms, $/ -36.9$ degrees
   C. 500 ohms, $/ 53.1$ degrees
   D. 500 ohms, $/ -53.1$ degrees

ANSWER D: Here is a series problem which is easily solved by first calculating the magnitude of the impedance which equals the square root of resistance squared plus reactance squared. The square root of 300 squared plus 400 squared is the square root of 250,000 or 500 ohms. Since the circuit is capacitive, the phase angle will have a minus sign in front of it. If you remember any trigonometry, you know that a right triangle with a 3-4-5 relationship of the sides is an easy one to determine the side values. With sides of 300 and 400, the hypotenuse will be 500, with a phase angle whose tangent is 3/4 or 4/3 depending on the relationship of the sides. In this case it's 4/3 and the angle is −53.1 degrees.

3D85 What is the impedance of a network composed of a 300-ohm-reactance capacitor, a 600-ohm-reactance inductor, and a 400-ohm resistor, all connected in series? (Specify your answer in polar coordinates.)
   A. 500 ohms, $/ 37$ degrees
   B. 400 ohms, $/ 27$ degrees
   C. 300 ohms, $/ 17$ degrees
   D. 200 ohms, $/ 10$ degrees

ANSWER A: Here's one that even though the answer is requested in polar coordinates, it is best to start out using rectangular coordinates. Z = (400 +j600 −j300) = (400 +j300). Now you can use the 3-4-5 ratio we talked about in the previous question. With resistance side of a right triangle at 400 ohms and the reactance side at 300 ohms, you know that the hypotenuse magnitude of the impedance is Z = 500. The angle is a positive one with an arctan of (an angle whose tangent is) 300/400, or 0.75. Grab a trig table or scientific calculator for an angle of 36.8° rounded to 37°.

3D86 What is the impedance of a network comprised of a 400-ohm-reactance inductor in parallel with a 300-ohm resistor? (Specify your answer in polar coordinates.)

A. 240 ohms, $\angle 36.9$ degrees
B. 240 ohms, $\angle -36.9$ degrees
C. 500 ohms, $\angle 53.1$ degrees
D. 500 ohms, $\angle -53.1$ degrees

ANSWER A: We know right off the bat that the answer is going to be either A or C because an inductive reactance will have a positive polar coordinate angle. To come up with 240 ohms at an angle of +36.9° as the correct impedance, let's start out by defining the two impedances that are in parallel:

$Z_1 = 300 \angle 0°$ since it is a resistance in **ohms**
$Z_2 = 400 \angle +90°$ since it is an inductive reactance in **ohms**

$$Z = \frac{Z_1 \times Z_2}{Z_1 + Z_2} \quad \text{or Special Case: } Z = \frac{RX_L \angle +90°}{\sqrt{R^2 + X_L^2} \angle \arctan \frac{X_L}{R}}$$

Additions and subtractions are easier in rectangular coordinates, and multiplication and division are easier in polar coordinates.

$$Z = \frac{300 \times 400 \angle 0° + 90°}{(300 + j0) + (0 + j400)} = \frac{120{,}000 \angle +90°}{500 \angle \arctan 400/300}$$

$$Z = 240 \angle +90° - 53.1°$$

$$Z = 240 \angle 36.9° \text{ ohms}$$

To solve for answer A, we used the 3-4-5 ratio, and both rectangular and polar coordinates.

3D87 What is the impedance of a network composed of a 1.0-millihenry inductor in series with a 200-ohm resistor, at 30 kHz? (Specify your answer in rectangular coordinates.)

A. 200 − j188
B. 200 + j188
C. 188 + j200
D. 188 − j200

ANSWER B: At last, another easy one. This is an inductive circuit with a resistance of 200 ohms; therefore, in rectangular coordinates the answer is 200 +j. Answer B is the only one 200 +j answer. Calculate inductive reactance to check your work. $X_L = 2\pi fL$ (where f = 30 kHz) $= 2 \times 3.14 \times 30 \times 10^3 \times 1 \times 10^{-3} = 188.4$.

Recall from algebra that $+2 \times +2 = +4$ and $-2 \times -2 = +4$ $\therefore \sqrt{4}$ has two roots $+2$ or -2. In ac circuit analysis there is a need to take the square root of a negative number. For example, $\sqrt{-4}$ has a root $\sqrt{-1} \times \sqrt{4} = j2$ and $-j2$. The root of a negative number is called an *imaginary number* and this is indicated by writing the operator j in front of the root. j is equal to $\sqrt{-1}$ and is the basic imaginary quantity. Since $j = \sqrt{-1}$, it is interesting to note the following because they will be encountered in ac circuit analysis:

$$j^2 = j \times j = \sqrt{-1} \times \sqrt{-1} = -1$$

$$j^3 = j \times j \times j = j^2 \times j = -1 \times j = -j$$

$$j^4 = j \times j \times j \times j = j^2 \times j^2 = -1 \times -1 = +1$$

$$\frac{1}{j} = \frac{1}{j} \times \frac{j}{j} = \frac{j}{j^2} = \frac{j}{-1} = -j$$

Summary
$j = \sqrt{-1}$
$j^2 = -1$
$j^3 = -j$
$j^4 = +1$
$\dfrac{1}{j} = -j$

j Operator as Vector Rotator

Note that an imaginary quantity can be considered as a vector being rotated by the operator j.

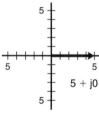

Vector starts at 0°.

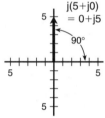

The j operator, when it multiplies a vector, rotates the vector by 90° counter clockwise.

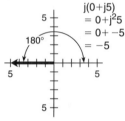

A second 90° rotation puts the vector at 180°.

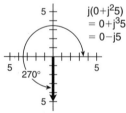

A third 90° rotation puts the vector at 270°.

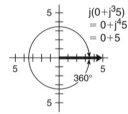

A fourth 90° rotation puts the vector back at 0°.

Complex Numbers (Real and Imaginary) and Operator j

3D88 What is the impedance of a network composed of a 0.01-microfarad capacitor in parallel with a 300-ohm resistor, at 50 kHz? (Specify your answer in rectangular coordinates.)
    A. $150 - j159$
    B. $150 + j159$
    C. $159 + j150$
    D. $159 - j150$
ANSWER D: Here is another one you will need to drag out the calculator on. Come on, isn't this fun? First find capacitive reactance. Refer back to question

3D81, if the capacitance is in **picofarads** and the frequency is in **megahertz**, then

$$X_C = \frac{10^{+6}}{2\pi fC}$$

50 kHz is 0.05 MHz and 0.01 microfarads is 10,000 picofarads, so you will divide 6.28 × .05 × 10,000 into 1 million, and end up with 318.47 ohms.

Now solve for impedance:

$$Z = \frac{Z_1 \times Z_2}{Z_1 + Z_2} \text{ or use Special Case: } Z = \frac{RX_C \, / -90°}{\sqrt{R^2 + X_C^2} \, / \arctan \frac{-X_C}{R}}$$

$$Z = \frac{300 \, / \, 0° \times 318.5 \, / -90°}{(300 + j0) + (0 - j318.5)}$$

$$Z = \frac{95{,}550 \, / -90°}{\sqrt{300^2 + 318.5^2} \, / \arctan 318.5/300} = \frac{95{,}550 \, / -90°}{438 \, / -46.5°}$$

$$Z = 218.3 \, / -90° + 46.5° = 218.3 \, / -43.5° \text{ ohms}$$

Your answer comes out 218 ohms. The answer needs to be in rectangular coordinates. To convert to rectangular coordinates, you will be finding the resistance side and the reactance side of a right triangle with a hypotenuse of 218.3 and a contained angle $\Theta = -43.5°$. The parts are:

R = 218.3 Cos Θ = 218.3 × 0.7253 = 158.3 (answer calls it 159)
X_C = 218.3 Sin Θ = 218.3 × 0.6883 = 150.2 (answer calls it 150)

Therefore, in rectangular coordinates:

Z = 158.3 −j150.2

Match this to the correct answer D of Z = 159 −j150.

3D89 Which of the following components is used in a power supply circuit to allow filter capacitors to discharge when power is turned off and aids in holding the voltage output more constant?
   A. Bleeder resistor
   B. Multiplier resistor
   C. Surge resistor
   D. Rectifying diodes
ANSWER A: Use extreme caution when working on any type of power supply, whether it's a transistorized switcher supply or a traditional transformer-type power supply. Those large capacitors can store lethal voltages for quite some time. The bleeder resistor ultimately drains the voltage after the supply is turned off, but the capacitors can hold a dangerous charge for as long as 2 minutes after a supply is turned off. Bleeder resistors also help in voltage regulation.

✱ 3D90 A 1-watt, 10-volt zener diode with the following characteristics: I min. = 5 mA, I max. = 95 mA, and Z = 8 ohms; is to be used as part of a voltage regulator. Approximately what size current-limiting resistor would set its bias to the midpoint of its operating range *(assume a 20V power supply)*?
   A. 100 ohms   C. 1 kilohms
   B. 200 ohms   D. 2 kilohms

ANSWER B: This is a classic Ohm's Law problem. Assume the supply voltage is 20V. Also assume that the midpoint current is 50 mA. This is calculated by 95 mA − 5 mA ÷ 2 = 45 + 5 = 50 mA. Now convert mA to amps, and apply Ohm's Law to get the answer as shown: 10 Volts ÷ 0.05 Amps = 200 Ohms. *[Ed. Note: Text in italics added.]*

---

* 3D91 A 1-watt, 10-volt zener diode with the following characteristics: I min. = 5 mA, I max. = 480 mA, and Z = 3 ohms; is to be used as part of a voltage regulator. Approximately what size current limiting resistor would set its bias to the midpoint of its operating range *(assume a 20V power supply)*?

A. 40 ohms
B. 100 ohms
C. 400 ohms
D. 1 kilohms

ANSWER A: Again assume the supply voltage is 20V. Calculate the mid-point current (242.5 mA) as in the last problem. Then, convert mA to amps and apply the formula (E ÷ I = R). 10 Volts ÷ 0.2425 = 41.24 The only answer that's close is 40. *[Ed. Note: Text in italics added.]*

[Based on real-world available zener diodes. this question is flawed. There are no 1W zener diodes with maximum currents of 480mA. It appears that a decimal point was misplaced, and the maximum current stated in the question should be 48mA. Based on 48mA, the correct answer would be C, 400 ohms. However, for purposes of the examination, base your calculations on the information provided in the question, and cite Answer A as the correct answer.]

3D92 A crowbar circuit is often added to a power supply to:

A. Prevent the circuit protective devices from being damaged.
B. Aide the filter section by increasing voltage regulation.
C. Protect the power supply by allowing the load to have as much current as it needs.
D. Protect the load by activating circuit protective devices.

ANSWER D: The crowbar circuit in power supplies immediately shuts down the output when a silicon-controlled rectifier senses too much current is being consumed. The SCR features a voltage-sensing trigger circuit at its gate, and a load short circuit immediately causes the power supply to safely shut down without damage. The power supply must be manually cycled off and back on to reset the crowbar circuitry after the short circuit has been remedied.

3D93 Given a power supply with a no-load voltage of 12 volts and a full-load voltage of 10 volts, what is the percentage of voltage regulation?

A. 17 %
B. 20 %
C. 80 %
D. 83 %

ANSWER B: E_{nl} = no-load voltage; E_{fl} = full-load voltage. The formula to calculate the percentage of voltage regulation is: $[(E_{nl} − E_{fl}) \times 100)] ÷ E_{fl}$ = % regulation. To work this out, multiply 100 times the difference of the voltages, and divide the answer by the full-load voltage. For our example, this would be $[(12 − 10) \times 100)] ÷ 10 = 20$ percent regulation.

3D94 Given a power supply with a full-load voltage of 200 volts and a regulation of 25%, what is the no-load voltage?

A. 150 volts
B. 160 volts
C. 240 volts
D. 250 volts

ANSWER D: Here you must rearrange the formula to compute the no-load voltage. The following 4 steps show how the formula is modified:

$$\frac{(E_{nl} - E_{fl})100}{E_{fl}} = \% \text{ regulation}$$

$$(E_{nl} - E_{fl})100 = \%(E_{fl})$$

$$(E_{nl} - E_{fl}) = \% \frac{E_{fl}}{100}$$

$$E_{nl} = \% \frac{E_{fl}}{100} + E_{fl}$$

Where:
E_{nl} = No-load voltage in **volts**
E_{fl} = Full-load voltage in **volts**
% = Regulation in **percent**

Now, plug in the numbers to arrive at the answer...250 volts.

$$E_{nl} = \% \frac{25(200)}{100} + E_{fl}$$

$$= \frac{5000}{100} = 50 + 200 = 250$$

3D95 Given a power supply with a no-load voltage of 200 volts and a regulation of 25%, what is the full-load voltage.
A. 150 volts
B. 160 volts
C. 240 volts
D. 250 volts

ANSWER B: To calculate the full-load voltage, you must rearrange the formula as follows:

$$\% \text{ reg} = \frac{(E_{nl} - E_{fl})100}{E_{fl}}$$

$$\%E_{fl} = 100E_{nl} - 100E_{fl}$$
$$\%E_{fl} + 100E_{fl} = 100E_{nl}$$
$$E_{fl}(\% + 100) = 100E_{nl}$$

Now, plug in the numbers to calculate the full-load voltage.

$$E_{fl} = \frac{100 \, E_{nl}}{(\% + 100)} = \frac{100 \times 200}{25 + 100} = \frac{20,000}{125} = 160 \text{ volts}$$

3D96 A 50-microampere meter movement has an internal resistance of 2 kilohms. What applied voltage is required to indicate half-scale deflection?
A. 0.01 Volts
B. 0.10 Volts
C. 0.005 Volts
D. 0.05 Volts

ANSWER D: In this Ohm's Law problem you multiply current by resistance, yielding full-scale deflection. Then, divide your answer by 2 to ge;t the final answer...0.05 volts. Remember that 50 microamperes is 6 places to the left for amps (0.000050 amps), and 2 kilohms is 3 decimal places to the right for ohms (2000 ohms). Check with: $50 \times 10^{-6} \times 2 \times 10^{3} = 100 \times 10^{-3} = 50 \times 10^{-3} = 0.05$ V.

3D97 The expression "voltage regulation" as it applies to a shunt-wound DC generator operating at a constant frequency refers to:
  A. Voltage output efficiency.
  B. Voltage in the secondary compared to the primary.
  C. Voltage fluctuations from load to no-load.
  D. Rotor winding voltage ratio.

ANSWER C: In shipboard installations, voltage regulation is extremely important. The term "voltage regulation" applies to shunt-wound DC generators, operating at a fixed frequency, where no-load voltage is compared to voltage under load (load to no-load), just as we demonstrated in the previous questions.

3D98 When an emergency transmitter uses 325 watts and a receiver uses 50 watts, how many hours can a 12.6 volt, 55 ampere-hour battery supply full power to both units?
  A. 6 hours
  B. 3 hours
  C. 1.8 hours
  D. 1.2 hours

ANSWER C: Operating power is 375 watts (325 + 50). Divide 375 by 12.6 volts to calculate the amount of current this emergency transmitter system is using (375 ÷ 12.6 = 29.76 amps at 12.6 VDC). Now, divide 29.76 amps into 55 ampere-hours to get the approximate operating time of 1.8 hours. Or you could multiply 12.6 × 55 = 693 watt hours and divide by 375 watts to get 1.85 hours.

3D99 The expression "voltage regulation" as it applies to a generator operating at a constant frequency refers to:
  A. Full load to no load.
  B. Limited load to peak load.
  C. Source input supply frequency.
  D. Field frequency.

ANSWER A: Voltage regulation in an AC generator with a constant frequency (constant revolutions per second) is the percentage of variation of the voltage from a full load to no load. As shown in question 3D93, the formula for calculating this percentage is:

$$[(E_{nl} - E_{fl}) \times 100] \div E_{fl} - \% \text{ regulation.}$$

3D100 The output of a separately-excited AC generator running at a constant speed can be controlled by:
  A. Armature.
  B. Brushes.
  C. Field current.
  D. Exciter.

ANSWER C: We control the output voltage of a separately-excited AC generator by varying the field current.

3D101 A transformer used to step up its input voltage must have:
  A. More turns of wire on its primary than on its secondary.
  B. More turns of wire on its secondary than on its primary.
  C. Equal number of primary and secondary turns of wire.
  D. None of the above statements are correct.

ANSWER B: Transformers that step up voltage have more turns on the secondary winding than on the primary. Isolation transformers have the same number of turns on both windings, and step-down transformers have fewer turns on the secondary winding.

3D102 A transformer used to step down its input voltage must have:
  A. More turns of wire on its primary than on its secondary.
  B. More turns of wire on its secondary than on its primary.
  C. Equal number of primary and secondary turns of wire.
  D. None of the above statements are correct.

ANSWER A: As stated in the previous question, a transformer used as a step-down device has fewer turns on the secondary winding that on the primary.

3D103 A 12.6-volt, 8-ampere-hour battery is supplying power to a receiver that uses 50 watts and a radar system that uses 300 watts. How long will the battery last?
  A. 100.8 hours
  B. 27.7 hours
  C. 1 hour
  D. 17 minutes or 0.3 hours

ANSWER D: In this question, first calculate the total power being consumed (300 watts + 50 watts = 350 watts). Then, divide 350 by 12.6 volts to determine the running current: a whopping 27 amps. Because 27 amps are being drawn from an emergency 8 ampere-hour battery, it should be obvious the system won't run for more than about a half hour.

Divide 27 *into* 8, and you end up with 0.29 hours of operation. Multiply this by 60 to calculate minutes, and the answer is 17 minutes, the same as 0.3 hours. Not much time for a radar system!

3D104 What is the total voltage when 12 Nickel-Cadmium batteries are connected in series?
  A. 12 volts.
  B. 12.6 volts.
  C. 15 volts.
  D. 72 volts.

ANSWER C: In this problem, you must remember that each nickel-cadmium battery delivers 1.25 volts. Batteries, when connected − to +, add together. Multiply 1.25 volts by 12 batteries, and you get a series voltage of 15 volts.

3D105 A ship radar unit uses 315 watts and a radio uses 50 watts. If the equipment is connected to a 50-ampere-hour battery rated at 12.6 volts, how long will the battery last?
  A. 28.97 hours
  B. 29 minutes
  C. 1 hour 43 minutes
  D. 10 hours 50 minutes

ANSWER C: Here is another easy problem. Divided the total watts used (315 + 50 = 365) by 12.6 volts to get 28.9 amps of constant current draw. Now, clear your calculator and divide 50 ampere-hours by 28.9 to get 1.7 hours. Finally, multiply 1.7 hours by 60 to get 103 minutes, which works out to be 1 hour and 43 minutes.

✱ 3D106 The turns ratio of a transformer is 1:20. When a 120-volt ac source is connected to its primary winding, the secondary voltage is:
   A. 120 volts.
   B. 1200 volts.
   C. 600 volts.
   D. 2400 volts.

ANSWER D: This ratio problem is solved by multiplying primary voltage (120) by the turns ratio (20), the number of turns on the secondary divided by the number on the primary. 120 volts × 20 = 2400 volts. *[Ed. Note: Standard practice for turns ratio should have stated it as 20:1, N_S/N_P. Secondary turns are 20 times the primary.]*

3D107 There is an improper impedance match between a 30-watt transmitter and the antenna and 5 watts is reflected. How much power is actually radiated?
   A. 35 watts
   B. 30 watts
   C. 25 watts
   D. 20 watts

ANSWER C: This is probably an *on-shore* VHF station that is permitted additional power above 25 watts. If 30 watts are coming out of the transmitter and 5 watts are reflected by a mismatched antenna, only 25 watts are actually being radiated by the antenna. The returning 5 watts would probably go up in heat.

3D108 How long will a 12.6-volt, 50-ampere-hour battery last if it supplies power to an emergency transmitter rated at 531 watts of plate input power and other emergency equipment with a combined power rating of 530 watts?
   A. 6 hours
   B. 4 hours
   C. 1 hour
   D. 35 minutes

ANSWER D: Be careful not to mis-read this question. When your author first worked it, he based his total power on a "combined" 530 watts. Incorrect. The total power being consumed is 1061 watts (531 + 530). Divide 1061 by 12.6 volts to get a current draw of 84 amps. Then, divide 84 amps *into* 50 ampere-hours, and you find that this system would run only 0.59 hours, which is 35 minutes. Or, you could simply take a look at the four possible answers and see there is only one answer less than 1 hour of operating time. Be careful that you don't read 0.59 hours decimal as 59 minutes—0.59 hours × 60 = 35 minutes.

3D109 A 12.6-volt, 55-ampere-hour battery is connected to a radar unit rated at 325 watts and a receiver that uses 20 watts. How long will radar unit and receiver be able to draw full power from the battery?
   A. 6 hours
   B. 4 hours
   C. 2.3 hours
   D. 2 hours

ANSWER D: Here is another easy problem—add the total watts (325 + 20 = 345), and divide that by 12.6 to get a current draw of 27.38 amps. Divide this *into* the 55 amp-hours, and you find the unit will run for about 2 hours.

 P(W) = E(V) × I(A) and I(A) × t(hrs) = ampere hours

3D110 A power transformer has a 120-volt primary winding and a 24-volt secondary winding. What is its turns ratio (referenced secondary to primary)?
  A. 10 : 1
  B. 1 : 10
  C. 5 : 1
  D. 1 : 5
ANSWER D: When dealing with voltage transfer in transformers, turns ratio is N_S/N_P and $N_S \div N_P = V_S \div V_P$. Therefore 24 ÷ 120 = 1 ÷ 5. The secondary is generating 24 V from an input of 120 V on the primary. The turns ratio is 1:5.

3D111 A 6-volt battery with 1.2 ohms internal resistance is connected across two 3-watt bulbs. What is the current flow?
  A. .57 amps
  B. .83 amps
  C. 1.0 amps
  D. 6.0 amps
ANSWER B: The circuit schematic has a battery in series with an internal resistance that feeds current to two light bulbs in parallel. It is assumed that the bulb resistance remains constant from cold to hot. To solve for the circuit current, you must know the total resistance; therefore, you must first find the equivalent resistance of the two light bulbs which are in parallel. The wattage of each light bulb is given as 3 watts, so you can find the resistance of each bulb by using R = $E^2 \div P$ = 36 ÷ 3 = 12 ohms for each bulb. Two 12-ohm resistors in parallel = 6 ohms. Thus, the total resistance across the battery is 1.2 ohms + 6 ohms = 7.2 ohms. To find the current, use I = E ÷ R = 6 ÷ 7.2 = 0.83 amp.

3D112 A power transformer has a primary winding of 200 turns of #24 wire and a secondary winding consisting of 500 turns of the same size wire. When 20 volts is applied to the primary winding, the expected secondary voltage is:
  A. 500 Volts.
  B. 25 Volts.
  C. 10 Volts.
  D. 50 Volts.
ANSWER D: This transformer has a turns ratio (N_S/N_P) of 500 ÷ 200 = 2.5 ÷ 1 or 2.5:1, so you simply multiply the primary voltage (20) by the ratio (2.5) to get the answer (50 volts).

3D113 The power input to a 52-ohm transmission line is 1,872 watts. The current flowing through the line is:
  A. 6 amps.
  B. 144 amps.
  C. 0.06 amps.
  D. 28.7 amps.
ANSWER A: Recall the formula for calculating current when power and resistance are known:

 $I^2R = P$ Where: I = Current in **amps**
                                        R = Resistance in **ohms**
$$I^2 = \frac{P}{R} = \frac{1872}{52} = 36$$ P = Power in **watts**

$$I = \sqrt{36} = 6 \text{ amps}$$

3D114 If a marine radiotelephone receiver uses 75 watts of power and a transmitter uses 325 watts, how long can they both operate before discharging a 50-ampere-hour 12-volt battery?
  A. 40 minutes
  B. 1 hour
  C. 1 1/2 hours
  D. 6 hours

ANSWER C: Add 325 watts to 75 watts for a total power comsumption of 400 watts. Divide that by 12.0 volts, to get the current draw of 33.33 amps. Divide this *into* 50-amp hours to find that the system will operate for exactly 1-1/2 hours.

3D115 A power transformer has a single primary winding and three secondary windings producing 5.0 volts, 12.6 volts, and 150 volts. Which of the three secondary windings will have the highest measured DC resistance.
  A. The 12.6-volt winding
  B. The 5.0-volt winding
  C. The 150-volt winding
  D. All will have equal resistance values

ANSWER C: You can determine which of three secondary windings may generate the highest voltage by metering each winding for its DC resistance—obviously, out of circuit. The 150-volt winding has the most turns of the secondary windings—the most wire—so it has the highest measured DC resistance.

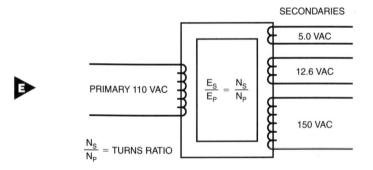

Power Transformer with Multiple Secondaries

3D116 Which of the following list is correct for listing common materials in order of descending conductivity:
  A. Silver, copper, aluminum, iron, and lead
  B. Lead, iron, silver, aluminum, and copper
  C. Iron, silver, aluminum, copper, and silver
  D. Silver, aluminum, iron, lead, and copper

ANSWER A: The acronym SCAIL stands for Silver, Copper, Aluminum, Iron, and Lead for conductivity. On a *SCAIL* of 1 to 10, you are going to do well on your examination, if you will use this simple reminder.

3D117 Under what condition may gas be a good conductor?
   A. When placed in an isotropic radiator
   B. When subjected to a vacuum
   C. When ionized
   D. When placed in a Leclanche' cell

ANSWER C: When gas is ionized in a tube, it becomes a good conductor. This is why our ionosphere, which is ionized, becomes a good conductor and refractor of electromagnetic energy.

3D118 Ship's power is generated 3-phase, ungrounded. On a delta-wound transformer with 120 Vac line-to-line secondary, the voltmeter reading from line to ground is:
   A. Approximately 67 Vac for a normal balanced system with no faults
   B. 0 Vac for a system with that phase faulted to ground
   C. 120 Vac when another phase is faulted to ground
   D. None of these

ANSWER D: On a ship, 3-phase, ungrounded, delta-wound transformer, each corner of the triangular power configuration is above ground, so there is no ground to measure the voltage.

3D119 The product of the number of turns and the current in amperes used to describe relative magnitude is:
   A. Ampere turns.
   B. Joules per second.
   C. Push-pull convergence.
   D. Dissipation collection.

ANSWER A: If you multiply the number of turns by the amount of current in amps, the result is in *ampere turns*; ampere turn is the unit of measurement for the *Gilbert.*

3D120 The factor by which the product of volts and amperes must be multiplied to obtain true power is:
   A. Apparent power.
   B. Power factor.
   C. Phase angle.
   D. Power angle.

ANSWER B: In an ac circuit, apparent power is multiplied by the power factor to get the value of true power.

   Volts × amps × *power factor* = *true power.*

3D121 Halving the cross-sectional area of a conductor will:
   A. Not affect the resistance.
   B. Quarter the resistance.
   C. Half the resistance.
   D. Double the resistance.

ANSWER D: If the cross-sectional area of a piece of wire is reduced by one-half, the resistance doubles.

3D122 If a resistance to which a constant voltage is applied is halved, what power dissipation will result?
   A. Doubled
   B. Halved

    C. Quadrupled

    D. Stay the same

ANSWER A: If the resistance in a piece of wire is reduced to half, the power dissipation doubles.

3D123 The effective value of an RF current and the heating value of the current are:

    A. Effective value divided by two equals the heating value.

    B. Effective value multiplied by two equals the heating value.

    C. The sum of the value of the divided parts multiplied by two equals

    D. The same.

ANSWER D: The effective (RMS) value of radio frequency current and its heating value are the same.

3D124 746 watts, roughly 3/4 kilowatt corresponding to the lifting of 550 pounds at the rate of one foot per second is:

    A. Quarter of a horsepower.

    B. Half of a horsepower.

    C. 3/4 of a horsepower.

    D. One horsepower.

ANSWER D: One horsepower is equal to 746 watts, which corresponds to the lifting of 550 pounds at the rate of one foot per second.

3D125 Assuming a power source to have a fixed value of internal impedance, maximum power will be transferred to the load when:

    A. The load impedance equals the internal impedance of the source.

    B. The load impedance is greater than the source impedance.

    C. The load impedance is less than the source impedance.

    D. The fixed values of internal impedance is not relative to the power source.

ANSWER A: When a load impedance equals the internal impedance of the source, maximum power is transferred. This is why we pay close attention to impedance matching in antenna and transceiver systems.

3D126 What is the conductance (G) of a circuit if 6 amperes of current flows when 12 Vdc is applied?

    A. 0.25 siemens(mho)

    B. 0.50 siemens(mho)

    C. 1.00 siemens(mho)

    D. 1.25 siemens(mho)

ANSWER B: The formula for determining conductance in mhos is $G = 1/R = I \div E$. So, 6 amps $\div$ 12 volts = 0.50 siemens. The siemens and the mho are units for the same amount of conductance (1 mho = 1 seimens).

3D127 If the voltage to a circuit is doubled and the resistance is tripled, what will be the final current?

    A. 1/3 the original current

    B. 2/3 the original current

    C. 1 1/3 the original current

    D. Double the original current

ANSWER B: The current in a circuit is $I = E/R$. When the voltage is doubled, and resistance is tripled, the current falls to 2/3 the original current ($I = 2E \div 3R$).

3D128 A relay coil has 500 ohms resistance, and operates on 125 mA. What value of resistance should be connected in series with it to operate from 110 Vdc?
  A. 150 ohms
  B. 220 ohms
  C. 380 ohms
  D. 470 ohms

ANSWER C: Here is another Ohm's Law problem where voltage divided by current equals resistance (E ÷ I = R). But you have to figure the voltage drop before you can apply the formula.

The voltage drop across the relay is 500 ohms × 0.125 amps = 62.5 volts. Now, subtract 62.5 from 110 to arrive at the remaining 47.5 volts that must be dropped across the series resistance. Finally, use Ohm's Law to solve for resistance to get the answer (47.5 Volts ÷ 0.125 Amps = 380 Ohms).

3D129 How many capacitors of 400 volts and 2.0 microfarads each would be necessary to obtain a combination rated at 1600 volts and 1.5 microfarads?
  A. 10
  B. 12
  C. 14
  D. 16

ANSWER B: The voltage ratings of capacitors add when capacitors are in series. In order to have a final capacitance to withstand 1600 volts, four of them must be in series. Four equal capacitors in series are just like four equal resistors in parallel. If all capacitors are equal, the equivalent capacitance is one fourth the value of one of them, or 0.5 microfarads. Capacitors in parallel add like resistors in series; therefore, there must be three 0.5 microfarad legs in parallel to end up with 1.5 microfarads in total. The equivalent 1.5 µF has a rating of 1600 volts.

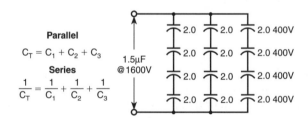

Series and Parallel Capacitor Combinations

3D130 The total inductance of two coils in parallel without any mutual coupling is:
  A. Equal to the product of the two inductances divided by their sum.
  B. Equal to the sum of the individual inductances.
  C. Equal to zero.
  D. Equal to the sum of the individual inductances divided by their product.

ANSWER A: Use the same basic formula to calculate the inductance of coils in parallel as your would to calculate the resistance of resistors in parallel. For coils, the formula is (L1 × L2) ÷ (L1 + L2).

3D131 Permeability is:

A. The magnetic field created by a conductor wound on a laminated core and carrying current.

B. The ratio of magnetic flux density in a substance to the magnetizing force that produces it.

C. Polarized molecular alignment in a ferromagnetic material while under the influence of a magnetizing force.

D. None of these.

ANSWER B: Permeability is a certain flux density that develops in a core for a certain value of current. If iron is inserted into the coil, more flux and flux density develops. Permeability is represented by the Greek letter μ (pronounced "mu",) and the formula is:

E $\quad \mu = \dfrac{\beta}{H}$ Where: β = Flux density in guass or **lines per square inch**

$\qquad\qquad\qquad\qquad\qquad$ H = Oersted of magnetizing force in **ampere turns per inch**

3D132 What is the total impedance of a series AC circuit having a resistance of 6 ohms, an inductive reactance of 17 ohms, and zero capacitive reactance?

A. 6.6 ohms

B. 11 ohms

C. 18 ohms

D. 23 ohms

ANSWER C: The formula for determining impedance is the square root of $R^2 + X^2$. $X = X_L$ when there is zero capacitive reactance. $6^2 = 36$ and $17^2 = 289$. The sum of the two equals 325. The square root of 325 is approximately 18 and the correct answer.

3D133 The opposition to the creation of magnetic lines of force in a magnetic circuit is known as:

A. Eddy currents.

B. Hysterisis.

C. Permeability.

D. Reluctance.

ANSWER D: The opposition to the creation of magnetic lines of force in a magnetic circuit is called *reluctance*. Iron has low reluctance because it has a very high μ. A segment of vacuum or air one centimeter long and one square centimeter of area with $\mu = 1$ has a reluctance of one unit.

The formula for reluctance is:

E $\quad \mathcal{R} = \dfrac{l}{\mu A}$ where: $\mathcal{R}$ = Reluctance in **reluctance units**

$\qquad\qquad\qquad\qquad\qquad$ l = Length in **cm** or **inches**

$\qquad\qquad\qquad\qquad\qquad$ A = Area in **cm²** or **in²**

3D134 Why is a center tap usually provided for vacuum tube plate and grid return circuits when an AC filament supply is used?

A. To prevent hum voltage from modulating the normal signals.

B. To allow more filament current.

C. All of these.

D. None of these.

ANSWER A: A center tap in a transformer is usually provided for vacuum tube plate and grid return circuits to reduce hum from modulating the normal signal. The hum is canceled at the source.

3D135 Given the following vacuum tube constants: Gp = 1000V, Ip = 150 mA, Ig = 10 mA, and grid leak = 5000 ohms; What would be the grid bias voltage?

    A. 25 V

    B. 50 V

    C. 100 V

    D. None of these

ANSWER B: Here we go with Ohm's Law again. Voltage (E) equals current (I) times resistance (R). Grid resistance is 5000 ohms, grid current is 10 mA. So, 5000 ohms $\times$ 0.01 amps = 50 volts.

3D136 A special type of power supply filter choke whose inductance is inversely proportional to the amount of current flowing through it is a:

    A. AF choke.

    B. RF choke.

    C. Smoothing choke.

    D. Swinging choke.

ANSWER D: The swinging choke is normally used in a vacuum tube rectifier power supply where it offers high inductance when low current is passing through it. When the current increases and saturates, the swinging choke has less inductance. This means there are different voltage drops across it, but the power supply voltage remains relatively constant under varying loads. This improves the voltage regulation of the entire circuit.

3D137 What turn ratio does a transformer need to match a source impedance of 500 ohms to a load of 10 ohms.

    A. 7.1 to 1

    B. 14.2 to 1

    C. 50 to 1

    D. None of these

ANSWER A: The ratio of turns on a primary to the number of turns on a secondary is equal to the square root of the ratio of the primary impedance to the secondary impedance. Divide 10 ohms into 500 ohms and you get a ratio of 50:1. You can approximate the square root of 50 in your head ($7 \times 7 = 49$, so the square root of 50 is slightly more than 7). Watch out for answer C; be sure to calculate the square root of 50 before you select an answer. Here's proof of the formula used: The power in the primary is equal to the power in the secondary; therefore, $P_P = P_S$

and $\dfrac{(V_P)^2}{Z_P} = \dfrac{(V_S)^2}{Z_S}$ and $\left(\dfrac{V_P}{V_S}\right)^2 = \dfrac{Z_P}{Z_S}$

Now $\dfrac{V_P}{V_S} = \dfrac{N_P}{N_S}$, therefore, $\left(\dfrac{N_P}{N_S}\right)^2 = \dfrac{Z_P}{Z_S}$ and $\dfrac{N_P}{N_S} = \sqrt{\dfrac{Z_P}{Z_S}}$

 $\dfrac{N_P}{N_S} = \sqrt{\dfrac{Z_P}{Z_S}}$

Where: N_P = Primary **turns**
N_S = Secondary **turns**
Z_P = Primary Z in **ohms**
Z_S = Secondary Z in **ohms**

3D138 What is the DC plate voltage of a resistance-coupled amplifier stage with a plate-supply voltage of 260 V, a plate current of 1 mA, and a plate load resistance of 100 kilohms.
 A. 60 volts
 B. 100 volts
 C. 160 volts
 D. 220 volts

ANSWER C: Ohm's Law again. Voltage (E) equals current (I) times resistance (R). Just be sure to convert the resistance with current through it into a voltage drop as follows: 100,000 ohms × 0.001 amps = 100 V. Subtract the 100 volts drop across the plate load resistor from the plate-supply voltage (260 volts) to get the DC plate voltage (160 volts).

3D139 The average fully charged voltage of an Edison storage cell is:
 A. 1 volt.
 B. 1.2 volts.
 C. 1.5 volts.
 D. 2 volts.

ANSWER B: A fully-charged Edison storage cell is rated at 1.2 volts.

3D140 The average fully charged voltage of a lead-acid storage cell is:
 A. 1 volt.
 B. 1.2 volts.
 C. 1.56 volts.
 D. 2.06 volts.

ANSWER D: The fully charged voltage of a lead-acid storage cell is 2.06 volts. Keep in mind they are talking about cells, not the entire series-configured battery.

3D141 A battery with a terminal voltage of 12.5 volts is to be trickle-charged at a 0.5 A rate. What resistance should be connected in series with the battery to charge it from a 110 Vdc line?
 A. 95 ohms
 B. 195 ohms
 C. 300 ohms
 D. None of these

ANSWER B: First calculate the voltage drop (110 − 12.5 volts = 97.5) that must occur across the resistor. Then, apply Ohm's Law (R = E ÷ I) as shown:

 97.5 ÷ 0.5 amps = 195 ohms.

3D142 A discharged storage cell of 3 cells has an open-circuit voltage of 1.8 volts per cell and an internal resistance of 0.1 ohms per cell. What voltage is needed to give an initial charging rate of 10 A?
 A. 8.4 volts
 B. 10 volts
 C. 12.5 volts
 D. 15 volts

ANSWER A: Ohm's Law again: The charging current is going to cause a voltage drop across the cell's internal resistance. Thus, 10 amps × 0.3 ohms (for 3 cells) = 3 volts.

 3 volts + 1.8 V + 1.8 V + 1.8 V = 8.4 volts.

3D143 What capacity in amperes does a storage battery need to be in order to operate a 50-watt transmitter for 6 hours? Assume a continuous transmitter load of 70% of the key-locked demand of 40 A, and an emergency light load of 1.5 A?
    A. 100 ampere-hours
    B. 177 ampere-hours
    C. 249 ampere-hours
    D. None of these
ANSWER B: You don't need to worry about how many watts the transmitter puts out because they give you no voltage to work with, but they do give you 40 amps on a 70-percent duty cycle. This works out to be 28 amps, plus an additional 1.5 amps, for a total of 29.5 amps. Because the desired battery life is 6 hours, multiply 29.5 times 6, and you end up with 177 ampere-hours. Some of the commercial questions provide information you don't really need to solve the problem.

3D144 What current will flow in a 6-volt storage battery with an internal resistance of 0.01 ohms, when a 3-watt, 6-volt lamp is connected?
    A. 0.4885 amps
    B. 0.4995 amps
    C. 0.5566 amps
    D. 0.5795 amps
ANSWER B: Ohm's Law is P (Power) = E^2(Volts) ÷ R, but you must modify the formula to solve for resistance of the lamp. R = E^2 ÷ P. 36 ÷ 3 = 12 ohms. 12 + 0.01 = 12.01 ohms total resistance. 6V ÷ 12.01 = 0.4995 amps.

3D145 What is the ratio of the output frequency to the input frequency of a single-phase full-wave rectifier?
    A. 1 : 1
    B. 1 : 2
    C. 2 : 1
    D. None of these
ANSWER C: A full-wave, single-phase rectifier has twice the output frequency compared to the input frequency. This is much easier to filter at 2:1.

3D146 A capacitor is sometimes placed in series with the primary of a power transformer to:
    A. Improve the power factor.
    B. Improve output voltage regulation.
    C. To rectify the primary windings.
    D. None of these.
ANSWER A: The ratio of true-to-apparent power in a circuit is known as the *power factor*. Power factor is a comparison of the power that is apparently being used in a circuit to the power the circuit is actually using. A capacitor placed in series with the primary of a power transformer improves the power factor.

3D147 What is the maximum allowable secondary voltage of a center-tapped transformer used in a full-wave rectifier with tubes having a peak inverse voltage rating of 10,000 volts?
    A. 5,000 volts
    B. 7,070 volts

C. 10,000 volts

D. 14,140 volts

ANSWER B: The full peak-to-peak voltage of the secondary will appear as peak inverse voltage across the rectifier. Therefore, since RMS voltage = 0.707 × peak voltage, the maximum secondary voltage is:

0.707 × 10,000 volts = 7070 volts.

3D148 A 3-horsepower, 100-Vdc motor is 85% efficient when developing its rated output. What is the current?

A. 8.545 amps

B. 20.345 amps

C. 26.3 amps

D. 25 amps

ANSWER C: The answer is C. The current is 26.3 amps. Here's the solution: The motor is rated at 3 horsepower; therefore, since:

$$1 \text{ Hp} = 746 \text{ watts}$$
$$3 \text{ Hp} = 3 \times 746 = 2{,}238 \text{ watts is } P_{OUT}$$

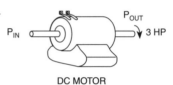

DC MOTOR

If the motor is only 85% efficient then the power input, P_{IN}, must be

$$P_{IN} \times 0.85 = P_{OUT}$$
$$P_{IN} = \frac{P_{OUT}}{0.85} = \frac{2238}{0.85} = 2632.9 \text{ watts}$$

To find current with V_{IN} = 100 VDC

$$V_{IN} \times I_{IN} = P_{IN}$$
$$I_{IN} = \frac{P_{IN}}{V_{IN}} = \frac{2632.9}{100} = 26.3 \text{ amps}$$

3D149 What is the line current of a 7-horsepower motor operating on 120 volts at full load, a power factor of 0.8, and 95% efficient?

A. 4.72 amps

B. 13.03 amps

C. 56 amps

D. 57.2 amps

ANSWER D: First convert the rated horsepower to output watts by multiplying 7 × 746 watts to get 5222 watts. Divide that by the power factor of 0.8 to get 6527.5 watts. Next, divide 6527.5 by 0.95, the efficiency, to get the actual power of 6871.05 watts. Finally, divide 6871.05 by the full load voltage (120 volts) to get the actual line current of 57.2 amps.

3D150 The second harmonic of a 380 kHz frequency is:

A. 2.

B. 190 kHz.

C. 760 kHz.

D. 144.4 GHz.

ANSWER C: To calculate the second harmonic of 380 kHz, simply multiply it by 2, and you get 760 kHz.

3D151 In a self-biased RF amplifier stage: plate voltage = 1250 volts, plate current = 150 mA, grid current = 15 mA, and grid-leak resistance = 4000 ohms. What is the operating grid bias voltage?
- A. 30 volts
- B. 60 volts
- C. 187.5 volts
- D. 540 volts

ANSWER B: In this problem, disregard the plate voltage and plate current. Use the following formula: voltage (grid bias voltage) = current (grid current) × resistance (grid-leak resistance). Don't forget to covert mA to amps before applying the formula: 0.015 amps × 4000 ohms = 60 volts.

3D152 What would be the dB change in field intensity at a given distance if the transmitter output power is doubled?
- A. 1.5 dB
- B. 6 dB
- C. 2 dB
- D. 3 dB

ANSWER A: Twice the current in an antenna quadruples the transmitter power output (I²R) and doubles the field strength at a distant point. Doubling the transmitter power output increases the field strength by √2 or 1.414, about 1.5 dB — Answer A.

3D153 If a field strength is 100 microvolts per meter at 100 miles, what is the field strength at 200 miles?
- A. 10 microvolts
- B. 25 microvolts
- C. 50 microvolts
- D. 150 microvolts

ANSWER C: As you travel further away from the antenna, field strength is inversely proportional to distance. Because the second distance is twice the original distance, field strength is cut in half to 50 microvolts. This holds true with medium frequencies, but there is greater attenuation of signals at higher frequencies.

3D154 What is the purpose of a multiplier resistor used with a voltmeter?
- A. A multiplier resistor is not used with a voltmeter.
- B. It is used to increase the voltage indicating range of the voltmeter.
- C. It is used to decrease the voltage indicating range of the voltmeter.
- D. It is used to increase the current indicating range of an ammeter not a voltmeter.

ANSWER B: The old Simpson 260 multimeter uses multiplier resistors to increase the voltage indicating range of the voltmeter part of the multimeter.

3D155 What is the purpose of a shunt resistor used with an ammeter?
- A. A shunt resistor is not used with an ammeter.
- B. It is used to increase the ampere indicating range of the ammeter.
- C. It is used to decrease the ampere indicating range of the ammeter.
- D. It is used to increase the voltage indicating range of the voltmeter not the ammeter.

ANSWER B: In the older Simpson meter, a shunt resistor increases the ampere indicating range of the ammeter by shunting some of the current around the meter movement, rather than through it.

3D156 The product of the readings of an ac voltmeter and ac ammeter is called:
   A. Apparent power.
   B. True power.
   C. Power factor.
   D. Current power.
ANSWER A: If you multiply ac volts by ac amps, the result is called *apparent power*. In this question, you must assume there is no phase angle.

3D157 An ac ammeter indicates:
   A. Effective (TRM) values of current.
   B. Effective (RMS) values of current.
   C. Peak values of current.
   D. Average values of current.
ANSWER B: An ac ammeter shows the effective [root-mean-square (RMS)] value of current in a circuit.

3D158 How may the range of a thermocouple ammeter be increased?
   A. By using a current transformer.
   B. By using a capacitor shunt.
   C. By using a current transformer and a capacitor shunt.
   D. By using a resistor shunt.
ANSWER C: You can increase the range of a thermocouple ammeter by using a current transformer and a capacitor shunt. Use the thermocouple ammeter for RF power output measurements and calculations.

3D159 By what factor must the voltage of an ac circuit, as indicated on the scale of an ac voltmeter, be multiplied to obtain the average voltage value?
   A. 0.707
   B. 0.9
   C. 1.414
   D. 3.14
ANSWER B: To calculate average voltage value when monitoring the voltage on the scale of an ac voltmeter, which measures the RMS value, multiply the actual voltage reading by 0.9 because $V_{AVG} = 0.9\ V_{RMS}$.

3D160 By what factor must the voltage of an ac circuit, as indicated on the scale of an ac voltmeter, be multiplied to obtain the peak voltage value?
   A. 0.707
   B. 0.9
   C. 1.414
   D. 3.14
ANSWER C: To calculate peak voltage from the voltage indicated on an ac voltmeter, which measures V_{RMS}, multiply by 1.414, because $V_{pk} = 1.414\ V_{RMS}$.

3D161 What is the energy consumed by a radio receiver drawing 50 watts of power for 10 hours?

A. 500 joules
B. 30000 joules
C. 1800000 joules
D. 30000 Ws

ANSWER C: One joule = 1 watt-second. First, convert the 10 hours to seconds. 60 minutes in an hour times 60 seconds in a minute gives 3600 seconds in an hour. Multiply 3600 × 10 hours to get 36,000 seconds, which times 50 watts gives 1,800,000 joules of energy.

3D162 What is the maximum rated current-carrying capacity of a resistor marked "2000 ohms, 200 watts"?

A. .316 amps
B. 3.16 amps
C. 10 amps
D. 100 amps

ANSWER A: The formula here is $P = I^2R$ or $I^2 = P \div R$. This gives us 200 ÷ 2000 = 0.1. Then, take the square root of 0.1 to get answer...0.316 amps.

OHM'S LAW   POWER

Unknown V   Unknown I   Unknown R   Power Circle

Ohm's Law and Power

3D163 A radio receiver rated at 50 watts draws 1.5 amps from the line. The effective resistance is:

A. 8.66 ohms.
B. 22.2 ohms.
C. 33.3 ohms.
D. 1.11 kilohms.

ANSWER B: The formula here is $P \div I^2 = R$. *Plug in* the values: 50 ÷ 2.25 = 22.22 ohms. Don't forget you have to square the 1.5 amps.

3D164 What is the maximum voltage that may be connected across a 20-watt, 2000-ohm resistor?

A. 10 volts
B. 100 volts
C. 200 volts
D. 10,000 volts

ANSWER C: Your formula here is $P \times R = E^2$. When you insert the values for this problem, this is what you have: 20 watts × 2000 ohms = 40,000. Then, take the square root of 40,000 to get the answer...200 volts.

3D165 What is the resistance of a 60-watt, 117-volt lamp?

A. 1.95 ohms
B. 2.76 ohms
C. 30.8 ohms
D. 228 ohms

ANSWER D: The formula is $E^2 \div P = R$. Square 117 volts to get 13,689. Then:

13689 ÷ 60 watts = 228.15 ohms.

3D166 How much energy is used in a week by a 117-volt clock having an internal resistance of 5000 ohms?
A. 19.2 watt-hours
B. 134 watt-hours
C. 460 watt-hours
D. 27.6 kilowatt-hours

ANSWER C: Because the current is not given, you must solve for it first, using

$E \div R = I$ (117 ÷ 5000 = 0.0234). E is in volts, R in ohms, and I in amperes.

Then, apply the formula for power, Volts × Amps = Power (117 × 0.0234 = 2.7378 watt). Finally, calculate the watt-hours by multiplying 2.7378 watts × 24 hours × 7 days to get 459.95 watt-hours.

3D167 If .8 coulombs pass a point in a circuit in .8 seconds, what is the average current value?
A. 1 ampere
B. .640 amperes
C. 1.28 amperes
D. .414 amperes

ANSWER A: Your formula is coulombs ÷ time = current. The answer is 0.8 coulombs ÷ 0.8 sec = 1 amp. Remember, amperes = coulombs per second.

3D168 How much power is developed when 117 volts forces 11.7 coulombs through a point in a circuit in 1.17 seconds?
A. 8.55 watts
B. 11.7 watts
C. 1.17 k watts
D. 1.6 kilowatts

ANSWER C: Because the current (amps) is not given, first solve for amps using the formula: Coulombs ÷ Seconds = Amps (11.7 Coulombs ÷ 1.17 Sec = 10 Amps). Then, apply the formula for power, $E \times I = P$ (117 Volts × 10 Amps = 1170 watts). Convert watts to kilowatts and you get 1.170 kW.

3D169 A 20-ohm, a 30-ohm, and an unknown-value resistor are connected in series across a 140-volt source with .5 amps flowing through the circuit. What is the unknown resistor's size?
A. 40 ohms
B. 90 ohms
C. 115 ohms
D. 230 ohms

ANSWER D: First use the formula $E \div I = R$ to figure the total resistance (140 ÷ 0.5 = 280 ohms). Because the total resistance of resistors in series is simply the sum of the resistances, all you have to do is subtract the known resistance (20 + 30 = 50) from the total resistance to find the value of the third resistor (280 ohms − 50 ohms = 230 ohms).

3D170 A 12-volt automotive battery with an internal resistance of .2 ohms is connected to a 2-ohm headlight lamp. What is the amount of current passing through the lamp?

A. 1.2 amperes
B. 5.45 amperes
C. 6 amperes
D. None of these

ANSWER B: Begin by combining the two resistances (2.0 + 0.2 = 2.2). Then apply the formula $E \div R = I$ to find the current (12 Volts ÷ 2.2 ohms = 5.45 amps).

3D171 What is the conductance of a circuit having three 300-ohm and two 200-ohm resistors connected in parallel?

A. 20 mS
B. 50 ohms
C. 8.33 mS
D. 120 ohms

ANSWER A: Conductance is equal to 1/R. In this case, R is the equivalent resistor of the resistors in parallel. The three 300-ohm resistors in parallel are equivalent to 100 ohms. The two 200-ohm resistors in parallel are equivalent to 100 ohms. Two 100-ohm equivalent resistances in parallel result in the total equivalent resistance which is 50 ohms. The conductance is then $1 \div 50 = 0.02$ mhos. The units of conductance are mhos, also called siemens. $0.02 = 20 \times 10^{-3}$ siemens, therefore, answer A is correct— 20 millisiemens.

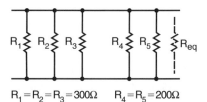

$R_1 = R_2 = R_3 = 300\Omega \quad R_4 = R_5 = 200\Omega$

3D172 A 500-ohm, 2-watt resistor and a 1500-ohm, 1-watt resistor are connected in parallel. What is the maximum voltage that can be applied across the parallel circuit without exceeding wattage ratings?

A. 22.4 volts
B. 31.6 volts
C. 38.7 volts
D. 875 volts

ANSWER B: To solve, you must test each resistor to determine the maximum voltage that can be applied before its power rating is exceeded. Use the formula $P \times R = E^2$. For the 500-ohm resistor, 2 × 500 = 1000. The square root of 1000 gives the maximum voltage of 31.6 volts. For the 1500-ohm resistor, 1 × 1500 = 1500. The square root of 1500 gives a maximum voltage of 38.72. So the 31.6 V for the 500-ohm resistor limits the voltage value.

★ 3D173 A 20-ohm and 30-ohm resistor form a parallel circuit connected to a 12-volt source with an internal resistance of .2 ohms. Which of the following correctly represents Kirchhoff's voltage laws for the circuit in question assuming A is the current through the 30-ohm resistor, and B is the current through the 20-ohm resistor *and C is the total current?*

A. 12 − .2A − 20A − 30B = 0.
B. 12 − .2A − 20B = 0.
C. 12 − .2A − 30A = 0.
D. 12 − .2C − 20B= 0 and 12 − .2C − 30A = 0.

ANSWER D: Assuming A is current through the 30-ohm resistor, B is the current through the 20-ohm resistor and C = A + B, the circuit is as follows:

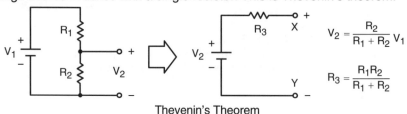

A is the current in Branch A
B is the current in Branch B
C is the total current

Kirchhoffs' Voltage Law: The sum of the voltages around a closed circuit loop is equal to zero.

Loop 1: $12 - 0.2C - R_2A = 0$
$12 - 0.2C - 30A = 0$
Loop 2: $12 - 0.2C - R_1B = 0$
$12 - 0.2C - 20B = 0$

By using C = A + B and the two voltage equations, A, B and C can be solved for: C = 0.983 A, A = 0.393 A, and B = 0.59 A. Check: C = 12V/12.25Ω = 0.984A. *[Ed. Note: Text in italics is added for understanding.]*

3D174 If a complex circuit is reduced to an equivalent circuit consisting of a single voltage source in series with a single resistor, this is an example of:
A. Norton's theorem.
B. Thevenin's theorem.
C. Ohm's law.
D. Kirchoff's law.

ANSWER B: You can replace a complex series/parallel circuit with a single voltage source in series with a single resistor. This is Thevenin's theorem.

$$V_2 = \frac{R_2}{R_1 + R_2} V_1$$

$$R_3 = \frac{R_1 R_2}{R_1 + R_2}$$

Thevenin's Theorem

3D175 If a complex circult is reduced to an equivalent circuit consisting of a single current source in parallel with a single resistor, this is an example of:
A. Norton's theorem.
B. Thevenin's theorem.
C. Ohm's law.
D. Kirchoff's law.

ANSWER A: To calculate a single current source in parallel with a single resistor, use Norton's theorem.

3D176 If a 20-ohm resistor, a 30-ohm resistor, and a 12-volt source with an internal resistance of 2 ohms are connected in a parallel circuit arrangement, how much current will flow from the source?
A. 0.240 amperes
B. 0.857 amperes
C. 0.750 amperes
D. 1.000 amperes

ANSWER B: Ohm's Law here again: current equals voltage divided by resistance. Solve for total resistance of two resistances in parallel: $(20 \times 30) \div (20 + 30) =$ 12 ohms. Then, add the internal 2 ohms resistance, and use 14 ohms as the total resistance. $E \div R = I$, therefore:

12 volts $\div$ 14 ohms = 0.857 amps

3D177 If a current flowing through a coil produces 200 lines of force in the core of the coil, it can be said that the core has:
  A. Flux of 14.14 lines.
  B. Flux of 100 lines.
  C. 200 gauss.
  D. 200 maxwells.
ANSWER D: One maxwell = 1 magnetic field line. So, 200 field lines yields 200 maxwells.

3D178 If the core of a coil is 2 centimeters squared and has 200 lines of force in the core, it can be said that the core has:
  A. Flux of 100.
  B. Flux of 400.
  C. 100 gauss.
  D. 100 maxwells.
ANSWER C: One line per square centimeter yields 1 gauss. 200 lines divided by 2 centimeters yields 100 gauss, answer C.

3D179 If a 50-turn coil has 2 amperes of current flowing through it, one can say it has:
  A. 126 gilberts.
  B. 100 flux.
  C. 100 gauss.
  D. 7.6 oersted.
ANSWER A: mmf (in gilberts) = 4πNI/10, where N = number of turns and I = current in amperes. If N = 1 and I = 1 (one ampere turn), then mmf = $4 \times 3.14/10 =$ 1.256 gilberts. One ampere turn produces 1.256 gilberts; therefore, for this problem, 100 ampere turns ($50 \times 2 = 100$) produces $100 \times 1.256 = 125.6$ gilberts.

3D180 If a 50 turn coil has 2 amperes of current flowing through it and a core length of 2 inches, one can say it has:
  A. 24.75 gilberts.
  B. 24.75 flux.
  C. 24.75 gauss.
  D. 24.75 oersted.
ANSWER D: First work on the core length. 2.54 centimeters $\times$ 2 inches = 5.08 centimeters. Field intensity, H, is in oersteds. The formula for the field intensity of a coil is:

$$H = \frac{4\pi NI}{10l}$$

Where: H = Field intensity in **Oersteds**
         N = Number of **turns**
         I = Current in **amperes**
         l = Length in **centimeters**
        4π = A constant, 4×3.14

$$H = \frac{4\pi \times 50 \times 2}{10 \times (2 \times 2.54)}$$

$$H = \frac{12.56 \times 50 \times 2}{50.8}$$

$$H = 24.72 \text{ oersteds}$$

3D181 In comparing an electric circuit with a magnetic circuit, volts is similar to ____, resistance is similar to ____, and amperes is similar to ____.

    A. Gilberts, reluctance, maxwells
    B. Reluctance, gilberts, maxwells
    C. Gilberts, maxwells, reluctance
    D. Maxwell, reluctance, gilberts

ANSWER A: In comparing an electric circuit with a magnetic circuit, gilberts are similar to volts, reluctance similar to resistance, and maxwells similar to amperes.

3D182 When magnetic lines are cut at a rate of ____ lines per second, an average EMF of 1 volt is produced.

    A. 10,000
    B. 100,000
    C. 10,000,000
    D. 100,000,000

ANSWER D: One weber is equal to 10^8 lines of flux. When one weber per second of flux is cut by a wire one volt is induced in the wire. One volt = 1 weber ÷ 1 second. The answer is 10^8 = 100,000,000.

3D183 Which of the following would shield a permanent-magnet field best?

    A. Copper
    B. Iron
    C. Lead
    D. Aluminum

ANSWER B: Iron is a permeable substance of magnetically soft material that best shields a permanent magnet field because it provides an easy path for the flux.

3D184 When induced currents produce expanding magnetic fields around conductors in a direction that opposes the original magnetic field, this is known as:

    A. Gilbert's law.
    B. Lenz's law.
    C. Maxwells' law.
    D. Norton's law.

ANSWER B: It is Lenz's law that describes induced currents producing expanding magnetic fields around conductors in a direction that actually opposes the original magnetic field.

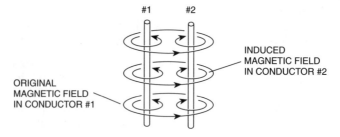

Illustration of Lenz's Law

3D185 At 240 degrees, what is the amplitude of sine-wave having a peak value of 5 volts?

 A. −4.3 volts
 B. −2.5 volts
 C. +2.5 volts
 D. +4.3 volts

ANSWER A: To work out the next few problems, I suggest you draw a complete sine-wave and note the degrees and recall the following formula $V = 5 \sin \Theta$.

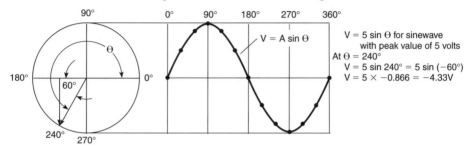

$V = 5 \sin \Theta$ for sinewave with peak value of 5 volts
At $\Theta = 240°$
$V = 5 \sin 240° = 5 \sin (−60°)$
$V = 5 \times −0.866 = −4.33V$

Vector of 5V Generates Sine Wave as It Rotates

3D186 At 150 degrees, what is the amplitude of a sine-wave having a peak value of 5 volts?

 A. −4.3 volts
 B. −2.5 volts
 C. +2.5 volts
 D. +4.3 volts

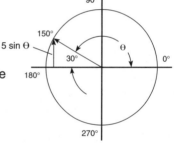

ANSWER C: Now the angle is 150°. The amplitude of the Vector $V = 5 \sin \Theta$ is now:

$V = 5 \sin 30°$
$= 5 \times 0.5 = 2.5 \ V$

3D187 At pi/3 radians, what is the amplitude of a sine-wave having a peak value of 5 volts?

 A. −4.3 volts
 B. −2.5 volts
 C. +2.5 volts
 D. +4.3 volts

ANSWER D: A radian is the angle within a circle whose arc length along the circumference is equal to the radius. There are 2π radians in a circle; therefore π radians = 180°. For this problem the angle is

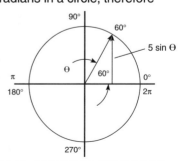

$$\frac{\pi}{3} = \frac{180°}{3} = 60°.$$

therefore, $V = 5 \sin 60°$
$V = 5 \times 0.866 = 4.33V$

3D188 If 4 amperes of current is flowing at 60 degrees, how much will flow at 150 degrees?
  A. 2.0 amperes
  B. 2.3 amperes
  C. 3.5 amperes
  D. 4.0 amperes

ANSWER B: This question relates to the three pervious questions, but now instead of voltage its a current sine wave I = A sin Θ. For this problem, 4 = A sin 60° where A is the peak value of I.

$$A = 4 \div \sin 60° = 4 \div 0.866 = 4.62$$

At 150°, I = 4.62 × sin 30° = 4.62 × 0.5 = 2.31 amps.

3D189 If 4 amperes of current is flowing at 30 degrees, how much will flow at 120 degrees?
  A. 2.3 amperes
  B. 4.0 amperes
  C. 6.9 amperes
  D. 8.0 amperes

ANSWER C: The same as question 3D188, but now I = A sin Θ equals 4 = A sin 30°. Thus, A = 4 ÷ sin 30° = 4 ÷ 0.5 = 8 at 120°, I = 8 sin 60° = 8 × 0.866

$$I = 6.93 \text{ amps.}$$

3D190 The frequency that is 2 octaves higher than 1000 Hz is:
  A. 2000 Hz.
  B. 3000 Hz.
  C. 4000 Hz.
  D. 5000 Hz.

ANSWER C: An octave represents twice the frequency. The first octave up from 1000 Hz is 2000 Hz, and the second octave is 2 × 2000 = 4000 Hz.

3D191 Which of the following would be considered a transducer?
  A. Light emitting diode
  B. Vacuum tube filament
  C. Speaker
  D. Microphone

ANSWER D: Technically, the light-emitting diode, the filament, and a speaker are also considered transducers, but for this specific question pool question, select answer D, the microphone, as the most obvious choice.

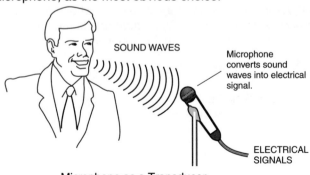

SOUND WAVES

Microphone converts sound waves into electrical signal.

ELECTRICAL SIGNALS

Microphone as a Transducer

3D192 What is the inductance in microhenries of a coil having a diameter of 1/2 inch, a length of 1 inch, and 200 turns?
    A. 25 microhenries
    B. 204 microhenries
    C. 305 microhenries
    D. 503 microhenries
ANSWER B: The *ARRL Handbook* gives the following formula for the approximate inductance of a single-layer air-core coil:

$$L = \frac{d^2 n^2}{18d + 40l}$$
    Where: d = diameter of coil in **inches**
                       n = number of **turns**
                       l = coil length in **inches**
                       L = inductance in **microhenries**

When d = 0.5″, l = 1.0″, n = 200, then,

$$L = \frac{d^2 n^2}{18d + 40l} = \frac{(0.5)^2 (200)^2}{18(0.5) + 40(1)} = 204 \ \mu H$$

3D193 If a 10-henry coil has 50 mA of current flowing through it, how much energy is stored in the magnetic field?
    A. 0.0125 Joules
    B. 12.5 Joules
    C. 0.500 Joules
    D. 500 Joules
ANSWER A: Our formula is $W = I^2 L \div 2$, where W is joules of energy, I is amps of current, and L is henrys of inductance. For this problem I = 0.05 A and L = 10 h, therefore $W = 0.05^2 \times 10 \div 2 = 0.0125$

3D194 If two coils are close enough to have a mutual inductance of 0.2 H and the coils have inductances of 2 H and 8 H, what is the coefficient of coupling?
    A. 2.5 %
    B. 5.0 %
    C. 10 %
    D. 25 %
ANSWER B: The coefficient (k) of coupling is represented by the formula:

$$k = \frac{M}{\sqrt{L_1 \times L_2}} = \frac{0.2}{\sqrt{2 \times 8}} = 0.05 (5\%)$$

Where: k = Coefficient of coupling in **units**
               M = Mutual inductance in **henries**
           L_1, L_2 = Inductance in **henries**

3D195 If a power transformer has a primary voltage of 120 Vac, a secondary voltage of 12 volts, and an efficiency of 95% when delivering 2 amperes of secondary current, what is the value of primary current?
    A. 190 milliamperes
    B. 200 milliamperes
    C. 211 milliamperes
    D. 2 amperes

ANSWER C: The efficiency formula is power output = 0.95 × power input. First, you must solve for power in watts on the secondary side (12 volts × 2 amps = 24 watts). Then, modify the formula to solve for power input (24 watts ÷ 0.95 = 25.26 watts power input)

Finally use the formula P ÷ E = I to get the current on the primary side (25.26 watts ÷ 120 volts = 0.211 amps). Move the decimal three place to the right to cover amperes to milliamperes, and you have the answer.

3D196 What is the capacitance of a capacitor with individual plate area of 2 square inches each, an air dielectric, and spacing between the two parallel plates of .01 inches?
   A. 45 picofarads
   B. 45 microfarads
   C. 200 picofarads
   D. 200 microfarads

ANSWER A: The formula for capacitance for a parallel plate capacitor is:

$$C = K\epsilon \frac{A}{d} \epsilon_o (n-1)$$

Where: C = Capacitance in **farads**
   A = Area in **square meters**
   d = Distance between plates in **meters**
   ϵ_o = permitivity of air (8.85×10^{-12} farads/m)
   $K\epsilon$ = Dielectric constant (1 for air)
   n = **Number** of parallel plates

If the conversion factors are included so that dimensions are in **inches** instead of **meters**, the formula is:

▶ $C = K\epsilon \, 0.225 \frac{A}{d} (n-1)$ in **pf**
   C = Capacitance in **picofarads**
   A = Area in **square inches**
   d = Distance between plates in **inches**

The problem solution is:

$$C = 1 \times 0.225 \frac{2}{0.01} (2-1) = \frac{4.5 \times 10^{-1}}{1 \times 10^{-2}} = 45 \text{ picofarads}$$

3D197 What is the capacitance of a capacitor with individual plate area of 1 square inch each, an air dielectric, and spacing between the two parallel plates of .01 inches?
   A. 2.25 picofarads
   B. 22.5 picofarads
   C. 2.25 microfarads
   D. 22.5 microfarads

ANSWER B: Using the formula from the previous question, C = 0.225(1) ÷ 0.01(2-1) = 22.5 picofarads.

3D198 Which of the following materials has a dielectric strength of approximately 80 volts?
   A. Air
   B. Bakelite
   C. Glass
   D. Mica

ANSWER A: This problem lacks the stated thickness between the two metal structures. In this case, choose air as having the best dielectric—all of the other answers have a dielectric slightly less than that of air.

3D199 The electron difference between the plates of a 5-microfarad capacitor connected to 10 Vdc will be:
A. 0.000005 coulombs
B. 0.00005 coulombs
C. 31.2 Terra electrons
D. 3.12 Terra electrons

ANSWER B: Electron difference in this question is charge. To determine a charge in coulombs where capacitance and voltage are given,

Q = CV	Where: Q = Charge in **coulombs**
Q = 5 × 10⁻⁶ × 10	C = Capacitance in **farads**
= 5 × 10⁻⁵	V = Voltage in **volts**
= 0.00005 coulombs, answer B.	

$$Q = CV$$
$$Q = 5 \times 10^{-6} \times 10$$
$$= 5 \times 10^{-5}$$
$$= 0.00005 \text{ coulombs, answer B.}$$

Where:
Q = Charge in **coulombs**
C = Capacitance in **farads**
V = Voltage in **volts**

Watch out for answer A—it looks good, but the decimal point is not correct.

3D200 If a series circuit consist of an inductor with an inductive reactance of 57.7 ohms and a resistance of 100 ohms, the phase angle between voltage and current will be approximately:
A. 30 degrees.
B. 35 degrees.
C. 55 degrees.
D. 60 degrees.

ANSWER A: Arctan Θ (phase angle) means an angle whose tangent is Θ. To calculate phase angle, the formula is:

Θ (Phase angle) = arctan (X_L/R)
Θ (Phase angle) = arctan $(57.7/100)$
Θ (Phase angle) = arctan (0.577)
Θ (Phase angle) = 29.98

3D201 If a series circuit consist of an inductor with an inductive reactance of 100 ohms and a resistance of 57.7 ohms, the phase angle between voltage and current will be approximately:
A. 30 degrees
B. 35 degrees
C. 55 degrees
D. 60 degrees

ANSWER D: Just like the previous question. Phase angle = arctan $(100 \div 57.7)$.
Arctan (1.733)
Arctan = 60

3D202 Approximately what capacitance value is needed to resonate a 2.5 millihenry coil to 2.146 MHz?
A. 2.2 picofarads
B. 2.2 microfarads
C. 87 picofarads
D. 87 microfarads

ANSWER A: Here is a problem that was a classic amateur radio examination question dealing with the capacitance value needed to resonate a coil at a specific frequency. The formula for resonance is:

$$f_r = \frac{1}{2\pi\sqrt{LC}}$$

Where: f_r = resonant frequency in **hertz**
L = Inductance in **henries**
C = Capacitance in **farads**
π = Constant equal to 3.14

If you solve for C, then

$$C = \frac{1}{(2\pi)^2 L f_r^2}$$

L must be in henries and f in hertz, when you plug in the values. Therefore,

$$C = \frac{1}{(2 \times 3.14)^2 \times 2.5 \times 10^{-3} \times (2.146 \times 10^6)^2}$$

$$C = \frac{1}{2.5 \times 10^{-3} \times 39.4 \times 4.61 \times 10^{12}}$$

$$C = \frac{1}{454.52 \times 10^{+9}}$$

$$C = \frac{1}{4.55 \times 10^{+11}}$$

$$C = 0.220 \times 10^{-11} = 2.20 \times 10^{-12}$$

$$C = 2.20 \text{ picofarads}$$

Subelement 3E – Circuit Components   (10 questions)

3E1 Structurally, what are the two main categories of semiconductor diodes?
  A. Junction and point contact
  B. Electrolytic and junction
  C. Electrolytic and point contact
  D. Vacuum and point contact
ANSWER A: The semiconductor diode is a common component found in mobile and base station radio equipment. Depending on how the diode is biased, it may look like an open circuit or a closed circuit. The junction and point contact are two main categories of semiconductor diodes.

3E2 What are the two primary classifications of Zener diodes?
  A. Hot carrier and tunnel
  B. Varactor and rectifying
  C. Voltage regulator and voltage reference
  D. Forward and reversed biased
ANSWER C: The Zener diode is found in both handheld transceivers as well as mobile radio equipment. The Zener may be used for voltage regulation, or as a voltage reference.

3E3 What is the principal characteristic of a Zener diode?
  A. A constant current under conditions of varying voltage
  B. A constant voltage under conditions of varying current
  C. A negative resistance region
  D. An internal capacitance that varies with the applied voltage
ANSWER B: Remember that a Zener diode is for voltage regulation, for constant voltage even though current changes.

3E4 What is the range of voltage ratings available in Zener diodes?
  A. 2.4 volts to 200 volts
  B. 1.2 volts to 7 volts
  C. 3 volts to 2000 volts
  D. 1.2 volts to 5.6 volts
ANSWER A: There is only one answer with two 2's in it, so 2.4 volts to 200 volts is the range of ratings of Zener diodes.

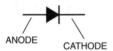

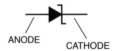

Here is the schematic symbol of a diode. Current will only flow ONE WAY in a diode. You can remember this diode diagram as a one-way arrow (key words).

Here is the schematic symbol of a Zener diode. Since a diode only passes energy in one direction, look for that one-way arrow, plus a "Z" indicating it is a Zener diode. Doesn't that vertical line look like a tiny "Z".

Semiconductor Diode   Zener Diode

3E5 What is the principal characteristic of a tunnel diode?

A. A high forward resistance
B. A very high PIV (peak inverse voltage)
C. A negative resistance region
D. A high forward current rating

ANSWER C: As a tunnel diode is conducting current, there is a spot where current increases as the voltage drop across the diode decreases. They call this a negative resistance region.

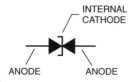

INTERNAL
CATHODE

ANODE   ANODE

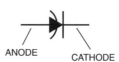

ANODE   CATHODE

Here is the schematic symbol of a tunnel diode. Memorize the schematic symbol for a tunnel diode. It has arrow-heads tip to tip, as if they are stuck in a tunnel.

Here is the schematic symbol for a varactor diode. The internal capacitance of a varactor diode varies as the voltage applied to its terminals changes. Your author builds microwave equipment with varactor diodes, so be sure to identify the proper symbol. The symbol is essentially a capacitor and a diode combined.

Tunnel Diode   Varactor Diode

3E6 What special type of diode is capable of both amplification and oscillation?

A. Point contact diodes
B. Zener diodes
C. Tunnel diodes
D. Junction diodes

ANSWER C: The tunnel diode is commonly used in both amplifier and oscillator circuits. The negative resistance region is particularly useful in oscillators.

3E7 What type of semiconductor diode varies its internal capacitance as the voltage applied to its terminals varies?

A. A varactor diode
B. A tunnel diode
C. A silicon-controlled rectifier
D. A Zener diode

ANSWER A: We use the varactor diode to tune VHF and UHF circuits by varying the voltage applied to the varactor diode. The symbol is shown at question 3E5.

3E8 What is the principal characteristic of a varactor diode?

A. It has a constant voltage under conditions of varying current.
B. Its internal capacitance varies with the applied voltage.
C. It has a negative resistance region.
D. It has a very high PIV(peak inverse voltage).

ANSWER B: Change the voltage, and the diode's internal capacitance will vary.

3E9 What is a common use of a varactor diode?
A. As a constant current source
B. As a constant voltage source
C. As a voltage controlled inductance
D. As a voltage controlled capacitance

ANSWER D: Change the voltage, and the internal capacitance will change—this is very useful in varying the resonant frequency of tuned circuits.

3E10 What is a common use of a hot-carrier diode?
A. As balanced inputs in SSB generation
B. As a variable capacitance in an automatic frequency control circuit
C. As a constant voltage reference in a power supply
D. As VHF and UHF mixers and detectors

ANSWER D: If your job requires servicing VHF and UHF equipment, you may find a circuit with a hot carrier diode because of its low noise-figure characteristics.

3E11 What limits the maximum forward current in a junction diode?
A. The peak inverse voltage(PIV)
B. The junction temperature
C. The forward voltage
D. The back EMF

ANSWER B: Guess what will kill most electronic components? High temperature! The limit of the maximum forward current in a junction diode is the junction temperature. This is why you will find these diodes mounted to the chassis of your equipment. The chassis acts as a heat sink to keep the junction temperature from exceeding its maximum limit.

3E12 How are junction diodes rated?
A. Maximum forward current and capacitance
B. Maximum reverse current and PIV(peak inverse voltage)
C. Maximum reverse current and capacitance
D. Maximum forward current and PIV(peak inverse voltage)

ANSWER D: When choosing a junction diode, you will need to know how much forward current will be passing through the diode, and the amount of peak inverse voltage (PIV) that the diode must stand in its non-conducting direction.

A ROW OF
TINY GLASS
SIGNAL DIODES

Diodes Mounted on a Circuit Board

3E13 What is a common use for point contact diodes?
A. As a constant current source
B. As a constant voltage source
C. As an RF detector
D. As a high voltage rectifier

ANSWER C: The little point contact diode, since it only conducts in one direction, may be used as an RF detector in some VHF and UHF equipment.

3E14 What type of diode is made of a metal whisker touching a very small semiconductor die?
A. Zener diode
B. Varactor diode
C. Junction diode
D. Point contact diode

ANSWER D: Your author remembers the early crystal radios. They used a point-contact diode made of a metal whisker touching sensitive areas of a very small semiconductor surface.

3E15 What is one common use for PIN diodes?
A. As a constant current source
B. As a constant voltage source
C. As an RF switch
D. As a high voltage rectifier

ANSWER C: PIN diodes may be used in small handheld transceivers for RF switching. This gets away from mechanical relays for switching.

3E16 What special type of diode is often used in RF switches, attenuators, and various types of phase shifting devices?
A. Tunnel diodes
B. Varactor diodes
C. PIN diodes
D. Junction diodes

ANSWER C: The PIN diode is a high-speed switching diode with an intrinsic layer of semiconductor between the P and N junction, thus, PIN. It may also be used in attenuators, along with RF switching.

3E17 What are the three terminals of a bipolar transistor?
A. Cathode, plate and grid
B. Base, collector and emitter
C. Gate, source and sink
D. Input, output and ground

ANSWER B: Almost like the alphabet, B for base, C for collector, (skip D), and E for emitter.

PNP Transistor   NPN Transistor

ROW OF
TRANSISTORS

Transistors Mounted on a Circuit Board

3E18 What is the meaning of the term alpha with regard to bipolar transistors?

A. The change of collector current with respect to base current
B. The change of base current with respect to collector current
C. The change of collector current with respect to emitter current
D. The change of collector current with respect to gate current

ANSWER C: In bipolar transistors, the term "alpha" is the variation of *COLLECTOR current* with respect to *EMITTER current* (CCEC, a key for remembering). This is an important consideration when designing a circuit that will utilize a bipolar transistor.

3E19 What is the term used to express the ratio of change in DC collector current to a change in emitter current in a bipolar transistor?

A. Gamma
B. Epsilon
C. Alpha
D. Beta

ANSWER C: A for alpha, (no B), C for collector current, (no D), E for emitter current. This is as simple as ACE. You will ace your examination!

3E20 What is the meaning of the term beta with regard to bipolar transistors?

A. The change of collector current with respect to base current
B. The change of base current with respect to emitter current
C. The change of collector current with respect to emitter current
D. The change in base current with respect to gate current

ANSWER A: When we talk about the term "beta" for a bipolar transistor, think of beta with a "B" for base current. This is the change of *COLLECTOR* current with respect to *BASE* current. B for base; B for beta. DC beta = $I_c \div I_b$.

3E21 What is the term used to express the ratio of change in the DC collector current to a change in base current in a bipolar transistor?

A. Alpha
B. Beta
C. Gamma
D. Delta

ANSWER B: Bipolar, base current, beta. AC beta = delta $I_c \div$ delta I_b.

3E22 What is the meaning of the term alpha cutoff frequency with regard to bipolar transistors?
- A. The practical lower frequency limit of a transistor in common emitter configuration
- B. The practical upper frequency limit of a transistor in common base configuration
- C. The practical lower frequency limit of a transistor in common base configuration
- D. The practical upper frequency limit of a transistor in common emitter configuration

ANSWER B: When we speak of a cutoff frequency, we speak of the upper frequency limit of a transistor. *Upper frequency limit* (key words)—in this case, for alpha which is the relation of collector current to emitter current in a common base circuit.

3E23 What is the term used to express that frequency at which the grounded base current gain has decreased to 0.707 of the gain obtainable at 1 kHz in a bipolar transistor?
- A. Corner frequency
- B. Alpha cutoff frequency
- C. Beta cutoff frequency
- D. Alpha rejection frequency

ANSWER B: *Grounded base* is the same as a common base configuration. Common base means *alpha*. Since 1 kHz is mentioned in this question, they are referring to the alpha cutoff frequency in a transistor.

3E24 What is the meaning of the term beta cutoff frequency with regard to a bipolar transistor?
- A. That frequency at which the grounded base current gain has decreased to 0.707 of that obtainable at 1 kHz in a transistor
- B. That frequency at which the grounded emitter current gain has decreased to 0.707 of that obtainable at 1 kHz in a transistor
- C. That frequency at which the grounded collector current gain has decreased to 0.707 of that obtainable at 1 kHz in a transistor
- D. That frequency at which the grounded gate current gain has decreased to 0.707 of that obtainable at 1 kHz in a transistor

ANSWER B: In a bipolar transistor, the *beta cutoff frequency* is that frequency at which the *grounded emitter current gain* (key words) has decreased to a certain value at a certain frequency.

3E25 What is the meaning of the term transition region with regard to a transistor?
- A. An area of low charge density around the P-N junction
- B. The area of maximum P-type charge
- C. The area of maximum N-type charge
- D. The point where wire leads are connected to the P- or N-type material

ANSWER A: When a transistor goes into saturation, transition region is the term for an area of low charge density around the P-N junction. When you go into saturation studying these questions and answers, put the book down, and get "recharged" from a *"low charge"* (key words) saturation!

3E26 What does it mean for a transistor to be fully saturated?
A. The collector current is at its maximum value.
B. The collector current is at its minimum value.
C. The transistor's Alpha is at its maximum value.
D. The transistor's Beta is at its maximum value.

ANSWER A: When a transistor is fully saturated, the collector current is maximum.

3E27 What does it mean for a transistor to be cut off?
A. There is no base current
B. The transistor is at its Class A operating point
C. There is no current between emitter and collector
D. There is maximum current between emitter and collector.

ANSWER C: At cutoff, in an NPN circuit, there is no current from emitter to the collector because the base-to-emitter junction is reverse biased.

3E28 What are the elements of a unijunction transistor?
A. Base 1, base 2, and emitter
B. Gate, cathode, and anode
C. Gate, base 1, and base 2
D. Gate, source, and sink

ANSWER A: The two horizontal lines coming out of the unijunction transistor represent *base 1 and base 2* (key words). Watch out for answer C. The angled arrowhead line is an *emitter* (key word).

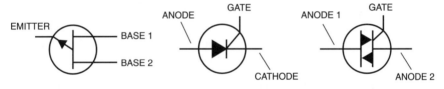

Notice there is an upper and lower line within the unijunction transistor diagram. Think of a uniform—upper and lower—part. There also is only one angled arrowhead line on the opposite side.	The SCR diagram looks a little bit like a switch, doesn't it? The SCR's gate element allows it to control other parts of a circuit. It is a diode with another gate control that turns on or off the SCR.	The triac is best visualized as two diodes connected in parallel, one in one direction, the other in the opposite direction, in a single package. Look at the diagram — this is the way it looks, doesn't it? It is really two SCRs connected head-to-toe.
Unijunction Transistor	Silicon Controlled Rectifier SCR	Bidirectional Triode Thyristor TRIAC

3E29 For best efficiency and stability, where on the load-line should a solid-state power amplifier be operated?
A. Just below the saturation point
B. Just above the saturation point
C. At the saturation point
D. At 1.414 times the saturation point

ANSWER A: Solid-state amplifiers are now common in commercial boat installations where high-power SSB equipment is desired. Most sideband sets run only 100 watts output, so a solid-state power amplifier, operated off your SSB set just below the saturation point, will put out a hefty signal. You must carefully monitor your drive power, as well as the voltage feeding the power amplifier circuit. Overdriving the SSB solid-state power amplifier could cause distortion, and possibly burn out one of the expensive output transistors.

3E30 What two elements widely used in semiconductor devices exhibit both metallic and non-metallic characteristics?
A. Silicon and gold
B. Silicon and germanium
C. Galena and germanium
D. Galena and bismuth

ANSWER B: Both silicon and germanium exhibit a combination of metallic and non-metallic properties. Think of *silicon* rectifiers and *germanium* diodes.

3E31 What are the three terminals of an SCR?
A. Anode, cathode, and gate
B. Gate, source, and sink
C. Base, collector, and emitter
D. Gate, base 1, and base 2

ANSWER A: Like any diode, it has an anode and a cathode, but it also has a special gate element. A control signal on the gate allows current from anode to cathode if the diode is forward biased. See figure at question 3E28.

3E32 What are the two stable operating conditions of an SCR?
A. Conducting and nonconducting
B. Oscillating and quiescent
C. Forward conducting and reverse conducting
D. NPN conduction and PNP conduction

ANSWER A: The SCR may be considered either conducting or not conducting. It's either on or off. Black or white. One or zero.

3E33 When an SCR is in the triggered or on condition, its electrical characteristics are similar to what other solid-state device (as measured between its cathode and anode)?
A. The junction diode
B. The tunnel diode
C. The hot-carrier diode
D. The varactor diode

ANSWER A: When the SCR is on, it's similar to a junction diode that will pass current in one direction.

3E34 Under what operating condition does an SCR exhibit electrical characteristics similar to a forward-biased silicon rectifier?
A. During a switching transition
B. When it is used as a detector
C. When it is gated "off"
D. When it is gated "on"

ANSWER D: When it is in the "on" position, as was stated in the previous question, the SCR looks like a forward-biased silicon junction diode. A rectifier is a junction diode designed to handle power circuitry.

3E35 What is the transistor called which is fabricated as two complementary SCRs in parallel with a common gate terminal?
  A. TRIAC
  B. Bilateral SCR
  C. Unijunction transistor
  D. Field effect transistor
ANSWER A: Since the TRIAC looks like two diodes, it could be compared to two complementary SCR's in parallel with a common gate terminal. See question 3E28.

3E36 What are the three terminals of a TRIAC?
  A. Emitter, base 1, and base 2
  B. Gate, anode 1, and anode 2
  C. Base, emitter, and collector
  D. Gate, source, and sink
ANSWER B: The triac has one gate and two anodes. The angled line is a gate. See figure at question 3E28. Its formal name is bidirectional triode thyristor.

3E37 What is the normal operating voltage and current for a light-emitting diode?
  A. 60 volts and 20 mA
  B. 5 volts and 50 mA
  C. 1.7 volts and 20 mA
  D. 0.7 volts and 60 mA
ANSWER C: The light-emitting diode (LED) is found on many types of marine and aviation two-way radio equipment. The LED draws approximately 20 milliamps of current, and requires only 1.7 volts to illuminate. Many manufacturers are staying away from LED readouts which may be used in an open cockpit of a boat. These are hard to see in the direct sunlight. There is nothing that can be done to make them more brilliant.

3E38 What type of bias is required for an LED to produce luminescence?
  A. Reverse bias
  B. Forward bias
  C. Zero bias
  D. Inductive bias
ANSWER B: The LED requires forward bias in order to illuminate. The LED is a diode, and it must be forward biased to make it give off light.

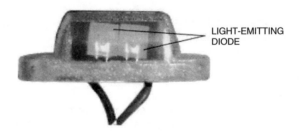

LIGHT-EMITTING DIODE

Two LEDs in a Single Package

3E39 What is the name of the semiconductor IC that has a fixed pattern of digital data stored in its memory matrix?
A. RAM—Random-Access Memory
B. ROM—Read-Only Memory
C. Register
D. Latch

ANSWER B: A ROM—Read-Only Memory is a semiconductor memory IC that has its digital data placed in the circuitry at the time of manufacture.

3E40 What colors are available in LEDs?
A. Yellow, blue, red, and brown
B. Red, violet, yellow, and peach
C. Violet, blue, orange, and red
D. Red, green, orange, and yellow

ANSWER D: Four of the popular colors for indicators are available in light-emitting-diodes. Although blue is also available, it is not listed here.

3E41 How can a neon lamp be used to check for the presence of RF?
A. A neon lamp will go out in the presence of RF.
B. A neon lamp will change color in the presence of RF.
C. A neon lamp will light only in the presence of very low frequency (VLF) signal.
D. A neon lamp will light in the presence of RF.

ANSWER D: A small neon lamp is handy when checking out marine radio, land mobile, or aeronautical radio installations. It should glow in the presence of strong RF. It may also be used to see whether or not a power amplifier has gone into oscillation. If the lamp continues to glow after you have cut off the power amplifier exciter, chances are you have an amplifier in oscillation. Shut it down—quick!

3E42 What would be the bandwidth of a good crystal lattice band-pass filter for a single-sideband phone emission?
A. 6 kHz at −6 dB
B. 2.1 kHz at −6 dB
C. 500 Hz at −6 dB
D. 15 kHz at −6 dB

ANSWER B: A single-sideband emission is normally about 3 kHz wide. Without making the voice sound pinched, a 2.1-kHz filter is what comes with most worldwide sets.

3E43 What would be the bandwidth of a good crystal lattice band-pass filter for a double-sideband phone emission?
A. 1 kHz at −6 dB
B. 500 Hz at −6 dB
C. 6 kHz at −6 dB
D. 15 kHz at −6 dB

ANSWER C: Both aeronautical VHF stations operate full carrier, double sideband, and so do many shortwave broadcast stations. Double sideband requires twice the bandwidth of single sideband. To receive the entire double-sideband signal, you would need a 6 kHz band-pass filter with 26 dB skirts to pass the signal.

3E44 What is a crystal lattice filter?

A. A power supply filter made with crisscrossed quartz crystals

B. An audio filter made with 4 quartz crystals at 1-kHz intervals

C. A filter with infinitely wide and shallow skirts made using quartz crystals

D. A filter with narrow bandwidth and steep skirts made using quartz crystals

ANSWER D: Marine and aircraft SSB transceivers usually have excellent SSB and CW filters built in. The filters have a nice narrow bandwidth and steep skirts. The filters are made up of quartz crystals. Some equipment may also offer accessory filters for a tighter response, such as for the reception of data or weather facsimile signals. Extremely sharp filters may sometimes be selected from the front panel of certain SSB receivers and transceivers.

3E45 What technique can be used to construct low cost, high performance crystal lattice filters?

A. Splitting and tumbling

B. Tumbling and grinding

C. Etching and splitting

D. Etching and grinding

ANSWER D: The process of etching and grinding crystal lattice filters allows you to custom-tailor their response. Your author remembers the old days of putting a penciled 3 on the face of the quartz crystal to lower its frequency, and using toothpaste to grind it down for a higher frequency. Anyone remember those days of changing frequency when "rock bound"?

3E46 What determines the bandwidth and response shape in a crystal lattice filter?

A. The relative frequencies of the individual crystals

B. The center frequency chosen for the filter

C. The amplitude of the RF stage preceding the filter

D. The amplitude of the signals passing through the filter

ANSWER A: The bandwidth and response of a crystal filter will relate to the frequency of the individual crystals within that filter unit.

3E47 What is an enhancement-mode FET?

A. An FET with a channel that blocks voltage through the gate

B. An FET with a channel that allows a current when the gate voltage is zero

C. An FET without a channel to hinder current through the gate

D. An FET without a channel; no current occurs with zero gate voltage

ANSWER D: Handheld manufacturers have *enhanced* their marine transceivers by using enhancement-mode field-effect transistors (FETs) that won't consume current when there is no voltage applied to the gate. As soon as a voltage of the right polarity is applied to the gate, a channel is formed, and there is current between source and drain. This type of "on demand" device is one more way that handheld transceivers try to conserve battery voltage.

3E48 What is a depletion-mode FET?

A. An FET that has a channel with no gate voltage applied; a current flows with zero gate voltage.

B. An FET that has a channel that blocks current when the gate voltage is zero.

C. An FET without a channel; no current flows with zero gate voltage.

D. An FET without a channel to hinder current through the gate.

ANSWER A: Your handheld battery may become *depleted* if depletion-mode FETs are used in many of the circuits. Even though there is no gate voltage applied, current still continues to flow.

3E49 Why do many MOSFET devices have built-in gate-protective Zener diodes?

A. The gate-protective Zener diode provides a voltage reference to provide the correct amount of reverse-bias gate voltage.

B. The gate-protective Zener diode protects the substrate from excessive voltages.

C. The gate-protective Zener diode keeps the gate voltage within specifications to prevent the device from overheating.

D. The gate-protective Zener diode prevents the gate insulation from being punctured by small static charges or excessive voltages.

ANSWER D: The "front-end" transistors of the modern transceiver must sometimes sustain major amounts of incoming signals from nearby transmitters, or maybe even a nearby lightning strike. Lightning *static* could destroy a MOSFET (Metal-Oxide-Semiconductor-Field-Effect Transistor) if it weren't for the built-in, gate-protective zener diode.

3E50 What do the initials CMOS stand for?

A. Common mode oscillating system

B. Complementary mica-oxide silicon

C. Complementary metal-oxide semiconductor

D. Complementary metal-oxide substrate

ANSWER C: When we talk about transistorized devices, they are in fact semiconductors. A CMOS is made from layers of metal-oxide semiconductor material.

3E51 Why are special precautions necessary in handling FET and CMOS devices?

A. They are susceptible to damage from static charges.

B. They have fragile leads that may break off.

C. They have micro-welded semiconductor junctions that are susceptible to breakage.

D. They are light sensitive.

ANSWER A: Any equipment using FET and CMOS transistors can become damaged easily from a nearby static discharge.

3E52 How does the input impedance of a field-effect transistor compare with that of a bipolar transistor?

A. One cannot compare input impedance without first knowing the supply voltage.

B. An FET has low input impedance; a bipolar transistor has high input impedance.

C. The input impedance of FETs and bipolar transistors is the same.

D. An FET has high input impedance; a bipolar transistor has low input impedance.

ANSWER D: An FET is much easier to work with in circuits than the simple bipolar transistor. Its higher input impedance is less likely to load down a circuit that it is attached to, and the FET can also handle big swings in signal levels.

3E53 What are the three terminals of a field-effect transistor?

A. Gate 1, gate 2, drain

B. Emitter, base, collector

C. Emitter, base 1, base 2

D. Gate, drain, source

ANSWER D: GaDS, better remember the three terminals of a field-effect transistor—gate, drain and source.

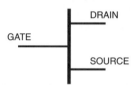

The field effect transistor (FET) is represented by a vertical bar with two lines coming in from the right and one coming in from the left.

Field-Effect Transistor

3E54 What are the two basic types of junction field-effect transistors?

A. N-channel and P-channel

B. High power and low power

C. MOSFET and GaAsFET

D. Silicon FET and germanium FET

ANSWER A: Be sure to note carefully the direction of the gate arrow to determine whether it is N-channel or P-channel.

3E55 What is an operational amplifier?

A. A high-gain, direct-coupled differential amplifier whose characteristics are determined by components external to the amplifier unit

B. A high-gain, direct-coupled audio amplifier whose characteristics are determined by components external to the amplifier unit

C. An amplifier used to increase the average output of frequency modulated signals

D. A program subroutine that calculates the gain of an RF amplifier

ANSWER A: The operational amplifier (op amp) is a direct-coupled *differential* amplifier that offers high gain and high input impedance. The "differential" wording refers to the op-amp input design where the output is determined by the difference voltage between the two inputs. Different op-amp circuits and characteristics are determined by the *external* components connected to the amplifier.

3E56 What would be the characteristics of the ideal op-amp?
A. Zero input impedance, infinite output impedance, infinite gain, flat frequency response
B. Infinite input impedance, zero output impedance, infinite gain, flat frequency response
C. Zero input impedance, zero output impedance, infinite gain, flat frequency response
D. Infinite input impedance, infinite output impedance, infinite gain, flat frequency response

ANSWER B: For an ideal amplifier, the output impedance of the op amp is just opposite of the input impedance—zero output impedance versus infinite input impedance. Zero output impedance contributes to a flat frequency response in audio amplifier circuits.

3E57 What determines the gain of a closed-loop op-amp circuit?
A. The external feedback network
B. The collector-to-base capacitance of the PNP stage
C. The power supply voltage
D. The PNP collector load

ANSWER A: The amplifier gain without feedback is very high. Feedback, applied to set the gain to particular amounts, is determined by an external feedback network. It may be one or more fixed resistors, or a type of variable circuit.

3E58 What is meant by the term op-amp offset voltage?
A. The output voltage of the op-amp minus its input voltage
B. The difference between the output voltage of the op-amp and the input voltage required in the following stage
C. The potential between the amplifier-input terminals of the op-amp in a closed-loop condition
D. The potential between the amplifier-input terminals of the op-amp in an open-loop condition

ANSWER C: If the op amp has been constructed properly, you should see almost no voltage between the amplifier input terminals when the feedback loop is *closed*.

3E59 What is the input impedance of a theoretically ideal op-amp?
A. 100 ohms
B. 1000 ohms
C. Very low
D. Very high

ANSWER D: Remember that the input impedance to an op amp is always very high, the higher the better.

3E60 What is the output impedance of a theoretically ideal op-amp?
A. Very low
B. Very high
C. 100 ohms
D. 1000 ohms

ANSWER A: Remember that the output impedance of an op amp is always very low, the lower the better.

3E61 What is a phase-locked loop circuit?

A. An electronic servo loop consisting of a ratio detector, reactance modulator, and voltage-controlled oscillator

B. An electronic circuit also known as a monostable multivibrator

C. An electronic circuit consisting of a precision push-pull amplifier with a differential input

D. An electronic servo loop consisting of a phase detector, a low-pass filter and voltage-controlled oscillator

ANSWER D: There are few circuits that have made such a big difference in communications as the phase-locked loop circuit. In fact, frequency selection from a transceiver with PLL synthesis came to a point where the operator actually had *too much control* of where the set might transmit and receive. Most land mobile equipment now locks out the consumer from getting into and making changes with the PLL. Same thing for marine radios—while mariners may tune in authorized ITU bands, the PLL circuitry has been locked out from the front panel so it can't be changed to go in any other region.

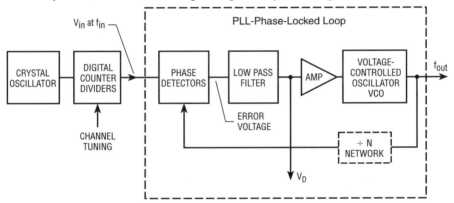

In a phase-locked loop the output frequency of the VCO is locked by phase and frequency to the frequency of the input signal V_{in}.

By using a very accurate crystal controlled source and dividing it down (or multiplying it up), the output frequency, f_{out}, of the PLL will be controlled very accurately and with rigid stability to the input frequency, f_{in}. The phase detector develops an error voltage that is determined by the frequencies and phase of the two inputs. The feedback system wants to reduce the error voltage to zero.

V_D is a voltage proportional to the frequency changes occuring in V_{in}; therefore, the PLL can be used as an FM demodulator.

With divide by N network, f_{out} will be f_{in} multiplied by N.

Channel tuning occurs in the Digital Counter Dividers.

Phase-Locked Loop (PLL)

3E62 What functions are performed by a phase-locked loop?

A. Wideband AF and RF power amplification

B. Comparison of two digital input signals, digital pulse counter

C. Photovoltaic conversion, optical coupling

D. Frequency synthesis, FM demodulation

ANSWER D: In building his first external frequency synthesizer, your author was able to "synthesize" hundreds of channels with just a couple of crystals. How many of you remember being "rock bound" with an old rig? See figure at 3E61 for FM demodulation.

3E63 A circuit compares the output from a voltage-controlled oscillator and a frequency standard. The difference between the two frequencies produces an error voltage that changes the voltage-controlled oscillator frequency. What is the name of the circuit?
A. A doubly balanced mixer
B. A phase-locked loop
C. A differential voltage amplifier
D. A variable frequency oscillator
ANSWER B: In a phase-locked loop, the *"difference between"* two frequencies will produce an error voltage to the voltage-controlled oscillator. *Difference* is a key word.

3E64 What do the initials TTL stand for?
A. Resistor-transistor logic
B. Transistor-transistor logic
C. Diode-transistor logic
D. Emitter-coupled logic
ANSWER B: If you regularly work with digital modes, transistor-transistor logic is commonplace within this type of transceiver. You will need a logic probe to troubleshoot TTL circuits.

3E65 What is the recommended power supply voltage for TTL series integrated circuits?
A. 12.00 volts
B. 50.00 volts
C. 5.00 volts
D. 13.60 volts
ANSWER C: If you ever worked on TTL circuits, you know the importance of those 5-volt regulators that always seem to burn up at the wrong time. TTL circuits run on +5.0 volts.

3E66 What logic state do the inputs of a TTL device assume if they are left open?
A. A high logic state
B. A low logic state
C. The device becomes randomized and will not provide consistent high or low logic states.
D. Open inputs on a TTL device are ignored.
ANSWER A: Here are several key words—inputs *open* and *high* state. Remember four letters make up the word *open*, and four letters make up the word *high*.

3E67 What is the range of input voltages considered to be a logic high input in a TTL device operating with a 5-volt power supply?
A. 2.0 to 5.5 volts
B. 1.5 to 3.0 volts
C. 1.0 to 1.5 volts
D. −5.0 to −2.0 volts
ANSWER A: If the input voltage is at the *high* level, depending on the number of circuits connected at the point, by specification the level will be between 2.0 volts and 5.5 volts. Logic low is 0.0 to 0.8 volts.

3E68 What is the range of input voltages considered to be a logic low in a CMOS device operating with an 18-volt power supply?
A. −0.8 to 0.0 volts
B. 0.0 to 5.4 volts
C. 0.0 to 0.8 volts
D. −0.8 to 0.4 volts
ANSWER B: A CMOS device operating with an 18-volt power supply is considered logic low with an input voltage of 0.0 to 5.4 volts. Don't forget, this is an 18-volt system.

3E69 Why do circuits containing TTL devices have several bypass capacitors per printed circuit board?
A. To prevent RFI to receivers
B. To keep the switching noise within the circuit, thus eliminating RFI
C. To filter out switching harmonics
D. To prevent switching transients from appearing on the supply line
ANSWER D: Since TTL devices are switching circuits with a rise and fall time so sharp they create troublesome transients, bypass capacitors will knock down the "spikes" that can create tremendous RFI. Remember, capacitors used as bypass want to keep the voltage across them constant.

3E70 What is a CMOS IC?
A. A chip with only P-channel transistors
B. A chip with P-channel and N-channel transistors
C. A chip with only N-channel transistors
D. A chip with only bipolar transistors
ANSWER B: The *C* in CMOS means *complementary,* which identifies an IC circuit that uses both P-channel and N-channel MOS transistors. MOS means Metal-Oxide-Semiconductor; therefore, this is MOS with both N and P transistors. Also, remember the problems associated with CMOS devices—you must make sure to minimize any static charges that could wipe them out in an instant.

3E71 What is one major advantage of CMOS over other devices?
A. Small size
B. Low current consumption
C. Low cost
D. Ease of circuit design
ANSWER B: That sensitive CMOS IC features low current consumption.

3E72 Why do CMOS digital integrated circuits have high immunity to noise on the input signal or power supply?
A. Larger bypass capacitors are used in CMOS circuit design.
B. The input switching threshold is about two times the power supply voltage.
C. The input switching threshold is about one-half the power supply voltage.
D. Input signals are stronger.
ANSWER C: The CMOS device is relatively immune to noise because its switching threshold is one-half the power supply voltage. As a result, power supply

noise transients or input transients much larger than other logic curcuit types will not cause a transition in the input state.

3E73 Signal energy is coupled into a traveling-wave tube at:
A. Collector end of helix.
B. Anode end of the helix.
C. Cathode end of the helix.
D. Focusing coils.

ANSWER C: The traveling-wave tube (TWT) is found in microwave telemetry equipment, and a low-power exciter signal is coupled to the TWT's cathode end of the helix. Care must be taken never to get near the waveguide output of a TWT system because there are dangerous microwaves when the circuit is switched on.

3E74 Permanent magnetic field that surrounds a traveling-wave tube (TWT) is intended to:
A. Provide a means of coupling.
B. Prevent the electron beam from spreading.
C. Prevent oscillations.
D. Prevent spurious oscillations.

ANSWER B: The permanent magnetic field that surrounds a TWT helps focus the electron beam.

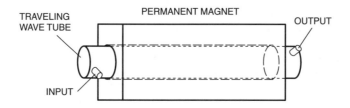

Travelling Wave Tube Assembly

3E75 Electromagnetic coils encase a traveling wave tube to:
A. Provide a means of coupling energy.
B. Prevent the electron beam from spreading.
C. Prevent oscillation.
D. Prevent spurious oscillation.

ANSWER B: The electromagnetic coils surrounding the traveling wave tube develop an axial magnetic focusing field to keep the electron beam diameter small and to keep the beam in the center of the helical coil inside the traveling wave tube.

3E76 When a doped semiconductor crystal is formed against a metal conductor, what type of diode is created?
A. Schottky diode
B. Tunnel diode
C. Varactor diode
D. Zener diode

ANSWER A: The Schottky diode requires little voltage to produce conduction, but offers excellent reverse-voltage opposition. The turn-on state and turn-off state of the Schottky diode is almost instantaneous. Schottky diodes are used in marine solar cells to minimize forward voltage drops and maximize reverse current opposition.

3E77 What type of diode contains no minority carriers in the junction region?
A. Tunnel diode
B. Varactor diode
C. Zener diode
D. Hot-carrier diode
ANSWER D: Hot carrier diodes, abbreviated HCDs, are quick acting and require little energy for conduction. They have no minority carriers in the junction region.

3E78 Mounting an LED facing a photodiode cell in a tiny light-tight enclosure produces a:
A. Seven segment LED.
B. Opto-isolator.
C. Opto-interrupter.
D. Photonic device.
ANSWER B: A light-emitting diode encased in a light-tight enclosure with a photodiode cell forms the opto-isolator stage found in many PLL synthesized marine radios. When you adjust the big knob, the opto-isolator detects an interrupted light source and counts the number of interruptions for the next digital stage to electronically process the command. It is important never to bang or bump the big knob on any opto-isolator system; poor alignment could cause erratic operation.

3E79 Most bipolar junction transistors have a ____ doped and ____ emitter region compared to the base and collector regions.
A. Heavily, thin
B. Heavily, small
C. Lightly, thin
D. Lightly, small
ANSWER B: The bipolar junction transistors have a heavily doped and very small emitter region, compared to the base and collector regions. In the bipolar, the amount of current that enters the base region controls the current between the emitter and the collector.

3E80 Most bipolar junction transistors have a ____ doped and ____ base region compared to the emitter and collector regions.
A. Heavily, thin
B. Heavily, small
C. Lightly, thin
D. Lightly, small
ANSWER C: This question asks about the bipolar junction transistor's base region, compared to its emitter and collector regions. The base is lightly doped, and there is a thin base region compared to emitter and collector regions.

3E81 Most bipolar junction transistors have a ____ doped and ____ collector region compared to the base and emitter regions.
  A. Heavily, large
  B. Heavily, small
  C. Lightly, thin
  D. Medium, large
ANSWER D: The collector on a bipolar transistor is medium doped and has a large collector region compared to the base and emitter regions. Be sure you understand this and the previous two questions—one will probably be on your upcoming test.

3E82 A common emitter amplifier has:
  A. More current gain than common base or common collector.
  B. More voltage gain than common base or common collector.
  C. More power gain than common base or common collector.
  D. Highest input impedance of the three amplifier configurations.
ANSWER C: The common emitter amplifier has more *power* gain than a common base or common collector amplifier. The audio current gain for a common emitter amplifier can be as much as 100 times.

3E83 A common base amplifier has:
  A. More current gain than common emitter or common collector.
  B. More voltage gain than common emitter or common collector.
  C. More power gain than common emitter or common collector.
  D. Highest input impedance of the three amplifier configurations.
ANSWER B: With the common base amplifier, the *voltage* gain is greater than that of a common emitter or common collector amplifier.

3E84 An emitter-follower amplifier has:
  A. More current gain than common emitter or common base.
  B. More voltage gain than common emitter or common base.
  C. More power gain than common emitter or common base.
  D. Lowest input impedance of the three amplifier configurations.
ANSWER A: A emitter-follower amplifier has more *current* gain than a common emitter that has power gain or a common base that has voltage gain. Remember ... common emitter has greater power gain, common base has greater voltage gain, and emitter-follower has greater current gain. One of these three will probably be on your test.

3E85 The JFET's _____ is the ratio of drain-source voltage change to gate-source voltage change with drain current constant.
  A. Amplification factor
  B. Dynamic drain resistance
  C. Transconductance
  D. Pinch-off voltage
ANSWER A: The JFET transistor is part of the field-effect transistor family and behaves similarly to those older-glass triode vacuum tubes found in the classic marine radiotelephones. There are three questions about JFET's. When drain current is constant, the ratio of drain-source voltage change to gate-source voltage change is called the JFET amplification factor.

3E86 The JFET's _____ is the ratio of drain-source voltage change to drain current change with the gate-source voltage constant.
A. Amplification factor
B. Dynamic drain resistance
C. Transconductance
D. Pinch-off voltage

ANSWER B: If the gate-source voltage is constant, the ratio of drain-source voltage change to drain current change is called *dynamic drain resistance*.

3E87 The JFET's _____ is the ratio of drain current change to gate-source voltage change with drain-source voltage constant.
A. Amplification factor
B. Dynamic drain resistance
C. Transconductance
D. Pinch-off voltage

ANSWER C: If the drain-source voltage is constant, the ratio of drain current change to gate-source voltage change is called the JFET's transconductance. Transconductance is the ability of the JFET to change the output circuit current when an input-voltage change is detected.

3E88 When working with JFETs, the following terms are used: amplification factor, dynamic drain resistance, and transconductance. Which of the following relationships is correct?
A. dynamic drain resistance = transconductance / amplification factor
B. transconductance = dynamic drain resistance / amplification factor
C. amplification factor = dynamic drain resistance / transconductance
D. amplification factor = dynamic drain resistance × tranconductance

ANSWER D: The JFET amplification factor (abbreviated µ) is equal to dynamic drain resistance times transconductance.

3E89 What describes a diode junction that is forward biased?
A. It is a high impedance.
B. It conducts very little current.
C. It is a low impedance.
D. It is an open circuit.

ANSWER C: A diode junction that is forward biased offers low impedance to the flow of electrons. Think of "forward" as easy movement.

3E90 What describes a diode junction that is reverse biased?
A. It is a short circuit.
B. It conducts a large current.
C. It is a low impedance.
D. It is a high impedance.

ANSWER D: A diode junction that is reverse biased has high impedance to the flow of electrons. Think of "reverse" as difficult movement.

3E91 What conditions exist when an NPN transistor is operating as a Class A amplifier?
A. The base-emitter junction is forward biased and the collector-base junction is reverse biased.
B. The base-emitter junction and collector-base junction are both forward biased.

C. The base-emitter junction and collector-base junction are both reverse biased.

D. The base-emitter junction is reverse biased and the collector-base junction is forward biased.

ANSWER A: In a Class A amplifier, the NPN transistor base-emitter junction is *forward* biased, and the collector-base junction is *reverse* biased. The biasing within the NPN transistor either adds or subtracts a specific amount of current from the signal at the input.

3E92 What conditions exist when a transistor is operating in saturation?

A. The base-emitter junction and collector-base junction are both forward biased.

B. The base-emitter junction and collector-base junction are both reverse biased.

C. The base-emitter junction is reverse biased and the collector-base junction is forward biased.

D. The base-emitter junction is forward biased and the collector-base junction is reverse biased.

ANSWER A: When a transistor goes into saturation, *both* the base-emitter junction and collector-base junction are *forward biased.* Schottky diodes are placed across the collector-base junction to allow the junction to switch out of saturation quickly.

3E93 What voltage is required on a silicon NPN switching transistor's base-emitter junction to cause current between collector and emitter?

A. The base must be at least 0.4 volts positive with respect to the emitter.

B. The base must be at a negative voltage with respect to the emitter.

C. The base must be at least 0.7 volts positive with respect to the emitter.

D. The base must be at least 0.7 volts negative with respect to the emitter.

ANSWER C: For the silicon NPN switching transistor to conduct current between the collector and emitter, the base must be at least 0.7 volts *positive* with respect to the emitter. Then the NPN switching transistor is considered "on."

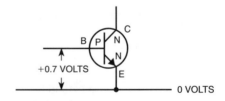

NPN Transistor

3E94 What voltage is required on a silicon PNP switching transistor's base-emitter junction to cause current between collector and emitter?
  A. The base must be at least 0.7 volts negative with respect to the emitter.
  B. The base must be at least 0.4 volts negative with respect to the emitter.
  C. The base must be positive with respect to the emitter.
  D. The base must be at least 0.4 volts positive with respect to the emitter.

ANSWER A: In contrast to an NPN transistor, the base of a PNP switching transistor must be at least 0.7 volts *negative* with respect to the emitter. The center letter of the transistor type can help you remember whether the forward bias voltage must be negative or positive.

  NPN = **P**ositive at 0.7 volts

  PNP = **N**egative at 0.7 volts

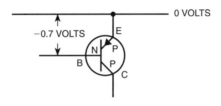

PNP Transistor

3E95 What semiconductor device controls current between source and drain due to a variable width channel controlled by a voltage applied between gate and source?
  A. A bipolar transistor (BJT)
  B. A field-effect transistor (FET)
  C. A gate-controlled diode
  D. A PNP transistor

ANSWER B: It is the field-effect transistor that is controlled by the *voltage* level of the input, as opposed to input current. The field-effect transistor is as close to a vacuum tube as you can get—even more so than a bipolar transistor.

3E96 What are the operating modes for field-effect transistors?
  A. Transition and depletion modes
  B. Enhancement and transition modes
  C. Transition and non-transition modes
  D. Depletion and enhancement modes

ANSWER D: The operating modes for an FET are *depletion* and *enhancement*. Depletion is normally on; enhancement is normally off and not conducting current.

3E97 What is an enhancement-mode FET?
  A. An FET with a channel that blocks voltage through the gate
  B. An FET with a channel that allows a current between source and drain when the gate voltage is zero

---

C. An FET without a channel to hinder current through the gate

D. An FET without a channel; no current between source and drain with zero gate voltage

ANSWER D: A field-effect transistor in the enhancement mode does not conduct current until gate voltage is increased. There is no current between source and drain when gate voltage is zero.

3E98 What is a depletion-mode FET?

A. An FET that has a channel with no gate voltage applied; there is current between source and drain when the gate voltage is zero.

B. An FET that has a channel that blocks current when the gate voltage is zero.

C. An FET without a channel; there is no current from source to drain with zero gate voltage.

D. An FET without a channel to hinder current through the gate.

ANSWER A: In the depletion mode, the field-effect transistor conducts current between source and drain when the gate voltage is zero. No gate voltage input is necessary for the depletion mode to conduct current between source and drain.

3E99 What is a silicon integrated circuit (IC)?

A. A complex semiconductor device containing within it all the circuit components interconnected on a single chip of silicon.

B. A large number of discrete components wired together on a silicon substrate.

C. Individual components integrated together on a printed wiring board.

D. A circuit of discrete individual silicon components.

ANSWER A: The integrated circuit is a complex device containing hundreds (even thousands) of semiconductor devices, all interconnected on and contained within a *single chip* of silicon. Some packaged ICs are soldered into place; others simply plug in. The key word in the answer is "chip."

3E100 What are the advantages of using an LED?

A. Low power consumption and long life.

B. High lumens per cm squared and low power consumption.

C. High lumens per cm squared and low voltage requirement.

D. A current flows when the device is exposed to a light source.

ANSWER A: The advantages of a light-emitting diode are extremely long life, few failures, and extremely low-power consumption.

3E101 What determines the visible color radiated by an LED?

A. The color of a lens in an eyepiece

B. The amount of voltage across the device

C. The amount of current through the device

D. The materials used to construct the device

ANSWER D: The popular colors of a light-emitting diode are red, green, orange, and yellow. Blue is the latest color. Color is determined by the material used to construct the diode.

3E102 What level of input voltage is considered a logic low in a TTL device operating with a 5-volt power supply?
  A. -2.0 to -5.5 volts
  B. 2.0 to 5.5 volts
  C. 0.0 to 0.8 volts
  D. -0.8 to 0.4 volts
ANSWER C: On a TTL device operating at 5 volts, logic low is considered 0.0 to 0.8 volts. Logic high is 2.0 to 5.5 volts.

3E103 Where is the external feedback network connected to control the gain of a closed-loop op-amp circuit?
  A. From output to inverting input
  B. From output to non-inverting input
  C. Across the output
  D. Across the input
ANSWER A: To control the gain of a closed-loop operational amplifier, external feedback is connected from the output to the inverting input.

3E104 What is the name of the semiconductor memory IC whose digital data can be written or read, and whose memory word address can be accessed randomly?
  A. ROM—Read-Only Memory
  B. PROM—Programmable Read-Only Memory
  C. RAM—Random-Access Memory
  D. EPROM—Electrically Programmable Read-Only Memory
ANSWER C: This question answers itself: Random-Access Memory (RAM).

3E105 What is the name of the random-accessed semiconductor memory IC that must be refreshed periodically to maintain reliable data storage in its memory matrix?
  A. ROM—Read-Only Memory
  B. PROM—Programmable Read-Only Memory
  C. PRAM—Programmable Random-Access Memory
  D. DRAM—Dynamic Random-Access Memory
ANSWER D: A RAM that must be periodically refreshed to maintain reliable data storage is called a dynamic random-access memory—DRAM.

3E106 What would be the bandwidth of a good crystal lattice band-pass filter for weather facsimile HF (high-frequency) reception?
  A. 1 kHz at -6 dB
  B. 500 Hz at -6 dB
  C. 6 kHz at -6 dB
  D. 15 kHz at -6 dB
ANSWER B: To receive good quality high-frequency weather facsimile imagery from a SSB transceiver, switch to the CW or AFSK mode to pull in tighter filters with steeper skirts. 500 Hz at -6 dB is a good choice.

3E107 The capacitance of a varactor _____ as the _____ bias is increased. Assume proper bias to use the diode as a varactor.
  A. Increases, forward
  B. Increases, reverse

C. Decreases, forward

D. Decreases, reverse

ANSWER D: The varactor decreases capacity as the reverse bias is increased. In a reverse biased diode, reverse-bias voltage increases the barrier between the materials, separating the two capacitor plates and reducing capacity.

3E108 The screen grid was added to vacuum tubes to reduce _____ and draw space-charge electrons from the cathode through the control grid wires.

A. Mu

B. Transconductance

C. Plate-impedance values

D. Grid-plate capacitance

ANSWER D: A screen grid in a vacuum tube decreases grid to plate capacitance, allowing the tube to draw electrons from the cathode through the control grid wires. This offers high plate current and high amplification, the main jobs of the tetrode tube. The screen grid normally runs at a 100-volt, or greater, dc potential.

3E109 When working with tetrodes, as long as plate-voltage is greater than screen-grid voltage, what effect on plate current would be noticed if you doubled plate voltage?

A. Plate current would be decreased by a factor of 1/2.

B. Plate current would be increased by a factor of 2.

C. Plate current would barely increase.

D. Plate current would be increased by a factor equal to Mu.

ANSWER C: Doubling the plate voltage on the tetrode causes little current increase. The plate current is relatively independent of the plate voltage. In contrast to the tetrode, doubling the plate voltage on a triode causes an almost two times increase in current.

3E110 Which of the following is *not* an advantage of using pentodes?

A. Usually requires no neutralization in high frequency circuits

B. Higher Mu than triodes or tetrodes

C. Higher secondary emission resulting in greater plate current

D. Better shielding between plate and control grid

ANSWER C: The pentode uses a suppressor grid between the plate and screen grid, which reduces secondary emission. The question asks, "Which of the following is NOT ... ," so answer C is correct.

3E111 Which of the following is *not* an advantage of gaseous tubes?

A. Little heat dissipation

B. Carry relatively low currents

C. High efficiency

D. Relatively constant voltage-drop across them

ANSWER B: The gaseous conductance can only carry relatively low current because there are no wires to carry higher currents.

3E112 A pulse width modulator IC would most likely be found in which of the following:

A. Ringing choke power supply.

B. Solid-state DC to DC converter.

C. Crowbar protection circuit.

D. Shunt regulator with error-signal amp.

ANSWER A: The pulse width modulator IC is normally found in the ringing choke power supply.

3E113 An 800-kHz crystal, calibrated at 40 degrees Celsius and having a temperature coefficient of -30 parts per million per degree Celsius, will resonate at what frequency when operated at 60 degrees Celsius?

A. 799.52 kHz

B. 799.40 kHz

C. 800.60 kHz

D. 800.48 kHz

ANSWER A: Thirty parts per million is 0.00003 per degree. A 20-degree change, multiplied by 0.00003, equals a -0.0006 total parts change of frequency. $-6 \times 10^{-4} \times 8 \times 10^5$ equals -480 Hz. If you deduct 480 Hz from 800 kHz, you end up with 799.52 kHz as the operating frequency.

3E114 An 800-kHz crystal, calibrated at 40 degrees Celsius and having a temperature coefficient of +30 parts per million per degree Celsius, will resonate at what frequency when operated at 60 degrees C?

A. 799.52 kHz

B. 799.40 kHz

C. 800.60 kHz

D. 800.48 kHz

ANSWER D: This is almost the same as the previous question, but this time the temperature coefficient is +30 parts per million. Add 480 Hz to the 800 kHz crystal, and you get 800.480 kHz, the correct answer.

3E115 Which of the following does not have a negative resistance region?

A. Dynatron

B. Tunnel diode

C. Unijunction transistor

D. Schottky diode

ANSWER D: The Schottky diode has no negative resistance region and many times is used aboard ships as a reverse polarity diode to the photo-cell charging system.

3E116 A phase-locked loop IC could be used in all of the following applications except:

A. Frequency-shift keying (FSK)

B. Horizontal sweep AFC

C. Frequency synthesis

D. Phase-shift oscillator

ANSWER D: The PLL chip could be used for frequency shift keying for a radioteleprinter, automatic frequency control in an oscilloscope, or frequency synthesis, but would normally not be used as a phase-shift oscillator. The PLL chip has a wide range of supply voltage capabilities, low power consumption, and extremely low-frequency drift with major changes in temperature.

3E117 Which of the following logic gates will provide an active high out when both inputs are active high?

A. AND

B. NAND

C. NOR

D. XOR

ANSWER A: As you can see from the *truth table*, the AND gate has a 1 or high output when all inputs are high. A and B must be "1" for C to be "1".

INPUTS		OUTPUT
A	B	C
0	0	0
0	1	0
1	0	0
1	1	1

2-Input AND Gate
Truth Table

INPUTS		OUTPUT
A	B	C
0	0	1
0	1	1
1	0	1
1	1	0

2-Input NAND Gate
Truth Table

3E118 Which of the following logic gates will provide an active low out when both inputs are active high?

A. AND

B. NAND

C. OR

D. XNOR

ANSWER B: A NAND gate is an AND gate with an inverter on the output. Look at the *truth table.*

3E119 Which of the following logic gates will provide an active high out when any input is active high?

A. AND

B. NAND

C. OR

D. NOR

ANSWER C: An OR gate outputs a 1 when any input is or both inputs are a 1. Refer to the *truth table.*

INPUTS		OUTPUT
A	B	C
0	0	0
0	1	1
1	0	1
1	1	1

2-Input OR Gate Truth Table

3E120 Which of the following logic gates will provide an active low out when any input is active high?

A. AND

B. NAND

C. OR

D. NOR

ANSWER D: A NOR gate is an OR gate with an inverter on the output. Again, look at the *truth table.*

INPUTS		OUTPUT
A	B	C
0	0	1
0	1	0
1	0	0
1	1	0

2-Input NOR Gate Truth Table

3E121 Which of the following logic gates will provide an active high out only when all inputs are different?
A. OR
B. NOR
C. XOR
D. XNOR

ANSWER C: As shown in the truth table below, an XOR (exclusive OR) gate outputs a logic 1 (high) only when one input or the other is a logic 1 (the inputs are not the same), not when both inputs are the same, either both logic 1 or logic 0.

INPUTS		OUTPUT
A	B	C
0	0	0
0	1	1
1	0	1
1	1	0

2-Input Exclusive OR (XOR)
Truth Table

INPUTS		OUTPUT
A	B	C
0	0	1
0	1	0
1	0	0
1	1	1

2-Input XNOR Gate
Truth Table

3E122 Which of the following logic gates will provide an active low out only when all inputs are different?
A. OR
B. NOR
C. XOR
D. XNOR

ANSWER D: And XNOR gate inverts the outputs from a XOR gate shown in the last question, so the outputs are logic 0 (low) when the inputs are not the same.

3E123 Which of the following inputs to a D flip-flop are considered to be asynchronous?
A. D, CLK
B. PRE, CLR
C. D, PRE
D. CLK, CLR

ANSWER B: An input that allows you to change the state of the flip-flop regardless of the state of the clock input is called an *asynchronous input.* They are usually direct reset (CLR) and direct set (PRE); PRE is usually a 1, and CLR is usually a 0.

3E124 Which of the following inputs to a J-K flip-flop are considered to be synchronous?
A. J, K
B. PRE, CLR
C. CLK, PRE
D. Q, CLR

ANSWER A: On a J-K flip-flop, the J-K inputs are considered synchronous. In synchronous operation, the outputs change only at specific times determined by the previous state, the clock, and the state of the inputs.

3E125 An R-S flip-flop is capable of doing all of the following except:
A. Accept data input into R-S inputs with CLK initiated.
B. Accept data input into PRE and CLR inputs without CLK being initiated.
C. Refuse to accept synchronous data if asynchronous data is being input at same time.
D. Operate in toggle mode with R-S inputs held constant and CLK initiated.

ANSWER D: An R-S flip-flop doesn't have a toggle mode, so answer D is correct. The R-S flip-flop is capable of all the other answers *except* D.

3E126 The toggle mode of operation, achieved by applying a string of CLK pulses, is a normal operation mode for which of the following?
A. D flip-flop
B. R-S flip-flop
C. J-K flip-flop
D. Bistable multivibrator

ANSWER C: A synchronous circuit may be used as a clock, and a J-K flip-flop yields toggle-mode operation when both J = 1 and K = 1, and a string of CLK pulses is applied.

3E127 How many R-S flip-flops would be required to construct an 8-bit storage register?
A. 2
B. 4
C. 8
D. 16

ANSWER C: One flip-flop holds one bit. You need 8 flip-flops for an 8-bit storage register. Some integrated circuits have four J-K flip-flops (FFs) inside the package, all operating in the toggle mode. To develop a MOD-8 (8-counts) counter, connect the output to the input on three of the J-K FFs. If all four FFs are used, you can count to 16.

3E128 How many J-K flip-flops would be required to construct a MOD-16 ripple counter?
A. 2
B. 4
C. 8
D. 16

ANSWER B: Each flip-flop requires 2 input pulses to generate 1 output pulse, so each time you add another flip-flop in series, you multiply the number of inputs required to generate 1 output by a factor of 2. Four flip-flops equals $2^4 = 16$.

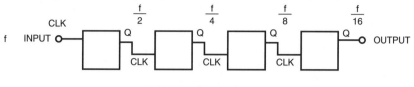

MOD-16 Ripple Counter

3E129 How many D flip-flops would be required to construct a MOD-16 ring counter?
> A. 2
> B. 4
> C. 8
> D. 16

ANSWER D: In a ring counter, each stage is a bit; therefore, a ring counter has the same number of stages as the MOD no., so MOD 16 is 16 flip-flops—one flip-flop per bit.

3E130 If an input CLK frequency of 160 kHz is applied to a four-bit ripple counter capable of achieving full count before roll-over, what frequency can you expect to measure at the MSB output?
> A. 10 kHz
> B. 20 kHz
> C. 40 kHz
> D. 320 kHz

ANSWER A: A 4-bit ripple counter consists of four flip-flops (1 bit = 1 flip-flop). As you learned in question 3E128, four flip-flops require 16 input pulses to generate 1 output pulse. So, divide 160 kHz by 16 and you get the answer—10 kHz.

3E131 If a CLK frequency of 160 kHz is applied to the Count-Up input pin of a BCD counting chip, what will be the frequency available at the Carry-Out pin of that same chip? (assume all other inputs are proper).
> A. 10 kHz
> B. 16 kHz
> C. 160 kHz
> D. 320 kHz

ANSWER B: The BCD (Binary-Coded Decimal) requires 10 input pulses to generate 1 output pulse, so the output would be 1/10 the input (160 kHz ÷ 10 = 16 kHz).

3E132 Which of the following are contained in an LCD display?
> A. Semiconductor P-N junctions
> B. Light emitting diodes
> C. Photovoltaic material
> D. Nematic fluid

ANSWER D: In an LCD display, the fluid that makes up the liquid crystal letters or numbers is called "nematic fluid." A polarizer forms the top and bottom of the sandwich, and the fluid is in between. You must be sure that an LCD display neither bakes in the sun or freezes in the cold.

3E133 Which of the following is *not* an analog-to-digital converter?
    A. Digital-ramp ADC
    B. Successive approximation ADC
    C. Flash ADC
    D. R/2R ladder ADC
ANSWER D: The ladder method is used in digital-to-analog converters, not analog-to-digital as the question asks. Answer D is digital to analog, so D is the correct answer for something that an analog-to-digital converter is *not!*

3E134 A — NOTed OR, bubbled OR, or negative OR — gate performs the same logic function as which of the following gates?
    A. AND
    B. NAND
    C. XOR
    D. XNOR
ANSWER B: NOTed OR, bubbled OR, and negative OR (NOR) are different names for the same logic gate — an OR gate with an inverter on the output. DeMorgan's theorem says that a NOR gate $\overline{A+B}$ is equal to $\overline{A} \cdot \overline{B}$, an AND gate of the complements of A and B. Table 1 is a truth table of the positive logic 2-input NOR gate, $\overline{A+B} = Q$. Table 2 is the truth table of $\overline{A} \cdot \overline{B} = Q$. Table 3 is the positive logic truth table for a 2-input NAND gate, $\overline{A \cdot B} = Q$. A truth table can be converted to negative logic by interchanging 1s and 0s resulting in the negative logic 2-input NAND gate of Table 4.

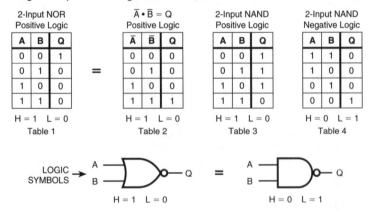

Not only is the 2-input NOR equivalent to Table 2 but also to a negative logic 2-input NAND gate.

Positive Logic NOR Equals Negative Logic NAND

3E135 A — NOTed AND, bubbled AND, or negative AND — gate performs the same logic function as which of the following gates?
    A. OR
    B. NOR
    C. XOR
    D. XNOR
ANSWER B: NOTed AND, bubbled AND, or negative AND (NAND) are different names for the same logic gate — an AND gate with an inverter on the output. DeMorgan's theorem says that a NAND gate $\overline{A \cdot B}$ is equal to $\overline{A} + \overline{B}$, an OR

gate of the complements of A and B. Table 1 is a truth table of a positive logic 2-input NAND gate, $\overline{A \cdot B} = Q$. Table 2 is the truth table of $\overline{A} + \overline{B} = Q$. Table 3 is the positive logic 2-input NOR gate truth table, $\overline{A+B} = Q$. Interchanging 1s and 0s in Table 3 produces the negative logic truth table for a 2-input NOR gate.

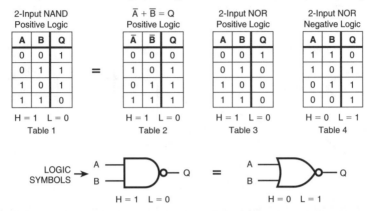

2-Input NAND Positive Logic				$\overline{A} + \overline{B} = Q$ Positive Logic				2-Input NOR Positive Logic				2-Input NOR Negative Logic		
A	B	Q		$\overline{A}$	$\overline{B}$	Q		A	B	Q		A	B	Q
0	0	1		0	0	0		0	0	1		1	1	0
0	1	1	=	1	0	1		0	1	0		1	0	1
1	0	1		0	1	1		1	0	0		0	1	1
1	1	0		1	1	1		1	1	0		0	0	1

H = 1 L = 0	H = 1 L = 0	H = 1 L = 0	H = 0 L = 1
Table 1	Table 2	Table 3	Table 4

LOGIC SYMBOLS

H = 1 L = 0   =   H = 0 L = 1

Not only is the 2-input NAND equivalent to Table 2 but also to a negative logic 2-input NOR gate.

Positive Logic NAND Equals Negative Logic NOR

3E136 Which of the following logic functions can not be duplicated when using a single IC containing quad 2-input NAND gates? (You can use all available gates contained in the one IC mentioned).
    A. AND
    B. XOR
    C. OR
    D. NOR
ANSWER B: You cannot have an XOR logic function when using a single IC containing quad 2-input NAND gates. You can use two of the gates for the logic and two for inverters, but an additional output gate would be required.

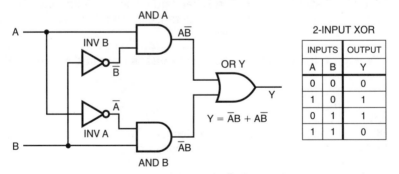

2-INPUT XOR		
INPUTS		OUTPUT
A	B	Y
0	0	0
1	0	1
0	1	1
1	1	0

An XOR Gate Using AND, OR and NOT Gate

3E137 Which of the following logic functions would require the fewest number of NOR gates to duplicate?
    A. AND
    B. NAND

    C. XOR
    D. XNOR

ANSWER A: The AND gate can be implemented by using three NOR gates, the fewest NOR gates to duplicate any of the functions shown. Two of the NOR gates are used as inverters, whose outputs feed the third NOR gate.

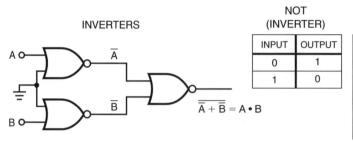

	NOT (INVERTER)	
INPUT	OUTPUT	
0	1	
1	0	

AND		
INPUTS		OUTPUT
A	B	C
0	0	0
0	1	0
1	0	0
1	1	1

AND Gate Using NOR Gates

3E138 Which of the following applications usually requires a parallel-to-serial conversion of data?
    A. Transfer a byte of data from a microcomputer's RAM to its CPU
    B. Transfer a byte of data from a microcomputer's ROM to its CPU
    C. Transfer a byte of data from a microcomputer's RAM to its monitor
    D. Transfer a byte of data from a microcomputer's RAM to its ALU

ANSWER C: To go from a microcomputer's RAM to its monitor, you need a parallel-to-serial conversion to transfer a byte of data.

3E139 Which of the following codes has gained the widest acceptance in modern times for exchange of data from one computer to another?
    A. ASCII code
    B. Baudot code
    C. Morse code
    D. Gray code

ANSWER A: ASCII stands for American Standard Code for Information Exchange, and it has gained the widest acceptance in modern times for exchange of data from one computer to another.

3E140 Which of the following op-amp circuits is operated open-loop?
    A. Comparator
    B. Non-inverting amp
    C. Inverting amp
    D. Active filter

ANSWER A: The comparator is a high-gain circuit with an open-loop configuration. The three wrong answers are closed-loop circuits.

3E141 Slew rate of an op-amp means:
    A. Output voltage change per nanosecond.
    B. Output voltage change per microsecond.
    C. Output voltage change per millisecond.
    D. Output voltage change per second.

ANSWER B: The slew rate limits the size of the output voltage at higher frequencies, and the slew rate of an op-amp is its output voltage change per microsecond. The key word in the answer is "microsecond."

3E142 RF chokes are sometimes constructed of universal wound pies to:
- A. Lower impedance to a relatively wide band of frequencies for which it was manufactured.
- B. Increase impedance to a relatively narrow band of frequencies for which it was manufactured.
- C. Increase the end-to-end distributed capacitance.
- D. Decrease the end-to-end distributed capacitance.

ANSWER D: Certain radios are troubled by the presence of RF near their circuits, so RF chokes are wound into a pie to decrease the end-to-end distributed capacitance that might form on a conventionally constructed choke.

3E143 Which of the following types of microphones is least likely to be used in broadcast applications?
- A. Condenser microphone
- B. Crystal microphone
- C. Dynamic microphone
- D. Magnetic microphone

ANSWER B: The old crystal microphone is no longer found in broadcast applications. After time, the crystals become fused, and the response of the microphone is severely degraded. The crystal microphone is also sensitive to heat. These limitations have made the crystal microphone just about obsolete, except for use in inexpensive radio projects.

3E144 How many individual memory cells would be contained in a memory IC that has 4 data bus input/output pins and 4 address pins for connection to the address bus?
- A. 8 memory cells
- B. 16 memory cells
- C. 32 memory cells
- D. 64 memory cells

ANSWER D: A 4-bit address means 16 memory addresses. Each memory address contains a 4-bit word. Sixteen words times 4 bits equals 64 individual memory cells.

3E145 When referring to digital IC's, which of the following contains between 100 to 9999 gates?
- A. MSI (medium-scale integration)
- B. LSI (large-scale integration)
- C. VLSI (very large-scale integration)
- D. ULSI (ultra large-scale integration)

ANSWER B: The large-scale integrated circuit has over 100 gates. The medium-scale integrated circuit has 10 to 99 gates, and the very large-scale integrated circuit has thousands of gates—well beyond 9999 gates.

3E146 Which of the following devices acts as two SCR's connected back to back, but facing in opposite directions and sharing a common gate?
- A. JFET
- B. Dual-gate MOSFET

C. DIAC

D. TRIAC

ANSWER D: The TRIAC acts like two silicon-controlled rectifiers in parallel to control current in either direction. See figure at question 3E28.

3E147 Which of the following devices is normally used to regulate the amount of AC current flowing to a load from approximately 0 degrees to no more than 180 degrees of the input signal?

A. DIAC

B. SCR

C. TRIAC

D. Class A BJT amp

ANSWER B: The silicon-controlled rectifier (SCR) allows ac currents from zero to 180 degrees to flow. See figure at question 3E28.

3E148 Which of the following devices is normally used to regulate the amount of AC current flowing to a load from approximately 0 degrees to 360 degrees of the input signal?

A. DIAC

B. SCR

C. TRIAC

D. Class B BJT amp

ANSWER C: The TRIAC allows ac current to the load to flow one direction while the input voltage is in its positive alternation (0 to 180 degrees) and in the other direction while the input voltage is in its negative alternation (180 to 360 degrees). The TRIAC is equivalent to back-to-back SCRs. See question 3E28.

3E149 Which of the following devices produces the least amount of noise when used as a part of a mixer stage in a superheterodyne receiver?

A. Bipolar-junction transistor

B. Dual-gate MOSFET

C. Duo-triode vacuum tube

D. P-N junction diode

ANSWER B: A dual-gate MOSFET is normally used as the mixer stage in a superhet receiver instead of diodes. Diodes and transistors are noisy. (Transistors are really just a combination of diodes.)

3E150 Which of the following devices is used principally as VHF and UHF parasitic suppressors?

A. Ferrite bead

B. Balun

C. Autotransformer

D. Swinging choke

ANSWER A: When designing VHF and UHF equipment, many times tiny ferrite beads are dressed on wires within the equipment to minimize RF flowing back into sensitive circuits.

3F1 What is a linear electronic voltage regulator?
A. A regulator that has a ramp voltage as its output
B. A regulator in which the pass transistor switches from the "off" state to the "on" state
C. A regulator in which the control device is switched on or off, with the duty cycle proportional to the line or load conditions
D. A regulator in which the conduction of a control element is varied in direct proportion to the line voltage or load current

ANSWER D: A linear electronic voltage regulator varies the conduction of a circuit in direct proportion to variations in the *line voltage* (key words) to, or the *load current* (key words) from, the device. You will find in your base station power supplies, which utilize a big heavy transformer, a sophisticated voltage regulation circuit.

3F2 What is a switching electronic voltage regulator?
A. A regulator in which the conduction of a control element is varied in direct proportion to the line voltage or load current
B. A regulator that provides more than one output voltage
C. A regulator in which the control device is switched on or off, with the duty cycle proportional to the line or load conditions
D. A regulator that gives a ramp voltage at its output

ANSWER C: The switching voltage regulator actually switches the control device completely *on or off* (key words).

3F3 What device is usually used as a stable reference voltage in a linear voltage regulator?
A. A Zener diode
B. A tunnel diode
C. An SCR
D. A varactor diode

ANSWER A: For voltage regulation and a stable reference voltage, Zener diodes are found in aeronautical handheld transceivers, aircraft radio, marine radio, and business band equipment. The most common cause of voltage regulator failure is a voltage transient or over-voltage. Whenever replacing a defective voltage regulator, monitor other voltage sources to see why the regulator failed.

3F4 What type of linear regulator is used in applications requiring efficient utilization of the primary power source?
A. A constant current source
B. A series regulator
C. A shunt regulator
D. A shunt current source

ANSWER B: A series regulator normally runs cool. It's more efficient than a shunt regulator. Everything passes through the series regulator.

3F5 What type of linear voltage regulator is used in applications where the load on the unregulated voltage source must be kept constant?

A. A constant current source
B. A series regulator
C. A shunt current source
D. A shunt regulator

ANSWER D: In circuits where the load on the unregulated input source must be kept constant, we use the slightly less-efficient shunt regulator. These can sometimes get quite warm, and are usually mounted to the chassis of the equipment. The chassis acts as a heat sink.

3F6 To obtain the best temperature stability, what should be the operating voltage of the reference diode in a linear voltage regulator?

A. Approximately 2.0 volts
B. Approximately 3.0 volts
C. Approximately 6.0 volts
D. Approximately 10.0 volts

ANSWER C: We normally run the operating voltage of the reference diode at around 6 volts. This gives the circuit good temperature stability.

3F7 What is the meaning of the term remote sensing with regard to a linear voltage regulator?

A. The feedback connection to the error amplifier is made directly to the load.
B. Sensing is accomplished by wireless inductive loops.
C. The load connection is made outside the feedback loop.
D. The error amplifier compares the input voltage to the reference voltage.

ANSWER A: Marine SSB transceivers, running 100 watts PEP output, must be run off of a stable power supply. Remote sensing of a portion of the output voltage directly at the load feeds back to the error amplifier for accurate conditions at the load. This keeps the output voltage from having wide variations due to an additional voltage drop between the regulator and the load.

3F8 What is a three-terminal regulator?

A. A regulator that supplies three voltages with variable current
B. A regulator that supplies three voltages at a constant current
C. A regulator containing three error amplifiers and sensing transistors
D. A regulator containing a voltage reference, error amplifier, sensing resistors and transistors, and a pass element

ANSWER D: The common voltage regulator found in mobile radio equipment is the 3-terminal regulator. When working on equipment, you can spot these regulators up alongside the metal chassis, using the chassis as a heat sink. Inside that small plastic IC is a voltage regulator containing a voltage reference, an error amplifier, sensing resistors and transistors, and a pass element. These regulators normally run relatively hot, so don't consider them necessarily bad if they seem hot to the touch. It's when you find these regulators stone cold that you may need to troubleshoot the circuit.

REGULATOR

Voltage Regulator IC Mounted in Circuit Board

3F9 What are the important characteristics of a three-terminal regulator?
A. Maximum and minimum input voltage, minimum output current and voltage
B. Maximum and minimum input voltage, maximum output current and voltage
C. Maximum and minimum input voltage, minimum output current and maximum output voltage
D. Maximum and minimum input voltage, minimum output voltage and maximum output current

ANSWER B: Two important characteristics of a 3-terminal regulator are maximum and minimum input voltage, which are given in all answers. The other three are minimum output current and minimum and maximum output voltage. Only one answer indicates *maximum output current and voltage* (key words), and these are important considerations when choosing a 3-terminal regulator for a particular circuit.

3F10 What is the distinguishing feature of a Class A amplifier?
A. Output for less than 180 degrees of the signal cycle
B. Output for the entire 360 degrees of the signal cycle
C. Output for more than 180 degrees and less than 360 degrees of the signal cycle
D. Output for exactly 180 degrees of the input signal cycle

ANSWER B: The Class A amplifier works during the entire 360-degree signal cycle. It offers good linearity, but with all its constant work, may have low efficiency. See figure at question 3F11.

3F11 What class of amplifier is distinguished by the presence of output throughout the entire signal cycle and the input never goes into the cutoff region?
A. Class A
B. Class B
C. Class C
D. Class D

ANSWER A: If it works off an entire signal cycle (key words), it's a Class A amplifier.

Various Classes of Transistorized Amplifiers

3F12 What is the distinguishing characteristic of a Class B amplifier?

A. Output for the entire input signal cycle

B. Output for greater than 180 degrees and less than 360 degrees of the input signal cycle

C. Output for less than 180 degrees of the input signal cycle

D. Output for 180 degrees of the input signal cycle

ANSWER D: The Class B amplifier only works during half of the input cycle, or for a total of 180 degrees. We usually run Class B amplifiers in push/pull pairs. Their efficiency is higher than Class A amplifiers because each tube or transistor gets to rest for half of the cycle. See figure above.

3F13 What class of amplifier is distinguished by the flow of current in the output essentially in 180 degree pulses?

A. Class A

B. Class B

C. Class C

D. Class D

ANSWER B: If it's working for 180 degrees, it's a Class B amplifier. See figure at question 3F11.

3F14 What is a Class AB amplifier?

A. Output is present for more than 180 degrees but less than 360 degrees of the signal input cycle.

B. Output is present for exactly 180 degrees of the input signal cycle.

C. Output is present for the entire input signal cycle.

D. Output is present for less than 180 degrees of the input signal cycle.

ANSWER A: The Class AB amplifier has some of the properties of both the Class A and the Class B amplifier. It has output for more than 180 degrees of the cycle, but less than 360 degrees of the cycle. See figure at question 3F11.

3F15 What is the distinguishing feature of a Class C amplifier?
A. Output is present for less than 180 degrees of the input signal cycle.
B. Output is present for exactly 180 degrees of the input signal cycle.
C. Output is present for the entire input signal cycle.
D. Output is present for more than 180 degrees but less than 360 degrees of the input signal cycle.

ANSWER A: A Class C amplifier gets to rest a lot! It has great efficiency, but its linearity is not as good as a Class B or Class A amp. Class C amplifiers are best for digital modes. The Class C amplifier output is always less than 180 degrees of the input signal cycle. See figure at question 3F11.

3F16 What class of amplifier is distinguished by the bias being set well beyond cutoff?
A. Class A
B. Class B
C. Class C
D. Class AB

ANSWER C: If the bias is set well beyond cutoff, the amplifier will only work less than 180 degrees of the cycle, and is a Class C amplifier. Class C amplifiers are very efficient, but not so great on linearity. See figure at question 3F11.

3F17 Which class of amplifier provides the highest efficiency?
A. Class A
B. Class B
C. Class C
D. Class AB

ANSWER C: The highest efficiency comes out of a Class C amplifier. Since it operates the minimum time out of the cycle, it dissipates less power, but at the same time, produces narrow pulses which can cause loss of linearity.

3F18 Which class of amplifier has the highest linearity and least distortion?
A. Class A
B. Class B
C. Class C
D. Class AB

ANSWER A: The best amplifier for a pure signal with the highest linearity and least distortion is a Class A amplifier. See figure at question 3F11.

3F19 Which class of amplifier has an operating angle of more than 180 degrees but less than 360 degrees when driven by a sine wave signal?
A. Class A
B. Class B
C. Class C
D. Class AB

ANSWER D: If it's more than 180, but less than 360, it must be a Class AB amplifier. See figure at question 3F11.

3F20 What is an L-network?

A. A network consisting entirely of four inductors
B. A network consisting of an inductor and a capacitor
C. A network used to generate a leading phase angle
D. A network used to generate a lagging phase angle

ANSWER B: When troubleshooting two-way radio equipment, make absolutely sure you have the technical manual before going out to the job. This will help you spot different types of networks, such as the L-network that has an inductor and capacitor in it.

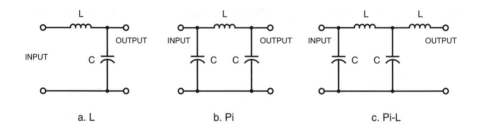

a. L   b. Pi   c. Pi-L

Matching Networks

3F21 What is a pi-network?

A. A network consisting entirely of four inductors or four capacitors
B. A Power Incidence network
C. An antenna matching network that is isolated from ground
D. A network consisting of one inductor and two capacitors or two inductors and one capacitor

ANSWER D: The pi-network consists of three components, looking like the Greek symbol pi ( ). The legs could be either capacitors or coils, and the horizontal line at the top could be either a coil or a capacitor.

3F22 What is a pi-L-network?

A. A Phase Inverter Load network
B. A network consisting of two inductors and two capacitors
C. A network with only three discrete parts
D. A matching network in which all components are isolated from ground

ANSWER B: The pi-L-network consists of two coils and two capacitors. Depending on how they are arranged, it could be a high-pass network or a low-pass network.

3F23 Which network provides the greatest harmonic suppression?

A. L-network
B. Pi-network
C. Inverse L-network
D. Pi-L-network

ANSWER D: The network with the most components will provide the greatest harmonic suppression. The pi-L-network is found in most worldwide marine SSB sets.

3F24 What are the three most commonly used networks to accomplish a match between an amplifying device and a transmission line?
>   A. M-network, pi-network and T-network
>   B. T-network, M-network and Q-network
>   C. L-network, pi-network and pi-L-network
>   D. L-network, M-network and C-network

ANSWER C: Worldwide and aeronautical long-range SSB equipment may have three different matching devices between the amplifier output and the transmission line—either pi, L, or pi-L networks.

3F25 How are networks able to transform one impedance to another?
>   A. Resistances in the networks substitute for resistances in the load.
>   B. The matching network introduces negative resistance to cancel the resistive part of an impedance.
>   C. The matching network introduces transconductance to cancel the reactive part of an impedance.
>   D. The matching network can cancel the reactive part of an impedance and change the value of the resistive part of an impedance.

ANSWER D: In older tube sets, you could vary the matching network to cancel the reactive (X) part of an impedance, and change the value of the resistive part (R) of an impedance. Newer transistorized sets have a fixed output, and there is no manual operator control of the fixed matching devices. However, manufacturers have come to the rescue in antenna matching by providing new worldwide SSB radios with built-in automatic antenna tuners.

3F26 Which type of network offers the greater transformation ratio?
>   A. L-network
>   B. Pi-network
>   C. Constant-K
>   D. Constant-M

ANSWER B: The pi-network is found in many worldwide transceivers. It offers the greatest transformation ratio.

3F27 Why is the L-network of limited utility in impedance matching?
>   A. It matches a small impedance range.
>   B. It has limited power handling capabilities.
>   C. It is thermally unstable.
>   D. It is prone to self resonance.

ANSWER A: The L-network alone can only match to a small impedance range. This is why we always prefer a pi-network.

3F28 What is an advantage of using a pi-L-network instead of a pi-network for impedance matching between the final amplifier of a vacuum-tube type transmitter and a multiband antenna?
>   A. Greater transformation range
>   B. Higher efficiency
>   C. Lower losses
>   D. Greater harmonic suppression

ANSWER D: Worldwide SSB equipment working into a multi-band antenna system, such as a quad antenna or a private coast station, could generate harmonics. The pi-L-network always gives us greater harmonic suppression.

3F29 Which type of network provides the greatest harmonic suppression?
A. L-network
B. Pi-network
C. Pi-L-network
D. Inverse-Pi network

ANSWER C: The greatest harmonic suppression is provided by a pi-L-network.

3F30 What are the three general groupings of filters?
A. High-pass, low-pass and band-pass
B. Inductive, capacitive and resistive
C. Audio, radio and capacitive
D. Hartley, Colpitts and Pierce

ANSWER A: High-pass filters are found in older TV sets. Low-pass filters are found in marine SSB sets. Band-pass filters may be found in VHF and UHF equipment.

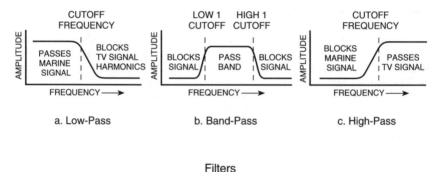

a. Low-Pass b. Band-Pass c. High-Pass

Filters

3F31 What is a constant-K filter?
A. A filter that uses Boltzmann's constant
B. A filter whose velocity factor is constant over a wide range of frequencies
C. A filter whose product of the series- and shunt-element impedances is a constant for all frequencies
D. A filter whose input impedance varies widely over the design bandwidth

ANSWER C: The constant K-filter is *constant for all frequencies* (key words), as opposed to a wide range of frequencies in answer B.

3F32 What is an advantage of a constant-k filter?
A. It has high attenuation for signals on frequencies far removed from the passband.
B. It can match impedances over a wide range of frequencies.
C. It uses elliptic functions.
D. The ratio of the cutoff frequency to the trap frequency can be varied.

ANSWER A: The constant K-filter continues to suppress harmonics and spurs far removed from its pass band.

3F33 What is an m-derived filter?

A. A filter whose input impedance varies widely over the design band-width

B. A filter whose product of the series- and shunt-element impedances is a constant for all frequencies

C. A filter whose schematic shape is the letter "M"

D. A filter that uses a trap to attenuate undesired frequencies too near cutoff for a constant-k filter

ANSWER D: The m-derived filter has a tuned *trap* (key word) to attenuate undesired spurious emissions too near the normal cutoff for a constant-K filter.

3F34 What are the distinguishing features of a Butterworth filter?

A. A filter whose product of the series- and shunt-element impedances is a constant for all frequencies.

B. It only requires capacitors.

C. It has a maximally flat response over its passband.

D. It requires only inductors.

ANSWER C: The Butterworth filter can be counted on for its flat response over the entire pass band of frequencies. Its cutoff, unfortunately, is not as sharp as other filters.

3F35 What are the distinguishing features of a Chebyshev filter?

A. It has a maximally flat response over its passband.

B. It allows ripple in the passband.

C. It only requires inductors.

D. A filter whose product of the series- and shunt-element impedances is a constant for all frequencies.

ANSWER B: It's a man's name who created the design. It sounds like "Chevy" connected to "Shev," but it's "Chebyshev." The Chebyshev filter may allow ripple to pass through on the pass band.

3F36 When would it be more desirable to use an m-derived filter over a constant-k filter?

A. When the response must be maximally flat at one frequency

B. When you need more attenuation at a certain frequency that is too close to the cut-off frequency for a constant-k filter

C. When the number of components must be minimized

D. When high power levels must be filtered

ANSWER B: Occasionally you need to knock down a close-in spur next to your transmitted signal. We would use an m-derived filter because it offers more attenuation at a certain frequency close to its cutoff.

3F37 What are three major oscillator circuits often used in radio equipment?

A. Taft, Pierce, and negative feedback

B. Colpitts, Hartley, and Taft

C. Taft, Hartley, and Pierce

D. Colpitts, Hartley, and Pierce

ANSWER D: Colpitts oscillators have a capacitor just like C in its name. Hartley is tapped, and a Pierce oscillator uses a crystal.

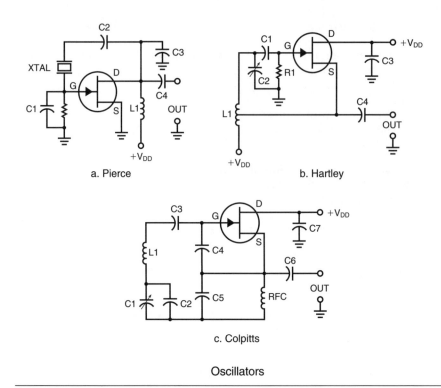

a. Pierce

b. Hartley

c. Colpitts

Oscillators

3F38 How is the positive feedback coupled to the input in a Hartley oscillator?
A. Through a neutralizing capacitor
B. Through a capacitive divider
C. Through link coupling
D. Through a tapped coil
ANSWER D: Hartley is always tapped. The tapped coil provides inductive coupling for positive feedback.

3F39 How is the positive feedback coupled to the input in a Colpitts oscillator?
A. Through a tapped coil
B. Through link coupling
C. Through a capacitive divider
D. Through a neutralizing capacitor
ANSWER C: On the Colpitts oscillator, we use a capacitive divider to provide feedback.

3F40 How is the positive feedback coupled to the input in a Pierce oscillator?
A. Through a tapped coil
B. Through link coupling
C. Through a capacitive divider
D. Through capacitive coupling

ANSWER D: On the Pierce oscillator, we also use capacitive coupling. Think of a quartz crystal as an element sandwiched between two plates. The two plates act a little bit like a capacitor.

3F41 Which of the three major oscillator circuits used in radio equipment utilizes a quartz crystal?
   A. Negative feedback
   B. Hartley
   C. Colpitts
   D. Pierce
ANSWER D: Quartz crystals are found in Pierce oscillators.

3F42 What is the piezoelectric effect?
   A. Mechanical vibration of a crystal by the application of a voltage
   B. Mechanical deformation of a crystal by the application of a magnetic field
   C. The generation of electrical energy by the application of light
   D. Reversed conduction states when a P-N junction is exposed to light
ANSWER A: If you apply a voltage to a quartz crystal, it will vibrate at a specific frequency. This is the reason crystal oscillators are so accurate. They oscillate at the frequency, or harmonics of the frequency, of the crystal, and continue to do so with great accuracy unless the temperature changes outside design limits.

3F43 What is the major advantage of a Pierce oscillator?
   A. It is easy to neutralize.
   B. It doesn't require an LC tank circuit.
   C. It can be tuned over a wide range.
   D. It has a high output power.
ANSWER B: The nice thing about a Pierce oscillator is that it doesn't require an inductive-capacitive tank circuit for excellent frequency stability. It operates on one frequency, not variable.

3F44 Which type of oscillator circuit is commonly used in a VFO?
   A. Pierce
   B. Colpitts
   C. Hartley
   D. Negative feedback
ANSWER B: Since the Colpitts oscillator uses a big capacitor in a variable frequency oscillator (VFO), it's the most common oscillator circuit for older VFO radios. Newer radios, even though they say they have a VFO, are really digitally controlled via an optical reader. They just look like they have a big capacitor behind that big tuning dial!

3F45 Why is the Colpitts oscillator circuit commonly used in a VFO?
   A. The frequency is a linear function of the load impedance.
   B. It can be used with or without crystal lock-in.
   C. It is stable.
   D. It has high output power.
ANSWER C: The Colpitts oscillator is relatively stable. After you let things warm up, the Colpitts oscillator normally stays put on a single frequency.

3F46 What is meant by the term: modulation?

A. The squelching of a signal until a critical signal-to-noise ratio is reached
B. Carrier rejection through phase nulling
C. A linear amplification mode
D. A mixing process whereby information is imposed upon a carrier

ANSWER D: When we speak into a microphone, we are putting information out on the air on a radio frequency carrier. In amplitude modulation, our modulation varies the carrier amplitude; in frequency modulation, our modulation alters the carrier frequency.

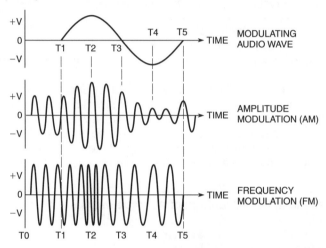

Modulation: from T0 to T1, carrier frequency is unmodulated; from T1 through T5, audio wave modulates carrier frequency.

3F47 How is a G3E FM-phone emission produced?

A. With a balanced modulator on the audio amplifier
B. With a reactance modulator on the oscillator
C. With a reactance modulator on the final amplifier
D. With a balanced modulator on the oscillator

ANSWER B: The *reactance modulator* (key words) causes the oscillator to vary frequency in accordance with the *phase* modulation. It's the oscillator (key word) that is modulated.

3F48 What is a reactance modulator?

A. A circuit that acts as a variable resistance or capacitance to produce FM signals
B. A circuit that acts as a variable resistance or capacitance to produce AM signals
C. A circuit that acts as a variable inductance or capacitance to produce FM signals
D. A circuit that acts as a variable inductance or capacitance to produce AM signals

ANSWER C: The reactance (key word) modulator is found in an FM transceiver, and varies the inductance or capacitance to produce the FM signal. The key words are *inductance, capacitance* and *FM*. Watch out for answer A!

3F49 What is a balanced modulator?
A. An FM modulator that produces a balanced deviation
B. A modulator that produces a double sideband, suppressed carrier signal
C. A modulator that produces a single sideband, suppressed carrier signal
D. A modulator that produces a full carrier signal

ANSWER B: The balanced modulator produces an upper and lower sideband, with the carrier suppressed. Remember, the double sidebands are *balanced* into an upper and lower sideband around a *suppressed* carrier.

3F50 How can a single-sideband phone signal be generated?
A. By driving a product detector with a DSB signal
B. By using a reactance modulator followed by a mixer
C. By using a loop modulator followed by a mixer
D. By using a balanced modulator followed by a filter

ANSWER D: In a single-sideband transceiver, a lower or upper sideband filter removes the unwanted sideband.

3F51 How can a double-sideband phone signal be generated?
A. By feeding a phase modulated signal into a low pass filter
B. By using a balanced modulator followed by a filter
C. By detuning a Hartley oscillator
D. By modulating the plate voltage of a class C amplifier

ANSWER D: For a double-sideband signal, we could modulate the plate voltage of an amplifier for amplitude modulation.

3F52 How is the efficiency of a power amplifier determined?
A. Efficiency = (RF power out / DC power in) × 100%
B. Efficiency = (RF power in / RF power out) × 100%
C. Efficiency = (RF power in / DC power in) × 100%
D. Efficiency = (DC power in / RF power in) × 100%

ANSWER A: The typical efficiency of a power amplifier is approximately half the power output for the amount of power input. The output power is RF power; the input power is DC power. The formula is: Eff(%) = P_{RF}/P_{DC} × 100. You can spot the answer by the key word *out*, followed by the key word *in*.

3F53 For reasonably efficient operation of a transistor amplifier, what should the load resistance be with 12 volts at the collector and 5 watts power output?
A. 100.3 ohms
B. 14.4 ohms
C. 10.3 ohms
D. 144 ohms

ANSWER B: To calculate the optimum load resistance for a transistor amplifier (assuming maximum efficiency of 50%), the following formula is used:

$$R_L = \frac{V_{CC}^2}{2P_O}$$ Where: R_L = Load resistance in **ohms**
V_{CC} = Transistor's collector voltage in **volts**
P_O = Amplifier's power output in **watts**

In this problem, 12 volts (V_{CC}) squared equals 144. Two times P_O is $2 \times 5 = 10$. Dividing 144 by 10 equals 14.4 ohms. The calculator keystrokes are: Clear, 12 $\times$ 12 = 144 $\div$ 10 = 14.4.

3F54 What is the flywheel effect?

A. The continued motion of a radio wave through space when the transmitter is turned off
B. The back and forth oscillation of electrons in an LC circuit
C. The use of a capacitor in a power supply to filter rectified AC
D. The transmission of a radio signal to a distant station by several hops through the ionosphere

ANSWER B: Some of you may remember the old farm machinery that had an external engine flywheel. The flywheel keeps the back and forth oscillations going. The flywheel effect is the way we describe how electrons flow in an LC circuit.

3F55 What order of Q is required by a tank-circuit sufficient to reduce harmonics to an acceptable level?

A. Approximately 120
B. Approximately 12
C. Approximately 1200
D. Approximately 1.2

ANSWER B: A nice tight tank circuit with enough Q to reduce harmonics will have a Q of at least 12. The higher the number, the greater the quality (Q) of the circuit. You might remember that you need "quality above 10" to help you choose the right answer.

3F56 How can parasitic oscillations be eliminated from a power amplifier?

A. By tuning for maximum SWR
B. By tuning for maximum power output
C. By neutralization
D. By tuning the output

ANSWER C: We can reduce parasitic oscillations, and clean up our signal, through the steps of neutralization in a power amplifier.

3F57 What is the process of detection?

A. The process of masking out the intelligence on a received carrier to make an S-meter operational
B. The recovery of intelligence from the modulated RF signal
C. The modulation of a carrier
D. The mixing of noise with the received signal

ANSWER B: Your radio's detector *recovers intelligence* (key words) from a modulated RF signal. You have modulation to put information on a carrier; you have detection to take information off a carrier.

3F58 What is the principle of detection in a diode detector?

A. Rectification and filtering of RF
B. Breakdown of the Zener voltage
C. Mixing with noise in the transition region of the diode
D. The change of reactance in the diode with respect to frequency

ANSWER A: Since a diode only conducts for half of the AC signal, it may be used as a rectifier. By filtering out the radio frequency energy after rectification, detection is accomplished—the modulating signal is what remains.

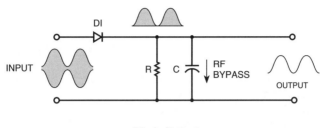

Diode Detector

3F59 What is a product detector?
A. A detector that provides local oscillations for input to the mixer
B. A detector that amplifies and narrows the band-pass frequencies
C. A detector that uses a mixing process with a locally generated carrier
D. A detector used to detect cross-modulation products

ANSWER C: A product detector is found in SSB receivers. It mixes the incoming signal with a beat frequency oscillator signal. The beat frequency oscillator signal is a locally generated carrier that is mixed with the incoming signal.

3F60 How are FM-phone signals detected?
A. By a balanced modulator
B. By a frequency discriminator
C. By a product detector
D. By a phase splitter

ANSWER B: FM signals must be detected differently than AM signals because frequency changes must be detected rather than amplitude changes. We use a reactance modulator to transmit FM. We use a *frequency discriminator* (key words) to detect an FM phone signal.

3F61 What is a frequency discriminator?
A. A circuit for detecting FM signals
B. A circuit for filtering two closely adjacent signals
C. An automatic bandswitching circuit
D. An FM generator

ANSWER A: Anytime you see the words *frequency discriminator* you know that you are dealing with frequency modulation and an FM transceiver.

3F62 What is the mixing process?
A. The elimination of noise in a wideband receiver by phase comparison
B. The elimination of noise in a wideband receiver by phase differentiation
C. Distortion caused by auroral propagation
D. The combination of two signals to produce sum and difference frequencies

ANSWER D: Inside the two-way radio equipment are several stages of mixers that combine two input signals to produce sum and different frequencies. These two inputs are the desired RF signal and the local oscillator (LO) signal. One of the output signals becomes the intermediate frequency (IF) signal inside the receiver section.

3F63 What are the principal frequencies that appear at the output of a mixer circuit?
A. Two and four times the original frequency
B. The sum, difference and square root of the input frequencies
C. The original frequencies and the sum and difference frequencies
D. 1.414 and 0.707 times the input frequency

ANSWER C: Out of the mixer comes your original two frequencies, and the sum and difference frequencies. The more elaborate the transceiver, the more mixing stages found in the sets. This helps filter out unwanted signals or phantom signals that could cause interference.

3F64 What are the advantages of the frequency-conversion process?
A. Automatic squelching and increased selectivity
B. Increased selectivity and optimal tuned-circuit design
C. Automatic soft limiting and automatic squelching
D. Automatic detection in the RF amplifier and increased selectivity

ANSWER B: Receivers with dual conversion are better than a set with only a single conversion. Triple conversion is better yet. The more stages of conversion, the more stages of increased selectivity and optimal tuned-circuit design.

3F65 What occurs in a receiver when an excessive amount of signal energy reaches the mixer circuit?
A. Spurious mixer products are generated.
B. Mixer blanking occurs.
C. Automatic limiting occurs.
D. A beat frequency is generated.

ANSWER A: Some VHF power amplifiers may incorporate a hefty pre-amplifier circuit for use on receiving. It boosts incoming signal levels. The pre-amp turned on could generate spurious mixer products as a result of the high signal level. The spurious products are not really signals that you are trying to receive. If you use a handheld with a power amp, turn the pre-amp off.

3F66 How much gain should be used in the RF amplifier stage of a receiver?
A. As much gain as possible short of self oscillation
B. Sufficient gain to allow weak signals to overcome noise generated in the first mixer stage
C. Sufficient gain to keep weak signals below the noise of the first mixer stage
D. It depends on the amplification factor of the first IF stage

ANSWER B: Very sensitive receiver transistors now give the signal amplifier stages all the gain they need. If you try to add a pre-amplifier on a modern two-way transceiver, you could create more interference than the possible weak-signal amplification would gain. You need enough gain to allow weak signals to overcome noise generated in the first mixer stage.

3F – PRACTICAL CIRCUITS

3F67 Why should the RF amplifier stage of a receiver only have sufficient gain to allow weak signals to overcome noise generated in the first mixer stage?
  A. To prevent the sum and difference frequencies from being generated
  B. To prevent bleed-through of the desired signal
  C. To prevent the generation of spurious mixer products
  D. To prevent bleed-through of the local oscillator
ANSWER C: Too much additional gain by an outside pre-amp could generate spurious mixer products and a terrible-sounding receiver.

3F68 What is the primary purpose of an RF amplifier in a receiver?
  A. To provide most of the receiver gain
  B. To vary the receiver image rejection by utilizing the AGC
  C. To improve the receiver's noise figure
  D. To develop the AGC voltage
ANSWER C: The RF amplifier, factory designed for your receiver, improves the *receiver's noise figure* (key words). Low noise transistors are now available for an improved noise-figure rating.

3F69 What is an i-f amplifier stage?
  A. A fixed-tuned pass-band amplifier
  B. A receiver demodulator
  C. A receiver filter
  D. A buffer oscillator
ANSWER A: The IF (intermediate frequency) amplifier is the next stage after the mixer and is tuned to the incoming intermediate frequency. It is fixed-tuned as a pass-band amplifier.

3F70 What factors should be considered when selecting an intermediate frequency?
  A. Cross-modulation distortion and interference
  B. Interference to other services
  C. Image rejection and selectivity
  D. Noise figure and distortion
ANSWER C: One of the key areas of receiver design is *image rejection* and *selectivity* (key words). Your author prefers a selective receiver any day over one that is slightly more sensitive.

3F71 What is the primary purpose of the first i-f amplifier stage in a receiver?
  A. Noise figure performance
  B. Tune out cross-modulation distortion
  C. Dynamic response
  D. Selectivity
ANSWER D: The first intermediate-frequency stage in an amplifier provides selectivity and image rejection. Look for the answer *selectivity* and *image rejection* (key words) on your examination. In this question, only *selectivity* is given.

3F72 What is the primary purpose of the final i-f amplifier stage in a receiver?

    A. Dynamic response
    B. Gain
    C. Noise figure performance
    D. Bypass undesired signals

ANSWER B: The purpose of the final IF amplifier stage in a receiver is both *gain* (key word) as well as selectivity. The answer for this question is gain.

3F73 What is a flip-flop circuit?

    A. A binary sequential logic element with one stable state
    B. A binary sequential logic element with eight stable states
    C. A binary sequential logic element with four stable states
    D. A binary sequential logic element with two stable states

ANSWER D: *"Flip flop"*, two words, two states. A flip-flop circuit is a form of bistable multivibrator—it remains in one of two states until triggered to change to the other state. It is a bistable sequential logic circuit.

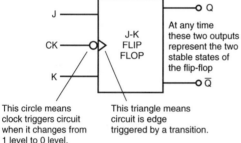

J_0	K_0	Q_1	$\overline{Q}_1$
0	0	Q_0	$\overline{Q}_0$
0	1	0	1
1	0	1	0
1	1	$\overline{Q}_0$	Q_0

0 Subscript is time before clock pulse
1 Subscript is time after clock pulse

Flip-Flop Circuit

3F74 How many bits of information can be stored in a single flip-flop circuit?

    A. 1
    B. 2
    C. 3
    D. 4

ANSWER A: One flip-flop circuit can only store a single bit of information. If semiconductor memories were made of flip-flops, one flip-flop would be required for each bit.

3F75 What is a bistable multivibrator circuit?

    A. An "AND" gate
    B. An "OR" gate
    C. A flip-flop
    D. A clock

ANSWER C: A bistable multivibrator is a flip-flop. Bistable means two stable states.

3F76 How many output changes are obtained for every two trigger pulses applied to the input of a bistable T flip-flop circuit?
    A. No output level changes
    B. One output level change
    C. Two output level changes
    D. Four output level changes
ANSWER C: Two trigger pulses, two output level changes.

3F77 The frequency of an AC signal can be divided electronically by what type of digital circuit?
    A. A free-running multivibrator
    B. An OR gate
    C. A bistable multivibrator
    D. An astable multivibrator
ANSWER C: A bistable multivibrator can be used to divide the frequency of an AC signal.

3F78 What type of digital IC is also known as a latch?
    A. A decade counter
    B. An OR gate
    C. A flip-flop
    D. An op-amp
ANSWER C: A flip-flop is used in a digital IC circuit that latches onto a bit and holds it until the time when it is triggered to latch onto the next bit.

3F79 How many flip-flops are required to divide a signal frequency by 4?
    A. 1
    B. 2
    C. 4
    D. 8
ANSWER B: Since a flip-flop has two stable states, you will need two flip-flops to divide a signal frequency by four.

3F80 What is an astable multivibrator?
    A. A circuit that alternates between two stable states
    B. A circuit that alternates between a stable state and an unstable state
    C. A circuit set to block either a 0 pulse or a 1 pulse and pass the other
    D. A circuit that alternates between two unstable states
ANSWER D: An *astable* multivibrator continuously switches back and forth between two states. It doesn't remain stable for a permanent time in either of its two states.

3F81 What is a monostable multivibrator?
    A. A circuit that can be switched momentarily to the opposite binary state and then returns after a set time to its original state
    B. A "clock" circuit that produces a continuous square wave oscillating between 1 and 0
    C. A circuit designed to store one bit of data in either the 0 or the 1 configuration
    D. A circuit that maintains a constant output voltage, regardless of variations in the input voltage

ANSWER A: This type of multivibrator may *momentarily* be *monostable* (*MO MO*). It stays in an original state until triggered to its other state, where it remains for a time usually determined by external components, after which it returns to the original state.

3F82 What is an AND gate?

A. A circuit that produces a logic "1" at its output only if all inputs are logic "1"

B. A circuit that produces a logic "0" at its output only if all inputs are logic "1"

C. A circuit that produces a logic "1" at its output if only one input is a logic "1"

D. A circuit that produces a logic "1" at its output if all inputs are logic "0"

ANSWER A: The AND gate has two or more inputs with a single output, and produces a logic "1" at its output only if *all* inputs are logic "1". Remember *AND*, *all*, and the number 1 and the number 1 looking like the letters ll in "all."

3F83 What is a NAND gate?

A. A circuit that produces a logic "0" at its output only when all inputs are logic "0"

B. A circuit that produces a logic "1" at its output only when all inputs are logic "1"

C. A circuit that produces a logic "0" at its output if some but not all of its inputs are logic "1"

D. A circuit that produces a logic "0" at its output only when all inputs are logic "1"

ANSWER D: The word *NAND* reminds me of the word "naw", meaning no or nothing. The *NAND* gate produces a logic "0" at its output only when all inputs are logic "1". N before AND means negative AND. If an AND gate outputs a "1" when all inputs are "1", a NAND gate outputs a "0", the negative of "1".

3F84 What is an OR gate?

A. A circuit that produces a logic "1" at its output if any input is logic "1"

B. A circuit that produces a logic "0" at its output if any input is logic "1"

C. A circuit that produces a logic "0" at its output if all inputs are logic "1"

D. A circuit that produces a logic "1" at its output if all inputs are logic "0"

ANSWER A: The OR gate will produce a logic "1" at its output if *any* input is, or all inputs are, a logic "1".

3F85 What is a NOR gate?

A. A circuit that produces a logic "0" at its output only if all inputs are logic "0"

B. A circuit that produces a logic "1" at its output only if all inputs are logic "1"

C. A circuit that produces a logic "0" at its output if any or all inputs are logic "1"

D. A circuit that produces a logic "1" at its output if some but not all of its inputs are logic "1"

ANSWER C: The NOR gate produces a logic "0" at its output if *ANY* or all inputs are logic "1". It's a negative OR. With OR gate or NOR gate, spot that word "ANY".

3F86 What is a NOT gate?

A. A circuit that produces a logic "O" at its output when the input is logic "1" and vice versa

B. A circuit that does not allow data transmission when its input is high

C. A circuit that allows data transmission only when its input is high

D. A circuit that produces a logic "1" at its output when the input is logic "1" and vice versa

ANSWER A: The NOT gate will produce a "0" or a "1" at the output—whichever is opposite from the level that is on the input.

3F87 What is a truth table?

A. A table of logic symbols that indicate the high logic states of an op-amp

B. A diagram showing logic states when the digital device's output is true

C. A list of input combinations and their corresponding outputs that characterizes a digital device's function

D. A table of logic symbols that indicates the low logic states of an op-amp

ANSWER C: We use a "truth table" in digital circuitry to characterize a digital device's function. If you buy the owner's technical manual on a piece of radio gear, you will usually find a page describing the truth table of digital devices used in the equipment.

3F88 In a positive-logic circuit, what level is used to represent a logic 1?

A. A low level

B. A positive-transition level

C. A negative-transition level

D. A high level

ANSWER D: In a positive-logic circuit, the logic "1" is a high level, or the most positive level. Think of the four letters in *high* and the four letters in *plus* (for positive).

3F89 In a positive-logic circuit, what level is used to represent a logic 0?

A. A low level

B. A positive-transition level

C. A negative-transition level

D. A high level

ANSWER A: In a positive-logic circuit, a logic "0" is a low level, or the least positive level.

3F90 In a negative-logic circuit, what level is used to represent a logic 1?
A. A low level
B. A positive-transition level
C. A negative-transition level
D. A high level
ANSWER A: In a negative-logic circuit, the logic "1" is now a low level. H=0, L=1. All 1s and 0s are interchanged from that in a postive logic truth table.

3F91 In a negative-logic circuit, what level is used to represent a logic 0?
A. A low level
B. A positive-transition level
C. A negative-transition level
D. A high level
ANSWER D: In a negative-logic circuit, the logic "0" is now a high level. H=0, L=1. All 1s and 0s are interchanged from that in a postive logic truth table.

3F92 What is a crystal-controlled marker generator?
A. A low-stability oscillator that "sweeps" through a band of frequencies
B. An oscillator often used in aircraft to determine the craft's location relative to the inner and outer markers at airports
C. A high-stability oscillator whose output frequency and amplitude can be varied over a wide range
D. A high-stability oscillator that generates a series of reference signals at known frequency intervals
ANSWER D: The crystal-controlled marker generator is a free-running oscillator that generates a series of reference signals at known frequency intervals.

3F93 What additional circuitry is required in a 100-kHz crystal-controlled marker generator to provide markers at 50 and 25 kHz?
A. An emitter-follower
B. Two frequency multipliers
C. Two flip-flops
D. A voltage divider
ANSWER C: We would use two flip-flops to take a 100-kHz signal and provide markers at 50 kHz and 25 kHz.

3F94 What is the purpose of a prescaler circuit?
A. It converts the output of a JK flip-flop to that of an RS flip-flop.
B. It multiplies an HF signal so a low-frequency counter can display the operating frequency.
C. It prevents oscillation in a low frequency counter circuit.
D. It divides an HF signal so a low-frequency counter can display the operating frequency.
ANSWER D: You will find a prescaler in frequency counters to divide down HF and VHF signals so they can be counted and displayed on a low-frequency counter.

3F95 What does the accuracy of a frequency counter depend on?
A. The internal crystal reference
B. A voltage-regulated power supply with an unvarying output

C. Accuracy of the AC input frequency to the power supply

D. Proper balancing of the power-supply diodes

ANSWER A: Every service shop should have a high frequency receiver that can tune in WWV and WWVH signals at 2.5 MHz, 5 MHz, 10 MHz, 15 MHz, or 20 MHz. If you hear the female voice, that is from WWVH in Hawaii. Either station may be used for frequency calibration of your frequency counter.

3F96 How many states does a decade counter digital IC have?

A. 6

B. 10

C. 15

D. 20

ANSWER B: Studying this material may seem like it is taking a decade, right? A decade is 10. Maybe a decade of days, or a decade of weeks, but, hopefully, not a decade of months.

3F97 What is the function of a decade counter digital IC?

A. Decode a decimal number for display on a seven-segment LED display

B. Produce one output pulse for every ten input pulses

C. Produce ten output pulses for every input pulse

D. Add two decimal numbers

ANSWER B: A decade counter digital IC gives one output pulse for every 10 input pulses.

3F98 What are the advantages of using an op-amp instead of LC elements in an audio filter?

A. Op-amps are more rugged and can withstand more abuse than can LC elements.

B. Op-amps are fixed at one frequency.

C. Op-amps are available in more styles and types than are LC elements.

D. Op-amps exhibit gain rather than insertion loss.

ANSWER D: An op amp exhibits high gain with negligible insertion loss because its input impedance is also high.

3F99 What determines the gain and frequency characteristics of an op-amp RC active filter?

A. Values of capacitances and resistances built into the op-amp

B. Values of capacitances and resistances external to the op-amp

C. Voltage and frequency of DC input to the op-amp power supply

D. Regulated DC voltage output from the op-amp power supply

ANSWER B: One of the nice features of an op amp is that it will operate as an RC filter—Resistors (R) and capacitors (C) are in a circuit outside the op amp. They are not built into the op amp; they are *external* (key word).

3F100 What are the principle uses of an op-amp RC active filter?

A. Op-amp circuits are used as high-pass filters to block RFI at the input to receivers.

B. Op-amp circuits are used as low-pass filters between transmitters and transmission lines.

C. Op-amp circuits are used as filters for smoothing power-supply output.

D. Op-amp circuits are used as audio filters for receivers.

ANSWER D: In radio transceivers, op-amp circuits are used in the audio section of receivers.

3F101 What type of capacitors should be used in an op-amp RC active filter circuit?

A. Electrolytic

B. Disc ceramic

C. Polystyrene

D. Paper dielectric

ANSWER C: The tiny op-amp filter should use polystyrene capacitors because they are very stable and will not vary when the circuit begins to heat up.

3F102 How can unwanted ringing and audio instability be prevented in a multisection op-amp RC audio filter circuit?

A. Restrict both gain and Q

B. Restrict gain, but increase Q

C. Restrict Q, but increase gain

D. Increase both gain and Q

ANSWER A: In order to keep an op amp from going into oscillation, the gain and the Q are restricted and set by the feedback circuit.

3F103 Where should an op-amp RC active audio filter be placed in a receiver?

A. In the IF strip, immediately before the detector

B. In the audio circuitry immediately before the speaker or phone jack

C. Between the balanced modulator and frequency multiplier

D. In the low-level audio stages

ANSWER D: We use an operational amplifier in the low-level audio stages in most mobile two-way radio transceivers.

3F104 What parameter must be selected when designing an audio filter using an op-amp?

A. Bandpass characteristics

B. Desired current gain

C. Temperature coefficient

D. Output-offset overshoot

ANSWER A: What frequencies do you want an op amp to amplify? This will help determine the bandpass characteristics required of the op amp.

3F105 What two factors determine the sensitivity of a receiver?

A. Dynamic range and third-order intercept

B. Cost and availability

C. Intermodulation distortion and dynamic range

D. Bandwidth and noise figure

ANSWER D: VHF and UHF transceivers with top sensitivity use low-noise front-end transistors. When choosing a receiver, look closely at the specifications for the bandwidth and the noise figure.

3F106 What is the limiting condition for sensitivity in a communications receiver?
A. The noise floor of the receiver
B. The power-supply output ripple
C. The two-tone intermodulation distortion
D. The input impedance to the detector

ANSWER A: Very weak signals get masked by the background noise of a receiver—the so-called noise floor of the receiver. The sensitivity of a receiver is limited by the noise generated by the devices inside, which contribute to the receiver's noise floor.

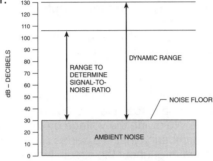

Noise Floor in a Receiver

Source: *Installing and Maintaining Sound Systems*, G. McComb, ©1996, Master Publishing, Inc.

3F107 What is the theoretical minimum noise floor of a receiver with a 400-hertz bandwidth?
A. −141 dBm
B. −148 dBm
C. −174 dBm
D. −180 dBm

ANSWER B: A receiver with a CW filter switched in at 400-Hz bandwidth will have approximately a −148 dBm minimum noise floor.
Noise threshold (dBn) = −174 + 10 LOG (bandwidth) IF (+ noise figure)

3F108 How can selectivity be achieved in the front-end circuitry of a communications receiver?
A. By using an audio filter
B. By using a preselector
C. By using an additional RF amplifier stage
D. By using an additional IF amplifier stage

ANSWER B: A preselector achieves selectivity in the front-end circuitry of a communications receiver. When you change bands, different preselectors are automatically brought on line.

3F109 A receiver selectivity of 2.4 kHz in the IF circuitry is optimum for what type of signals?
A. CW
B. SSB voice
C. Double-sideband AM voice
D. FSK RTTY

ANSWER B: A good 2.4-kHz filter in the IF circuitry is perfect for SSB voice. A 1.8 kHz filter would even be better, but might make the audio sound pinched.

3F110 What occurs during CW reception if too narrow a filter bandwidth is used in the IF stage of a receiver?
 A. Undesired signals will reach the audio stage
 B. Output-offset overshoot
 C. Cross-modulation distortion
 D. Filter ringing

ANSWER D: While working CW, if you adjust the shaping of your filter network in the IF stage of a receiver, you can only pinch the bandpass so tight. Any further tightening creates filter ringing.

3F111 (Refer to figure EL3-F1) To find the supply power with only a voltmeter, measure between:
 A. Z to W voltage times 150,000 divided by 2000.
 B. Y to Z voltage squared divided by 2000.
 C. W to Y voltage divided by 150.
 D. W to X voltage divided by 150,000 times W to Z voltage.

ANSWER D: If we measure the voltage with our voltmeter between points W and X, we will read the voltage drop across the 150,000 ohm resistor. Now to solve for current, use I = E ÷ R. We divide the voltage by 150,000, to determine the current for the circuit, remembering that in a series circuit like this, we will have the same amount of current throughout the circuit. Now that we have calculated the current, use P = E × I to find the power. Multiply the current by the voltage between W and Z, to calculate the total supply power.

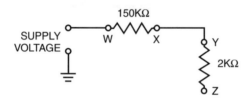

Figure EL3-F1

3F112 A receiver selectivity of 10 kHz in the IF circuitry is optimum for what type of signals?
 A. SSB voice
 B. Double-sideband AM
 C. CW
 D. FSK RTTY

ANSWER B: A very wide 10-kHz filter would be appropriate for tuning in short-wave broadcast stations on double-sideband AM.

3F113 What degree of selectivity is desirable in the IF circuitry of a single-sideband phone receiver?
 A. 1 kHz
 B. 2.4 kHz
 C. 4.2 kHz
 D. 4.8 kHz

ANSWER B: For SSB, a 2.4-kHz filter bandwidth is the most popular for good fidelity audio.

3F114 What is an undesirable effect of using too wide a filter bandwidth in the IF section of a receiver?
  A. Output-offset overshoot
  B. Undesired signals will reach the audio stage
  C. Thermal-noise distortion
  D. Filter ringing
ANSWER B: By using an AM 10-kHz IF filter for pulling in SSB signals, it would result in undesired signals beyond the normal bandwidth of the desired signal.

3F115 How should the filter bandwidth of a receiver IF section compare with the bandwidth of a received signal?
  A. Filter bandwidth should be slightly greater than the received-signal bandwidth
  B. Filter bandwidth should be approximately half the received-signal bandwidth
  C. Filter bandwidth should be approximately two times the received-signal bandwidth
  D. Filter bandwidth should be approximately four times the received-signal bandwidth
ANSWER A: For best fidelity, filter bandwidth should be slightly greater than the received signal bandwidth. If you are using one of those new SSB base stations, you have plenty of choices!

3F116 What degree of selectivity is desirable in the IF circuitry of an FM-phone receiver?
  A. 1 kHz
  B. 2.4 kHz
  C. 4.2 kHz
  D. 15 kHz
ANSWER D: In your FM equipment, the filter must accommodate a minimum of ± 5 kHz, a total of 10-kHz bandwidth. The 15-kHz filter is your best choice.

3F117 How can selectivity be achieved in the IF circuitry of a communications receiver?
  A. Incorporate a means of varying the supply voltage to the local oscillator circuitry
  B. Replace the standard JFET mixer with a bipolar transistor followed by a capacitor of the proper value
  C. Remove AGC action from the IF stage and confine it to the audio stage only
  D. Incorporate a high-Q filter
ANSWER D: Circuits with high-Q filters will give you the best selectivity in the IF section.

3F118 What is meant by the dynamic range of a communications receiver?
A. The number of kHz between the lowest and the highest frequency to which the receiver can be tuned
B. The maximum possible undistorted audio output of the receiver, referenced to one milliwatt
C. The ratio between the minimum discernible signal and the largest tolerable signal without causing audible distortion products
D. The difference between the lowest-frequency signal and the highest-frequency signal detectable without moving the tuning knob
ANSWER C: The dynamic range of a communications receiver is a *ratio* of the maximum signal to the minimum signal without distortion. Look for the key word *ratio* in the answer to this question.

3F119 What is the term for the ratio between the largest tolerable receiver input signal and the minimum discernible signal?
A. Intermodulation distortion
B. Noise floor
C. Noise figure
D. Dynamic range
ANSWER D: If your client's older two-way radio equipment sounds distorted when receiving extremely strong signals, chances are it has poor dynamic range. Good dynamic range allows the receiver to capture and send to the amplifier stage good, clean audio on extremely strong signal inputs all the way down to the minimum discernible signal.

3F120 What type of problems are caused by poor dynamic range in a communications receiver?
A. Cross-modulation of the desired signal and desensitization from strong adjacent signals
B. Oscillator instability requiring frequent retuning, and loss of ability to recover the opposite sideband, should it be transmitted
C. Cross-modulation of the desired signal and insufficient audio power to operate the speaker
D. Oscillator instability and severe audio distortion of all but the strongest received signals
ANSWER A: In a receiver that does not have good dynamic range, cross-modulation of the desired signal can be encountered, and the receiver can drop in sensitivity (be desensitized) by strong adjacent frequency signals.

3F121 The ability of a communications receiver to perform well in the presence of strong signals outside the band of interest is indicated by what parameter?
A. Noise figure
B. Blocking dynamic range
C. Signal-to-noise ratio
D. Audio output
ANSWER B: Look for a receiver that has good "blocking" to keep unwanted signals out of the pass band of the radio band you are working.

3F122 What is meant by the term noise figure of a communications receiver?

A. The level of noise entering the receiver from the antenna

B. The relative strength of a received signal 3 kHz removed from the carrier frequency

C. The level of noise generated in the front end and succeeding stages of a receiver

D. The ability of a receiver to reject unwanted signals at frequencies close to the desired one

ANSWER C: For VHF and UHF satellite gear, always look for the lowest noise figure with a hot front-end complement of transistors.

3F123 Which stage of a receiver primarily establishes its noise figure?

A. The audio stage

B. The IF strip

C. The RF stage

D. The local oscillator

ANSWER C: It's the RF amplifier stage where we are most concerned about a low noise figure.

3F124 What is an inverting op-amp circuit?

A. An operational amplifier circuit connected such that the input and output signals are 180 degrees out of phase

B. An operational amplifier circuit connected such that the input and output signals are in phase

C. An operational amplifier circuit connected such that the input and output signals are 90 degrees out of phase

D. An operational amplifier circuit connected such that the input impedance is held at zero, while the output impedance is high

ANSWER A: On an inverting op amp, the input and output signals are 180 degrees out of phase. Remember *inverting* means a 180° phase shift.

3F125 What is a noninverting op-amp circuit?

A. An operational amplifier circuit connected such that the input and output signals are 180 degrees out of phase

B. An operational amplifier circuit connected such that the input and output signals are in phase

C. An operational amplifier circuit connected such that the input and output signals are 90 degrees out of phase

D. An operational amplifier circuit connected such that the input impedance is held at zero while the output impedance is high

ANSWER B: In a *non-inverting* op amp circuit, the input and output signals are *in phase. Non-inverting* means *no phase shift; inverting* means 180° *phase shift.*

Ideal Operational Amplifier

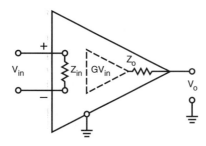

Z_{in} = Infinity

G = Gain = Infinity

Z_o = Zero

Bandwith = Infinity

V_o has no offset
    (V_o=0 when V_{in}=0)

IC Operational Ampliifer

Even though IC operational amplifiers do not meet all the ideal specifications, G and Z_{in} are very large. Because G is very large, any small input voltage would drive the output into saturation (for practical supply voltages); therefore, normal operational amplifier operation is with feedback to set the gain. Here's an example for an inverting amplifier:

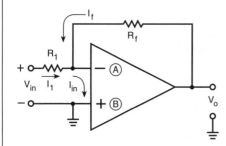

Since Z_{in} is very large, $I_{in} = 0$

$$\therefore I_1 + I_f = 0 \qquad \therefore \frac{V_{in}}{R_1} + \frac{V_o}{R_f} = 0$$

$$I_1 = \frac{V_{in}}{R_1} \qquad \frac{V_o}{R_f} = -\frac{V_{in}}{R_1}$$

$$I_f = \frac{V_o}{R_f} \qquad \therefore \frac{V_o}{V_{in}} = -\frac{R_f}{R_1}$$

(A) Inverting Input — An input voltage that is more positive on this input will cause the output voltage to be less positive.

(B) Non-Inverting Input — An input voltage that is more positive on this input will cause the output voltage to be more positive.

The gain of an *inverting IC operational amplifier* is:

$$G = -\frac{R_f}{R_1}$$

The minus sign means the output is out of phase with the input.

Operational Amplifier Basics

3F126 How does the gain of a theoretically ideal operational amplifier vary with frequency?

A. The gain increases linearly with increasing frequency.

B. The gain decreases linearly with increasing frequency.

C. The gain decreases logarithmically with increasing frequency.

D. The gain does not vary with frequency.

ANSWER D: The gain on an *ideal* op amp should not vary with frequency.

3F127 What determines the input impedance in a FET common-source amplifier?
   A. The input impedance is essentially determined by the resistance between the drain and substrate.
   B. The input impedance is essentially determined by the resistance between the source and drain.
   C. The input impedance is essentially determined by the gate biasing network.
   D. The input impedance is essentially determined by the resistance between the source and substrate.
ANSWER C: Since the input impedance of the device itself is very high, the input impedance to a FET common source amplifier is determined by the *gate biasing network*.

3F128 What determines the output impedance in a FET common-source amplifier?
   A. The output impedance is essentially determined by the drain resistor.
   B. The output impedance is essentially determined by the input impedance of the FET.
   C. The output impedance is essentially determined by the drain supply voltage.
   D. The output impedance is essentially determined by the gate supply voltage.
ANSWER A: The resistance connected to the drain (drain resistor) determines the output impedance of a FET common-source amplifier.

3F129 What is the purpose of a bypass capacitor?
   A. It increases the resonant frequency of the circuit.
   B. It removes direct current from the circuit by shunting DC to ground.
   C. It removes alternating current by providing a low impedance path to ground.
   D. It acts as a voltage divider.
ANSWER C: Bypass capacitors shunt AC through a low impedance path to ground.

3F130 What is the purpose of a coupling capacitor?
   A. It blocks direct current and passes alternating current.
   B. It blocks alternating current and passes direct current.
   C. It increases the resonant frequency of the circuit.
   D. It decreases the resonant frequency of the circuit.
ANSWER A: Coupling capacitors block direct current, and pass the AC signal. Put an ohmmeter across a capacitor and you read an open circuit (after the capacitor charges), yet holding onto one lead of a capacitor and putting a scope on the other lead produces a signal on the scope.

3F131 What condition must exist for a circuit to oscillate?
   A. It must have a gain of less than 1.
   B. It must be neutralized.
   C. It must have positive feedback sufficient to overcome losses.
   D. It must have negative feedback sufficient to cancel the input.
ANSWER C: The circuit must have just enough positive feedback to overcome losses in order to continue to properly oscillate.

3F132 (Refer to figure EL3-F2) What is the voltage drop across R1?
A. 9 volts
B. 7 volts
C. 5 volts
D. 3 volts

ANSWER C: Look at Figure EL3-F2 and notice that the diode is reverse biased, and will look like an open circuit to this series resistance circuit problem. Current will be the same throughout all of the circuit, and voltage drops will equal the source voltage. Combine R3 and R2 as a total of 300 ohms, R3 + R2 equals the resistance of R1, 300 ohms, so each will drop 5 volts. The voltage drop across R1 equals 5 volts.

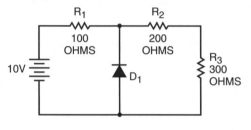

Figure EL3-F2

3F133 (Refer to figure EL3-F3) What is the voltage drop across R1?
A. 1.2 volts
B. 2.4 volts
C. 3.7 volts
D. 9 volts

ANSWER D: Look at Figure EL3-F3 and note that the Zener diode will drop 3 volts from the supply 12 volts, and 12 minus 3 volts equals a 9-volt drop at R1.

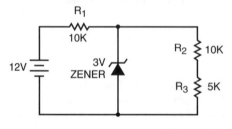

Figure EL3-F3

3F134 In a properly operating marine transmitter, if the power supply bleeder resistor opens:
A. Short circuit of supply voltage due to overload.
B. Regulation would decrease.
C. Next stage would fail due to short circuit.
D. Filter capacitors might short from voltage surge.

ANSWER D: Bleeder resistors assist in voltage regulation in the filter capacitor section of a power supply. If the bleeder resistors should open, the filter capacitors would no longer be protected, and they could short from a voltage surge.

3F135 (Refer to figure EL3-F4) Which of the following can occur that would least affect this circuit?
   A. C1 shorts.
   B. C1 opens.
   C. C3 shorts.
   D. C18 opens.
ANSWER D: It is assumed that the circuit is an operating circuit (e.g., Class C operation) and doesn't need base biasing. Based on this, C18 is a bypass capacitor, and if it opens, the circuit would probably continue to function. However, if it shorts, the circuit will not operate properly.

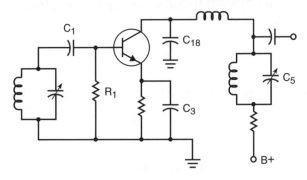

Figure EL3-F4

3F136 (Refer to figure EL3-F5) When S1 is closed, lights L1 and L2 go on. What is the condition of both lamps when both S1 and S2 are closed?
   A. Both lamps stay on.
   B. L1 turns off; L2 stays on.
   C. Both lamps turn off.
   D. L1 stays on; L2 turns off.
ANSWER D: When both switch 1 and switch 2 are closed, they will provide a positive voltage on the bases of both NPN transistors. Both transistors will conduct, and lamp L1 will be in series and will illuminate. With both switches closed, lamp L2 is shunted by the lower transistor, and there will not be enough current through it to light.

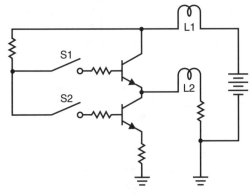

Figure EL3-F5

3F137 (Refer to figure EL3-F5) If S1 is closed both lamps light, what happens when S1 and S2 are closed?
A. L1 and L2 are off.
B. L1 is on and L2 is flashing.
C. L1 is off and L2 is on.
D. L1 is on and L2 is off.

ANSWER D: This is almost the same question as above, but worded slightly different. The top lamp, L1, in Figure EL3-F5 will stay illuminated, and L2 will be off due to insufficient current.

3F138 (Refer to figure EL3-F6) How can we correct the defect, if any, in this voltage doubler circuit?
A. Omit C1.
B. Reverse polarity signs.
C. Ground X.
D. Reverse polarity on C1.

ANSWER B: Look at Figure EL3-F6, and notice the load polarity—this is not correct. You will normally find the positive leg on top; and if you look at the components within the diagram, you will see that the load polarity has reversed polarity signs.

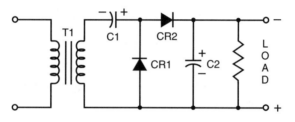

Figure EL3-F6

3F139 (Refer to figure EL3-F7) What change is needed in order to correct the grounded emitter amplifier shown?
A. No change is necessary.
B. Polarities of emitter-base battery should be reversed.
C. Polarities of collector-base battery should be reversed.
D. Point A should be replaced with a low value capacitor.

ANSWER A: Figure EL3-F7 illustrates a grounded emitter amplifier, and no change is necessary because this figure is accurately drawn.

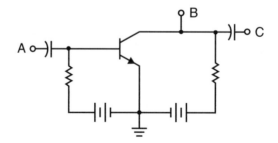

Figure EL3-F7

Subelement 3G – Signals and Emissions

(10 questions)

3G1 What is emission A3C?
A. Facsimile
B. RTTY
C. ATV
D. Slow Scan TV

ANSWER A: Three character signals and emissions may still show up on paperwork going to and from the FCC, or paperwork to local frequency coordinators. When you see the number 3 in the middle of an emission designator, it usually means voice or other analog information. The A means amplitude modulation, and the C is what gives away this answer—C for faCsimile. See chart at 3G71.

3G2 What type of emission is produced when an amplitude-modulated transmitter is modulated by a facsimile signal?
A. A3F
B. A3C
C. F3F
D. F3C

ANSWER B: Remember, C for faCsimile. Put this together with A for Amplitude modulated and you have the correct answer, A3C. Most marine SSB sets will accept weather facsimile programs when tied into a computer, and read out FAXed weather charts quite well. See chart at 3G71.

3G3 What is emission F3C?
A. Voice transmission
B. Slow Scan TV
C. RTTY
D. Facsimile

ANSWER D: Think of the letter "C" in the word faCsimile. F for FM.

3G4 What type of emission is produced when a frequency-modulated transmitter is modulated by a facsimile signal?
A. F3C
B. A3C
C. F3F
D. A3F

ANSWER A: If it is a facsimile signal then remember the C, but frequency modulated means the new designator will then be F3C, the F for *frequency* modulation.

3G5 What is emission A3F?
A. RTTY
B. Television
C. SSB
D. Modulated CW

ANSWER B: You will get an F on your exam if all you do is watch television, and not study this book. A3F is amplitude modulated TV.

3G6 What type of emission is produced when an amplitude-modulated transmitter is modulated by a television signal?
A. F3F
B. A3F
C. A3C
D. F3C

ANSWER B: Television, A3F. A for Amplitude modulated and F for TV.

3G7 What is emission F3F?
A. Modulated CW
B. Facsimile
C. RTTY
D. Television

ANSWER D: Same thing, F for television, but this time F for the frequency modulated part of the TV signal.

3G8 What type of emission is produced when a frequency-modulated transmitter is modulated by a television signal?
A. A3F
B. A3C
C. F3F
D. F3C

ANSWER C: Television, F3F—F for FM, F for TV. Since it is frequency modulated, answer A is incorrect.

3G9 How can an FM-phone signal be produced?
A. By modulating the supply voltage to a class-B amplifier
B. By modulating the supply voltage to a class-C amplifier
C. By using a reactance modulator on an oscillator
D. By using a balanced modulator on an oscillator

ANSWER C: They are coming back around. Here's that word again, reactance modulator. An FM phone signal is produced by using a reactance modulator.

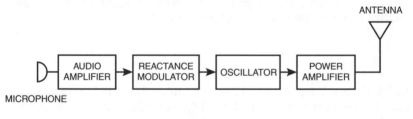

FM Transmitter

3G10 How can a double-sideband phone signal be produced?
A. By using a reactance modulator on an oscillator
B. By varying the voltage to the varactor in an oscillator circuit
C. By using a phase detector, oscillator, and filter in a feedback loop
D. By modulating the plate supply voltage to a class C amplifier

ANSWER D: Since we're talking about double-sideband phone, it means AM with both sidebands present. Answer A is wrong because it's talking about FM. Answer D is the correct answer because we're talking about a Class C amplifier, an okay amplifier to plate modulate with a double-sideband phone signal.

3G11 How can a single-sideband phone signal be produced?

A. By producing a double sideband signal with a balanced modulator and then removing the unwanted sideband by filtering

B. By producing a double sideband signal with a balanced modulator and then removing the unwanted sideband by heterodyning

C. By producing a double sideband signal with a balanced modulator and then removing the unwanted sideband by mixing

D. By producing a double sideband signal with a balanced modulator and then removing the unwanted sideband by neutralization

ANSWER A: On single sideband, the signal is first produced by a balanced modulator with both sidebands present, and a filter then removes the unwanted upper or lower sideband. It is the *filtering* (key word) that removes the unwanted sideband.

3G12 What is meant by the term deviation ratio?

A. The ratio of the audio modulating frequency to the center carrier frequency

B. The ratio of the maximum carrier frequency deviation to the highest audio modulating frequency

C. The ratio of the carrier center frequency to the audio modulating frequency

D. The ratio of the highest audio modulating frequency to the average audio modulating frequency

ANSWER B: It's important to set the deviation ratio properly on your FM transceiver to prevent splatter, and a signal bandwidth that is too wide. Deviation ratio is the ratio of the maximum carrier swing to your highest audio modulating frequency when you speak into the microphone.

$$\text{Deviation Ratio} = \frac{\text{Maximum Carrier Frequency Deviation in kHz}}{\text{Maximum Modulation Frequency in kHz}}$$

3G13 In an FM-phone signal, what is the term for the maximum deviation from the carrier frequency divided by the maximum audio modulating frequency?

A. Deviation index

B. Modulation index

C. Deviation ratio

D. Modulation ratio

ANSWER C: Simple division of the maximum audio frequency into the maximum deviation of the carrier frequency will lead you to the deviation ratio. Division produces a ratio.

3G14 What is the deviation ratio for an FM-phone signal having a maximum frequency swing of plus or minus 5 kHz and accepting a maximum modulation rate of 3 kHz?

A. 60

B. 0.16

C. 0.6

D. 1.66

ANSWER D: Use formula at 3G12. Divide the maximum carrier frequency swing by the maximum modulation frequency. 3 kHz modulation frequency into 5 kHz frequency swing gives you a deviation ratio of 1.66, answer D.

3G15 What is the deviation ratio of an FM-phone signal having a maximum frequency swing of plus or minus 7.5 kHz and accepting a maximum modulation rate of 3.5 kHz?

A. 2.14
B. 0.214
C. 0.47
D. 47

ANSWER A: Divide 3.5 (maximum modulation frequency) into 7.5 (maximum carrier frequency deviation), and you end up with 2.14, your deviation ratio.

3G16 What is meant by the term modulation index?

A. The processor index
B. The ratio between the deviation of a frequency modulated signal and the modulating frequency
C. The FM signal-to-noise ratio
D. The ratio of the maximum carrier frequency deviation to the highest audio modulating frequency

ANSWER B: With modulation index, instead of looking at the highest audio frequency, we look at a specific modulating frequency. This is an important consideration for the amount of deviation on a packet station on FM.

3G17 In an FM-phone signal, what is the term for the ratio between the deviation of the frequency-modulated signal and the modulating frequency?

A. FM compressibility
B. Quieting index
C. Percentage of modulation
D. Modulation index

ANSWER D: Be careful with this one. Since they don't say the highest modulating frequency, but say only "modulating frequency", they are looking for the modulation index. Notice that there is not any answer that says "deviation ratio" to get you in trouble.

3G18 How does the modulation index of a phase-modulated emission vary with the modulated frequency?

A. The modulation index increases as the RF carrier frequency (the modulated frequency) increases.
B. The modulation index decreases as the RF carrier frequency (the modulated frequency) increases.
C. The modulation index varies with the square root of the RF carrier frequency (the modulated frequency).
D. The modulation index does not depend on the RF carrier frequency (the modulated frequency).

ANSWER D: This is one of those answers that has a "no" in it that makes it correct. When you calculate modulation index, you *do not* (key words) look at the actual RF carrier frequency as part of your calculations.

3G19 In an FM-phone signal having a maximum frequency deviation of 3.0 kHz either side of the carrier frequency, what is the modulation index when the modulating frequency is 1.0 kHz?
    A. 3
    B. 0.3
    C. 3000
    D. 1000
ANSWER A: 3000 divided by 1000 gives you a modulation index of 3.

3G20 What is the modulation index of an FM-phone transmitter producing an instantaneous carrier deviation of 6 kHz when modulated with a 2 kHz modulating frequency?
    A. 6000
    B. 3
    C. 2000
    D. 1/3.
ANSWER B: 6 divided by 2 gives us a modulation index of 3.

3G21 What are electromagnetic waves?
    A. Alternating currents in the core of an electromagnet
    B. A wave consisting of two electric fields at right angles to each other
    C. A wave consisting of an electric field and a magnetic field at right angles to each other
    D. A wave consisting of two magnetic fields at right angles to each other
ANSWER C: When you think of an electromagnetic radio wave, think of it as a combination of two fields at right angles traveling in waves through the atmosphere at the same time. One field is vertically polarized and the other is horizontally polarized. If the antenna is vertically polarized, the electric field will be vertical, and the magnetic field will be horizontal. See figure at 3G24.

3G22 What is a wave front?
    A. A voltage pulse in a conductor
    B. A current pulse in a conductor
    C. A voltage pulse across a resistor
    D. A fixed point in an electromagnetic wave
ANSWER D: When we analyze propagation of signals, we take into account the wave front which is a fixed point at the beginning of an electromagnetic wave.

3G23 At what speed do electromagnetic waves travel in free space?
    A. Approximately 300 million meters per second
    B. Approximately 468 million meters per second
    C. Approximately 186,300 feet per second
    D. Approximately 300 million miles per second
ANSWER A: Radio waves travel at the rate of 300 million meters per second in free space. That's where the number 300 comes from when we convert frequency in MHz to wavelength in meters, or visa versa, wavelength in meters to frequency in MHz.

3G24 What are the two interrelated fields considered to make up an electromagnetic wave?

A. An electric field and a current field
B. An electric field and a magnetic field
C. An electric field and a voltage field
D. A voltage field and a current field

ANSWER B: Radio waves consist of an electric field and a magnetic field traveling together at right angles to each other.

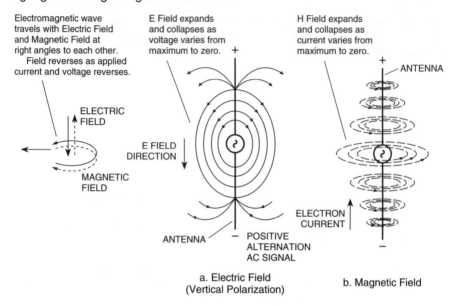

Electromagnetic wave travels with Electric Field and Magnetic Field at right angles to each other. Field reverses as applied current and voltage reverses.

ELECTRIC FIELD

E FIELD DIRECTION

MAGNETIC FIELD

E Field expands and collapses as voltage varies from maximum to zero.

ANTENNA — POSITIVE ALTERNATION AC SIGNAL

a. Electric Field (Vertical Polarization)

H Field expands and collapses as current varies from maximum to zero.

ANTENNA

ELECTRON CURRENT

b. Magnetic Field

Propagating Electromagnetic Waves

Source: *Antennas – Selection and Installation,* A.J. Evans, ©1986 Master Publishing, Inc., Richardson, Texas

3G25 Why do electromagnetic waves not penetrate a good conductor to any great extent?

A. The electromagnetic field induces currents in the insulator.
B. The oxide on the conductor surface acts as a shield.
C. Because of eddy currents
D. The resistivity of the conductor dissipates the field.

ANSWER C: Radio waves travel best on the outside of a conductor. The reason they do not travel inside a conductor, to any great extent, is because of *eddy* currents which tend to cancel the inside travel of the radio wave.

3G26 What is meant by referring to electromagnetic waves as horizontally polarized?

A. The electric field is parallel to the earth.
B. The magnetic field is parallel to the earth.
C. Both the electric and magnetic fields are horizontal.
D. Both the electric and magnetic fields are vertical.

ANSWER A: When we talk about polarization, we always refer to the electric lines of force, not the magnetic lines of force. On a horizontally polarized wave, the electric field is *parallel to the earth* (key words).

3G27 What is meant by referring to electromagnetic waves as having circular polarization?
A. The electric field is bent into a circular shape.
B. The electric field rotates.
C. The electromagnetic wave continues to circle the earth.
D. The electromagnetic wave has been generated by a quad antenna.

ANSWER B: Some satellite enthusiasts will use circular polarization to improve their transmission and reception through an orbiting satellite. With circular polarization, *the electric field rotates* (key words). It could be right-hand circular, or left-hand circular, depending on the antenna feed system.

3G28 When the electric field is perpendicular to the surface of the earth, what is the polarization of the electromagnetic wave?
A. Circular
B. Horizontal
C. Vertical
D. Elliptical

ANSWER C: If the electric field is perpendicular to the earth, polarization is vertical.

3G29 When the magnetic field is parallel to the surface of the earth, what is the polarization of the electromagnetic wave?
A. Circular
B. Horizontal
C. Elliptical
D. Vertical

ANSWER D: Watch out for this one—here they are asking about the magnetic field which is parallel to the surface of the earth. If the magnetic field is parallel, the *electric field is vertical*, so polarization would be *vertical*.

3G30 When the magnetic field is perpendicular to the surface of the earth, what is the polarization of the electromagnetic field?
A. Horizontal
B. Circular
C. Elliptical
D. Vertical

ANSWER A: If the magnetic field is perpendicular (vertical) to the surface of the earth, the *electric* lines of force would be *horizontal*, and this would give us *horizontal polarization*.

3G31 When the electric field is parallel to the surface of the earth, what is the polarization of the electromagnetic wave?
A. Vertical
B. Horizontal
C. Circular
D. Elliptical

ANSWER B: Here the *electric* field is *parallel*, so the polarization is *horizontal*.

3G32 What is a sine wave?
A. A constant-voltage, varying-current wave
B. A wave whose amplitude at any given instant can be represented by a point on a wheel rotating at a uniform speed
C. A wave following the laws of the trigonometric tangent function
D. A wave whose polarity changes in a random manner

ANSWER B: A sine wave is as smooth as a turning wheel, giving us a uniform cycle of energy. The point on a wheel is like a vector rotating through 360°. If the sine of the angle is plotted against degrees rotation, we will plot a sinewave.

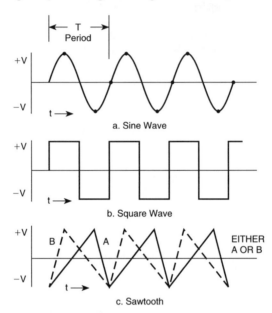

a. Sine Wave

b. Square Wave

c. Sawtooth

Sine, Square and Sawtooth Waveforms

3G33 If a sine wave begins from above or below the zero axis, how many times will it cross the zero axis in one complete cycle?
A. 180 times
B. 4 times
C. 2 times
D. 360 times

ANSWER C: In a complete cycle, the sine wave begins at the axis, goes up to a maximum, then down through the axis once to a maximum in the opposite direction, and then back to the zero axis for a total of two crossings, completing one cycle.

3G34 How many degrees are there in one complete sine wave cycle?
A. 90 degrees
B. 270 degrees
C. 180 degrees
D. 360 degrees

ANSWER D: One complete sine wave cycle is 360 degrees, just like a wheel.

3G35 What is the period of a wave?

A. The time required to complete one cycle
B. The number of degrees in one cycle
C. The number of zero crossings in one cycle
D. The amplitude of the wave

ANSWER A: The period of a wave is expressed and measured in time. Just like school periods for classes, and periods of studying this book, the period of the wave is the time required to complete one cycle.

3G36 What is a square wave?

A. A wave with only 300 degrees in one cycle
B. A wave that abruptly changes back and forth between two voltage levels and which remains an equal time at each level
C. A wave that makes four zero crossings per cycle
D. A wave in which the positive and negative excursions occupy unequal portions of the cycle time

ANSWER B: Square waves have abrupt changes back and forth between two voltage levels. The *abrupt change* (key words) of a square wave is not desirable as a CW key-attack and key-release wave form. Square waves on CW generate key clicks because of their abrupt changes. See figure at question 3G32.

3G37 What is a wave called which abruptly changes back and forth between two voltage levels and which remains an equal time at each level?

A. A sine wave
B. A cosine wave
C. A square wave
D. A rectangular wave

ANSWER C: A square waveform has steep waveform edges, in which the waveform changes rapidly from one voltage level to another. The times that the waveform is at each level are equal. The key words are abruptly changes back and forth; when the waveform does, then it is a square wave. See figure at question 3G32.

3G38 Which sine waves make up a square wave?

A. 0.707 times the fundamental frequency
B. The fundamental frequency and all odd and even harmonics
C. The fundamental frequency and all even harmonics
D. The fundamental frequency and all odd harmonics

ANSWER D: The square wave is a tough one to get rid of harmonics, because the fundamental frequency and all odd harmonics may be found in a square wave. But only the odd harmonics—1, 3, 5, 7, etc. A square wave gives odd results!

3G39 What type of wave is made up of sine waves of the fundamental frequency and all the odd harmonics?

A. Square wave
B. Sine wave
C. Cosine wave
D. Tangent wave

ANSWER A: Again, you get *odd* harmonics from *square* waves.

3G40 What is a sawtooth wave?

A. A wave that alternates between two values and spends an equal time at each level
B. A wave with a straight line rise time faster than the fall time (or vice versa)
C. A wave that produces a phase angle tangent to the unit circle
D. A wave whose amplitude at any given instant can be represented by a point on a wheel rotating at a uniform speed

ANSWER B: A sawtooth wave has a straight line rise, and an immediate fall, or vice versa. It looks like a saw blade if you look at it on an oscilloscope. See figure at question 3G32.

3G41 What type of wave is characterized by a rise time significantly faster than the fall time (or vice versa)?

A. A cosine wave
B. A square wave
C. A sawtooth wave
D. A sine wave

ANSWER C: The sawtooth wave can be backwards with immediate rise time significantly faster than the fall time. See figure at question 3G32.

3G42 Which sine waves make up a sawtooth wave?

A. The fundamental frequency and all prime harmonics
B. The fundamental frequency and all even harmonics
C. The fundamental frequency and all odd harmonics
D. The fundamental frequency and all harmonics

ANSWER D: The sawtooth wave may be used in frequency multipliers because the fundamental frequency is also rich in all harmonics, both odd and even.

3G43 What type of wave is made up of sine waves at the fundamental frequency and all the harmonics?

A. A sawtooth wave
B. A square wave
C. A sine wave
D. A cosine wave

ANSWER A: If it has *all the harmonics* (key words), the answer is a sawtooth wave.

3G44 What is the meaning of the term, root mean square value of an AC voltage?

A. The value of an AC voltage found by squaring the average value of the peak AC voltage
B. The value of a DC voltage that would cause the same heating effect in a given resistor as a peak AC voltage
C. The value of an AC voltage that would cause the same heating effect in a given resistor as a DC voltage of the same value
D. The value of an AC voltage found by taking the square root of the average AC value

ANSWER C: You sometimes see this as RMS, also called effective value of an ac voltage. Since ac voltage is continually changing in amplitude, we must look at its root mean square value before we can come up with the same heating effect in a given resistor as a constant dc voltage.

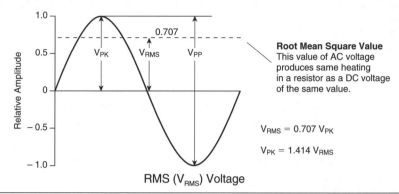

RMS (V_{RMS}) Voltage

Root Mean Square Value
This value of AC voltage produces same heating in a resistor as a DC voltage of the same value.

$V_{RMS} = 0.707\ V_{PK}$

$V_{PK} = 1.414\ V_{RMS}$

3G45 What is the term used in reference to a DC voltage that would cause the same heating in a resistor as a certain value of AC voltage?
A. Cosine voltage
B. Power factor
C. Root mean square
D. Average voltage

ANSWER C: When we compare a dc voltage to an ac voltage that produces the same heating in a resistor, we need to choose the root-mean-square ac voltage.

3G46 What would be the most accurate way of determining the RMS voltage of a complex waveform?
A. By using a grid dip-meter
B. By measuring the voltage with a D'Arsonval meter
C. By using an absorption wavemeter
D. By measuring the heating effect in a known resistor

ANSWER D: In a very complex wave form, one way to determine its effective value is to measure the heating effect in a known resistor, compared to a dc known voltage.

3G47 What is the RMS voltage at a common household electrical power outlet?
A. 117-V AC
B. 331-V AC
C. 82.7-V AC
D. 165.5-V AC

ANSWER A: Household alternating current electrical power is rated at 117 VAC RMS.

3G48 What is the peak voltage at a common household electrical outlet?
A. 234 volts
B. 165.5 volts
C. 117 volts
D. 331 volts

ANSWER B: When you measure house power with a voltmeter, you are measuring RMS. If you were to look at it on an oscilloscope, and look at the electrical peaks, it would be 1.414 times higher than 117 volts, or a total of 165 volts peak.

3G49 What is the peak-to-peak voltage at a common household electrical outlet?
    A. 234 volts
    B. 117 volts
    C. 331 volts
    D. 165.5 volts
ANSWER C: If you really want to impress your friends about your house power, tell them that you have 331 volts coming out of the socket. If they bet you don't, tell them you are stating the voltage as peak-to-peak, which is twice as much as peak voltage.

3G50 What is RMS voltage of a 165-volt peak pure sine wave?
    A. 233-V AC
    B. 330-V AC
    C. 58.3-V AC
    D. 117-V AC
ANSWER D: If your scope shows 165 volts peak pure sine wave, the RMS value is 117 VAC, normal RMS house power. Multiply the peak voltage times 0.707 to find the RMS volts.

3G51 What is the RMS value of a 331-volt peak-to-peak pure sine wave?
    A. 117-V AC
    B. 165-V AC
    C. 234-V AC
    D. 300-V AC
ANSWER A: 331 volts peak-to-peak divided in half is 165.5 volts peak. Multiply the peak voltage by 0.707, and you end up with house power, 117 volts RMS.

3G52 For many types of voices, what is the ratio of PEP to average power during a modulation peak in a single-sideband phone signal?
    A. Approximately 1.0 to 1
    B. Approximately 25 to 1
    C. Approximately 2.5 to 1
    D. Approximately 100 to 1
ANSWER C: When you talk about power output from your SSB worldwide set, you normally refer to power as peak envelope power. Depending on your modulation, the ratio of PEP to average power is about 2.5 to 1. You can remember 2.5 because that's very close to 2.5 kHz of band occupancy for a properly modulated SSB signal.

3G53 In a single-sideband phone signal, what determines the PEP-to-average power ratio?
    A. The frequency of the modulating signal
    B. The degree of carrier suppression
    C. The speech characteristics
    D. The amplifier power
ANSWER C: It's not all that easy to determine the ratio of peak envelope power (PEP) to average power in your worldwide SSB set because your speech characteristics may vary from those when another operator is using the same equipment.

    $PEP = P_{DC} \times Eff$ Where: PEP = Peak Envelope Power in **watts**
                                             P_{DC} = Input dc power in **watts**

3G54 What is the approximate DC input power to a Class B RF power amplifier stage in an FM-phone transmitter when the PEP output power is 1500 watts?

A. Approximately 900 watts
B. Approximately 1765 watts
C. Approximately 2500 watts
D. Approximately 3000 watts

ANSWER C: You can determine input power by dividing PEP output power by the efficiency of the amplifier. Because efficiency is normally rated in a percentage, divide percentage efficiency by 100 to get the decimal value. In this problem, divide 1500 watts PEP output power by 0.6 because the efficiency of a Class B amplifier is approximately 60 percent. They don't tell you that—you must remember that a Class B amp is a bit more efficient than a class AB amplifier that has an efficiency of 50%.

1500 divided by 0.6 gives a power input of approximately 2500 watts, and for the common Class B amplifier, 2500 watts in will normally provide about 1500 watts out.

3G55 What is the approximate DC input power to a Class C RF power amplifier stage in a RTTY transmitter when the PEP output power is 1000 watts?

A. Approximately 850 watts
B. Approximately 1250 watts
C. Approximately 1667 watts
D. Approximately 2000 watts

ANSWER B: The Class C amplifier is more efficient—sometimes as good as 80 percent. If our output is 1000 watts, our 80 percent efficient amplifier would need only 1250 watts for input power.

You can do this in your head, but double check your work using the following steps: Divide the 80% by 100 to get 0.8. Divide the PEP power of 1000 by 0.8 to get input power of 1250 watts.

3G56 What is the approximate DC input power to a Class AB RF power amplifier stage in an unmodulated carrier transmitter when the PEP output power is 500 watts?

A. Approximately 250 watts
B. Approximately 600 watts
C. Approximately 800 watts
D. Approximately 1000 watts

ANSWER D: The Class AB amplifier is a real work horse, and is one of the most common amps found in worldwide radio shacks. It's 50 percent efficient. If you're putting out 500 watts, your input power is approximately 1000 watts.

Remember, Class C amplifier efficiency is about 80 percent; Class B is approximately 60 percent; and Class AB is approximately 50 percent. These values will not be given on your test, so be sure to have them memorized before the big examination.

3G57 Where is the noise generated which primarily determines the signal-to-noise ratio in a VHF (150 MHz) marine band receiver?

A. In the receiver front end
B. Man-made noise

C. In the atmosphere

D. In the ionosphere

ANSWER A: Back in the early days of marine band on 2 MHz, most noise was generated in the atmosphere and ionosphere, plus plenty onboard. Now that we have marine VHF frequencies, there is little noise that can be detected by an FM receiver, so the only noise that we really need to work with is noise found in the receiver front-end transistorized circuitry.

3G58 In a pulse-width modulation system, what parameter does the modulating signal vary?

A. Pulse duration

B. Pulse frequency

C. Pulse amplitude

D. Pulse intensity

ANSWER A: It is the *duration* of the pulse that the modulating signal varies.

3G59 What is the type of modulation in which the modulating signal varies the duration of the transmitted pulse?

A. Amplitude modulation

B. Frequency modulation

C. Pulse-width modulation

D. Pulse-height modulation

ANSWER C: When the modulating signals varies the *duration* of the transmitted pulse, it is pulse-width modulation.

3G60 In a pulse-position modulation system, what parameter does the modulating signal vary?

A. The number of pulses per second

B. Both the frequency and amplitude of the pulses

C. The duration of the pulses

D. The time at which each pulse occurs

ANSWER D: This question is about a pulse-position modulation system. The position is the time at which each pulse occurs. The modulating signal varies the *time* when pulses occur.

3G61 Why is the transmitter peak power in a pulse modulation system much greater than its average power?

A. The signal duty cycle is less than 100%.

B. The signal reaches peak amplitude only when voice modulated.

C. The signal reaches peak amplitude only when voltage spikes are generated within the modulator.

D. The signal reaches peak amplitude only when the pulses are also amplitude modulated.

ANSWER A: Since pulse modulation is not a continuous burst of power, the duty cycle is always less than 100 percent.

3G62 What is one way that voice is transmitted in a pulse-width modula-tion system?

A. A standard pulse is varied in amplitude by an amount depending on the voice waveform at that instant.

B. The position of a standard pulse is varied by an amount depending on the voice waveform at that instant.

C. A standard pulse is varied in duration by an amount depending on the voice waveform at that instant.

D. The number of standard pulses per second varies depending on the voice waveform at that instant.

ANSWER C: Remember *pulse-width* and *duration*. A standard pulse is varied *in duration* by an amount depending on the modulating waveform of the voice signal.

3G63 The International Organization for Standardization has developed a seven-level reference model for a packet-radio communications structure. What level is responsible for the actual transmission of data and handshaking signals?

A. The physical layer

B. The transport layer

C. The communications layer

D. The synchronization layer

ANSWER A: The data and handshaking signals are in the *physical layer.*

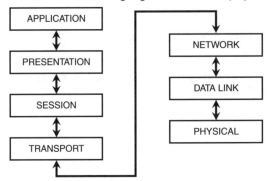

ISO Model for Open Systems Interconnection

3G64 The International Organization for Standardization has developed a seven-level reference model for a packet-radio communications structure. What level arranges the bits into frames and controls data flow?

A. The transport layer

B. The link layer

C. The communications layer

D. The synchronization layer

ANSWER B: It is the *link layer* that arranges the bits into frames, and also controls data flow.

3G65 What is one advantage of using the ASCII code, with its larger character set, instead of the Baudot code?

A. ASCII includes built-in error-correction features

B. ASCII characters contain fewer information bits than Baudot characters

C. It is possible to transmit upper and lower case text

D. The larger character set allows store-and-forward control characters to be added to a message

ANSWER C: ASCII code allows upper and lower case text to be transmitted; Baudot does not.

3G66 What is the duration of a 45-baud Baudot RTTY data pulse?
A. 11 milliseconds
B. 40 milliseconds
C. 31 milliseconds
D. 22 milliseconds
ANSWER D: The very slow 45-baud RTTY data pulse is 22 milliseconds long.

3G67 What is the duration of a 45-baud Baudot RTTY start pulse?
A. 11 milliseconds
B. 22 milliseconds
C. 31 milliseconds
D. 40 milliseconds
ANSWER B: The 45-baud RTTY start pulse is also 22 milliseconds long.

3G68 What is the duration of a 45-baud Baudot RTTY stop pulse?
A. 11 milliseconds
B. 18 milliseconds
C. 31 milliseconds
D. 40 milliseconds
ANSWER C: The 45-baud stop pulse is slightly longer than the data and start pulse—31 milliseconds.

3G69 What is the necessary bandwidth of a 170-hertz shift, 45-baud Baudot emission F1B transmission?
A. 45 Hz
B. 249 Hz
C. 442 Hz
D. 600 Hz
ANSWER B: The formula for bandwidth for a Baudot transmission is:

 BW = baud rate + (1.2 × frequency shift).

In this problem the baud rate is 45 plus (1.2 × 170). This works out to be 249 Hz, the correct answer B.

3G70 What is the necessary bandwidth of a 170-hertz shift, 45-baud Baudot emission J2B transmission?
A. 45 Hz
B. 249 Hz
C. 442 Hz
D. 600 Hz
ANSWER B: Use the same formula as in question 3G69:

BW = 45 plus (1.2 × 170) = 249 Hz, which is answer B.

3G71 What is the necessary bandwidth of a 170-hertz shift, 74-baud Baudot emission F1B transmission?
A. 250 Hz
B. 278 Hz
C. 442 Hz
D. 600 Hz
ANSWER B: Use the same formula as in question 3G69:

BW = 74 plus (1.2 × 170) = 278 Hz, which is answer B.

First symbol – type of main modulation
(Note: Modulation is the process of varying the ratio to convey information)
- N – Emission of an unmodulated carrier
- A – Amplitude Modulation
- J – Single sideband, suppressed carrier (the emission you are authorized on ten meters between 28.3 and 28.5 MHz)
- F – Frequency modulation
- G – Phase modulation
- P – Sequence of unmodulated pulses

Second symbol – Nature of signals modulating the main carrier
- 0 – No modulating signal
- 1 – Single channel digital information, no modulation
- 2 – Single channel digital information with modulation
- 3 – Single channel carrying analog information
- 7 – Two or more channels carrying digital information
- 8 – Two or more channels carrying analog information
- 9 – Combination of digital and analog information

Third symbol – Type of Information to be transmitted
- N – No information transmitted
- A – Telegraphy to be received by human hearing
- B – Telegraphy to be received by automatic equipment
- C – Facsimile, transmission of pictures by radio
- D – Data transmission, telemetry, telecommand
- E – Telephony (voice information)
- F – Video, television
- W – Combination of above

ITU Emissions

3G72 What is the necessary bandwidth of a 170-hertz shift, 74-baud Baudot emission J2B transmission?
- A. 250 Hz
- B. 278 Hz
- C. 442 Hz
- D. 600 Hz

ANSWER B: Use the same formula as in question 3G69:

BW = 74 plus (1.2 $\times$ 170) = 278 Hz, which is answer B.

3G73 What is the necessary bandwidth of a 1000-hertz shift, 1200-baud SCII emission F1D transmission?
- A. 1000 Hz
- B. 1200 Hz
- C. 440 Hz
- D. 2400 Hz

ANSWER D: Now the question is back to digital bandwidth. Use the formula in question 3G69. Take your baud rate at 1200 and add it to (1.2 $\times$ 1000). This gives you 1200 plus 1200, or 2400 Hz which matches answer D.

3G74 What is the necessary bandwidth of a 4800-hertz frequency shift, 9600-baud ASCII emission F1D transmission?
- A. 15.36 kHz
- B. 9.6 kHz
- C. 4.8 kHz
- D. 5.76 kHz

ANSWER A: Digital bandwidth again—use the formula at question 3G69. In this problem, the baud rate is 9600 and the frequency shift is 4800. Bandwidth = 9600 + (1.2 × 4800) = 15,360. The answer is 15,360 Hz, and when the decimal point is moved three places to the left, the bandwidth works out to be 15.36 kHz, matching answer A.

3G75 What is the necessary bandwidth of a 4800-hertz frequency shift, 9600-baud ASCII emission J2D transmission?
    A. 15.36 kHz
    B. 9.6 kHz
    C. 4.8 kHz
    D. 5.76 kHz
ANSWER A: Digital bandwidth again; use the same formula again. Same calculation: Bandwidth = 9600 + (1.2 × 4800) = 15,360. You get the same answer of 15.36 kHz.

3G76 What is the necessary bandwidth of a 170-hertz shift, 110-baud ASCII emission F1B transmission?
    A. 304 Hz
    B. 314 Hz
    C. 608 Hz
    D. 628 Hz
ANSWER B: Use the formula at question 3G69. The calculations are: Bandwidth = 110 + (1.2 × 170) = 314 Hz.

3G77 What is the necessary bandwidth of a 170-hertz shift, 110-baud ASCII emission J2B transmission?
    A. 304 Hz
    B. 314 Hz
    C. 608 Hz
    D. 628 Hz
ANSWER B: Same digital transmission, same formula as question 3G69. Even though the the emission changed from F1B to J2B, the bandwidth solution is the same.

3G78 What is the necessary bandwidth of a 170-hertz shift, 300-baud ASCII emission F1D transmission?
    A. 0 Hz
    B. 0.3 kHz
    C. 0.5 kHz
    D. 1.0 kHz
ANSWER C: Same digital transmission, same formula, same solution: In words this time, bandwidth equals the baud rate of 300 plus the product of 1.2 times the frequency shift of 170 Hz. Bandwidth works out to be 504 Hz, and moving the decimal point three places to the left gives 0.5 kHz.

3G79 What is the necessary bandwidth for a 170-hertz shift, 300-baud ASCII emission J2D transmission?
    A. 0 Hz
    B. 0.3 kHz
    C. 0.5 kHz
    D. 1.0 kHz

ANSWER C: No change here from the last question, other than the emission, which does not effect the calculation of the bandwidth. It's the same: Bandwidth = 300 + (1.2 × 170) = 504 Hz = 0.504 kHz.

3G80 What is amplitude compandored single sideband?
    A. Reception of single sideband with a conventional CW receiver
    B. Reception of single sideband with a conventional FM receiver
    C. Single sideband incorporating speech compression at the transmitter and speech expansion at the receiver
    D. Single sideband incorporating speech expansion at the transmitter and speech compression at the receiver
ANSWER C: Compandoring a single-sideband signal will compress it at the transmitter, and the receiver will then expand it at the other end of the circuit. ACSB may soon become popular on land mobile VHF frequencies where it's now still considered an experimental emission. ACSB compression provides more efficient use of transmitter power; the expansion provides better signal-to-noise ratio at the receiver.

3G81 What is meant by compandoring?
    A. Compressing speech at the transmitter and expanding it at the receiver
    B. Using an audio-frequency signal to produce pulse-length modulation
    C. Combining amplitude and frequency modulation to produce a single-sideband signal
    D. Detecting and demodulating a single-sideband signal by converting it to a pulse-modulated signal
ANSWER A: We compress the speech to compandor our transmitted signal. It is expanded again at the receiver. That's where the name compandoring comes from—COMpression and exPANDing.

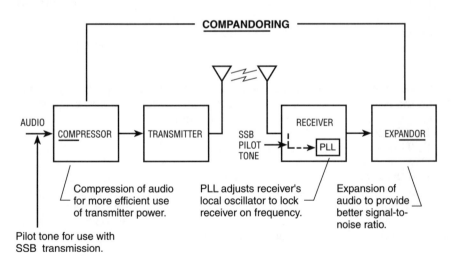

Compandoring

3G82 What is the purpose of a pilot tone in an amplitude compandored single-sideband system?
- A. It permits rapid tuning of a mobile receiver.
- B. It replaces the suppressed carrier at the receiver.
- C. It permits rapid change of frequency to escape high-powered interference.
- D. It acts as a beacon to indicate the present propagation characteristic of the band.

ANSWER A: The pilot tone, used only for ACSB transmissions, allows an automatic mobile receiver to lock onto a signal. This allows rapid tuning of the mobile receiver in the automatic mode.

3G83 What is the approximate frequency of the pilot tone in an amplitude compandored single-sideband system?
- A. 1 kHz
- B. 5 MHz
- C. 455 kHz
- D. 3 kHz

ANSWER D: The pilot tone is a 3000 Hz (3 kHz) signal. After the pilot tone is used for tuning the receiver, it is filtered out so it will not be heard in the speaker.

3G84 How many more voice transmissions can be packed into a given frequency band for amplitude-compandored single-sideband systems over conventional FM-phone systems?
- A. 2
- B. 4
- C. 8
- D. 16

ANSWER B: In the VHF land mobile radio service, ACSB gives us four channels in the place of one conventional, narrow-band, FM channel. Although there is no tremendous use of ACSB on VHF 150-MHz land mobile frequencies, there promises to be plenty of ACSB up on the newly created 220-222-MHz land mobile band.

3G85 What term describes a wide-bandwidth communications system in which the RF carrier varies according to some predetermined sequence?
- A. Amplitude compandored single sideband
- B. SITOR
- C. Time-domain frequency modulation
- D. Spread spectrum communication

ANSWER D: Spread spectrum communications are becoming more popular. You probably don't realize it, but your 900-MHz cordless phone uses spread spectrum communications. The frequency hops in a pre-arranged sequence, eliminating interference and eavesdropping. The new global positioning system (GPS) at 1500 MHz uses spread-spectrum techniques.

3G86 What is the term used to describe a spread spectrum communications system where the center frequency of a conventional carrier is altered many times per second in accordance with a pseudo-random list of channels?
A. Frequency hopping
B. Direct sequence
C. Time-domain frequency modulation
D. Frequency compandored spread spectrum
ANSWER A: Spread spectrum communications alters the center frequency of an RF carrier in psuedo-random manner, causing the frequency to "hop" around from channel to channel. This is termed *frequency hopping.*

3G87 What term is used to describe a spread spectrum communications system in which a very fast binary bit stream is used to shift the phase of an RF carrier?
A. Frequency hopping
B. Direct sequence
C. Binary phase-shift keying
D. Phase compandored spread spectrum
ANSWER B: The term "direct sequence" is used to describe a system where a fast binary stream shifts the phase of an RF carrier.

3G88 What is the term for the amplitude of the maximum positive excursion of a signal as viewed on an oscilloscope?
A. Peak-to-peak voltage
B. Inverse peak negative voltage
C. RMS voltage
D. Peak positive voltage
ANSWER D: The maximum positive excursion for the amplitude of a signal on an oscilloscope is called the peak positive voltage. See figure at question 3G90.

3G89 What is the term for the amplitude of the maximum negative excursion of a signal as viewed on an oscilloscope?
A. Peak-to-peak voltage
B. Inverse peak positive voltage
C. RMS voltage
D. Peak negative voltage
ANSWER D: AC signals alternate from positive to negative. The maximum negative excursion is called the peak negative voltage, which is identified as a negative peak amplitude in the figure at question 3G90.

3G90 What is the easiest voltage amplitude dimension to measure by viewing a pure sine wave signal on an oscilloscope?
A. Peak-to-peak voltage
B. RMS voltage
C. Average voltage
D. DC voltage
ANSWER A: Looking at a pure sine wave signal on a scope, it is easy to identify the peak-to-peak voltage measured from the maximum positive excursion of the signal to the maximum negative excursion of the signal.

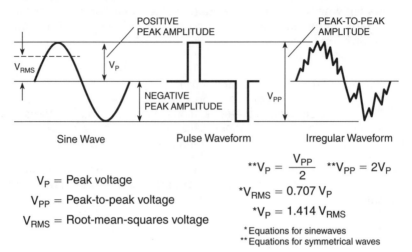

Peak, Peak-to-Peak and RMS Voltages

3G91 What is the relationship between the peak-to-peak voltage and the peak voltage amplitude in a symmetrical wave form?
　　A. 1:1
　　B. 2:1
　　C. 3:1
　　D. 4:1

ANSWER B: The relationship between peak-to-peak voltage and peak voltage is 2:1. It is important to realize that the waveform must be symmetrical—as in a sine wave. See figure above.

3G92 What input-amplitude parameter is valuable in evaluating the signal-handling capability of a Class A amplifier?
　　A. Peak voltage
　　B. Average voltage
　　C. RMS voltage
　　D. Resting voltage

ANSWER A: In a class A amplifier, peak voltage provides a double check on the capability and linearity of that amplifier. Many times, distortion occurs when the amplifier is handling a peak voltage.

3G93 If 480 kHz is radiated from a 1/4 wavelength antenna, what is the 7th harmonic?
　　A. 3.360 MHz
　　B. 840 kHz
　　C. 3350 kHz
　　D. 480 kHz

ANSWER A: Disregard the electrical wavelength of the antenna. Multiple 7 × 480 = 3360 kHz. To change kHz to MHz, remember KLM (kilohertz-left-megahertz), and convert kilohertz to megahertz by moving the decimal point 3 places to the left.

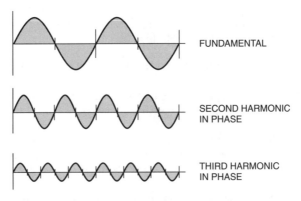

FUNDAMENTAL

SECOND HARMONIC
IN PHASE

THIRD HARMONIC
IN PHASE

Fundamental Radio Wave and Harmonics

3G94 What is the seventh harmonic of a 100 MHz quarter-wavelength antenna?
A. 14.28 MHz
B. 107 MHz
C. 149 MHz
D. 700 MHz
ANSWER D: The seventh harmonic of 100 MHz is simply 7 × 100 = 700 MHz.

3G95 What is the seventh harmonic of 2182 kHz when the transmitter is connected to a half-wave Hertz antenna?
A. 2182 kHz
B. 15.27 MHz
C. 311.7 kHz
D. 7.64 MHz
ANSWER B: Here again, disregard the length of the antenna. The harmonics are actually generated within the transmitter section. It's just that some antennas will radiate harmonics more than others. 7 × 2182 = 15,274 kHz = 15.27 MHz.

3G96 What is the fifth harmonic frequency of a transmitter operating on 480 kHz with a 1/4 wavelength antenna?
A. 2.4 MHz
B. 240 MHz
C. 600 kHz
D. 1.2 MHz
ANSWER A: One more time, disregard the length of the antenna. Multiply the frequency by the harmonics: 480 kHz × 5 = 2400 kHz, (KLM) 2.4 MHz. Just keep in mind that a multi-band trap antenna is prone to making these harmonics sound stronger over the airwaves.

3G97 What is the frequency range of the ground-based Very-High-Frequency Omnidirectional Range (VOR) stations used for aircraft navigation?
A. 108.00 kHz to 117.95 kHz
B. 329.15 MHz to 335.00 MHz
C. 329.15 kHz to 335.00 kHz
D. 108.00 MHz to 117.95 MHz

ANSWER D: The frequency range for the Very-High-Frequency Omnidirectional Range stations is from 108 MHz to 117.95 MHz. Watch out for answer A—it is incorrect because it is in kHz; VHF is in MHz.

3G98 What is the function of a commercial aircraft's SELCAL installation?

    A. SELCAL is a type of aircraft communications device that allows an aircraft's receiver to be continuously CALibrated for signal SELectivity.

    B. SELCAL is a type of aircraft communications system where a ground-based transmitter can CALL a SELected aircraft or group of aircraft without the flight crew monitoring the ground-station frequency.

    C. SELCAL is a type of aircraft transmission that uses SEquential LogiC ALgorithm encryption to prevent public "eavesdropping" of crucial aircraft flight data.

    D. SELCAL is a type of aircraft communications system where an airborne transmitter can SELectively CALculate the line-of-sight distance to several ground-station receivers.

ANSWER B: Selective calling that allows base stations to contact an individual aircraft is preceded by a series of tones that trigger a specific aircraft's SELCAL decoder. The decoder normally remains silent so the flight crew does not need to audibly monitor for the ground-station call.

3G99 What is the main underlying operating principle of the Very-High-Frequency Omnidirectional Range (VOR) aircraft navigational system?

    A. A definite amount of time is required to send and receive a radio signal.

    B. The difference between the peak values of two DC voltages may be used to determine an aircraft's altitude above a selected VOR station.

    C. A phase difference between two AC voltages may be used to determine an aircraft's azimuth position in relation to a selected VOR station.

    D. A phase difference between two AC voltages may be used to determine an aircraft's distance from a selected VOR station.

ANSWER C: The VOR navigation receiver in an aircraft compares the phase differences between two AC voltages to determine an aircraft's azimuth position in relation to a selected VOR station. The key word to look for in the answer is **azimuth**.

3G100 The amplitude modulated variable phase signal and the frequency modulated reference phase signal of a Very-High-Frequency Omnidirectional Range (VOR) station used for aircraft navigation are synchronized so that both signals are in phase with each other at:

    A. 180 degrees South, true bearing position of the VOR station

    B. 360 degrees North, magnetic bearing position of the VOR station

    C. 180 degrees South, magnetic bearing position of the VOR station

    D. 0 degrees North, true bearing position of the VOR station

ANSWER B: When both VOR signals are in phase, they indicate exactly 360 degrees magnetic north. Not true north, but magnetic north.

3G101 Choose the only correct statement about the effective range of a Very-High-Frequency Omnidirectional Range (VOR) station used for aircraft navigation.
　　A. Its reception range is based on both the aircraft's altitude and the aircraft's line-of-sight to the VOR station.
　　B. Its reception range is not a function of the aircraft's altitude.
　　C. Its reception range is not a function of the aircraft's longitude and latitude position in relation to the VOR station's position.
　　D. Its reception range is greatly affected by atmospheric effects and propagation anomalies.
ANSWER A: The reception of a distant VOR station varies in signal strength and range, based on both the aircraft's altitude and the aircraft line-of-sight to the VOR station. Low altitude reduces range because the VOR could actually be past the radio horizon. Mountains in between the aircraft and the VOR can also reduce reception range.

3G102 What is the name of the mechanically rotatable variable RF transformer device of a Very-High-Frequency Omnidirectional Range (VOR) station that is used to generate the amplitude modulated variable phase signal used in aircraft navigation?
　　A. A ghandimeter
　　B. A goniometer
　　C. A gondolameter
　　D. A gorgonzolameter
ANSWER B: The device within the VOR transmitting station that creates the amplitude modulated variable phase signal in 360 one-degree increments is called a *goniometer*. Watch out for the wrong ANSWERs—they look similar.

3G103 What is the frequency range of the localizer beam system used by aircraft to find the centerline of a runway during an Instrument Landing System (ILS) approach to an airport?
　　A. 108.10 kHz to 111.95 kHz
　　B. 329.15 MHz to 335.00 MHz
　　C. 329.15 kHz to 335.00 kHz
　　D. 108.10 MHz to 111.95 MHz
ANSWER D: The VHF frequency for the localizer beam system for an instrument landing system approach is 108.1 MHz to 111.95 Mhz (just above the FM radio band). If you're near an airport, crank your FM receiver dial all the way to the right, and you can usually hear the localizer. Watch out for answer A; these are the right numbers, but answer A is in kHz instead of MHz.

3G104 Choose the only correct statement about the localizer beam system used by aircraft to find the centerline of a runway during an Instrument Landing System (ILS) approach to an airport.
　　A. The localizer beam system operates within the assigned frequency range of 108.10 GHz to 111.95 GHz.
　　B. The localizer beam system produces two amplitude modulated antenna patterns; one pattern above and one pattern below the normal 2.5 degree approach glidepath of the aircraft.

C. The localizer beam system's frequencies are automatically tuned-in when the proper glideslope frequency is selected on the aircraft's Navigation and Communication (NAV/COMM) transceiver.

D. The localizer beam system produces two amplitude modulated antenna patterns; one pattern with an audio frequency of 90 Hz and one pattern with an audio frequency of 150 Hz.

ANSWER D: The two audio frequencies that are amplitude modulated for the localizer beam are 90 Hz and 150 Hz. No other answer has 90 and 150 in it.

3G105 Choose the only correct statement about the localizer beam system used by aircraft to find the centerline of a runway during an Instrument Landing System (ILS) approach to an airport.

A. If the ratio of the 90-Hz audio signal strength to the 150-Hz audio signal strength of the antenna patterns is equal; the aircraft is on the proper 2.5-degree approach glidepath.

B. If the strength of the 90-Hz audio signal is greater than the strength of the 150-Hz audio signal of the antenna patterns; the aircraft is to the left of the centerline of the runway.

C. If the strength of the 150-Hz audio signal is greater than the strength of the 90-Hz audio signal of the antenna patterns; the aircraft is to the left of the centerline of the runway.

D. If the strength of the 150-Hz audio signal is greater than the strength of the 90-Hz audio signal of the antenna patterns; the aircraft is above the proper 2.5-degree approach glidepath.

ANSWER B: The localizer beam system is used by an aircraft to find the center line of a runway during ILS approaches. The signal strength of the 90-Hz audio signal is greater than the strength of the 150-Hz audio signal if the aircraft is to the left of the center line of the runway. (Remember, *localizer-lower-left.*)

✱ 3G106 What is the frequency range of the glideslope beam system used by aircraft to maintain the proper ascent angle to the surface of a runway during an Instrument Landing System (ILS) approach to an airport?

A. 108.00 MHz to 117.95 MHz

B. 329.15 kHz to 335.00 kHz

C. 329.15 MHz to 335.00 MHz

D. 108.10 kHz to 117.95 kHz

ANSWER C: The frequency range for a glideslope beam used in an ILS approach is 329.15 MHz to 335 MHz. Watch out for answer B—it's the right numbers, but the answer is incorrect because it is stated in kHz, not MHz. *[Ed. Note: The question is stated incorrectly: it should ask about the descent angle.]*

✱ 3G107 Choose the only correct statement about the glideslope beam system used by aircraft to maintain the proper ascent angle to the surface of a runway during an Instrument Landing System (ILS) approach to an airport.

A. Two antenna patterns are produced; one to the left of the runway's centerline at an audio modulated tone of 90 Hz and one to the right of the runway's centerline at an audio modulated tone of 150 Hz.

B. Two antenna patterns are produced; one to the left of the runway's centerline at an audio modulated tone of 150 Hz and one to the right of the runway's centerline at an audio modulated tone of 90 Hz.

C. Two antenna patterns are produced; one above the normal 2.5 degree ascent angle to the runway's surface at an audio modulated tone of 150 Hz and one below the normal 2.5-degree ascent angle to the runway's surface at an audio modulated tone of 90 Hz.

D. Two antenna patterns are produced; one above the normal 2.5 degree ascent angle to the runway's surface at an audio modulated tone of 90 Hz and one below the normal 2.5-degree ascent angle to the runway's surface at an audio modulated tone of 150 Hz.

ANSWER D: The glideslope beam pattern consists of two antenna patterns, one above the normal 3-degree descent angle to the runway's surface at an audio modulated tone of 90 Hz, and one BELOW the normal 3-degree descent angle to the runway's surface at an audio modulated tone of 150 Hz. Remember, this question is about glideslope, not localizer left and right. The 90-Hz tone is heard above the glideslope, and the 150-Hz tone is heard below the glideslope. *[Ed. Note: The question is stated incorrectly: it should ask about the <u>descent</u> angle.]*

3G108 What is the frequency range of the marker beacon system used to indicate an aircraft's position during an Instrument Landing System (ILS) approach to an airport's runway?

A. The outer, middle, and inner marker beacons' UHF frequencies are unique for each ILS equipped airport to provide unambiguous frequency protected reception areas in the 329.15 MHz to 335.00 MHz range.

B. The outer marker beacon's carrier frequency is 400 MHz, the middle marker beacon's carrier frequency is 1300 MHz, and the inner marker beacon's carrier frequency is 3000 MHz.

C. The outer, the middle, and the inner marker beacon's carrier frequencies are all 75 MHz but the marker beacons are 95% tone-modulated at 400 Hz (outer), 1300 Hz (middle), and 3000 Hz (inner).

D. The outer marker beacon's carrier frequency is 3000 kHz, the middle marker beacon's carrier frequency is 1300 kHz, and the inner marker beacon's carrier frequency is 400 kHz.

ANSWER C: The marker beacon system that illustrates an aircraft's position during an ILS approach uses an outer marker, middle marker, and inner marker, all on 75 MHz. A rising tone indicates the aircraft is coming in for a landing. The outer marker is modulated at 400 Hz, the middle marker at 1300 Hz, and the inner marker at 3000 Hz.

Watch out for those incorrect answers that list the tone-modulated audio MHz or kHz. There is only one answer with the correct 75-MHz carrier frequency.

3G109 What is the frequency range of the Distance Measuring Equipment (DME) used to indicate an aircraft's slant range distance to a selected ground-based navigation station?

A. 962 MHz to 1213 MHz

B. 108.10 MHz to 111.95 MHz

C. 108.00 MHz to 117.95 MHz

D. 329.15 MHz to 335.00 MHz

ANSWER A: The distant measuring equipment (DME) that indicates an aircraft's slant range to a selected ground-based navigation station uses frequencies from 962 MHz to 1213 MHz.

3G110 What is the main underlying operating principle of an aircraft's Distance Measuring Equipment (DME)?

A. A measurable amount of time is required to send and receive a radio signal through the earth's atmosphere.
B. The difference between the peak values of two DC voltages may be used to determine an aircraft's distance to another aircraft.
C. A measurable frequency compression of an AC signal may be used to determine an aircraft's altitude above the earth.
D. A phase inversion between two AC voltages may be used to determine an aircraft's distance to the exit ramp of an airport's runway.

ANSWER A: DME equipment uses the principle of radar and time delays between a transmit and received signal. DME equipment measures the amount of time required to send and receive a radio signal through the earth's atmosphere and converts that time to distance. The key word is *time*.

3G111 What is the slant range distance of an aircraft's Distance Measuring Equipment (DME)?

A. It is the distance between two aircraft of different altitudes.
B. It is the distance between two ground-based navigation stations having differences in their elevations above mean sea level.
C. It is the line-of-sight distance between an aircraft and a selected ground-based navigation station.
D. It is the radius-of-the-earth distance between two ground-based navigation stations having the same elevations above mean sea level.

ANSWER C: The slant range of an aircraft DME readout is the line-of-sight distance between an aircraft and a selected ground-based navigation station. The key words are *"line-of-sight."*

✱ 3G112 What is the distance to a selected Distance Measuring Equipment (DME) station if an aircraft receives the ground station's reply 175 microseconds after it transmits its airborne interrogation signal? Use the standard 50 microsecond DME reply delay.

A. 20.2 nautical miles
B. 11.6 statute miles
C. 14.2 nautical miles
D. 10.1 statute miles

ANSWER B: Radio signals travel at the speed of light, and take 12.36 microseconds to travel one *nautical* mile. To calculate this problem, subtract the 50 microsecond reply delay from the total reply time of 175 microseconds. Then, divide the remaining 125 microseconds by the speed-of-light constant to get the distance (125 microsec ÷ 12.36 = 10.1 *nautical* miles). This doesn't match any of the given answers, so multiply 10.1 *nautical* miles by 1.15 to convert to *statute* miles and you have it ...11.6 statute miles.

[Ed. Note: As asked, none of the answers listed are correct. Answer B provides the round trip distance the radar signal travels to and from the DME station — not the distance to the station. The correct answer is one-half of Answer B, or 5.8 statute miles. If this question appears on your examination, use Answer B — it is considered the correct answer for this question. After the examination, question your examination team to make sure Answer B was used.]

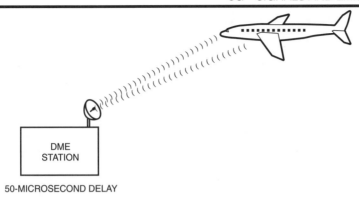

DME
STATION

50-MICROSECOND DELAY

Measuring Aircraft's Distance to Station

3G113 What are the transmission and the reception frequencies of an aircraft's mode-C transponder operating in the Air Traffic Control Radar Beacon System (ATCRBS)?

A. Transmit at 1090 MHz and receive at 1030 MHz
B. Transmit at 1030 kHz and receive at 1090 kHz
C. Transmit at 1090 kHz and receive at 1030 kHz
D. Transmit at 1030 MHz and receive at 1090 MHz

ANSWER A: The mode-C transponder transmits at 1090 MHz, and receives 60 MHz lower at 1030 MHz. Watch out for incorrect answers that have this backwards, and incorrect answers in kHz instead of Mhz!

3G114 What type of information is derived from an aircraft's mode-C transponder transmission operating in the Air Traffic Control Radar Beacon System (ATCRBS)?

A. Range, aircraft weight, and fuel aboard information
B. Range, aircraft weight, and altitude information
C. Range, fuel aboard, and altitude information
D. Range, bearing, and altitude information

ANSWER D: The mode-C transponder squawk allows an air traffic control radar beacon system to determine aircraft range, bearing, and altitude. Read over the incorrect answers to see how ridiculously wrong they are.

3G115 What type of encoding is used in an aircraft's mode-C transponder transmission to a ground station of the Air Traffic Control Radar Beacon System (ATCRBS)?

A. Differential phase shift keying
B. Pulse position modulation
C. Doppler effect compressional encryption
D. Amplitude modulation at 95%

ANSWER B: The encoding for a mode-C transponder is PPM. This stands for *pulse position modulation*, not parts per million!

3G116 What type of encryption is used in the P6 informational pulse of an aircraft's mode-S transponder transmission to a ground station of the Air Traffic Control Radar Beacon System (ATCRBS)?

A. Differential phase shift keying
B. Pulse position modulation

C. Doppler effect compressional encryption

D. Amplitude modulation at 95%

ANSWER A: A mode-S transponder use different phase shift keying.

3G117 What is the function of the P2 side lobe suppression (SLS) pulse that is received by an aircraft's mode-C or mode-S transponder operating in the Air Traffic Control Radar Beacon System (ATCRBS)?

A. The P2 SLS pulse is radiated from an ATCRBS omnidirectional antenna and is used to suppress an aircraft's transponder reply if it is located in the spurious side lobes of the ATCRBS interrogation directional beam pattern.

B. The P2 SLS pulse is radiated from an ATCRBS directional antenna and is used to ensure that an aircraft's transponder will continuously reply to all transmitted ATCRBS interrogation signals while suppressing all other aircraft from replying.

C. The P2 SLS pulse is radiated from an ATCRBS omnidirectional antenna to suppress the reply of non-transponder equipped aircraft within the ATCRBS' reception area.

D. The P2 SLS pulse is transmitted by the aircraft's transponder to suppress other aircrafts' transponders from replying to an ATCRBS interrogation signal thereby eliminating reply transmission confusion.

ANSWER A: Spurious side lobes of an air traffic control radar beacon system may yield unwanted information, and the P2 SLS pulse from the omnidirectional antenna helps suppress spurious side lobe error reception.

3G118 In addition to duplicating the functions of a mode-C transponder, an aircraft's mode-S transponder can also provide:

A. Primary radar surveillance capabilities

B. Long range lightning detection

C. Mid-air collision avoidance capabilities

D. Backup VHF voice communication abilities

ANSWER C: The mode-S transponder have all the functions of a mode-C transponder; however, plus mode-S transponders also has mid-air collision avoidance capabilities.

3G119 In the Air Traffic Control Radar Beacon System (ATCRBS), what is an aircraft's mode-S squitter transmission?

A. It is a 1090 MHz transmission that announces the unique presence of a mode S capable aircraft by encrypting the aircraft's tail number in the signal.

B. It is a 1030 MHz transmission that is only used to determine the distance to an aircraft that is not transponder equipped.

C. It is a 1090 MHz transmission that is used to actively track other aircraft that are not transponder equipped.

D. It is a 1030 MHz transmission that is used as a safety distress signal when an aircraft is in danger of a mid-air collision.

ANSWER A: The 1090 MHz "squitter" transmission announces the presence of a mode-S transponder that contains an the encrypted tail number of the aircraft.

3G120 What is the frequency range of an aircraft's radio altimeter?
A. 962 MHz to 1213 MHz
B. 329.15 MHz to 335.00 MHz
C. 4250 MHz to 4350 MHz
D. 108.00 MHz to 117.95 MHz
ANSWER C: Here is one more frequency to file in memory before the big test— radio altimeter from 4250 MHz to 4350 MHz. The answers are the only four-digit numbers (up in the radar range). The radio altimeter signal is a frequency-modulated continuous wave.

3G121 What type of transmission is radiated from an aircraft's radio altimeter antenna?
A. An amplitude modulated continuous wave
B. A pulse position modulated UHF signal
C. A differential phase shift keyed UHF signal
D. A frequency modulated continuous wave
ANSWER D: A frequency-modulated continuous wave makes up the transmitting portion of the 4250 MHz - 4350 MHz radio altimeter.

3G122 What is the normal Above Ground Level (AGL) range of an aircraft's radio altimeter?
A. 0 feet AGL to 2,500 feet AGL
B. 0 feet AGL to 25,000 feet AGL
C. 0 feet AGL to 75,000 feet AGL
D. 0 feet AGL to the inner boundaries of the troposphere
ANSWER A: The radio altimeter is only used for low-level flying, up to 2,500 feet above ground level.

3G123 What occurs when an aircraft's radio altimeter experiences the double bounce phenomenon?
A. One aircraft's radio altimeter transmission is received by another aircraft's radio altimeter antenna, giving erroneous indications.
B. An aircraft's radio altimeter transmission is first bounced-off the earth's troposphere, then bounced-off the earth's surface, then received.
C. An aircraft's radio altimeter transmission is bounced-off the earth, then bounced-off the aircraft's skin, then bounced-off the earth again, then received.
D. One aircraft's radio altimeter transmission is bounced-off another aircraft's skin, then received, giving erroneous indications.
ANSWER C: Over sea water and other smooth terrain, an aircraft radio altimeter could experience a double bounce. The signal is received twice; from the earth, up to the aircraft skin, back to the earth, and then back to the aircraft receiver.

3G124 What is the frequency range of an aircraft's Automatic Direction Finding (ADF) equipment?
A. 190 MHz to 1750 MHz
B. 108.10 MHz to 111.95 MHz
C. 190 kHz to 1750 kHz
D. 108.00 MHz to 117.95 MHz

ANSWER C: Automatic direction finders work on low-frequency and medium-frequency bands from 190 kHz to 1750 kHz. Watch out for answer A—it is incorrectly stated in MHz.

3G125 Choose the only correct statement about an aircraft's Automatic Direction Finding (ADF) equipment.
  A. An aircraft's ADF transmission exhibits primarily a line-of-sight range to the ground-based target station and will not follow the curvature of the earth.
  B. Only a single omnidirectional sense antenna is required to receive an NDB transmission and process the signal to calculate the aircraft's bearing to the selected ground station.
  C. All frequencies in the ADF's operating range except the commercial standard broadcast stations (550 kHz to 1660 kHz) can be utilized as a navigational Non Directional Beacon (NDB) signal.
  D. An aircraft's ADF antennas can receive transmissions that are over the earth's horizon (sometimes several hundred miles away) since these signals will follow the curvature of the earth.

ANSWER D: An aircraft reception of older non-directional beacons on low frequency and medium frequency sometimes propagates well beyond the earth's horizon because these lower frequencies tend to follow the curvature of the earth and are sometimes refracted by the ionosphere.

3G126 What is the cause of quadrantal error when using an aircraft's Automatic Direction Finding (ADF) equipment?
  A. Quadrantal error is caused by the altitude of an aircraft in relation to the curvature of the earth.
  B. Quadrantal error is caused by the geographic position of an aircraft in relation to the North, South, East, or West quadrant of the earth.
  C. Quadrantal error is caused by the presence of the aircraft in the electromagnetic field of the NDB transmission.
  D. Quadrantal error is caused when the aircraft's ADF transmission is not attenuated sufficiently and is then received at an elevated power level.

ANSWER C: The quadrantal error is caused by the presence of the aircraft ADF system and the strong electromagnetic field of a non-directional beacon transmitting station.

3G127 What is meant by the term, "night effect", when using an aircraft's Automatic Direction Finding (ADF) equipment?
  A. Night effect refers to the fact that all Non-Directional Beacon (NDB) transmitters are turned-off at dusk and turned-on at dawn.
  B. Night effect refers to the fact that Non-Directional Beacon (NDB) transmissions can bounce-off the earth's ionosphere at night and be received at almost any direction.
  C. Night effect refers to the fact that an aircraft's ADF transmissions will be slowed at night due to the increased density of the earth's atmosphere after sunset.
  D. Night effect refers to the fact that an aircraft's ADF antennas usually collect dew moisture after sunset which decreases their effective reception distance from an NDB transmitter.

ANSWER B: At nighttime, the ionosphere can refract and *bounce* NDB signals, and through back scatter could be received from a direction different than where the beacon is actually located.

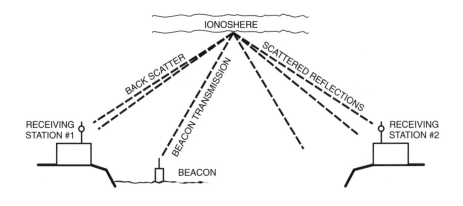

"Night Effect" Using ADF Equipment

3G128 What is the frequency range of an aircraft's High-Frequency (HF) communications?
  A. 118.000 MHz to 136.975 MHz (worldwide up to 151.975 MHz)
  B. 108.00 MHz to 117.95 MHz
  C. 329.15 MHz to 335.00 MHz
  D. 2.000 MHz to 29.999 MHz

ANSWER D: High-frequency communications span 3 MHz to 30 MHz, but the correct answer is 2 MHz to 29.999 MHz, and this is plenty close enough.

3G129 Why is an antenna coupler used in an aircraft's High-Frequency (HF) communications installation?
  A. The antenna coupler is a TNC connector that electrically and mechanically bonds the HF receiver and, or box to the antenna's coaxial lead.
  B. The antenna coupler is a device that is used to secure an HF antenna to the aircraft's fuselage.
  C. The antenna coupler is a device that is used to provide an efficient voltage standing-wave ratio impedance match between the HF transmitter load and the transmission line antenna.
  D. The antenna coupler is a device that is used to "daisy chain" several HF antennas on an aircraft to increase its HF reception range.

ANSWER C: An aeronautical antenna coupler hooked into the aeronautical HF SSB transceiver is used to provide an efficient voltage standing-wave ratio impedance match between the HF SSB equipment and the antenna at the end of the transmission line.

3G130 What is the frequency range of an aircraft's Very-High-Frequency (VHF) communications?

A. 118.000 MHz to 136.975 MHz (worldwide up to 151.975 MHz)
B. 108.00 MHz to 117.95 MHz
C. 329.15 MHz to 335.00 MHz
D. 2.000 MHz to 29.999 MHz

ANSWER A: Aeronautical Very-High-Frequency channels are found from 118 MHz to 136.975 Mhz (and sometimes higher) throughout the world.

3G131 Choose the only correct statement about an aircraft's High Frequency (HF) and Very High Frequency (VHF) communication transmissions.

A. VHF transmissions usually follow the curvature of the earth or are reflected by the ionosphere to provide long-range communications.
B. HF transmission antennas are smaller and more aerodynamic than are VHF transmission antennas.
C. VHF transmissions experience less channel interference than do HF transmissions.
D. HF transmissions are primarily a line-of-sight system between the transmitter station and the receiver station.

ANSWER C: High frequency has a greater range than VHF, but VHF transmissions experience less channel interference from atmospheric noise and onboard aircraft electrical noise than do HF transmissions.

Subelement 3H – Antennas and Feedlines   (10 questions)

3H1 What is meant by the term antenna *gain*?

A. The numerical ratio relating the radiated signal strength of an antenna to that of another antenna
B. The ratio of the signal in the forward direction to the signal in the back direction
C. The ratio of the amount of power produced by the antenna compared to the output power of the transmitter
D. The final amplifier gain minus the transmission line losses (including any phasing lines present)

ANSWER A: We normally rate antenna gain figures in dB. This is the numerical ratio relating the radiated signal strength of your antenna to that of another antenna down the street. Generally, the higher gain antenna will get better signal reports.

3H2 What is the term for a numerical ratio that relates the performance of one antenna to that of another real or theoretical antenna?

A. Effective radiated power
B. Antenna gain
C. Conversion gain
D. Peak effective power

ANSWER B: Antenna gain is an important consideration when choosing an antenna system.

3H3 What is meant by the term antenna bandwidth?
A. Antenna length divided by the number of elements
B. The frequency range over which an antenna can be expected to perform well
C. The angle between the half-power radiation points
D. The angle formed between two imaginary lines drawn through the ends of the elements

ANSWER B: As a ship station on worldwide frequencies switches to lower bands, such as the 4-MHz, 6-MHz, or 8-MHz band, antenna bandwidth becomes important. The associated automatic antenna tuner will probably seek a new LC setting with frequency excursions as little as 50 kHz.

3H4 What is the wavelength of a shorted stub used to absorb even harmonics?
A. 1/2 wavelength
B. 1/3 wavelength
C. 1/4 wavelength
D. 1/8 wavelength

ANSWER C: We would use a 1/4 wavelength shorted stub to absorb even harmonics for a specific frequency antenna system.

3H5 What is a trap antenna?
A. An antenna for rejecting interfering signals
B. A highly sensitive antenna with maximum gain in all directions
C. An antenna capable of being used on more than one band because of the presence of parallel LC networks
D. An antenna with a large capture area

ANSWER C: For worldwide frequencies aboard a boat, 2 MHz through 22 MHz, a multi-band trap antenna is a good performer. This will allow the ship station to use a 50-ohm SSB transceiver without an auxiliary antenna tuner. The trap antenna consists of parallel resonant LC networks to trap out the higher bands from getting up to the top of the antenna. These traps also provide loading for the lower bands. Trap antennas use parallel LC networks which prevent the signal from traveling beyond the trap.

3H6 What is an advantage of using a trap antenna?
A. It has high directivity in the high-frequency bands.
B. It has high gain.
C. It minimizes harmonic radiation.
D. It may be used for multiband operation.

ANSWER D: The trap antenna may be used for multiband operation, from a single feed line. There are trap dipoles, trap verticals, and trap beam antennas for multiband operation.

3H7 What is a disadvantage of using a trap antenna?
A. It will radiate harmonics.
B. It can only be used for single band operation.
C. It is too sharply directional at lower frequencies.
D. It must be neutralized.

ANSWER A: Trap antennas have one big disadvantage—on older equipment they can radiate harmonics. Since marine band harmonics relate to bands twice the operating frequency, you can see that the second harmonic of 8 MHz ends up on sensitive frequencies at 16 MHz from a trap antenna. If your ship station is using a 50-ohm trap antenna, pay particular attention to unintentional harmonics that trap antennas are well known for!

3H8 What is the principle of a trap antenna?

A. Beamwidth may be controlled by non-linear impedances.
B. The traps form a high impedance to isolate parts of the antenna.
C. The effective radiated power can be increased if the space around the antenna "sees" a high impedance.
D. The traps increase the antenna gain.

ANSWER B: The trap antenna will use loading coils and outside capacitive coil sleeves to develop a high-impedance network to isolate longer parts of the antenna to those frequencies requiring a shorter radiating element.

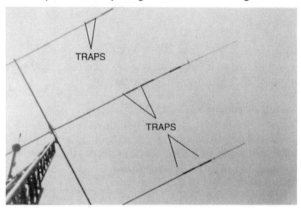

Yagi Antenna with Traps

3H9 What is a parasitic element of an antenna?

A. An element polarized 90 degrees opposite the driven element
B. An element dependent on the antenna structure for support
C. An element that receives its excitation from mutual coupling rather than from a transmission line
D. A transmission line that radiates radio-frequency energy

ANSWER C: You can build a beam antenna on a broomstick. Even though the elements are not electrically connected, they receive their excitation from *mutual coupling* (key words).

3H10 How does a parasitic element generate an electromagnetic field?

A. By the RF current received from a connected transmission line
B. By interacting with the earth's magnetic field
C. By altering the phase of the current on the driven element
D. By currents induced into the element from a surrounding electric field

ANSWER D: Radio frequency voltages, which are induced into the parasitic elements from a surrounding electric field, produce currents in the parasitic elements which radiate supporting fields.

3H11 How does the length of the reflector element of a parasitic element beam antenna compare with that of the driven element?

A. It is about 5% longer.
B. It is about 5% shorter.
C. It is twice as long.
D. It is one-half as long.

ANSWER A: The beam antenna reflector element is normally 5 percent longer than the driven element. The reflector helps form the front lobe of the radiation pattern. See also figure at question 3H79.

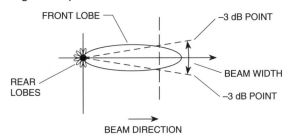

Directional Radiation Pattern of a Yagi Beam

Source: *Antennas – Selection and Installation,* A.J. Evans, ©1986 Master Publishing, Inc., Richardson, Texas

3H12 How does the length of the director element of a parasitic-element beam antenna compare with that of the driven element?

A. It is about 5% longer.
B. It is about 5% shorter.
C. It is one-half as long.
D. It is twice as long.

ANSWER B: The director of a beam antenna is approximately 5 percent shorter than the driven element. This is why most beams have a tapered look when you view them from the ground. If receiving, the shorter end always points toward the transmitting stations. See figure at question 3H79.

3H13 What is meant by the term radiation resistance for an antenna?

A. Losses in the antenna elements and feed line
B. The specific impedance of the antenna
C. An equivalent resistance that would dissipate the same amount of power as that radiated from an antenna
D. The resistance in the trap coils to received signals

ANSWER C: Radiation resistance is the amount of power radiated from an antenna as electromagnetic waves. It is equal to the amount of DC power that would be radiated as heat if the antenna were replaced by a resistor of equal ohmic value. Radiation resistance is a positive measurement of an antenna's efficiency, as opposed to the DC losses of ohmic resistance.

3H14 What are the factors that determine the radiation resistance of an antenna?

A. Transmission line length and height of antenna
B. The location of the antenna with respect to nearby objects and the length/diameter ratio of the conductors
C. It is a constant for all antennas since it is a physical constant
D. Sunspot activity and the time of day

ANSWER B: Radiation resistance of a mobile antenna may be severely compromised if the antenna is mounted too low. The location of the antenna with respect to nearby objects, as well as the length to diameter ratio of the conductors, is very important.

3H15 What is a driven element of an antenna?
   A. Always the rearmost element
   B. Always the forwardmost element
   C. The element fed by the transmission line
   D. The element connected to the rotator

ANSWER C: The driven element of a Yagi antenna is the element fed by the *transmission line* (key words). See figure at question 3H79.

3H16 What is the usual electrical length of a driven element in a HF beam antenna?
   A. 1/4 wavelength
   B. 1/2 wavelength
   C. 3/4 wavelength
   D. 1 wavelength

ANSWER B: The usual electrical length of a driven element on a high-frequency beam antenna is like a dipole, one-half wavelength long.

3H17 What is the term for an antenna element that is supplied power from a transmitter through a transmission line?
   A. Driven element
   B. Director element
   C. Reflector element
   D. Parasitic element

ANSWER A: The transmission line supplies power from the transmitter to the driven element.

3H18 How is antenna "efficiency" computed?
   A. Efficiency = (radiation resistance/transmission resistance) × 100%
   B. Efficiency = (radiation resistance/total resistance) × 100%
   C. Efficiency = (total resistance/radiation resistance) × 100%
   D. Efficiency = (effective radiated power/transmitter output) × 100%

ANSWER B: The efficiency of your antenna is an important consideration when you plan to upgrade your antenna system. Efficiency equals the radiation resistance divided by the total resistance, multiplied by 100 percent. You want to keep your ohmic resistance as low as possible to make your antenna more efficient. Generally, bigger antenna elements have greater radiation resistance, lower ohmic resistance, higher efficiency, and increased bandwidth. Bigger is always better!

3H19 What is the term for the ratio of the radiation resistance of an antenna to the total resistance of the system?
   A. Effective radiated power
   B. Radiation conversion loss
   C. Antenna efficiency
   D. Beamwidth

ANSWER C: The ratio of the radiation resistance of an antenna to the total resistance of the system times 100 is the percent antenna efficiency.

3H20 What is included in the total resistance of an antenna system?
A. Radiation resistance plus space impedance
B. Radiation resistance plus transmission resistance
C. Transmission line resistance plus radiation resistance
D. Radiation resistance plus ohmic resistance

ANSWER D: You want the *ohmic resistance* (key words) as small as possible for improved antenna efficiency.

3H21 How can the antenna efficiency of a HF grounded vertical antenna be made comparable to that of a half-wave antenna?
A. By installing a good ground radial system
B. By isolating the coax shield from ground
C. By shortening the vertical
D. By lengthening the vertical

ANSWER A: Shipboard vertical antennas are 1/4 wavelength long, and use a mirror image counterpoise to compare to that of a half-wave antenna. A good ground radial system, interconnected to bonded through-hulls, will give the 1/4 wavelength radiator good performance, and the good ground will also lower the noise floor and lower the take-off angle, which leads to longer range results.

3H22 Why does a half-wave antenna operate at very high efficiency?
A. Because it is non-resonant
B. Because the conductor resistance is low compared to the radiation resistance
C. Because earth-induced currents add to its radiated power
D. Because it has less corona from the element ends than other types of antennas

ANSWER B: Half-wave antennas operate at high efficiency because their ohmic conductor resistance is low compared to their radiation resistance. For marine applications, insulated stainless steel stays supporting the mast may make good antenna elements with extremely low ohmic resistance.

3H23 What is a folded dipole antenna?
A. A dipole that is one-quarter wavelength long
B. A ground plane antenna
C. A dipole whose ends are connected by another one-half wavelength piece of wire
D. A fictional antenna used in theoretical discussions to replace the radiation resistance

ANSWER C: The folded dipole is a fun one to build out of television twin-lead cable. As shown in the diagram, the twin-lead folded dipole is constructed by using a half-wavelength piece of twin lead. Scrape the insulation from the wire ends and twist the wires together on each end. In the center of one of the leads in the twin-lead, cut the lead and scrape some of the insulation off to expose bare wires on the end of each quarter-wave piece. Feed these wire with standard 300-ohm twin-lead, or connect a balun transformer to the bare wires to permit 50-ohm or 75-ohm coax cable to feed the folded dipole.

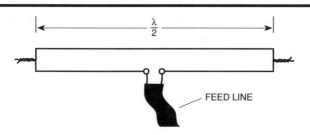

Folded Dipole Antenna Made from 300-ohm Twin-lead

3H24 How does the bandwidth of a folded dipole antenna compare with that of a simple dipole antenna?
    A. It is 0.707 times the simple dipole bandwidth.
    B. It is essentially the same.
    C. It is less than 50% that of a simple dipole.
    D. It is greater.
ANSWER D: The advantage of a folded dipole over a simple dipole is *greater bandwidth* (key words). The bandwidth is greater because twice the added half-wavelength of wire connecting the driven-element dipole makes the folded dipole look like a "thick" antenna, and thus gives it a wider bandwidth.

3H25 What is the input terminal impedance at the center of a folded dipole antenna?
    A. 300 ohms
    B. 72 ohms
    C. 50 ohms
    D. 450 ohms
ANSWER A: The impedance at the center of a folded dipole may be a whopping 300 ohms. You'll need an impedance-matching transformer to feed it with 50-ohm coax; however, ordinary TV twin-lead would match very well.

3H26 What is the meaning of the term "velocity factor" of a transmission line?
    A. The ratio of the characteristic impedance of the line to the terminating impedance
    B. The index of shielding for coaxial cable
    C. The velocity of the wave on the transmission line multiplied by the velocity of light in a vacuum
    D. The velocity of the wave on the transmission line divided by the velocity of light in a vacuum
ANSWER D: Radio waves travel in free space at 300 million meters per second. But in coaxial cable transmission lines, and other types of feed lines, the radio waves move slower. Velocity factor is the velocity of the radio wave on the transmission line divided by the velocity in free space.

3H27 What is the term for the ratio of actual velocity at which a signal travels through a line to the speed of light in a vacuum?
    A. Velocity factor
    B. Characteristic impedance
    C. Surge impedance
    D. Standing wave ratio

ANSWER A: Comparing the velocity of radio signals through a piece of feed line to the speed of light in a vacuum is called the velocity factor of the feed line.

3H28 What is the velocity factor for non-foam dielectric 50- or 75-ohm flexible coaxial cable such as RG 8, 11, 58 and 59?

    A. 2.70
    B. 0.66
    C. 0.30
    D. 0.10

ANSWER B: Most marine grade coax cable has a velocity factor of 0.66. This is not stated on the outside jacket. You may need to check the manufacturer's coax cable specification sheet to determine the velocity factor.

3H29 What determines the velocity factor in a transmission line?

    A. The termination impedance
    B. The line length
    C. Dielectrics in the line
    D. The center conductor resistivity

ANSWER C: There are some great low-loss coaxial cable types out there in radio land with different dielectrics separating the center conductor and the outside shield. It's the dielectric that makes a big difference in the velocity factor.

3H30 Why is the physical length of a coaxial cable transmission line shorter than its electrical length?

    A. Skin effect is less pronounced in the coaxial cable.
    B. RF energy moves slower along the coaxial cable.
    C. The surge impedance is higher in the parallel feed line.
    D. The characteristic impedance is higher in the parallel feed line.

ANSWER B: Radio frequency energy moves slightly *slower* (key word) inside coaxial cable than it would in free space.

3H31 What would be the physical length of a typical coaxial transmission line that is electrically one-quarter wavelength long at 14.1 MHz?

    A. 20 meters
    B. 3.51 meters
    C. 2.33 meters
    D. 0.25 meters

ANSWER B: When you work on antenna phasing harnesses, you will need to calculate the velocity factor for a piece of coax in order to know where to place the connectors. You must be accurate down to a fraction of an inch! You can calculate the physical length of a coax cable which is electrically one-quarter wavelength long using the formula:

▶ $$L \text{ (in feet)} = \frac{984 \lambda V}{f}$$ Where: L = Coax cable length in **feet**
                                                   λ = Wavelength in **meters**
                                                 V = Velocity factor
                                                 f = Frequency in **MHz**

For this problem, multiply 984 × 0.25, because the coax is one quarter wavelength. Then multiply the result by 0.66, the velocity factor. Remember, 0.66 is the velocity factor for coax. We are assuming that value for these problems. Divide the total by 14.1, the frequency in MHz. The length comes out to about

11.51 feet. The answer needs to be in meters, so divide 11.5 by 3.28 (3.28 feet = 1 meter). The result is 3.51 meters. (For a quick approximation, divide 11.5 by 3. The result is 3.83, which gets close enough to pick an answer.)

3H32 What would be the physical length of a typical coaxial transmission line that is electrically one-quarter wavelength long at 7.2 MHz?
   A. 10.5 meters
   B. 6.88 meters
   C. 24 meters
   D. 50 meters

ANSWER B: The coax cable is again one-quarter wavelength, so multiple 984 × 0.25, and the result by 0.66, the velocity factor. We again will assume this value. They won't give you 0.66 on the exam. You must know it. Now divide these three multiplied figures by the frequency, 7.2 MHz, and you end up with 22.55 feet of coax. Dividing by 3 to get approximate length of 7.5 meters, leads you to 6.88 meters of answer B.

3H33 What is the physical length of a parallel antenna feedline that is electrically one-half wavelength long at 14.10 MHz? (Assume a velocity factor of 0.82.)
   A. 15 meters
   B. 24.3 meters
   C. 8.7 meters
   D. 70.8 meters

ANSWER C: In this problem, the parallel feedline is a half-wavelength long, and the velocity factor is given as 0.82. No problem—multiply 984 × 0.5 × 0.82, and divide by 14.1, the frequency in MHz. The answer is 28.61 feet. This works out to be about 8.7 meters long.

3H34 What is the physical length of a twin lead transmission feedline at 36.5 MHz? (Assume a velocity factor of 0.80.)
   A. Electrical length times 0.8
   B. Electrical length divided by 0.8
   C. 80 meters
   D. 160 meters

ANSWER A: This problem is slightly different—all they want to know is the physical length of a feed line at 36.5 MHz with a velocity factor of 0.8. They don't ask anything about how many wavelengths long, so simply take electrical length and multiply it times 0.8. Be careful, times 0.8! Your answer is A.

3H35 In a half-wave antenna, where are the current nodes?
   A. At the ends
   B. At the center
   C. Three-quarters of the way from the feed point toward the end
   D. One-half of the way from the feed point toward the end

ANSWER A: On a half-wave dipole, current nodes are areas of minimum current on the tip ends of a dipole. Current is maximum in the center. See figure at question 3H36.

3H36 In a half-wave antenna, where are the voltage nodes?
A. At the ends
B. At the feed point
C. Three-quarters of the way from the feed point toward the end
D. One-half of the way from the feed point toward the end

ANSWER B: In a half-wave dipole, the voltage node is just opposite of the current—the voltage node of minimum voltage is at the feed point where current is a maximum.

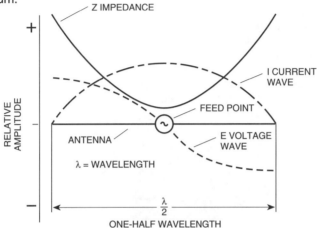

Voltage, Current and Impedance on a Half-wave Dipole

3H37 At the ends of a half-wave antenna, what values of current and voltage exist compared to the remainder of the antenna?
A. Equal voltage and current
B. Minimum voltage and maximum current
C. Maximum voltage and minimum current
D. Minimum voltage and minimum current

ANSWER C: There is maximum voltage on the tip ends of a dipole, but minimum current. This is why you can sometimes spark a lead pencil (be careful) off of the tip ends of a dipole.

3H38 At the center of a half-wave antenna, what values of voltage and current exist compared to the remainder of the antenna?
A. Equal voltage and current
B. Maximum voltage and minimum current
C. Minimum voltage and minimum current
D. Minimum voltage and maximum current

ANSWER D: At the center of a halfwave dipole, voltage is minimum, but current is maximum. Since current at the feed point is maximum, we always want to insure a tight feed-point connection.

3H39 What happens to the base feed point of a fixed length mobile antenna as the frequency of operation is lowered?
A. The resistance decreases and the capacitive reactance decreases
B. The resistance decreases and the capacitive reactance increases
C. The resistance increases and the capacitive reactance decreases
D. The resistance increases and the capacitive reactance increases

---

ANSWER B: As we operate lower on the bands, the base feed point resistance decreases (lower decreases), and the capacitive reactance (X_c) increases. You can visualize the correct answer by thinking the lower you operate in frequency, the more coil (inductive reactance, X_L) you will need to offset and equal the added capacitive reactance, which increases.

3H40 Why should an HF mobile antenna loading coil have a high ratio of reactance to resistance?
- A. To swamp out harmonics
- B. To maximize losses
- C. To minimize losses
- D. To minimize the Q

ANSWER C: When you see those giant open-air coils halfway up the mast on a mobile whip antenna, you are looking at high Q coils which minimize losses. The bigger the coil, the lower the losses.

Mobile Antenna Loading Coil

3H41 Why is a loading coil often used with an HF mobile antenna?
- A. To improve reception
- B. To lower the losses
- C. To lower the Q
- D. To tune out the capacitive reactance

ANSWER D: Since X_L must equal X_c in order for a whip antenna to achieve resonance, we need to add a loading coil on an antenna for lower frequency operation to tune out capacitive reactance.

3H42 For a shortened vertical antenna, where should a loading coil be placed to minimize losses and produce the most effective performance?
- A. Near the center of the vertical radiator
- B. As low as possible on the vertical radiator
- C. As close to the transmitter as possible
- D. At a voltage node

ANSWER A: *Center loading* (key words) is one of the best ways to produce an effective high-frequency mobile antenna system. Also, the larger the coil, the lower the losses.

3H43 What happens to the bandwidth of an antenna as it is shortened through the use of loading coils?
A. It is increased.
B. It is decreased.
C. No change occurs.
D. It becomes flat.

ANSWER B: The shorter your mobile whip, the more loading you need, and bandwidth is decreased. Also, your efficiency usually goes down—so for high efficiency, and more bandwidth, go for a longer whip with less loading.

3H44 What type of antenna is used in an aircraft's Instrument Landing System (ILS) glideslope installation?
A. A vertically polarized antenna that radiates an omnidirectional antenna pattern
B. A balanced loop reception antenna
C. A folded dipole reception antenna
D. An electronically steerable phased-array antenna that radiates a directional antenna pattern

ANSWER C: The type of antenna used in an aircraft instrument landing system glideslope installation is a folded dipole reception antenna.

3H45 What is an advantage of using top loading in a shortened HF vertical antenna?
A. Lower Q
B. Greater structural strength
C. Higher losses
D. Improved radiation efficiency

ANSWER D: Helical top-loaded whips have *improved radiation efficiency* (key words) over shorter whips with big fat loading coils in the center. While they are not as aerodynamic on the top of your mobile unit, the top helical-loaded whip has improved radiation efficiency.

3H46 What is an isotropic radiator?
A. A hypothetical, omnidirectional antenna
B. In the northern hemisphere, an antenna whose directive pattern is constant in southern directions
C. An antenna high enough in the air that its directive pattern is substantially unaffected by the ground beneath it
D. An antenna whose directive pattern is substantially unaffected by the spacing of the elements

ANSWER A: The isotropic radiator radiates equally well in all directions. It is a hypothetical antenna, used as a standard. The isotropic radiator is at least 2 dB down from the simple dipole antenna that is many times also used as a standard.

3H47 When is it useful to refer to an isotropic radiator?
A. When comparing the gains of directional antennas
B. When testing a transmission line for standing wave ratio
C. When (in the northern hemisphere) directing the transmission in a southerly direction
D. When using a dummy load to tune a transmitter

ANSWER A: When the gain of directional antennas and collinear arrays are compared, they are usually compared to an isotropic radiator.

3H48 What theoretical reference antenna provides a comparison for antenna measurements?
A. Quarter-wave vertical
B. Yagi-Uda array
C. Bobtail curtain
D. Isotropic radiator

ANSWER D: The isotropic radiator is a hypothetical antenna used for a theoretical reference.

3H49 What purpose does an isotropic radiator serve?
A. It is used to compare signal strengths (at a distant point) of different transmitters.
B. It is used as a reference for antenna gain measurements.
C. It is used as a dummy load for tuning transmitters.
D. It is used to measure the standing-wave-ratio on a transmission line.

ANSWER B: Marine antenna manufacturers may advertise their VHF antenna gain as compared to an isotropic radiator. This elevates the gain figure, and makes their antennas more competitive. However, in the land mobile industry, most antenna specifications meet EIA (Electronics Industry of America) guidelines, and they are usually several dB less than the same type of marine antenna. Most land mobile VHF and UHF collinear array gain figures are compared to a dipole (dBd).

3H50 How much gain does a 1/2-wavelength dipole have over an isotropic radiator?
A. About 1.5 dB
B. About 2.1 dB
C. About 3.0 dB
D. About 6.0 dB

ANSWER B: The halfwave dipole has about 2.1 dB advantage in gain over an isotropic radiator.

3H51 How much gain does an antenna have over a 1/2-wavelength dipole when it has 6 dB gain over an isotropic radiator?
A. About 3.9 dB
B. About 6.0 dB
C. About 8.1 dB
D. About 10.0 dB

ANSWER A: If an antenna has 6 dB gain over an isotropic radiator, it will have 2.1 dB less over a dipole, resulting in a true 3.9 (6 - 2.1) dBd—that little "d" at the end means over a dipole. dBi means gain over an isotropic radiator.

3H52 How much gain does an antenna have over a 1/2-wavelength dipole when it has 12 dB gain over an isotropic radiator?
A. About 6.1 dB
B. About 9.9 dB
C. About 12.0 dB
D. About 14.1 dB

ANSWER B: An antenna with 12 dBi gain is only 9.9 dBd. Since the dipole has a 2.1 dB gain over an isotropic antenna, the dBi of the antenna in the question is reduced to 9.9 dBd when referenced to the dipole.

3H53 What is the antenna pattern for an isotropic radiator?
A. A figure-8
B. A unidirectional cardioid
C. A parabola
D. A sphere

ANSWER D: The isotropic radiator radiates in all directions, like a sphere.

3H54 What type of directivity pattern does an isotropic radiator have?
A. A figure-8
B. A unidirectional cardioid
C. A parabola
D. A sphere

ANSWER D: The isotropic antenna (radiator) radiates in a pattern that is equal at any point the same distance from the antenna in all directions. Connecting the points would form a perfect sphere.

3H55 What factors determine the receiving antenna gain required at a station in earth operation?
A. Height, transmitter power and antennas of satellite
B. Length of transmission line and impedance match between receiver and transmission line
C. Preamplifier location on transmission line and presence or absence of RF amplifier stages
D. Height of earth antenna and satellite orbit

ANSWER A: To determine your receiving station antenna requirements from a satellite, you will need to take in satellite range information, satellite transmitter ERP, and satellite antenna polarity.

3H56 What factors determine the EIRP required by a station in earth operation?
A. Satellite antennas and height, satellite receiver sensitivity
B. Path loss, earth antenna gain, signal-to-noise ratio
C. Satellite transmitter power and orientation of ground receiving antenna
D. Elevation of satellite above horizon, signal-to-noise ratio, satellite transmitter power

ANSWER A: To determine the amount of gain you will need for your earth station, review all of the parameters of the satellite and its antenna system.

3H57 What factors determine the EIRP required by a station in telecommand operation?

    A. Path loss, earth antenna gain, signal-to-noise ratio

    B. Satellite antennas and height, satellite receiver sensitivity

    C. Satellite transmitter power and orientation of ground receiving antenna

    D. Elevation of satellite above horizon, signal-to-noise ratio, satellite transmitter power

ANSWER B: VHF or UHF frequencies are used for telecommand operation, so satellite antennas and height, plus satellite receiver sensitivity, are important factors for determining EIRP.

3H58 How does the gain of a parabolic dish type antenna change when the operating frequency is doubled?

    A. Gain does not change.

    B. Gain is multiplied by 0.707.

    C. Gain increases 6 dB.

    D. Gain increases 3 dB.

ANSWER C: From 1270 MHz on up, you may wish to use a parabolic dish antenna for microwave work. If you double the frequency, the gain of the dish increases by 6 dB, a 4 times increase.

Parabolic Antenna

3H59 What happens to the beamwidth of an antenna as the gain is increased?

    A. The beamwidth increases geometrically as the gain is increased.

    B. The beamwidth increases arithmetically as the gain is increased.

    C. The beamwidth is essentially unaffected by the gain of the antenna.

    D. The beamwidth decreases as the gain is increased.

ANSWER D: The parabolic dish concentrates that beam width as gain increases. The higher the gain of a dish, the narrower the beam width. Beam width *decreases* as gain *increases*.

3H60 What is the beamwidth of a symmetrical-pattern antenna with a gain of 20 dB as compared to an isotropic radiator?
    A. 10.1 degrees
    B. 20.3 degrees
    C. 45.0 degrees
    D. 60.9 degrees
ANSWER B: Beam width (in degrees) is equal to a fixed value of 203 divided by the square root of 10, raised to a power represented by the antenna gain, in dB, divided by 10:

$$\text{Beamwidth} = \frac{203}{(\sqrt{10})^x} \text{ in } \textbf{degrees} \qquad x = \frac{\text{Antenna gain in } \textbf{dB}}{10}$$

In this first question, the antenna gain is 20 dB, and x = 20 ÷ 10 = 2

$$\text{Beamwidth} = \frac{203}{(\sqrt{10})^2} = \frac{203}{10} = 20.3 \text{ degrees}$$

The beamwidth for this antenna is 20.3° matching answer B.

3H61 What is the beamwidth of a symmetrical-pattern antenna with a gain of 30 dB as compared to an isotropic radiator?
    A. 3.2 degrees
    B. 6.4 degrees
    C. 37 degrees
    D. 60.4 degrees
ANSWER B: Let's use the same formula as the previous question. In this question the antenna gain is 30 dB and x = 30 ÷ 10 = 3; therefore,

$$\text{Beamwidth} = \frac{203}{(\sqrt{10})^3} = \frac{203}{10\sqrt{10}} = \frac{203}{31.63} = 6.42°$$

The beamwidth for this antenna is 6.42°, matching answer B.

3H62 What is the beamwidth of a symmetrical-pattern antenna with a gain of 15 dB as compared to an isotropic radiator?
    A. 72 degrees
    B. 52 degrees
    C. 36.1 degrees
    D. 3.61 degrees
ANSWER C: Same formula as question 3H60. The gain is 15 dB and x = 15 ÷ 10 = 1.5. The beamwidth is:

$$\text{Beamwidth} = \frac{203}{(\sqrt{10})^{1.5}} = \frac{203}{3.16^{1.5}} = \frac{203}{5.62} = 36.1°$$

You may not know how to find $3.16^{1.5}$. What you can do is estimate. 3.16^2 is 10; therefore, the value you are looking for is between 3.16 and 10. Let's call it Y; therefore, Y = 3.16 × Z. Z is somewhere between 0 and 3.16. Since the power is 1.5, you could choose half of 3.16 or 1.58. However, the power function is logarithmic so 0.5 is weighted past the linear one-half point. Let's choose Z = 1.7, therefore, Y = 3.16 × 1.7 = 5.37. With 5.37, the beamwidth = 203 ÷ 5.37 = 37.8°. In this question, you would be close enough to choose the correct answer C of 36.1°.

3H63 What is the beamwidth of a symmetrical-pattern antenna with a gain of 12 dB as compared to an isotropic radiator?
  A. 34.8 degrees
  B. 45.0 degrees
  C. 58.0 degrees
  D. 51.0 degrees
ANSWER D: Same formula, but now the gain is 12 dB and **X** = 12 ÷ 10 = 1.2.

$$\text{Beamwidth} = \frac{203}{\sqrt{10}^{1.2}} = \frac{203}{3.16^{1.2}} = \frac{203}{3.98} = 51°$$

If you estimate $3.16^{1.2} = 3.16 \times 1.2 = 3.79$, the beamwidth would be 53.6°. This would get you close enough to choose answer D, 51 degrees.
  Let's review what's been done in the last four questions. A big 20 dB dish will provide a tight pattern of 20 degrees—20/20. A giant 30 dB dish will narrow the beam down to about 6.5 degrees. But a much smaller 15 dB dish will widen the beam to a rather broad pattern of 36 degrees, and a 12 dB dish even a wider pattern of 51 degrees. You can almost visualize these patterns.

3H64 How is circular polarization produced using linearly-polarized antennas?
  A. Stack two yagis, fed 90 degrees out of phase, to form an array with the respective elements in parallel planes
  B. Stack two yagis, fed in phase, to form an array with the respective elements in parallel planes
  C. Arrange two yagis perpendicular to each other, with the driven elements in the same plane, fed 90 degrees out of phase
  D. Arrange two yagis perpendicular to each other, with the driven elements in the same plane, fed in phase
ANSWER C: To get circular polarization out of a pair of Yagis, arrange the Yagis perpendicular to each other, with the driven elements in the same plane, and fed 90 degrees out of phase.

3H65 Why does an antenna system for earth operation (for communications through a satellite) need to have rotators for both azimuth and elevation control?
  A. In order to point the antenna above the horizon to avoid terrestrial interference
  B. Satellite antennas require two rotators because they are so large and heavy
  C. In order to track the satellite as it orbits the earth
  D. The elevation rotator points the antenna at the satellite and the azimuth rotator changes the antenna polarization
ANSWER C: You need both an azimuth as well as an elevation rotor control to track the satellites up in orbit. This will be useful if you plan to track orbiting weather satellites at 137 MHz for weather facsimile reception.

3H66 What term describes a method used to match a high-impedance transmission line to a lower impedance antenna by connecting the line to the driven element in two places, spaced a fraction of a wavelength on each side of the driven element center?

A. The gamma matching system
B. The delta matching system
C. The omega matching system
D. The stub matching system

ANSWER B: The delta matching system is a popular way to change the imped-ance of the transmission line to the impedance of the antenna input.

3H67 What term describes an unbalanced feed system in which the driven element is fed both at the center of that element and a fraction of a wavelength to one side of center?
A. The gamma matching system
B. The delta matching system
C. The omega matching system
D. The stub matching system

ANSWER A: A gamma match is used with coax cable that forms an unbalanced feed system going to a tube that acts as a capacitor in series with the connec-tion point.

3H68 What term describes a method of antenna impedance matching that uses a short section of transmission line connected to the antenna feed line near the antenna and perpendicular to the feed line?
A. The gamma matching system
B. The delta matching system
C. The omega matching system
D. The stub matching system

ANSWER D: When a short section of transmission line is connected to the antenna feed line near the antenna, it is called stub matching.

3H69 What kind of impedance does a 1/8-wavelength transmission line present to a generator when the line is shorted at the far end?
A. A capacitive reactance
B. The same as the characteristic impedance of the line
C. An inductive reactance
D. The same as the input impedance to the final generator stage

ANSWER C: If a 1/8-wavelength transmission line is shorted at the far end, it will look like an inductive reactance. See figure at question 3H70.

3H70 What kind of impedance does a 1/8-wavelength transmission line present to a generator when the line is open at the far end?
A. The same as the characteristic impedance of the line
B. An inductive reactance
C. A capacitive reactance
D. The same as the input impedance of the final generator stage

ANSWER C: If that 1/8-wavelength transmission line is open at the far end, it will look like a capacitive reactance.

Quarter wavelength ($\frac{\lambda}{4}$) stubs

(or multiples of odd numbers of $\frac{\lambda}{4}$ s)

Less than $\frac{\lambda}{4}$ stub

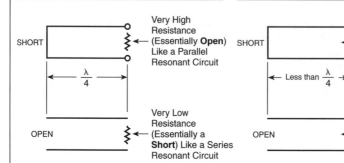

SHORT — Very High Resistance (Essentially **Open**) Like a Parallel Resonant Circuit — $\frac{\lambda}{4}$

OPEN — Very Low Resistance (Essentially a **Short**) Like a Series Resonant Circuit

SHORT — INDUCTIVE [1] REACTANCE — Less than $\frac{\lambda}{4}$

OPEN — CAPACITIVE [2] REACTANCE

Half-wavelength ($\frac{\lambda}{2}$) stubs

(or multiples of $\frac{\lambda}{2}$ s)

Greater than $\frac{\lambda}{4}$, less than $\frac{\lambda}{2}$

Think of it as two $\frac{\lambda}{4}$ stubs in series.

Think of it as a $\frac{\lambda}{4}$ stub plus one less than $\frac{\lambda}{4}$.

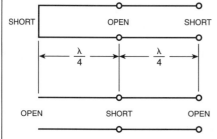

SHORT — OPEN — SHORT — $\frac{\lambda}{4}$ — $\frac{\lambda}{4}$

OPEN — SHORT — OPEN

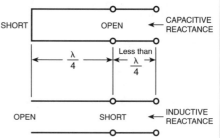

SHORT — OPEN — CAPACITIVE REACTANCE — $\frac{\lambda}{4}$ — Less than $\frac{\lambda}{4}$

OPEN — SHORT — INDUCTIVE REACTANCE

Notes

1. Since a short at end makes the diagram look like a long wire, this should remind reader that impedance is an inductive reactance.
2. Since an open at end makes the diagram look like two parallel plates, this should remind reader that impedance is a capacitive reactance.

Impedance of Matching Stubs

3H71 What kind of impedance does a 1/4-wavelength transmission line present to a generator when the line is shorted at the far end?

    A. A very high impedance

    B. A very low impedance

    C. The same as the characteristic impedance of the transmission line

    D. The same as the generator output impedance

ANSWER A: On a quarter-wavelength transmission line, shorting it at the far end will look like a very high impedance. See figure above.

3H72 What kind of impedance does a 1/4-wavelength transmission line present to a generator when the line is open at the far end?
A. A very high impedance
B. A very low impedance
C. The same as the characteristic impedance of the line
D. The same as the input impedance to the final generator stage
ANSWER B: If the far end of a quarter-wavelength transmission line is open, it will look like a very low impedance. See figure at question 3H70.

3H73 What kind of impedance does a 3/8-wavelength transmission line present to a generator when the line is shorted at the far end?
A. The same as the characteristic impedance of the line
B. An inductive reactance
C. A capacitive reactance
D. The same as the input impedance to the final generator stage
ANSWER C: On a 3/8-wavelength transmission line, if it is shorted at the far end, it looks like a capacitive reactance. See figure at question 3H70.

3H74 What kind of impedance does a 3/8-wavelength transmission line present to a generator when the line is open at the far end?
A. A capacitive reactance
B. The same as the characteristic impedance of the line
C. An inductive reactance
D. The same as the input impedance to the final generator stage
ANSWER C: If that same 3/8-wavelength transmission line is now open at the far end, it will look like an inductive reactance. See figure at question 3H70.

3H75 What kind of impedance does a 1/2-wavelength transmission line present to a generator when the line is shorted at the far end?
A. A very high impedance
B. A very low impedance
C. The same as the characteristic impedance of the line
D. The same as the output impedance of the generator
ANSWER B: If a half-wavelength transmission line is shorted at the far end, it will have a very low impedance. See figure at question 3H70.

3H76 What kind of impedance does a 1/2-wavelength transmission line present to a generator when the line is open at the far end?
A. A very high impedance
B. A very low impedance
C. The same as the characteristic impedance of the line
D. The same as the output impedance of the generator
ANSWER A: If that half-wavelength transmission line is open at the far end, it will have a very high impedance. See figure at question 3H70.

3H77 What is the term used for an equivalent resistance that would dissipate the same amount of energy as that radiated from an antenna?
A. Space resistance
B. Loss resistance
C. Transmission line loss
D. Radiation resistance

ANSWER D: Radiation resistance is that equivalent resistance which would dissipate the same amount of energy as that radiated from the antenna.

3H78 Why is the value of the radiation resistance of an antenna important?

A. Knowing the radiation resistance makes it possible to match impedances for maximum power transfer.
B. Knowing the radiation resistance makes it possible to measure the near-field radiation density from a transmitting antenna.
C. The value of the radiation resistance represents the front-to-side ratio of the antenna.
D. The value of the radiation resistance represents the front-to-back ratio of the antenna.

ANSWER A: If we know the radiation resistance of a specific antenna, it will allow us to match impedances to that antenna for maximum power transfer. Most coaxial cables have a characteristic impedance of 50 to 52 ohms, and this is what is commonly used in land mobile, air, and marine installations.

3H79 Adding parasitic elements to an antenna will:

A. Decrease its directional characteristics
B. Decrease its sensitivity
C. Increase its directional characteristics
D. Increase its sensitivity

ANSWER C: Adding parasitic elements to a directional antenna for worldwide marine use, or VHF aeronautical use, will increase its directional characteristics. See also questions 3H11, 3H12 and 3H15.

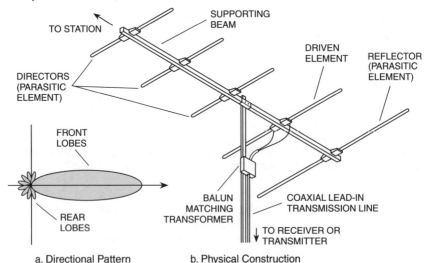

a. Directional Pattern   b. Physical Construction

A Beam Antenna — The Yagi Antenna

Source: *Antennas — Selection and Installation,* © 1986 Master Publishing, Inc., Richardson Texas

3H80 What ferrite rod device prevents the formation of reflected waves on a waveguide transmission line?

A. Reflector
B. Isolator

C. Wave-trap

D. SWR refractor

ANSWER B: At microwave frequencies, the ferrite rod isolator prevents reflected waves on a waveguide transmission line.

3H81 Frequencies most affected by knife-edge refraction are:

A. Low and medium frequencies.

B. High frequencies.

C. Very high and ultra high frequencies.

D. 100 kHz. to 3.0 MHz.

ANSWER C: VHF and UHF signals can be bent over mountains, and around and over buildings with a beam antenna and the effect of knife-edge refraction.

3H82 When measuring I and V along a 1/2-wave Hertz antenna, where would you find the points where I and V are maximum and minimum?

A. V and I are high at the ends

B. V and I are high in the middle

C. V and I are uniform throughout the antenna

D. V is maximum at both ends, I is maximum in the middle

ANSWER D: Voltage is always maximum at the ends of a halfwave dipole. Current is maximum in the middle and minimum at the ends. Remember volts and the letter "S" on its side at the dipole, and current like the setting sun on the dipole. See figure at question 3H36.

3H83 To increase the resonant frequency of a 1/4-wavelength antenna:

A. Add a capacitor.

B. Lower capacitor value.

C. Cut antenna.

D. Add an inductor.

ANSWER A: We add capacitance to raise the resonant frequency of a 1/4-wavelength antenna. Answer C is close, but a better answer would be "trim antenna" rather than to just cut it.

3H84 Why are concentric transmission lines sometimes filled with nitrogen?

A. Reduces resistance at high frequencies

B. Prevent water damage underground

C. Keep moisture out and prevent oxidation

D. Reduce microwave line losses

ANSWER C: In high power commercial land radio stations, we sometimes will back fill concentric transmission line with dry nitrogen to drive out moisture and to prevent oxidation on the center conductor.

3H85 A vertical 1/4-wave antenna receives signals:

A. In the microwave band.

B. In one vertical direction.

C. In one horizontal direction.

D. Equally from all horizontal directions.

ANSWER D: While this answer sounds strange, the 1/4-wavelength vertical antenna receives its energy from other vertical antennas in all horizontal directions around the antenna. You have to think about this for a few seconds to get the picture—it receives in all directions—all horizontal directions.

3H86 Which of the following represents the best standing wave ratio (SWR)?
  A. 1:1
  B. 1:1.5
  C. 1:3
  D. 1:4
ANSWER A: A 1:1 is the best SWR.

3H87 What is the purpose of stacking elements on an antenna?
  A. Sharper directional pattern
  B. Increased gain
  C. Improved bandpass
  D. All of these
ANSWER D: A collinear antenna is normally used for land mobile and marine applications. The collinear antenna has stacked elements inside a white radome. We normally do not use a collinear antenna in the aviation service because it might not pick up well at high elevations.

3H88 What type of antenna is used in an aircraft's Instrument Landing System (ILS) marker beacon installation?
  A. An electronically steerable phased-array antenna that radiates a directional antenna pattern
  B. A folded dipole reception antenna
  C. A balanced loop reception antenna
  D. A horizontally polarized antenna that radiates an omnidirectional antenna pattern
ANSWER C: For the instrument landing system marker beacon, a balanced loop reception antenna is used.

3H89 On a half-wave Hertz antenna:
  A. Voltage is maximum at both ends and current is maximum at the center of the antenna
  B. Current is maximum at both ends and voltage in the center
  C. Voltage and current are uniform throughout the antenna
  D. Voltage and current are high at the ends
ANSWER A: On the halfwave dipole, voltage is maximum at both ends, and current is maximum at the center. See figure at question 3H36.

3H90 What type of antenna is designed for minimum radiation?
  A. Dummy antenna
  B. Quarter-wave antenna
  C. Half-wave antenna
  D. Directional antenna
ANSWER A: We use a dummy antenna when working on two-way radio equipment to keep it from radiating out on the airwaves.

3H91 What type of antenna is used in an aircraft's Very-High-Frequency Omnidirectional Range (VOR) and Localizer (LOC) installations?
  A. A vertically polarized antenna that radiates an omnidirectional antenna pattern
  B. A folded dipole reception antenna

C. A balanced loop transmission antenna

D. A horizontally polarized omnidirection reception antenna

ANSWER D: The antenna for VOR and Localizer installations is a horizontally polarized omnidirection reception antenna.

3H92 Adding parasitic elements to a quarter-wavelength antenna will:

A. Reduce its directional characteristics.

B. Increase its directional characteristics.

C. Increase its sensitivity.

D. Increase its directional characteristics and increase its sensitivity.

ANSWER D: Adding parasitic elements in front of and behind a quarter-wavelength antenna will increase its directional characteristics and add to the sensitivity of the system *in one direction.* Therefore, the answer is D because both B and C are correct.

3H93 Ignoring line losses, voltage at a point on a transmission line without standing waves is:

A. Equal to the product of the line current and impedance.

B. Equal to the product of the line current and power factor.

C. Equal to the product of the line current and the surge impedance.

D. Zero at both ends.

ANSWER C: Measuring voltage on a transmission line is the product of the line current times the surge impedance.

3H94 Stacking antenna elements:

A. Will suppress odd harmonics.

B. Decrease signal to noise ratio.

C. Increases sensitivity to weak signals.

D. Increases selectivity.

ANSWER C: Stacking antenna elements in a collinear array increases the sensitivity to extremely weak signals down close to the horizon.

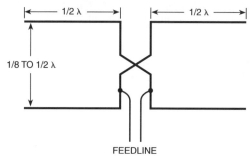

A 4-Element Antenna Array

Courtesy of *The ARRL Antenna Book,* ©1991, The American Radio Relay League, Inc.

3H95 What allows microwaves to pass in only one direction?

A. RF emitter

B. Ferrite isolator

C. Capacitor

D. Varactor-triac

ANSWER B: We use a ferrite isolator to channel microwaves to pass in only one direction to protect the receiver section hooked up to the same isolator.

3H96 What would be added to make a receiving antenna more directional?

A. Inductor
B. Capacitor
C. Parasitic elements
D. Height

ANSWER C: Adding parasitic antenna elements to a receiving antenna will make it more directional.

3H97 Nitrogen is placed in transmission lines to:

A. Improve the "skin-effect" of microwaves.
B. Reduce arcing in the line.
C. Reduce the standing wave ratio of the line.
D. Prevent moisture from entering the line.

ANSWER D: Nitrogen inside a transmission line prevents moisture from getting into the line to protect the center conductor from oxidation.

3H98 Neglecting line losses, the voltage at any point along a transmission line, having no standing waves, will be equal to the:

A. Transmitter output.
B. Product of the line voltage and the surge impedance of the line.
C. Product of the line current and the surge impedance of the line.
D. Product of the resistance and surge impedance of the line.

ANSWER C: To calculate voltage in a transmission line, multiply line current by surge impedance.

3H99 Adding a capacitor in series with a Marconi antenna:

A. Increases the antenna circuit resonant frequency.
B. Decreases the antenna circuit resonant frequency.
C. Blocks the transmission of signals from the antenna.
D. Increases the power handling capacity of the antenna.

ANSWER A: Adding a capacitor in series with a Marconi antenna will raise the antenna's resonant frequency. Adding an inductor will lower the antenna's resonant frequency.

3H100 An antenna is carrying an unmodulated signal, when 100% modulation is impressed, the antenna current:

A. Goes up 50%.
B. Goes down one half.
C. Stays the same.
D. Goes up 22.5%.

ANSWER D: You can calculate 100 percent modulated antenna current from the following formula:

$$\% \text{ increase in modulated} = \left(\sqrt{1 + \frac{m^2}{2}} - 1 \right) \times 100$$

Where m = a value between 0 and 1 representing the % modulation, e.g., 1 = 100%

Out in the field, there may be many different types of modulation, so this formula is an approximate way to calculate 100 percent modulated antenna current. Typical is 20 percent to 25 percent.

3H101 An excited 1/2-wavelength antenna produces:
A. Residual fields.
B. An electro-magnetic field only.
C. Both electro-magnetic and electro-static fields.
D. An electro-flux field sometimes.

ANSWER C: All antennas radiate two fields—an electric field and a magnetic field. They are at right angles to each other. It is the electric field that is used in describing polarization.

3H102 An antenna which intercepts signals equally from all horizontal directions is:
A. Parabolic.
B. Vertical loop.
C. Horizontal marconi.
D. Vertical 1/4 wave.

ANSWER D: An antenna that receives signals equally well from all horizontal directions is the vertical quarter-wavelength dipole.

3H103 A Loop-antenna:
A. Is bi-directional.
B. Is usually vertical.
C. Is more often used as a receiving antenna.
D. Any of these.

ANSWER D: The loop antenna may be used for direction-finding on a radio receiver. It is bi-directional, and usually mounted vertical.

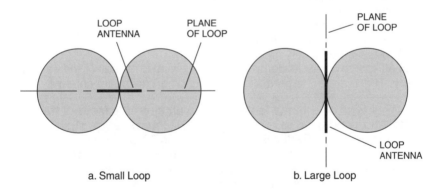

a. Small Loop   b. Large Loop

Radiation Patterns of Loop Antennas

3H104 Referred to the fundamental frequency, a shorted stub line attached to the transmission line to absorb even harmonics could have a wavelength of:
A. 1.41 wavelengths.
B. 1/2 wavelengths.
C. 1/4 wavelengths.
D. 1/6 wavelengths.

ANSWER C: The 1/4-wavelength shorted stub line is attached to a transmitter to absorb even harmonics on a single frequency antenna system.

3H105 Nitrogen gas in concentric RF transmission lines is used to:
   A. Keep moisture out.
   B. Prevent oxidation.
   C. Act as an insulator.
   D. Keep moisture out and prevent oxidation.
ANSWER D: Nitrogen in an RF transmission hardline acts as an insulator and helps drive out moisture. Moisture causes unwanted oxidation.

3H106 "Stacking" elements on an antenna:
   A. Makes for better reception.
   B. Makes for poorer reception.
   C. Decreases antenna current.
   D. Decreases directivity.
ANSWER A: Stacking elements in a collinear array increases reception.

3H107 What type of antenna is attached to an aircraft's Mode-C transponder installation and used to receive 1030 MHz interrogation signals from the Air Traffic Control Radar Beacon System (ATCRBS)?
   A. An L-band monopole blade-type omnidirectional antenna
   B. An electronically steerable phased-array directional antenna
   C. A folded dipole reception antenna
   D. An internally mounted, mechanically rotatable loop antenna
ANSWER A: An L-band monopole blade-type omnidirectional antenna is used to receive 1030-MHz mode-C transponder interrogations.

3H108 The parasitic elements on a receiving antenna:
   A. Increase its directivity.
   B. Decrease its directivity.
   C. Have no effect on its impedance.
   D. Make it more nearly omnidirectional.
ANSWER A: Adding parasitic elements on a receiving antenna will increase its directivity. This is similar to parasitic elements on a Yagi beam antenna.

3H109 What type of antenna is attached to an aircraft's Mode-S transponder installation and used to receive 1030-MHz interrogation signals from the Air Traffic Control Radar Beacon System (ATCRBS)?
   A. An L-band monopole blade-type omnidirectional antenna
   B. An electronically steerable phased-array directional antenna
   C. A folded dipole reception antenna
   D. An internally mounted, mechanically rotatable loop antenna
ANSWER A: For the mode-S transponder, an L-band monopole blade-type omnidirectional antenna is used.

3H110 The resonant frequency of a Hertz antenna can be lowered by:
   A. Lowering the frequency of the transmitter.
   B. Placing a condenser in series with the antenna.
   C. Placing a resistor in series with the antenna.
   D. Placing an inductance in series with the antenna.
ANSWER D: If we add an inductance (coil) in series with a Hertz antenna, it will lower its resonant frequency.

3H111 Parasitic elements are useful in a receiving antenna for:
A. Increasing directivity.
B. Increasing selectivity.
C. Increasing sensitivity.
D. Increasing directivity and increasing sensitivity.
ANSWER D: Using parasitic elements on a receiving antenna increases directivity and increase sensitivity to weak incoming radio waves.

3H112 What type of antenna radiates the 1030-MHz interrogation signals from an aircraft's Mode-S Traffic alert and Collision Avoidance System (TCAS) installation?
A. An L-band monopole blade-type omnidirectional antenna
B. An electronically steerable phased-array directional antenna
C. A folded dipole antenna
D. An internally mounted, mechanically rotatable loop antenna
ANSWER B: For the aircraft mode-S Traffic alert and Collision Avoidance System, an electronically steerable phased-array directional antenna is used.

3H113 Concerning shipboard satellite dish antenna systems, azimuth is:
A. Vertical aiming of the antenna.
B. Horizontal aiming of the antenna.
C. 0 - 90 degrees.
D. North to east.
ANSWER B: Azimuth is the horizontal aiming of the antenna. A radar antenna has 360-degrees azimuth when it is rotating.

3H114 What is the effect of adding a capacitor in series to an antenna?
A. Resonant frequency will decrease
B. Resonant frequency will increase
C. Resonant frequency will remain same
D. Electrical length will be longer
ANSWER B: Adding a capacitor in series with an antenna will raise the resonant frequency.

3H115 If a transmission line has a power loss of 6 dB per 100 feet, what is the power at the feed point to the antenna at the end of a 200-foot transmission line fed by a 100-watt transmitter?
A. 70 watts
B. 50 watts
C. 25 watts
D. 6 watts
ANSWER D: You can do this one in your head. If there is 6 dB loss for 100 feet of coax, there will be 12 dB loss for 200 feet. 12 dB loss is more than 10 dB which is easy to remember as a multiplier of 10. Knowing that your loss is greater than 10 times, 6 watts is the only answer that will agree with 12 dB.

3H116 Waveguides are:
A. Used exclusively in high frequency power supplies.
B. Ceramic couplers attached to antenna terminals.
C. High-pass filters used at low radio frequencies.
D. Hollow metal conductors used to carry high frequency current.

ANSWER D: The key words are *hollow metal conductors*. None of the other answers is even close, but answer D has an unfortunate wording by using "high-frequency" current. It does not refer to the designated HF band of 3-30 MHz. We normally use waveguides at frequencies above 500 MHz, mostly above 1 GHz. See 3H124 for an example of a waveguide.

3H117 What type of antenna pattern is radiated from a phased-array directional antenna when transmitting the P1 or P3 Pulse-Position-Modulated pulses in a Mode-S interrogation signal of an aircraft's Traffic alert and Collision Avoidance System (TCAS) installation?
  A. A 1090 MHz directional pattern
  B. A 1030 MHz omnidirectional pattern
  C. A 1090 MHz omnidirectional pattern
  D. A 1030 MHz directional pattern
ANSWER D: A 1030 MHz directional pattern is radiated from a phased-array directional antenna when P1 or P3 signals are transmitted.

3H118 A 520-kHz signal is fed to a 1/2-wave Hertz antenna. The fifth harmonic will be:
  A. 2.65 MHz
  B. 2650 kHz
  C. 2600 KHz
  D. 104 kHz
ANSWER C: $5 \times 520 = 2600$ kHz. This is near 2638 kHz, one of the early AM medium-frequency, ship-to-ship channels.

3H119 When a capacitor is connected in series with a Marconi antenna:
  A. An inductor of equal value must be added.
  B. No change occurs to antenna.
  C. Antenna open circuit stops transmission.
  D. Antenna resonant frequency increases.
ANSWER D: If you add a capacitor in series with a Marconi antenna, it will raise the resonant frequency.

3H120 How do you increase the electrical length of an antenna?
  A. Add an inductor in parallel
  B. Add an inductor in series
  C. Add a capacitor in series
  D. Add a resistor in series
ANSWER B: To lower the resonant frequency of a Marconi antenna, you will need to add inductance in series. This is the same as increasing the electrical length of the antenna.

3H121 A coaxial cable has 7 dB of reflected power when the input is 5 watts. What is the output of the transmission line?
  A. 5 watts
  B. 2.5 watts
  C. 1.25 watts
  D. 1 watt
ANSWER D: This one you can do in your head. 6 dB reflection is a 4 times loss in this circuit. Since it's a little bit more than a 4 times loss at 5 watts, you would only have 1 watt getting out of the transmission line into the antenna circuit.

3H122 What is the 7th harmonic of 450 kHz when fed through a 1/4-wavelength vertical antenna?

- A. 3150 Hz
- B. 3150 MHz
- C. 787.5 kHz
- D. 3.15 MHz

ANSWER D: Disregard the length of the antenna. 7 × 450 kHz = 3150 kHz; and if you move the decimal point 3 places to the left to calculate MHz, the answer is 3.15 MHz.

3H123 What is the 5th harmonic of a 450 kHz transmitter carrier fed to a 1/4-wave antenna?

- A. 562.5 MHz
- B. 1125 kHz
- C. 2250 MHz
- D. 2.25 MHz

ANSWER D: Here again, disregard the length of the antenna. 5 × 450 kHz = 2250 kHz; and moving the decimal point 3 places to the left (KLM), you end up with 2.25 MHz.

3H124 Waveguide construction:

- A. Should not use silver plating.
- B. Should not use copper.
- C. Should have short vertical runs.
- D. Should not have long horizontal runs.

ANSWER D: With any waveguide installation, you should not have any long runs.

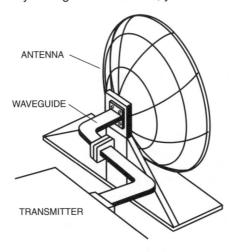

Waveguide Between Transmitter and Antenna

3H125 What type of antenna radiates the 1090-MHz "squitter" signals from an aircraft's Mode-S transponder installation of the Air Traffic Control Radar Beacon System (ATCRBS)?

- A. An L-band monopole blade-type omnidirectional antenna
- B. An electronically steerable phased-array antenna radiating an omnidirectional pattern

C. A folded dipole antenna

D. An internally mounted, mechanically rotatable loop antenna

ANSWER A: The "squitter" signals from mode-S transponder are radiated by an L-band monopole blade-type omnidirectional antenna.

3H126 To lengthen an antenna electrically, add a:

A. Coil.

B. Resistor.

C. Battery.

D. Conduit.

ANSWER A: To lengthen an antenna electrically, add an inductor, such as a coil, in series with the antenna at the base.

3H127 How do you electrically decrease the length of an antenna?

A. Add an inductor in series

B. Add a capacitor in series

C. Add an inductor in parallel

D. Add a resistor in series

ANSWER B: To electrically shorten an antenna, add a capacitor in series at the base.

3H128 If the length of an antenna is changed from 1.5 feet to 1.6 feet its resonant frequency will:

A. Decrease.

B. Increase.

C. Be 6.7% higher.

D. Be 6% lower.

ANSWER A: Lower, longer, giving the antenna better performance at a decrease in frequency. Answer D cannot be verified unless you know more details about how the antenna is loaded or matched.

3H129 To couple energy into and out of a waveguide:

A. Use wide copper sheeting.

B. Use an LC circuit.

C. Use capacitive coupling.

D. Use a thin piece of wire as an antenna.

ANSWER D: With the transmitter turned off, and the waveguide feed assembly out of circuit, take a look inside and spot the tiny thin piece of wire that is acting as the antenna element. This will couple energy into and out of a waveguide.

★ 3H130 An isolator:

A. Acts as a buffer between a microwave oscillator coupled to a waveguide.

B. Acts as a buffer to protect a microwave oscillator from variations in line load changes.

C. Shields UHF circuits from RF transfer.

D. Acts as a buffer between a microwave oscillator coupled to a waveguide and acts as a buffer to protect a microwave oscillator from variations in line load changes.

ANSWER D: The isolator in an antenna circuit will act as a buffer between the microwave oscillator coupled to the waveguide and load changes. *[Ed. Note: While A and B are both partially correct, Answer D is the correct answer.]*

3H131 A high SWR creates losses in transmission lines. A high standing-wave ratio might be caused by:
A. Improper turns ratio between primary and secondary in the plate tank transformer.
B. Screen grid current flow.
C. An antenna electrically too long for its frequency.
D. An impedance mismatch.

ANSWER D: An impedance mismatch in the antenna or a deformed transmission line could lead to high SWR. Answer C is good, but not necessarily correct.

3H132 A properly installed shunt-fed, 1/4-wave Marconi antenna:
A. Has zero resistance to ground.
B. Has high resistance to ground.
C. Should be cut to 1/2 wave.
D. Should not be shunt-fed.

ANSWER A: The perfect Marconi antenna, dc shunt-fed, should have zero resistance to ground.

3H133 When a capacitor is connected in series with a Marconi antenna:
A. An inductor of equal value must be added.
B. No change occurs to antenna.
C. Antenna open circuit stops transmission.
D. Antenna resonant frequency increases.

ANSWER D: Adding capacitance to an antenna will raise its resonant frequency.

3H134 When excited by RF, a half-wave antenna will radiate:
A. A space wave.
B. A ground wave.
C. Electromagnetic fields.
D. Both electromagnetic and electrostatic fields.

ANSWER D: When an antenna emits radio waves, it emits magnetic fields and electric fields at right angles to each other.

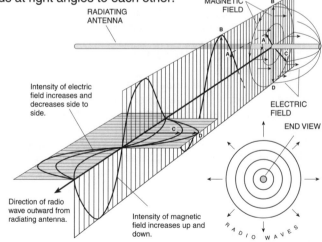

Radiation Pattern of Radio Waves Around Antenna

Source: *Basic Electronics,* G. McWhorter, A.L. Evans, ©1994, Master Publishing, Inc.

3H135 Waveguides are:
A. Used exclusively in high frequency power supplies.
B. Ceramic couplers attached to antenna terminals.
C. High-pass filters used at low radio frequencies.
D. Hollow metal conductors used to carry high frequency current.

ANSWER D: This question is identical to question 3H116. Key words are *hollow metal conductors.*

3H136 A 520-kHz signal is fed to a 1/2-wave Hertz antenna. The fifth harmonic will be:
A. 2.65 MHz.
B. 2650 kHz.
C. 2600 kHz.
D. 1300 kHz.

ANSWER C: Question 3H118 is identical to this question. 5 × 520 kHz = 2600 kHz. Disregard the length of the antenna.

3H137 The voltage produced in a receiving antenna is:
A. Out of phase with the current if connected properly.
B. Out of phase with the current if cut to 1/3 wavelength.
C. Variable depending on the station's SWR.
D. Always proportional to the received field strength.

ANSWER D: The voltage produced in a receiving antenna is always proportional to the received field strength at the antenna. Electromagnetic radio waves induce a voltage in the receiving antenna which produces a current in the antenna and input circuit to which the antenna is connected.

3H138 A properly connected transmission line:
A. Is grounded at the transmitter end.
B. Is cut to a harmonic of the carrier frequency.
C. Is cut to an even harmonic of the carrier frequency.
D. Has a SWR as near as 1:1 as possible.

ANSWER D: A properly connected transmission line whose impedance matches the load will have an SWR as near as 1:1 as possible.

3H139 Conductance takes place in a waveguide:
A. By interelectron delay.
B. Through electrostatic field reluctance.
C. In the same manner as a transmission line.
D. Through electromagnetic and electrostatic fields in the walls of the waveguide.

ANSWER D: In a waveguide, the conductance takes place in the walls of the waveguide through magnetic and electric fields.

3H140 Concerning shunt-fed 1/4-wavelength Marconi antenna:

A. DC resistance of the antenna to ground is zero.

B. RF resistance from antenna feed point to ground is zero.

C. Harmonic radiation is zero under all conditions.

D. It must be grounded at both feed and far ends.

ANSWER A: In a shunt-fed, 1/4-wavelength Marconi antenna, dc resistance of the antenna to ground is zero.

3H141 What type of antenna pattern is radiated from a phased-array directional antenna when transmitting the P2 Side-Lobe-Suppression (SLS) pulse in a Mode-S interrogation signal of an aircraft's Traffic alert and Collision Avoidance System (TCAS) installation?

A. A 1090 MHz directional pattern

B. A 1030 MHz omnidirectional pattern

C. A 1090 MHz omnidirectional pattern

D. A 1030 MHz directional pattern

ANSWER B: The phased-array directional antenna transmitting the P2-Side-Lobe-Suppression pulse in Mode-S generates an omnidirectional pattern on 1030 MHz.

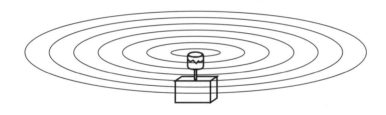

Omnidirectional Pattern From Phased-Array Antenna

3H142 If a 3/4-wavelength transmission line is shorted at one end, impedance at the open will be:

A. Zero.

B. Infinite.

C. Decreased.

D. Increased.

ANSWER B: With a 3/4-wavelength transmission line shorted at one end, impedance at the opposite end, open end, will always be high. Refer to the figure at 3H70 and consider the 3/4-wavelength transmission line as a 1/2-wavelength line with a 1/4-wavelength added to it. A short at the end of the 1/2-wavelength reflects a short to the 1/4-wavelength. A short at the end of the 1/4-wavelength reflects a high impedance at the open end.

3H143 A dummy antenna is a:

A. Non-directional receiver antenna.

B. Wide bandwidth directional receiver antenna.

C. Transmitter test antenna designed for minimum radiation.

D. Transmitter non-directional narrow-band antenna.

ANSWER C: A dummy antenna is one that is designed for minimal radiation and output.

7

Radar – Radio Detection and Ranging

Radar is *an electronic instrument used to detect and locate moving or fixed objects.* It is a system of remote sensing that exploits the fact that radio waves produce echoes as they are reflected from objects. These objects can be almost anything including mountains, a shoreline, ships, aircraft, automobiles — even a baseball traveling at 100 miles-per-hour.

There are many advantages of radar. It can operate long range, is very accurate and has the ability to penetrate darkness, fog and adverse weather.

HISTORY OF RADAR

It had been known for more than a century that radio waves could be reflected back from solid objects. In 1904, a German patent was granted for an anti-collision device that used radio frequency to detect ships and trains in fog.

In 1922, scientists working for the U.S. Naval Research Laboratory noted a difference in received signals caused by reflections from a wooden vessel on the Potomac River. They reasoned that these different echoes could be used to detect and track enemy ships.

A few years later, physicists at Washington's Carnegie Institution measured the distance to the ionosphere by "pulse ranging." By beaming a series of short signals of slightly different frequencies to the ionosphere, they were able to determine its height by measuring the time that each pulse took to return to earth.

The first use of radar was during the 1930's for military use in detecting aircraft and surface vessels. France, Germany and the United States developed experimental early warning radar systems. Radar was later installed in airplanes to detect enemy surface vessels.

Radar came into its own, however, during World War II when it became a vital weapon. Anti-aircraft guns were controlled by radar. Airborne radar played a large part in strategic bombing. In 1941, radar-directed gunfire sank many enemy ships. Using radar, Naval battleships could maneuver close to enemy coasts during low visibility.

Radar has now grown into a major industry and is particularly indispensable to civil aviation, maritime shipping, National defense systems, and weather forecasting. Over-the-horizon (OTH) search radar has a range of 2,500 miles using ionosphere-reflected waves operating in the HF range.

HOW DOES RADAR WORK?

Modern radar systems use an extremely wide range of frequencies, from the HF to infrared range. Microwave frequencies are the most commonly used since they can travel through clouds and bad weather without absorption and noise. Radar operates on two principles. (1) objects illuminated by a radio frequency (RF) beam reflects radiation in all directions and, (2) the speed of RF is constant.

While the reflected energy is scattered in all directions, a detectable portion is also reflected back to where it originated. To detect and locate objects, a radar transmits radio frequency (RF) energy and picks up the reflections with a radio receiver. The receiver is so sensitive that power levels of less than a millionth of a millionth of a watt can be detected and amplified.

Radio signals, like all electromagnetic waves, travel at the speed of light. That is 186,282 statute miles per second or 0.186 miles (982 feet) in a microsecond. By dividing by two (reflected signals must make a two-way trip) one can accurately gauge an object's distance. Therefore, an echo returned in one microsecond indicates that a target is 491 feet away from the radar set. A "radar mile" refers to nautical miles which are 15 percent longer than statute miles. 1.15 statute miles equals 1 nautical mile; therefore, radio waves travel 0.162 nautical miles in a microsecond.

Like light, RF can be focused in a certain direction. An object is located by scanning a region and the beam position that yields the strongest echoes indicates the existence of a reflecting surface of the object.

TYPES OF RADAR

There are two general types of radar — pulsed and continuous wave (cw). A pulsed radar transmits its RF energy in short bursts or pulses. A cw radar operates constantly instead of in short bursts. *Magnetrons* are the vacuum tube oscillators that generate high-power microwave pulses. The energy can be focused into a narrow beam with the so-called radar "dish" — an antenna with a parabolic shape.

Pulse Radar

Simple pulse radars transmit a beam of powerful on-off bursts of microwave radiation. The radar burst lasts only a few millionths of a second. During the interval between the pulses, the return echo is received and interpreted as shown in *Figure 7-1*. A high-speed electronic switch called a duplexer switches the transmitter on and off, and the antenna from "transmit" to "receive." This permits the same antenna to be used for transmitting and receiving. The pulse duration is determined by the minimum range of the remote object and must be short enough so as not to mask the return echo. Pulse radars are used primarily to determine distance and direction.

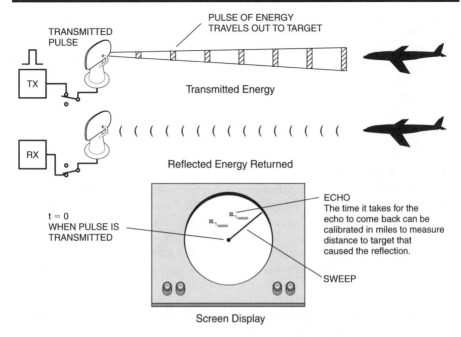

Figure 7-1. RADAR – RAdio Detection And Ranging

Distance (range) can be determined by measuring the time between the transmission of the pulse and the reception of the echo. A pulse radar set automatically converts the round trip time into distance to the object. Direction is determined by synchronizing a scan on a display screen to a sweeping antenna which focuses a narrow beam of energy out ahead of it. An object in the path of the highly directional beam reflects an echo that appears in position on the display. Pulse radar can track a moving object by continuously measuring distance and direction at regular intervals as the antenna sweeps an area.

Radar maps can be produced by scanning many pulses over an area and plotting the echoes from each direction on a radar screen.

Continuous-Wave Radar

Pulse radar provides no information about target velocity. Continuous wave (CW) radar transmits and receives at the same time. There are no pulses. The problem of interference between the transmitter and receiver is overcome by isolating them as much as possible and by designing the receiving system so that it can distinguish differences between the transmitted and reflected signals. Systems that can do this are of two types; (1) Doppler radar and (2) frequency-modulated (FM) radar.

Doppler Radar

If the object is moving, the transmitted and reflected signals will have different frequencies. This is known as a Doppler shift. When the outgoing wave strikes an object that is approaching the radar antenna, the wave is reflected at a slightly higher frequency. The reflection is bounced back at a lower frequency when the target moves away from the antenna. The greater the frequency difference between the transmitted and reflected waves, the faster the object is moving. Doppler radar does not determine distance, but rate of change of distance. Doppler is used by police to detect speeding motorists, by the military to direct weapon fire at moving targets, and by weather stations to track tornadoes.

Frequency-Modulated Radar

If the target is stationary, differences may be created by continuously varying the transmitter's frequency, as with frequency modulation. FM radar, unlike Doppler radar, can determine distance. The return echo will have a slightly different frequency than the outbound burst. The difference between the transmitter frequency and the reflection is converted into distance. The farther away the object is, the greater the frequency difference.

Because of the difficulty in isolating the transmitter and receiver and of the ease of measuring distance with pulses, cw radar is not as widely used as pulsed radar.

Radar Displays

Radar sets come in all sizes, from a small four-pound hand-held police version to those with giant parabolic dishes weighing tons and a thousand feet in diameter! But they all have somewhat similar parts — a transmitter, a receiver, an antenna and a display. If it is a pulsed radar, a TX/RX electronic switch is required.

The received radar information can be displayed in a number of ways. Some sets have analog meters and digital read-outs. Others have a circular cathode-ray tube (CRT) screen, as shown in *Figure 7-1,* commonly called a PPI — for Plan Position Indicator.

The most common CRT PPI display is a circular map-like picture of the area scanned by the radar beam. The observing station is at the center of the display. Each pulse appears as an electron beam from the center of the tube to the outer edge. Radar echoes appear as blips on the screen. Concentric rings indicate distance. Land based radars are calibrated in statute miles, ship radar sets in nautical miles. Speed is determined by noting how fast the blips move across the screen.

Most recently, PPI phosphorous cathode ray tubes are being replaced by digital technology with computer-driven raster scan displays. As a result, all types of digital display techniques can be used.

AVIATION AND MARITIME RADAR

The largest non-military users of radar are police departments for speed-detection, weather-forecasting (rain reflects radar signals), civil aviation and ship navigation. Air controllers are able to determine the position of planes in the air at least 50 miles from an airport. Planes map the areas over which they fly. Radar altimeters show how high a plane is flying. Weather radar detects storms in the area.

SHIP RADAR

Radar is used by vessels of all sizes; from small pleasure boats to huge cargo ships and ocean liners. Radar provides visibility in bad weather, spots other vessels, shows water depth and highlights possible obstacles to navigation. A ship's position can be determined by radar echoes from reflector buoys.

Harbor masters keep track of ships using radar and the Coast Guard uses radar for search and rescue. Radar beacons (Racons) operate somewhat like a radar set in reverse. A radar beacon is triggered by pulses from a radar set and sends back a distinctive reply which provides precise location and bearing information.

Ship radar is one of the most important pieces of navigation equipment in the wheelhouse. Radar, combined with depth sounding and a sharp lookout, is an absolute requirement when entering congested shipping lanes. Although the Global Positioning System (GPS) receiver and associated electronic chart display is important to stay on track in the vessel traffic system (VTS), only radar will show targets all around the vessel not necessarily printed on paper charts or stored in electronic chart displays.

CAUTION

A word of CAUTION! Radar waves are DANGEROUS! Never expose yourself to the output of the waveguide assembly which radiates radar microwaves. Be extremely cautious of thousands of volts inside the radar transceiver system. Never allow a turned-on radar to sweep you with radiated transmit energy. Never stand close nor look into a transmitting radar system. These are microwaves, and they are DANGEROUS.

RADAR ENDORSEMENT FOR THE COMMERCIAL LICENSES

The importance of shipboard radar is so great that the Federal Communications Commission requires passing the commercial radio operator's Element 8 ship radar endorsement as part of the commercial radio licensing process.

Since radar equipment contains a powerful transmitter, the FCC requires specific commercial radio operator licenses (those obtained by passing either written electronics Element 3 or 6) as a prerequisite to obtaining a ship radar endorsement. This endorsement is required to make internal adjustments to the radar transmitter circuits even if the radar unit is installed aboard a non-commercial vessel such as a pleasure boat.

Ship Radar Endorsement

The ship radar endorsement is not a stand-alone license. It must be placed on the General Radiotelephone Operator License (GROL), the Global Maritime Distress and Safety System Radio Maintainer's license (GMDSS/M), or on the First or Second Class Radiotelegraph Operator Certificates (T-1 or T-2). Only those technicians whose commercial radio operator license bears this radar endorsement may open up and repair, maintain or internally adjust ship radar equipment. Refer to Chapter 2 for the eligibility requirements to obtain a ship radar endorsement. A ship radar endorsement is not needed to routinely operate an aircraft or ship radar set, but the operation must be limited to manipulation of external controls.

EXAMINATION REQUIREMENTS

The examination is on specialized theory and practice concerning proper installation, servicing and maintenance of ship radar equipment. It also will contain questions covering special rules applicable to ship radar stations, and the technical fundamentals of radar.

In this book, each question in the pool has the question, its four multiple-choice answers, the right answer identified, and a detailed explanation of why the right answer is correct.

The COLEM organization administering the examination will select questions that will be identical to those in this book, although the A-B-C-D order of the multiple choices may be changed. No rewording of questions or changes in the four potential answers are allowed by the FCC.

8

Question Pool – Element 8

The questions asked in each Element 8 examination must be taken from the most recent pool. The release date of the following question pool was October 1, 1995, and its effective date was January 1, 1996. All examinations administered after January 1, 1996 must use questions from this pool.

An Element 8 examination is used to prove that the examinee possesses the operational and technical qualifications to repair, maintain or internally adjust ship radar equipment. The Element 8 examination is not for a license itself, but for an *endorsement* placed only on a General Radiotelephone Operator License (GROL) [PG], GMDSS Radio Maintainer's License (DM) or a First or Second Class Radiotelegraph Operator's Certificate (T1 or T2.)

A GROL is held by any person who adjusts, maintains, or internally repairs a radiotelephone transmitter at a station licensed by the FCC in aviation, maritime, or international fixed public radio services. A GMDSS/M is held by any person who is required to repair and maintain satellite-based marine emergency subsystems and equipment. A Second Class Radiotelegraph Operator's Certificate (T2) is held by any person who is required to operate, repair and maintain ship and coast radiotelegraph stations in the maritime services. A T1 license is held by any person required to do everything a T2 license holder does, and who serves as chief radio operator on a U.S. passenger ship.

Each Element 8 examination is administered by a Commercial Operator License Examination Manager (COLEM). The COLEM must construct the examination by selecting 50 questions from this 321 question pool. You must have 38 or more answers correct from the 50 in order to pass the examination.

The COLEM may change the order of the answers and distracters (incorrect choices) of questions from the order that appears in this pool.

Suggestions concerning improvement to the questions in the pool or submittal of new questions for consideration in updated pools may be sent to technical author Gordon West, 2414 College Drive, Costa Mesa, CA 92626.

Subelement 8A – Ship Radar Techniques

(50 questions)

8A1 To install and maintain a radar unit, you must have:

A. Permission of the master but no license.
B. A General Radiotelephone Operator License (GROL) with a Radar Endorsement or a GMDSS Radio Maintainer License (GMDSS/M) with a Radar Endorsement.
C. An Amateur Extra Class License.
D. A General Radiotelephone Operator License or a GMDSS Radio Maintainer License (GMDSS/M) only.

ANSWER B: To remove the cover and install, fix, adjust, and maintain a radar unit, you must be the holder of a GROL, GMDSS/M, a T-1, or a T-2 license, endorsed with radar. If you are just manipulating external controls or making external adjustments the radar endorsement is not necessary.

8A2 Who may replace fuses and receiving-type tubes and circuits in a radar unit?

A. An unlicensed person
B. A holder of a GMDSS Radio Maintainer License (GMDSS/M)
C. A holder of a General Radiotelephone Operator License with a Radar Endorsement
D. Any of these

ANSWER C: Anytime the cover of a radar system is removed to do repair or maintenance a radar endorsement is required on a GROL, GMDSS/M, T-1, or T-2 license.

8A3 Under what circumstances can an unlicensed person operate a radar set?

A. Only if the transmitter is a non-tunable type
B. When the master of the ship designates him to operate it
C. Never
D. Only if the transmitter is a non-tunable type and when the master of the ship designates him to operate it

ANSWER D: The modern radar is fixed tuned on the transmitter, so the operator does not need any special license to run the equipment. The master of the ship may designate anyone qualified to operate a non-tunable radar set.

8A4 What radar maintenance work may be done by unlicensed workers?

A. None
B. Replacement of magnetron and klystron tubes
C. Replacement tubes and circuits
D. Minor frequency adjustments only

ANSWER A: The answer is essentially none because if an equipment cover needs to be removed, a radar endorsement is required. Cleaning a radar radome would not require a license.

8A5 Who is permitted to operate a ship radar unit?

A. Only qualified FCC Licensed persons with a Radar Endorsement
B. Only the engineer or someone under his or her direct supervision
C. The master of the ship, or anyone designated by the master
D. Anyone who is knowledgeable

ANSWER C: The operation of a ship radar unit is under the control of the master of the ship, and the master may designate a specific person to operate the radar.

8A6 What component of a radar receiver is represented by block 46 in Diagram EL8-A?
A. The ATR box
B. The TR box
C. The RF Attenuator
D. The Crystal Detector

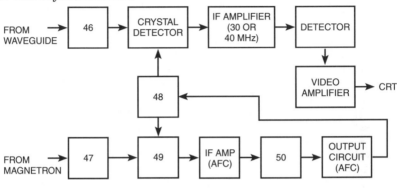

Diagram EL8-A

ANSWER B: Block 46 in Diagram EL8-A is the TR box.

8A7 What component of a radar receiver is represented by block 47 in Diagram EL8-A?
A. The ATR box
B. The TR box
C. The RF Attenuator
D. The Crystal Detector

ANSWER C: Refer to the diagram in question 8A6. Block 47 in Diagram EL8-A is the RF attenuator.

8A8 What component of a radar receiver is represented by block 48 in Diagram EL8-A?
A. The Discriminator
B. The IF Amplifier
C. The Klystron
D. The Crystal Detector

ANSWER C: Refer to the diagram in question 8A6. Block 48 in Diagram EL8-A is the klystron.

8A9 What component of a radar receiver is represented by block 49 in Diagram EL8-A?
A. The Discriminator
B. The IF Amplifier
C. The Klystron
D. The Crystal Detector

ANSWER D: Refer to the diagram in question 8A6. This is the crystal detector section of the radar receiver.

8A10 What component is block 50 in Diagram EL8-A?
 A. The Discriminator
 B. The IF amplifier
 C. The AFC amplifier
 D. The Crystal detector
ANSWER A: Block 50 in Diagram EL8-A is the discriminator stage and is fed by the intermediate frequency amplifier and its output is the output AFC circuit. Refer to the diagram in question 8A6.

8A11 The minimum range of a RADAR is primarily determined by:
 A. The pulse width and TR (TRL) cell recovery time.
 B. The ATR cell recovery time.
 C. The overall height of the antenna.
 D. The peak power output of the radar transmitter.
ANSWER A: For a radar to detect close-in targets, the pulse width must be extremely narrow because the return echo is arriving right after the transmitted pulse. The pulse width and the transmit/receive cell recovery time is the major determining factor on how close in a radar can see. Small marine radars typically see in as close as 75 yards, but no less than that.

8A12 An RF mixer has what purpose in a radar system?
 A. It mixes the CW transmitter output to form pulsed waves.
 B. It converts a low-level signal to a different frequency.
 C. It prevents microwave oscillations from reaching the antenna.
 D. It combines audio tones with RF to produce the radar signal.
ANSWER B: To obtain the required gain and bandwidth of a signal, a crystal mixer is employed to lower the frequency of the received signal to the 30-60-MHz band.

8A13 Pulse transformers and pulse-forming networks are commonly used to shape the microwave energy burst radar transmitter. The switching devices most often used in such pulse forming circuits are:
 A. Power MOSFETS and Triacs.
 B. Diacs and SCR's.
 C. Thyratrons and BJT's.
 D. SCR's and Thyratrons.
ANSWER D: The switching device found in the pulse-forming circuit of most radars is silicon-controlled rectifiers (SCR's) and Thyratrons.

8A14 Good bearing resolution largely depends upon:
 A. A high transmitter output reading.
 B. A high duty cycle.
 C. A narrow antenna beam in the vertical plane.
 D. A narrow antenna beam in the horizontal plane.
ANSWER D: The resolution of the radar beam is proportional to the length of the rotating antenna. The narrower the beam, the better the ability to differentiate multiple targets.

8A15 Bearing resolution is:
 A. The ability to distinguish two targets of different distances.
 B. The ability to distinguish two targets of different elevations.

C. The ability to distinguish two adjacent targets of equal distance.

D. The ability to distinguish two targets of different size.

ANSWER C: Bearing resolution is the ability to differentiate between two targets that are separated horizontally. If the radar antenna is relatively small, a wider beamwidth illustrates two individual targets as one echo.

8A16 An artificial transmission line is used for:

A. The transmission of radar pulses.

B. Testing the radar unit, when actual targets are not available.

C. Determining the shape and duration of pulses.

D. Testing the delay time for artificial targets.

ANSWER C: Artificial transmission lines are used to simulate the actual lines and to allow determination of the shapes and durations of the pulses.

8A17 What device is located between the magnetron and the mixer and prevents received signals from entering the magnetron?

A. The TR Box

B. The ATR Box

C. The RF Attenuator

D. A resonant cavity

ANSWER B: The ATR circuit keeps received echoes out of the transmitter.

8A18 The ATR box:

A. Protects the receiver from strong radar signals.

B. Prevents the received signal from entering the transmitter.

C. Turns off the receiver when the transmitter is on.

D. All of these.

ANSWER B: The ATR circuit keeps received echoes out of the transmitter.

8A19 When a pulse radar is radiating, which elements in the TR box are energized?

A. The TR tube only

B. The ATR tube only

C. Both the TR and ATR tubes

D. Neither the TR nor ATR tubes

ANSWER C: As the radar is transmitting a pulse, gas switches in both the TR and the ATR tubes becomes ionized; therefore, both elements are energized.

8A20 In the AFC system, the discriminator compares the frequencies of the:

A. Magnetron and klystron.

B. PRR generator and magnetron.

C. Magnetron and crystal detector.

D. Magnetron and video amplifier.

ANSWER A: In the automatic frequency control system, the discriminator compares the frequencies of both the magnetron and the klystron.

8A21 The purpose of the discriminator circuit in a radar set is to:

A. Discriminate against nearby objects.

B. Discriminate against two objects with very similar bearings.

C. Generate a corrective voltage for controlling the frequency of the klystron local oscillator.

D. Demodulate or remove the intelligence from the FM signal.

ANSWER C: The discriminator circuit in a radar is part of the automatic frequency control circuitry. It detects any change of the magnetron or klystron frequency so that their difference in frequency maintains the intermediate frequency.

8A22 Where is a RF attenuator used in a radar unit?

A. Between the antenna and the receiver

B. Between the magnetron and the antenna

C. Between the magnetron and the AFC section of the receiver

D. Between the AFC section and the klystron

ANSWER C: RF attenuators are staged between the magnetron and the automatic frequency control section of the receiver to prevent overload.

8A23 What frequency is the discriminator tuned to?

A. The magnetron frequency

B. The local oscillator frequency

C. The 30 MHz or 60 MHz IF

D. The pulse repetition frequency

ANSWER C: A 30-MHz or 60-MHz intermediate frequency makes possible the necessary gain as well as the right amount of band width to pull the signal from the mixing process at the local oscillator crystal mixer.

8A24 The usual intermediate frequency of a shipboard radar unit is:

A. 455 kHz.

B. 10.7 MHz.

C. 30 or 60 MHz.

D. 120 MHz.

ANSWER C: 30 or 60 MHz is the common intermediate frequency of most radar units.

8A25 The error voltage from the discriminator is applied to:

A. The repeller (reflector) of the klystron.

B. The grids of the IF amplifier.

C. The grids of the RF amplifiers.

D. The magnetron.

ANSWER A: The error correction voltage from the discriminator is usually applied to the repeller (reflector) of the klystron.

8A26 How may the frequency of the klystron be varied?

A. Small changes can be made by adjusting the repeller voltage.

B. Large changes can be made by adjusting the size of the resonant cavity.

C. By changing the phasing of the buncher grids

D. Small changes can be made by adjusting the repeller voltage and large changes can be made by adjusting the size of the resonant cavity.

ANSWER D: You have seen this question before; you can slightly change the frequency of the klystron by adjusting the repeller voltage, and you can make major changes in frequency by adjusting the actual size of the resonant cavity.

8A27 Fine adjustments of a reflex klystron are accomplished by:
A. Adjusting the flexible wall of the cavity.
B. Varying the repeller voltage.
C. Adjusting the AFC control system.
D. Varying the cavity grid potential.

ANSWER B: Fine adjustments of a reflex klystron are accomplished using small changes in the control voltage on the repeller plates. Large adjustment are made by changing the cavity volume.

8A28 The pulse repetition rate (PRR) refers to:
A. The reciprocal of the duty cycle.
B. The pulse rate of the local oscillator tube.
C. The pulse rate of the magnetron.
D. The pulse rate of the klystron.

ANSWER C: The pulse repetition rate (PRR) is the number of pulses per second sent to the target. It is also the number of pulses per second sent to the magnetron.

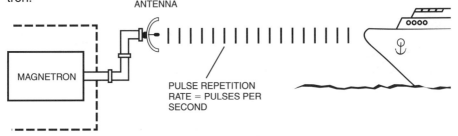

ANTENNA

MAGNETRON

PULSE REPETITION
RATE = PULSES PER
SECOND

Pulse Repetltion Rate from Antenna is Same as from Magnetron

8A29 The timer circuit:
A. Determines the pulse repetition rate (PRR).
B. Determines range markers.
C. Provides blanking and unblanking signals for the CRT.
D. All of these.

ANSWER D: The timer circuit determines pulse repetition rate and range markers, and provides blanking and unblanking signals for the cathode-ray tube in a raster scan display.

8A30 On what frequency is radar expected to cause interference?
A. On the pulse repetition frequency
B. On the klystron oscillator frequency
C. On the magnetron frequency
D. On most any communications frequency

ANSWER D: Radar interference may be noticed on many communications receivers onboard. This is because of the rich harmonics produced by the pulsing, the chopper circuit in the power supply, and the relatively long cable runs between the antenna TR unit and the radar presentation scope. Good grounding techniques on both the communications receiver and the radar helps minimize interference.

8A31 If the magnetron is allowed to operate without the magnetic field in place:
   A. Its output will be somewhat distorted.
   B. It will quickly destroy itself from excessive current flow.
   C. Its frequency will change slightly.
   D. Nothing serious will happen.
ANSWER B: If the magnetic field in the magnetron gets weak, the current flow to the anode is unrestrained, and the tube overheats and destroys itself.

8A32 What radar circuit determines the pulse repetition rate (PRR)?
   A. The discriminator
   B. The timer (synchronizer circuit)
   C. The artificial transmission line
   D. The pulse-rate-indicator circuit
ANSWER B: The pulse repetition rate is determined by the radar timer synchronizer circuit.

8A33 Range markers are determined by:
   A. The CRT.
   B. The magnetron.
   C. The timer.
   D. The video amplifier.
ANSWER C: It's the timer circuit that determines range rings.

8A34 The timer circuit:
   A. Determines the pulse repetition rate (PRR).
   B. Determines range markers.
   C. Provides blanking and unblanking signals for the CRT.
   D. All of these.
ANSWER D: The timer circuit determines pulse repetition rate and range markers, and provides blanking and unblanking signals for the cathode-ray tube in a raster scan display.

8A35 Unblanking pulses are produced by the timer circuit. Where are they sent?
   A. To the IF amplifiers
   B. To the CRT
   C. To the mixer
   D. To the discriminator
ANSWER B: The timer circuit determines pulse repetition rate and range markers, and provides blanking and unblanking signals for the cathode-ray tube in a raster scan display.

8A36 Accurate range markers must be developed using very narrow pulses. A circuit that could be used to provide these high-quality pulses for the CRT is a:
   A. Ringing oscillator.
   B. Blocking oscillator.
   C. Monostable multivibrator.
   D. Triggered bi-stable multivibrator.

ANSWER B: A blocking oscillator, like all oscillators, makes use of feedback. That is, some of the output of the circuit is sent back to the input. But, as its name suggests, the blocking oscillator "blocks out" most of the input and feedback, and allows only a very sharp, narrow pulse at its output. Of course, this is perfect for generating the thin, unobtrusive range rings that allow you to see the distances to targets on a radar scope without "blocking out" the targets.

8A37 What determines the maximum unambiguous range of a radar set?
  A. The duty cycle
  B. The peak power output
  C. The time between the transmitted pulses
  D. The sensitivity of the discriminator

ANSWER C: The range, or distance, for maximum radar echo detection is determined by the time between the transmitted pulses.

8A38 The final stage of the sweep circuits of an electromagnetic CRT is usually a power amplifier. The reason for using a power amplifier rather than a voltage amplifier is:
  A. A voltage amplifier is likely to develop excessive sweep nonlinearity.
  B. To maintain a constant level of output current.
  C. To provide a relatively high output voltage to drive the deflection coils.
  D. To provide a relatively high output current to drive the deflection coils.

ANSWER D: A power amplifier is used rather than a voltage amplifier because the high current required by the deflection coils demands maximum power. Voltage amplifiers do not develop enough current.

8A39 Using tuned circuits for selectivity, radar IF amplifier stages would normally be biased to operate:
  A. Class A amplifiers.
  B. Class B amplifiers.
  C. Class C amplifiers.
  D. Class AB amplifiers.

ANSWER A: Class A biasing introduces the least distortion to the signal.

8A40 The AFC system is used to:
  A. Control the frequency of the magnetron.
  B. Control the frequency of the klystron.
  C. Control the receiver gain.
  D. Control the frequency of the incoming pulses.

ANSWER B: The automatic frequency control system controls the frequency of the klystron.

8A41 The output of an AFC Discriminator is:
  A. A sinewave at the IF frequency.
  B. Zero volts if the IF is "on frequency".
  C. AC if "on frequency"; DC if "off frequency".
  D. Always filtered with a high-pass filter.

ANSWER B: The AFC discriminator produces a voltage only if the IF is *off frequency.* If the unit is *on frequency,* no error voltage output is produced.

8A42 The primary operating frequency of a reflex klystron is controlled by the:
  A. Dimensions of the resonant cavity.
  B. Level of voltage on the control grid.
  C. Voltage applied to the cavity grids.
  D. Voltage applied to the repeller plate.

ANSWER A: The larger the resonant cavity of the klystron, the lower the frequency. The resonant cavity determines the primary operating frequency of a reflex klystron transmitter stage.

8A43 Electrons provide regeneration to oscillating cavities when they are:
  A. Accelerating.
  B. Decelerating.
  C. Entering the cavity.
  D. Oscillations involve electric and magnetic fields, so electrons have no control function.

ANSWER B: In the magnetron, for example, when some of the electrons from the cathode are caused to slow down, these electrons curve away from the anode and give up some of their energy into the resonant cavity. This type of electron is often referred to as a working electron. The working electrons, by staying in space longer, cause the RF cavity to receive more energy than it gives up, and thus sustains oscillation.

8A44 Power losses in cavity resonators are very low because:
  A. Conducting surfaces are large.
  B. Cavities only require low operating currents.
  C. Heavy insulation is used for all interconnections.
  D. Only small conducting surfaces are used.

ANSWER A: To minimize power loss, resonant cavities in microwave equipment are extremely large providing large conducting surfaces.

8A45 In a circular resonant cavity with flat ends, the E-field and the H-field form with specific relationships. The:
  A. E-lines are parallel to the top and bottom walls.
  B. E-lines are perpendicular to the end walls.
  C. H-lines are perpendicular to the side walls.
  D. H-lines are parallel to the end walls.

ANSWER B: The E-field and the H-field are perpendicular to each other in the resonant cavity. *(*NOTE: Distracter D also is technically correct, but B is the correct answer for the examination.)*

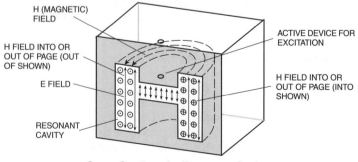

Cross Section of a Resonant Cavity

8A – SHIP RADAR TECHNIQUES **8**

8A46 Practical radar operation requires the use of microwave frequencies so that:

A. Stronger target echoes will be produced.

B. Ground clutter interference will be minimized.

C. Interference to other communication systems will be eliminated.

D. Antennas will be more efficient for both transmitting and receiving.

ANSWER A: Radars operate *line-of-sight.* Microwave frequencies faithfully reflect off of metal and land, and are partially reflected off fiberglass. Lower frequencies penetrate these materials, and do not produces the return echoes necessary for radar operation.

8A47 Short range radars would most likely transmit:

A. Narrow pulses at a fast rate.

B. Narrow pulses at a slow rate.

C. Wide pulses at a fast rate.

D. Wide pulses at a slow rate.

ANSWER A: To detect an echo when the target is close by, you would need a narrow pulse with a fast pulse repetition rate. A wide, slow pulse would actually cover up the return echo.

8A48 The microwave energy developed by the radar transmitter and radiated into space:

A. May be refracted by the ionosphere.

B. Can "see" targets beyond the horizon.

C. Travels through space "line of sight" only.

D. Is not usually reflected by the earth or the sea.

ANSWER C: Microwave radar transmits *line-of-sight* energy. This energy is not reflected by the ionosphere, and does not travel much beyond the horizon. Sometimes you can get an echo off of thunderstorms, and there are some rare cases of even receiving echoes off of the moon.

8A49 In comparing the overall operation of a pulse radar and a CW radar, it is determined that:

A. The pulse radar provides more accurate long-range and short-range measurements.

B. The CW radar produces spurious ground clutter on the display scope.

C. The CW radar has better short-range detection capabilities.

D. The CW radar develops a lower average output power.

ANSWER A: Long- and short-range measurements are best determined with a radar that has short-duration RF pulses. A CW radar system, commonly called Doppler Radar, is best for determining the velocity of a target, rather than its range.

8A50 The characteristic of the magnetron output pulse that relates to accurate range measurement is its:

A. Amplitude.

B. Decay time.

C. Rise time.

D. Duration.

ANSWER C: The rise time of the magnetron output must be as short as possible to produce the desired rapid pulse rates. If the rise time is too long, the measured target distance is inaccurate.

8A51 Magnetrons may be frequency modulated by:
A. Nutating the transmitter antenna.
B. Varying the magnetron load impedance.
C. Varying the power supply regulation.
D. Varying the amplitude of the magnetron input pulses.
ANSWER B: One way of varying the frequency of a magnetron is to vary the load impedance.

8A52 Some aircraft radars and avionics suites operate with a prime power line frequency of 400 Hz. What is the principle advantage of a higher line frequency?
A. 400-Hz power supplies draw less current than 60-Hz supplies, allowing more current available for other systems on the aircraft.
B. The magnetic devices in a 400-Hz power supply such as transformers, chokes and filters are smaller and lighter in weight than those used in 60-Hz power supplies.
C. A 400-Hz power supply generates less heat and operates much more efficiently than a 60-Hz power supply.
D. 400-Hz power supplies are much less expensive to produce than power supplies with lower line frequencies.
ANSWER B: Many high-voltage aircraft avionics operate with a power line frequency of 400-Hz, rather than lower ac frequencies. This allows the magnetic devices like filters, chokes, and transformers to be much smaller and lighter than those in 60-Hz devices. Smaller and lighter components are necessary for the basic aerodynamic designs of most aircraft.

8A53 When a radar signal is sent to an object, the Doppler effect is:
A. Objects moving towards you reflect back a lower frequency.
B. Objects moving away from you reflect back a higher frequency.
C. Stationary objects reflect back a slightly lower frequency.
D. Objects moving towards you reflect back a higher frequency.
ANSWER D: In addition to detecting the target, Doppler radar indicates whether the target is stationary, moving away, or moving toward the transmitting radar. Doppler shifts cause the reflected energy to come back at a slightly higher frequency if the target is moving toward your radar transmitter. Normally you see this motion as a variation in colors on your radar scope.

8A54 Airport wind shear radars depend upon:
A. The doppler effect to track rapid shifts in wind patterns.
B. High resolution to track moisture content cells in the atmosphere.
C. Phase measurements and FM & CW waveforms.
D. Large, high-gain antenna systems.
ANSWER A: Airports' wind-shear radars operate CW and often use a frequency near 450 MHz. The signals are transmitted straight up to detect shifts in wind patterns.

8A55 Choose the most correct statement containing the parameters which control the size of the target echo.
A. Transmitted power, antenna effective area, transmit and receive losses, radar cross section of the target, range to target
B. Height of antenna, power radiated, size of target, receiver gain, pulse width

C. Power radiated, antenna gain, size of target, shape of target, pulse width, receiver gain

D. Magnetron gain, antenna gain, size of target, range to target, waveguide loss

ANSWER A: The key word in the question is "size" of the target echo. Size of the target echo is determined by the target cross-sectional area, range to the target, transmit and receive path losses, antenna effective area, and transmitter power. Improvements in low-noise receivers now allow small targets to appear even bigger as a target echo.

8A56 Frequencies generally used for marine radar are in the ___ part of the radio spectrum.
A. UHF
B. EHF
C. SHF
D. VHF
ANSWER C:

VHF = 30 - 300 MHz
UHF = 300 - 3,000 MHz
SHF = 3,000 - 30,000 MHz
EHF = 30,000 - 300,000 MHz

The three marine radar bands are 2,900 - 3,100 MHz, 9,300 - 9,500 MHz, and 14,000 - 14,050 MHz. All of these bands fall in the super-high frequency (3,000-30,000 MHz) marine radar bands. The most popular band, X-band (9,300 - 9,500 Mhz), is sometimes noted in gigahertz (9.3 - 9.5 GHz).

8A57 The minimum range of a radar is determined by:
A. The frequency of the radar transmitter.
B. The pulse repetition rate.
C. The transmitted pulse width.
D. The pulse repetition frequency.
ANSWER C: The minimum range at which a radar can detect a target is determined by the pulse duration, sometimes called *pulse width or pulse length.* As the radar operator switches down to a close-in scale, the radar automatically shortens its pulse width to as little as 0.1 microsecond. A 0.1 microsecond pulse covers approximately 160 yards of range on the display. Small-boat marine radars use even shorter pulses for target detection as close as 75 feet.

8A58 An S-band radar operates in which frequency band?
A. 1-2 GHz
B. 2-4 GHz
C. 4-8 GHz
D. 8-12 GHz
ANSWER B: An S-band radar operates on marine assigned frequencies 2,900-3,100 MHz. The S-band range is sometimes abbreviated 2 - 4 GHz.

8A59 An X band radar operates in which frequency band?
A. 1-2 GHz
B. 2-4 GHz
C. 4-8 GHz
D. 8-12 GHz

ANSWER D: An X-band radar operates from 9,300 MHz to 9,500 MHz. On FCC radio station Form 506, this is Item 22R, which must be checked to insure the radar onboard is properly licensed.

8A60 The major advantage of an S-band radar over an X-band radar is:
- A. It has greater bearing resolution.
- B. It is less affected by weather conditions.
- C. It is mechanically less complex.
- D. It has greater power output.

ANSWER B: Lower frequency radars on the S-band exhibit poor reflectivity to nearby thunderstorms, and thus have the ability to "see" better through light snow and rain.

8A61 A major consideration for the use of a switching regulator power supply over a linear regulator is:
- A. The switching regulator has better regulation.
- B. The linear regulator does not require a transformer to step down AC line voltages to a usable level.
- C. The switching regulator can be used in nearly all applications requiring regulated voltage.
- D. The overall efficiency of a switching regulator is much higher than a linear power supply.

ANSWER D: Switching power supplies offer incredible efficiency. Typical efficiency is 75 percent. For a linear power supply using a heavy transformer, efficiency is not great—only about 30 percent.

8A62 The major advantage of digitally processing a radar signal is:
- A. Digital readouts appear on the radar display.
- B. Enhancement of weak target returns.
- C. An improved operator interface.
- D. Rectangular display geometry is far easier to read on the CRT.

ANSWER B: Digital signal processing, abbreviated DSP, allows enhancement of return echoes by the use of sampling and *Fast Fourier* analysis. This allows reduction of interference that may come in with the return echo. A DSP circuit strips out the interference, and accentuates the real echo.

8A63 In order to ensure that a practical filter is able to remove undesired components from the output of an analog-to-digital converter, the sampling frequency should be:
- A. The same as the lowest component of the analog frequency.
- B. Two times the highest component of the analog frequency.
- C. Greater than two times the highest component of the sampled frequency.
- D. The same as the highest component of the sampled frequency.

ANSWER C: According to the *Nyquist Theorem,* to correctly represent the analog signal, the sampling frequency must be at least *two* times the highest frequency of the analog signal being sampled.

8A64 Which of the following is the most practical means of increasing the range of a ship radar installation?
- A. Increase the height of the radar antenna and the height of the target
- B. Increase the height of the target and use a corner reflector
- C. Use a metallic corner reflector and increase pulse width
- D. Increase transmitter power, increase pulse width, and increase the time between transmitted pulses

ANSWER D: To increase target echo detection to a greater range, a radar must be more powerful, have a longer pulse width, and have a longer delay between transmitted pulses.

8A65 A shipboard radar uses a PFN driving a magnetron cathode through a step-up transformer. This results in which type of modulation?
- A. Frequency modulation.
- B. Amplitude modulation.
- C. Continuous Wave (CW) modulation.
- D. Pulse modulation.

ANSWER D: The term "PFN" stands for pulse frequency network, and this is a type of pulse modulation.

8A66 In a solid-state radar modulator, the duration of the transmitted pulse is determined by:
- A. The thyratron.
- B. The magnetron voltage.
- C. The pulse forming network.
- D. The trigger pulse.

ANSWER C: The pulse forming network determines the pulse width, which in turn determines the duration of the transmitted pulse.

8A67 What device is used as a transmitter in a marine radar system?
- A. A Klystron
- B. A beam-powered pentode
- C. A Magnetron
- D. A Thyratron

ANSWER C: The magnetron produces high-power outputs for marine radars. The klystron (incorrect Answer A) is generally used as the oscillator for the system.

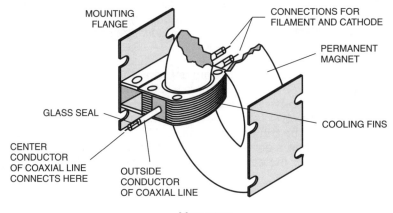

Magnetron

8A68 What device(s) could be used as the local oscillator in a radar receiver?
A. A Thyratron
B. A Klystron
C. A Klystron and a Gunn Diode
D. A Gunn diode

ANSWER C: Klystrons and Gunn diodes are used as oscillators in radar receivers.

8A69 The klystron local oscillator is constantly kept on frequency by:
A. Constant manual adjustments.
B. The Automatic Frequency Control circuit.
C. A feedback loop from the crystal detector.
D. A feedback loop from the TR box.

ANSWER B: Because the magnetron's frequency can *drift,* the automatic frequency control circuit causes the klystron to track any changes in magnetron frequency and uses the difference between the two frequencies to maintain a constant intermediate frequency.

8A70 What device(s) may act as the modulator of a radar system?
A. A magnetron
B. A thyratron
C. A silicon-controlled rectifier (SCR)
D. A thyratron and a silicon-controlled rectifier

ANSWER D: A Thyratron or an SCR is used in the STC (sensitivity time control) circuit to provide modulation for the radar system's pulses.

8A71 A circuit used to develop AFC voltage in a radar receiver is called the:
A. Peak detector.
B. Crystal mixer.
C. Second detector.
D. Discriminator.

ANSWER D: The discriminator is tuned to the IF frequency. When operating properly, the difference in magnetron and klystron frequency is the IF frequency. As long as the difference remains the same as the IF frequency, no error voltage is produced. If there is a difference, an error voltage is produced and is applied to the klystron to change the frequency difference back to the IF frequency. This corrective action is called *automatic frequency control.*

8A72 The TR box:
A. Prevents the received signal from entering the transmitter.
B. Protects the receiver from the strong radar pulses.
C. Turns off the receiver when the transmitter is on.
D. Protects the receiver from the strong radar pulses and turns off the receiver when the transmitter is on.

ANSWER D: In the TR box, a solid-state circuit acts as a mechanical T/R relay. The transmit/receive box protects the receiver from the high-powered signals produced during transmit.

8A73 A DC keep-alive potential:
A. Is applied to a TR tube to make it more sensitive.
B. Partially ionizes the gas in a TR tube, making it very sensitive to transmitter pulses.
C. Fully ionizes the gas in a TR tube.
D. Is applied to a TR tube to make it more sensitive and partially ionizes the gas in a TR tube, making it very sensitive to transmitter pulses.

ANSWER D: The TR tube is a gas switch. When the gas is not ionized, there is a high impedance to the signal. When the gas is ionized, it tends to short circuit the line by presenting a low-impedance path. An additional electrode is inserted to provide a *keep-alive* voltage to keep the gas partially ionized. This allows the gas to be more quickly and easily ionized by the transmit pulse.

8A74 What determines the minimum range of a radar set?
A. The duty cycle
B. The average power
C. The peak power
D. The transmitted pulse width and the T/R cell recovery time

ANSWER D: Here is that same question about minimum radar range. Minimum range is determined by transmitted pulse width and the transmit/receive cell recovery time.

8A75 The ARPA term CPA refers to:
A. The closest point a ship or target will approach your own ship.
B. The furthest point a ship or target will get to your own ship's bow.
C. Direction of target relative to your own ship's direction.
D. The combined detection and processing of targets.

ANSWER A: CPA stands for *closest point of approach. (NOTE: The acronym ARPA stands for Automatic Radio Plotting Aid.)*

8A76 What is the purpose or function of the "Trial Mode" used in most ARPA equipment?
A. It selects trial dost for targets' recent past positions.
B. It is used to display target position and your own ship's data such as TCPA, CPA, etc.
C. It is used to allow results of proposed maneuvers to be assessed.
D. None of these.

ANSWER C: Using the trial mode, you can execute intended maneuvers that cause the screen to display an intended course change. This allows you to judge the results of a proposed maneuver before you actually engage the steering system and engine speed for that maneuver.

8A77 On most ARPA equipment, this is a line on the PPI display which indicates a target's position. The speed of the target is shown by the length of the line. The course of the target is shown by the direction of the line. This statement best describes a/an:
A. Vector.
B. Electronic Bearing Line.
C. Range Marker.
D. Heading Marker.

ANSWER A: A vector is defined as a line representing course and position.

8A78 Raster-scan displays are frequently used in the newer and the more sophisticated radars. These displays are most like those found in:
  A. Television sets.
  B. A-scan Radars.
  C. B-scan Radars.
  D. PPI displays.

ANSWER A: Raster scan displays are similar to television sets. The old phosphorous displays are called "PPI." *Oldtimers* like the PPI display because of the slow decay in echoes.

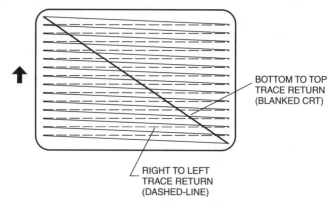

BOTTOM TO TOP
TRACE RETURN
(BLANKED CRT)

RIGHT TO LEFT
TRACE RETURN
(DASHED-LINE)

Raster Scan in Radar Similar to TV Scan

8A79 Voltages used in CRT anode-circuits are in what range of value?
  A. 0.5-10 mV
  B. 20-50 mV
  C. 200-1000 V
  D. 10-50 kV

ANSWER D: The 10,000 to 50,000 volts used in CRTs can produce *lethal results* if a technician touches the anode of a radar cathode-ray tube.

8A80 In older radar sets, how far from the waveguides are the spark gaps located?
  A. One quarter wavelength
  B. One half wavelength
  C. One wavelength
  D. Two wavelengths

ANSWER A: Spark-gap tubes are placed one-quarter wavelength from the waveguide.

8A81 The echo box is used for:
  A. Testing and tuning of the radar unit by providing artificial targets.
  B. Testing the wavelength of the incoming echo signal.
  C. Amplification of the echo signal.
  D. Detection of the echo pulses.

ANSWER A: The echo box simulates artificial targets so you can check radar system performance against a known base line.

8A82 The heading flash is a momentary intensification of the sweep line on the PPI presentation. Its function is to:
  A. Alert the operator when a target is within range.
  B. Alert the operator when shallow water is near.
  C. Inform the operator of the dead-ahead position on the PPI scope.
  D. Inform the operator when the antenna is pointed to the rear of the ship.
ANSWER C: The heading flash reveals a dead-ahead position of the bow of the vessel. A microswitch and a cam trigger send the heading flash to the scope. The heading flash should be adjusted about once a year.

8A83 The magnetron is:
  A. A type of diode that requires an internal magnetic field.
  B. A triode that requires an external magnetic field.
  C. A type of diode that requires an external magnetic field.
  D. Used as the local oscillator in the radar unit.
ANSWER C: The magnetron is a diode with a magnetic field between the cathode and anode; the magnetic field controls the tube. When the magnet gets weak, the magnetron must be replaced.

8A84 In a radar unit, the local oscillator is:
  A. A hydrogen thyratron.
  B. A klystron.
  C. A pentagrid converter tube.
  D. A reactance tube modulator.
ANSWER B: The local oscillator is the klystron (LOK).

8A85 What type of video output tube is used in most marine radar units?
  A. A standard CRT
  B. A plan position indicator (PPI)
  C. An oscilloscope
  D. None of these
ANSWER B: For this examination, your best answer for the video output tube in radar sets is the plan position indicator, abbreviated PPI. In old radars, the PPI was a phosphorous coating on the inside of the tube. On newer radars, it is a raster scan display similar to—but not exactly like—a television cathode ray tube. See figure at 8A78. On newer radars, a regular CRT can be used as a repeater, but for the test, the letters "PPI" are the plan position indicator.

8A86 How are ranges changed so that range markers represent different distances?
  A. By changing the electrical potential on the video amplifier
  B. By changing the PPI screen
  C. By changing the oscillating frequency of the ringing oscillator
  D. By changing the peak power
ANSWER C: An external variable range-ring control knob adjusts the ringing oscillator frequency and, consequently, varies the range represented by the markers.

8A87 Sea return is:
  A. Sea water that gets into the antenna system.
  B. The return echo from a target at sea.

C. The reflection of radar signals from nearby waves.

D. None of these.

ANSWER C: Sea return is echoes reflected off the face of nearby waves. You can even see the wake of certain ships as sea return.

8A88 The sensitivity-time control (STC) circuit:

A. Increases the sensitivity of the receiver for close objects.

B. Increases the sensitivity of the receiver for distant objects.

C. Decreases the sensitivity of the receiver for close objects.

D. Decreases the sensitivity of the transmitter for close objects.

ANSWER C: The sensitivity time control is used for minimizing false echoes from sea return during heavy weather. It minimizes the pick up of echoes from nearby targets.

8A89 The primary tube used in the STC circuit is:

A. A pentode.

B. A triode.

C. A diode.

D. A hydrogen thyratron.

ANSWER D: The hydrogen thyratron tube is used in the STC circuit.

8A90 In a radar unit, the mixer uses:

A. A silicon crystal or PIN diode.

B. A pentagrid converter tube.

C. A field-effect transistor.

D. A microwave transistor.

ANSWER A: The PIN diode is a point-contact, silicon-crystal diode with a fine tungsten, bronze-phosphor, or beryllium-copper "cat whisker" between two N- and P-type silicon crystals. These diodes must be treated carefully to avoid a static electricity burnout. They normally plug into the T/R unit, and many times require replacement when another powerful radar operating nearby causes the diode to overload and immediately burnout.

8A91 Why is hydrogen gas used in thyratron tubes?

A. It ionizes and deionizes slowly.

B. It ionizes and deionizes quickly.

C. Because it is much lighter than other gases.

D. Because it does not ionize at all.

ANSWER B: The hydrogen gas in thyratron tubes ionizes and deionizes more quickly than any other gas.

8A92 In video amplifiers, compensation for the input and output stage capacitances must be accomplished to prevent distorting the video pulses. This compensation is normally accomplished by:

A. Connecting inductors in parallel with both the input and output capacitances.

B. Connecting resistances in parallel with both the input and output capacitances.

C. Connecting an inductor in parallel with the input capacitance and an inductor in series with the output capacitance.

D. Connecting an inductor in series with the input capacitance and an inductor in parallel with the output capacitance.

ANSWER D: An inductor placed in series with the capacitance at the amplifiers input slightly expands the input pulse. This lowers the resonant frequency and spreads out the signal, which helps to ensure that weaker signals are included with the stronger signals in the input. You could use resistors to do this, but they would slow the pulse and introduce unwanted distortion.

At the amplifier's output, the impedance is changed by placing an inductor in parallel with the capacitance. This compressed the pulse and raises the frequency of the signal going into the CRT.

All of this happens so rapidly that your eye can't see the flicker on the CRT.

8A93 The video (second) detector in a pulse modulated radar system would most likely use a (an):
- A. Discriminator detector.
- B. Diode detector.
- C. Ratio detector.
- D. Infinite impedance detector.

ANSWER B: A crystal detector diode is most often used to extract the video modulation from the carrier.

8A94 In the receive mode, frequency conversion is generally accomplished by a:
- A. Tunable waveguide section.
- B. Pentagrid converter.
- C. Crystal diode.
- D. Ferrite device.

ANSWER C: Frequency conversion is usually accomplished by the crystal diode.

8A95 When the receiver employs an MTI circuit:
- A. The receiver gain increases with time.
- B. Only moving targets will be displayed.
- C. The receiver AGC circuits are disabled.
- D. Ground clutter will be free of "rabbits".

ANSWER B: In the "moving target" mode, the MTI circuit eliminates all echoes except those from vessels that are *underway.*

8A96 Video amplifiers in pulse radar receivers must have a broad bandwidth because:
- A. Weak pulses must be amplified.
- B. High frequency sinewaves must be amplified.
- C. The radars operate at PRFs above 100.
- D. The pulses produced are normally too wide for video amplification.

ANSWER A: The broad bandwidth capabilities in the video amp of a pulse radar receiver allow detection of weak pulses. A video amplifier with a narrow bandwidth allows weak pulses to fall into the noise threshold so that they are not seen on the CRT.

8A97 To minimize video display saturation caused by RF energy from external sources, such as jamming, a receiver might employ:
- A. STC circuits.
- B. GTC circuits.
- C. FTC circuits.
- D. AGC circuits.

ANSWER C: The fast time control circuit, FTC, minimizes the "spoked wheel" effect of interference coming in from other nearby radar units.

8A98 A magnetron has a cathode-anode potential of 20 kilovolts. Electrons emitted by the cathode:
    A. Never reach the anode.
    B. Enter the cavities to sustain oscillations.
    C. Travel to the anode in straight line paths.
    D. Sustain oscillation by their cycloidal paths.
ANSWER D: The electrons follow a cycloidal path as they travel from cathode to anode. This electron motion produces oscillations in the resonant cavities.

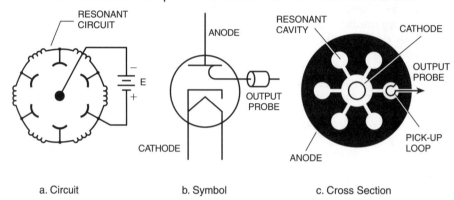

a. Circuit   b. Symbol   c. Cross Section

A Magnetron Generates Transmitter Power

8A99 How many active elements are contained in a magnetron?
    A. 2
    B. 3
    C. 4
    D. 5
ANSWER A: The cathode and anode are the active elements in the magnetron, and all other elements are passive.

8A100 The shape and duration of the high-voltage pulse delivered to the magnetron is established by:
    A. An RC network in the keyer stage.
    B. The duration of the modulator input trigger.
    C. An artificial delay line.
    D. The time required to saturate the pulse transformer.
ANSWER C: It is the delay line that shapes and delays the high-voltage pulse delivered to the magnetron.

8A101 The pulse developed by the modulator may have an amplitude greater than the supply voltage. This is possible by:
    A. Using a voltage multiplier circuit.
    B. Employing a resonant charging choke.
    C. Discharging a capacitor through an inductor.
    D. Discharging two capacitors in series and combining their charges.
ANSWER B: At resonance, the amplitude of the pulse can be well above the supply voltage. This is made possible by employing a resonant charging choke.

8A102 To minimize video display saturation caused by RF energy from external sources, such as jamming, a receiver might employ:
    A. STC circuits.
    B. GTC circuits.
    C. FTC circuits.
    D. AGC circuits.
ANSWER C: The fast time control circuit minimizes the "spoked wheel" effect of interference coming in from other nearby radar units.

8A103 An advantage of resonant charging is that it:
    A. Eliminates the need for a reverse current diode.
    B. Guarantees perfectly square output pulses.
    C. Reduces the high-voltage power supply requirements.
    D. Maintains a constant magnetron output frequency.
ANSWER C: Because a resonant charging circuit can develop extremely high voltages above the normal B+ voltage, it reduces the high-voltage power supply requirements.

8A104 Pulse radars require precise timing for their operation. Which type circuit below might best be used to provide these accurate timing pulses?
    A. A single-swing blocking oscillator
    B. An AFC controlled sinewave oscillator
    C. A non-symmetrical astable multivibrator
    D. A triggered flip-flop type multivibrator
ANSWER A: The blocking oscillator oscillates for a period of time, followed by a period of time where no oscillation occurs. The circuit then oscillates again. This oscillator is often used to form the pulse in a radar transmitter.

8A105 Standard vacuum tubes are not used at radar frequencies because:
    A. Their interelectrode capacitance is too small.
    B. Transit time is so short that oscillation may result.
    C. Pentode gain is too high and reduces sensitivity.
    D. The inductance of the leads is excessive.
ANSWER D: Vacuum tubes have been phased out of most radar sets because of the length of the leads to the tube. These leads create additional unwanted inductance in the circuit.

8A106 Transit time might be defined as the time required for:
    A. RF energy to travel through the waveguide.
    B. A pulse to travel a wavelength inside a waveguide.
    C. One cycle of operation to be completed.
    D. Electrons to travel from cathode to anode.
ANSWER D: The time it takes for electrons to travel from the cathode to the anode is called *transit time.* In a standard vacuum tube, this time is too long to allow the tube to function properly at radar frequencies.

8A107 A CW radar may be used to determine the rate of travel of a target and whether it is moving toward or away from the radar position. To make this determination:
    A. The carrier must be frequency modulated.
    B. The AFC control circuits must be disabled.

C. The carrier must be pulse modulated.

D. The principles of doppler effect are employed.

ANSWER D: The Doppler principle is in use here. If the target is moving toward the radar, the return frequency is higher than the transmitted signal; if the target is moving away from the radar, the return frequency is lower. This frequency change is called *Doppler shift.*

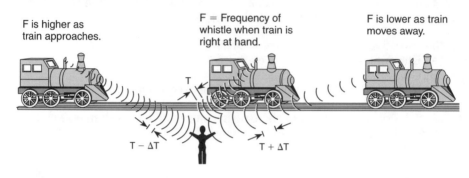

F is higher as train approaches.

F = Frequency of whistle when train is right at hand.

F is lower as train moves away.

T

T − ΔT

T + ΔT

Dopper Effect

8A108 A CW radar is being frequency modulated with a low-frequency sinewave to provide range measurements. To develop a range indication, the receiver would most likely use a:

A. Diode detector, tuned to the IF.

B. Diode detector, tuned to the modulating frequency.

C. Phase comparator, tuned to the IF.

D. Phase comparator, tuned to the modulating frequency.

ANSWER D: This CW radar is based on Doppler shift, and, therefore, a phase comparator tuned to the modulating frequency is used to detect frequency changes in the return echoes and, thereby, determine the speed of a target.

8A109 To obtain an indication of target movement using the doppler principle, a CW radar receiver would most likely employ a:

A. Discriminator, tuned to the IF.

B. Discriminator, tuned to the transmitter frequency.

C. Discriminator, tuned to an audio frequency.

D. Slope detector, tuned slightly above the IF.

ANSWER A: For Doppler detection, the discriminator operates at the intermediate frequency.

8A110 A negative voltage is commonly applied to the magnetron cathode rather than a positive voltage to the magnetron anode because:

A. The cathode must be made negative to force electronics into the drift area.

B. A positive voltage would tend to nullify or weaken the magnetic field.

C. The anode can be operated at ground potential for safety reasons.

D. The cavities might not be shock-excited into oscillation by a positive voltage.

ANSWER C: The magnetron anode is normally operated at ground potential to reduce the likelihood of an accidental shock when technicians are working on a turned-on radar T/R unit.

8A111 Solid-state microwave amplifier devices operating at C-Band and above are typically made from:
- A. Germanium.
- B. Silicon Dioxide.
- C. Silicon Carbide.
- D. Gallium Arsenide.

ANSWER D: The GaAsFET strip-line, surface-mounted transistor offers high gain with an extremely low noise floor, which makes it suitable as a microwave amplifier device.

8A112 The SSR subsystem of an Airport Surveillance Radar System uses:
- A. Multiple waveforms to resolve range and doppler ambiguities.
- B. A much higher frequency than the PSR subsystem.
- C. Coded response information from aircraft transponders.
- D. A high azimuth scan rate and peak transmitter power.

ANSWER C: Aeronautical SSR systems, part of airport surveillance radar systems, receive coded response information from a specifically-addressed aircraft transponder.

8A113 What is the transmit frequency of an aircraft transponder?
- A. 5250 MHz
- B. 1.09 GHz
- C. 1030 MHz
- D. 2950 MHz

ANSWER B: Aeronautical transponders transmit at 1090 MHz, which is the same as 1.09 GHz.

8A114 What is the receive frequency of an aircraft transponder?
- A. 5250 MHz
- B. 1.09 GHz
- C. 1030 MHz
- D. 2950 MHz

ANSWER C: Aircraft transponders receive and listen to signals at 1030 MHz, sometimes abbreviated as 1.03 GHz.

8A115 Choose the most correct statement with respect to component damage from electrostatic discharge:
- A. ESD damage occurs primarily in passive components which are easily identified and replaced.
- B. ESD damage occurs primarily in active components which are easily identified and replaced.
- C. The technician will feel a small static shock and recognize that ESD damage has occurred to the circuit.
- D. ESD damage may cause immediate circuit failures, but may also cause failures much later at times when the radar set is critically needed.

ANSWER D: The sensitive microwave receivers in aeronautical radar are constantly subjected to electrostatic discharge (ESD), either from the aircraft flying through the dry air and building up a charge-to-discharge state, or from the aircraft flying through heavily charged clouds within a severe weather system.

Sometimes ESD damage is subtle, and you won't realize you have an onboard failure until you begin to use a certain parameter of a piece of microwave equipment.

8A116 The display of a 2-D Airport Surveillance radar typically provides:
A. Range, elevation, and target speed.
B. Range and azimuth only.
C. Range, azimuth, and elevation.
D. Range, azimuth, and SSR responses.

ANSWER D: A two-dimensional Airport Surveillance radar shows range and azimuth (bearing) to a target, and shows coded responses from a transponder.

8A117 In a conventional PPI display, the electron beam is scanned:
A. From the center of the display to the outer edges.
B. From the center of the display to the outer edges and in a rotating pattern which follows the antenna position.
C. In a rotating pattern which follows the antenna position.
D. From one specified X-Y coordinate to the next.

ANSWER B: A PPI radar screen may be either an older phosphor screen or a newer raster scan display. In a conventional PPI display, the electron beam moves from the center of the display to the outer edge and rotates in a pattern following the antenna's rotation.

8A118 What does the term ARPA/CAS refer to?
A. The basic radar system in operation
B. The device which displays the optional U.S. Coast Guard Acquisition and Search radar information on a CRT display
C. The device which acquires and tracks targets which are displayed on the radar indicator's CRT
D. The device which allows the ship to automatically steer around potential hazards

ANSWER C: Big-ship radar systems feature ARPA/CAS which allows automatic tracking of any target and leaves an electronic "wake" that is easily monitored on the radar's cathode-ray tube.

8A119 The characteristics of a Field-Effect Transistor (FET) used in a modern radar switching power supply can be compared as follows:
A. "On" state compares to a bipolar transistor. "Off" state compares to a 1-Megohm resistor.
B. "On" state compares to a pure resistor. "Off" state compares to a mechanical relay.
C. "On" state compares to an low resistance inductor. "Off" state compares to a 10-Megohm resistor.
D. "On" state compares to a resistor. "Off" state compares to a capacitor.

ANSWER B: The Field-Effect Transistor (FET) is the popular choice in switching power supplies where it is desirable for the ON state to be low resistance and the OFF state to offer infinite resistance similar to an open mechanical relay.

8A120 Which of the following characteristics are true of a power MOSFET used in a radar switching supply?
A. Low input impedance; failure mode can be gate punch-through
B. High input impedance; failure mode can be gate punch-through
C. High input impedance; failure mode can be thermal runaway
D. Low input impedance; failure mode can be gate breakdown

ANSWER B: The MOSFET power transistor commonly found in radar switcher power supplies has a high input impedance, but when the device fails, the failure is in the gate "punch-through" current (breakdown at gate insulation.)

8A121 On a basic syncro system, the angular information is carried on:
A. The DC feedback signal.
B. The stator lines.
C. The deflection coils.
D. The rotor lines.

ANSWER B: Sawtooth current pulses make up the angular information carried on the radar stator lines.

8A122 In a fixed-frequency switching power supply, the pulse width of the switching circuit will increase when:
A. The load current decreases.
B. The output voltage increases.
C. The input voltage increases.
D. The load impedance decreases.

ANSWER D: A switching-mode power supply drops and regulates the output voltage by chopping the dc input to a square wave, and then recovering the voltage with a low-pass filter. As the input voltage or the load changes, the duty cycle of the on/off cycle changes and keeps the voltage constant.

8A123 The output of an RC integrator, when driven by a square wave with a period of much less than one time constant is:
A. A Sawtooth wave.
B. A Sine wave.
C. A series of narrow spikes.
D. A Triangle wave.

ANSWER D: As the square wave goes positive, the capacitor in the RC circuit charges. With the square wave period much less than the time constant, the square wave returns to zero before the capacitor can charge or discharge, generating a triangular wave output. The amplitude of the wave is less than the square wave amplitude.

8A124 A pulse-width modulator in a switching power supply is used to:
A. Provide the reference voltage for the regulator.
B. Vary the frequency of the switching regulator to control the output voltage.
C. Vary the duty cycle of the regulator switch to control the output voltage.
D. Compare the reference voltage with the output voltage sample and produce an error voltage.

ANSWER C: As the duty cycle (on time ÷ period of cycle) of the power supply increases, the voltage increases. Therefore, you can use the duty cycle to regulate the output voltage.

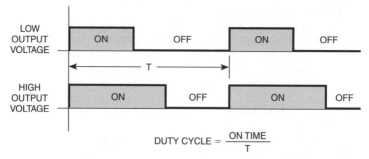

DUTY CYCLE = $\dfrac{\text{ON TIME}}{T}$

Duty Cycle Varies to Control Output Voltage in a Switching Power Supply

---

8A125 In video amplifiers, compensation for the input and output stage capacitances must be accomplished to prevent distorting the video pulses. This compensation is normally accomplished by:
  A. Connecting inductors in parallel with both the input and output capacitances.
  B. Connecting resistances in parallel with both the input and output capacitances.
  C. Connecting an inductor in parallel with the input capacitance and an inductor in series with the output capacitance.
  D. Connecting an inductor in series with the input capacitance and an inductor in parallel with the output capacitance.

ANSWER D: An inductor placed in series with the capacitance at the amplifiers input slightly expands the input pulse. This lowers the resonant frequency and spreads out the signal, which helps to ensure that weaker signals are included with the stronger signals in the input. You could use resistors to do this, but they would slow the pulse and introduce unwanted distortion.

At the amplifier's output, the impedance is changed by placing an inductor in parallel with the capacitance. This compressed the pulse and raises the frequency of the signal going into the CRT.

[Ed. Note: This question is a duplicate of 8A92, page 358 and has been removed from the question pool by the FCC. It should not appear on any examination.]

8A126 When comparing TTL and CMOS logic families, which of the following is true:
  A. CMOS logic requires a supply voltage of 5 volts ±20%, whereas TTL logic requires 5 volts ±5%.
  B. Unused inputs should be tied high or low as necessary only in the CMOS family.
  C. At higher operating frequencies, CMOS circuits consume almost as much power as TTL circuits.
  D. When a CMOS input is held low, it sources current into whatever it drives.

ANSWER C: One of the features of CMOS is the low power consumption at low frequencies compared to TTL. However, due to increased capacitance at high frequencies, the power consumption is about the same.

8A127 When comparing a TTL and a CMOS NAND gate:

A. Both have active pull-up characteristics.
B. Both have three output states.
C. Both have comparable input power sourcing.
D. Both employ Schmitt diodes for increased speed capabilities.

ANSWER A: Both the TTL and the CMOS NAND gate have an active pull-up characteristic.

8A128 A basic sample-and-hold circuit contains:

A. An analog switch and an amplifier.
B. An analog switch, a capacitor, and an amplifier.
C. An analog multiplexer and a capacitor.
D. An analog switch, a capacitor, amplifiers and input and output buffers.

ANSWER D: A capacitor samples the input voltage until the amplifier is placed in the hold mode by an analog switch. The input and outputs are buffered.

8A129 Which of the following is not a method of analog-to-digital conversion?

A. Delta-sigma conversion
B. Dynamic-range conversion
C. Switched-capacitor conversion
D. Dual-slope integration

ANSWER B: Dynamic range is specifically related to the amplitude levels of multiple signals that a receiver can accommodate. This is not an indication of an analog-to-digital converter.

8A130 Some hand shaking signals you might find on a EIA-232-D interface are:

A. NRFD and NDAC.
B. ACKNLG and DATA STROBE.
C. PIO and LIFO.
D. RTS and CTS.

ANSWER D: The confirmation signals "request to send" (RTS) and "clear to send" (CTS) are two of the important signals in the EIA-232-D datastream.

8A131 In standard CMOS logic IC databooks, the parameter t_{TLH} is considered to be:

A. Negligible in CMOS logic.
B. Measured from the leading edge to the trailing edge.
C. Measured from the 10% to 90% points of the leading edge.
D. Measured at the 50% point of the leading edge.

ANSWER C: t_{TLH} means time to transition from low level to high.

8A132 Which of the following would not be considered an input to the computer of a collision avoidance system?

A. Own ship's exact position from navigation satellite receiver.
B. Own ship's gyrocompass heading.
C. Own ship's speed from Doppler log.
D. Own ship's wind velocity from an anemometer.

ANSWER D: Wind velocity is not considered a normal input to the computer collision avoidance system.

8A133 High voltage is applied to what element of the magnetron?
A. The waveguide
B. The anode
C. The plate cap
D. The cathode

ANSWER D: High voltage is found within the heart of the magnetron on the sealed cathode. Cathode current is several amps, and the cathode is large so that it can withstand the tremendous amount of heat that develops. Keep metal tools well away from the magnetron because of its large permanent magnet. The exposed anode is at ground potential, and a large negative voltage of several kilovolts is applied to the cathode during the magnetron's oscillating state. See figure at 8A67 and 8A98.

8A134 An ion discharge (TR) cell is used to:
A. Protect the transmitter from high SWRs.
B. Lower the noise figure of the receiver.
C. Protect the receiver mixer during the transmit pulse.
D. Tune the local oscillator of the radar receiver.

ANSWER C: The TR cell is similar to that of a T/R switch on a transceiver. The TR cell protects the receiver input during the transmit pulse output.

8A135 In a radar unit, the mixer uses:
A. PIN diodes and silicon crystals.
B. PIN diodes.
C. Boettcher crystals.
D. Silicon crystals.

ANSWER A: A radar mixer almost always uses a PIN diode and silicon crystals.

8A136 An AFC system keeps the receiver tuned to the transmitted signal by varying the frequency of:
A. The magnetron.
B. The IF amplifier stage.
C. The local oscillator.
D. The cavity duplexer.

ANSWER C: The automatic frequency control function takes place in the local oscillator.

8A137 A Gunn diode oscillator takes advantage of what effect?
A. Negative resistance
B. Avalanche transit time
C. Bulk-effect
D. Negative resistance and bulk-effect

ANSWER D: Gunn diodes are not really diodes in the normal rectification sense. They are used as oscillators and only retain the diode name because they have two elements. The Gunn diode operates because of the bulk semiconductor material (bulk effect) and capacitance and negative resistance.

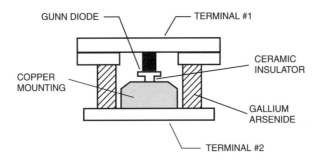

GUNN DIODE — TERMINAL #1

CERAMIC
INSULATOR

COPPER
MOUNTING

GALLIUM
ARSENIDE

TERMINAL #2

Gunn Diode

8A138 The basic frequency determining element in a Gunn oscillator is:
A. The power supply voltage.
B. The type of semiconductor used.
C. The resonant cavity.
D. The loading of the oscillator by the mixer.

ANSWER C: A varactor can be used to vary the frequency for narrow bandwidth applications. But for wide bandwidths applications, such as radar, YIG tuning may be employed. In YIG tuning, the resonance of a cavity is changed by using a magnetic coil.

8A139 A logarithmic IF amplifier is preferable to a linear IF amplifier in a radar receiver because:
A. It has higher gain.
B. It is more easily aligned.
C. It has a lower noise figure.
D. It has a greater dynamic range.

ANSWER D: The characteristic of a log amplifier provides an output voltage that is proportional to the log of the input signal. This characteristic allows greater dynamic ranges for the receiver.

8A140 Silicon crystals:
A. Are very sensitive to static electric charges.
B. Should be wrapped in lead foil for storage.
C. Tolerate very low currents.
D. All of these.

ANSWER D: Working with silicon crystals requires you to take great precautions not to accidentally *zap* them with a static discharge. They can tolerate only an extremely low current, and for protection when shipping, they are wrapped in lead foil to protect against a static discharge.

8A141 The TWT (traveling wave tube) is:
A. A type of waveguide.
B. Not really a vacuum tube but a semiconductor device.
C. A microwave amplifier tube.
D. A microwave tube that requires a very strong external magnetic field for proper operation.

ANSWER C: The traveling wave tube (TWT) is a microwave amplifier tube. A beam of electrons is sent from the cathode, through a focusing coil (helix), and to the anode at the other end of the tube. The input is applied at the cathode end of the helix coil.

8A142 The input signal to a TWT is inserted at:
A. The anode end of the helix.
B. The grid end of the helix.
C. The collector end of the helix.
D. The cathode end of the helix.

ANSWER D: The input to a TWT is applied at the cathode.

8A143 Which is typical current for a silicon crystal used in a radar mixer or detector circuit?
A. 3 mA
B. 15 mA
C. 50 mA
D. 100 mA

ANSWER A: The typical current for a silicon crystal in a radar mixer or detector is no more than 3 or 4 milliamps...maximum. Any more would burn it out.

8A144 Oscillations in a klystron local oscillator tube are maintained by:
A. The grid-feedback loop.
B. Bunches of electrons passing the cavity grids.
C. The circulation of electrons.
D. The LC circuit.

ANSWER B: The klystron is used as the local oscillator for the receiver. The cathode emits a steady stream of electrons directed toward a resonant cavity, and accelerated by a grid. In the resonant cavity, the electrons are placed in oscillation at the cavity's resonant frequency. The electrons are then bunched together and pass through the cavity grids. The repeller plate repels them back into the cavity in the proper phase to sustain oscillation.

8A145 The STC circuit:
A. Increases the sensitivity of the receiver for close targets.
B. Decreases sea return on the PPI scope.
C. Helps to increase the bearing resolution of targets.
D. Increases sea return on the PPI scope.

ANSWER B: The short time control (sensitivity time control) decreases return echoes from the sea in the direction of the incoming wind. It is the front side of the ocean waves that leads to the greatest number of false-echo sea returns.

8A146 In a radar receiver, the RF power amplifier:
A. Is high gain.
B. Is low gain.
C. Requires a wide bandwidth.
D. Does not exist.

ANSWER D: In radar receivers, RF receive amplifiers really don't exist. A crystal mixer stage is used as the receiver "front end," and the crystal is a small point-contact diode.

8A147 What is the average range of a ship's surface search radar unit?
A. 20 miles
B. 40 miles
C. 80 miles
D. 100 miles

ANSWER B: Most marine radars are 36 to 40 mile, 3 KW - 5 KW equipment. Further range would simply be lost in space, because microwave energy travels 4/3 radio range to the optical horizon. Extremely small boat radars might see only 16 miles, but 40 miles is your best answer in most cases.

8A148 The anode of a magnetron is normally maintained at ground potential:
A. Because it operates more efficiently that way.
B. For safety purposes.
C. Never. It must be highly positive to attract the electrons.
D. Because greater peak-power ratings can be achieved.

ANSWER B: The anode is at ground potential to protect operations personnel from shocks and to eliminate the problem of insulating the magnetron from the chassis.

8A149 A keep-alive voltage is applied to:
A. The crystal detector.
B. The ATR tube.
C. The TR tube.
D. The magnetron.

ANSWER C: A keep-alive voltage is always kept on the TR tube to insure it has the fastest response time for the changeover.

8A150 How does a TWT amplify?
A. Through a positive feedback circuit
B. Through a negative feedback circuit
C. Through a capacitive feedback circuit
D. By transfer of energy from the signal to the electron beam

ANSWER D: The traveling wave tube (TWT) is an amplifier that is similar to the klystron. The beam-generating portion of the tube is identical to the klystron; however, the modulation of the signal is accomplished differently. The TWT applies the signal to an element that is positioned parallel to the electron beam.

8A151 A wavelength is equal to how many degrees of a sinewave?
A. 360
B. 180
C. 90
D. 45

ANSWER A: A sine wave is generated by a vector with angle Θ as the vector rotates through a 360° circle. See figure on next page. A wavelengh is the distance the wave travels in the time of one cycle.

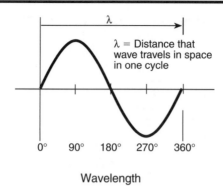

λ = Distance that wave travels in space in one cycle

0°   90°   180°   270°   360°

Wavelength

8A152 A raster scan radar display maintains the display presentation on the CRT face by the use of:
A. Long-persistence phosphors in the CRT face.
B. Analog to digital conversion.
C. Fast sweeping of the radar video.
D. Video RAM.

ANSWER D: It is video RAM that drives the display on the screen of a raster scan radar. On the old PPI scopes, long-persistence phosphorous performs this function. Almost all new radars use a raster scan display.

8A153 In a raster-type display, the electron beam is scanned:
A. From the center of the display to the outer edges.
B. Horizontally and vertically across the CRT face.
C. In a rotating pattern which follows the antenna position.
D. From one specified X-Y coordinate to the next.

ANSWER B: In a raster scan display, the electron beam scans horizontally and vertically across the CRT face, similar to the scanning in the picture tube of a standard TV set. See figure at 8A78.

8A154 The position of the PPI scope sweep must indicate the position of the antenna. The sweep and antenna positions are frequently kept in synchronism by the use of:
A. Servosystems.
B. Synchro systems.
C. DC positioning motors.
D. Differential amplifiers.

ANSWER B: It is the job of the synchro system to keep the radar sweep and antenna in alignment. After several years of operation, mechanical slippage could require you to get into the antenna unit and readjust the synchro settings and heading marker microswitch.

8A155 What is the main difference between an analog and a digital receiver?
A. Special amplification circuitry
B. The presence of decision circuitry to distinguish between "on" and "off" signal levels
C. The two cannot be compared
D. Digital receivers produce no distortion

ANSWER B: The digital radar receiver has a much lower noise floor, thanks to digital signal processing. Digital circuitry distinguishes only on-and-off signal levels.

8A156 The purpose of the aquadag coating on the CRT is:
A. To protect the electrons from strong electric fields.
B. To act as a second anode.
C. To attract secondary emissions from the CRT screen.
D. All of these.

ANSWER D: The aquadag coating on a CRT acts as a second anode and attracts secondary emissions from the CRT. It also provides shielding from strong electric fields that may be present near the CRT tube.

8A157 When you examine the RADAR you notice that there is no target video in the center of the CRT. The blank spot gets smaller in diameter as you increase the range scale. What operator front panel control could be misadjusted?
A. TUNE
B. Sensitivity Time Control (STC)
C. Anti-Clutter Rain (ACR)
D. False Target Elimination (FTE)

ANSWER B: One way of minimizing sea echo return is to switch on the STC circuit. If the control is turned too high, you won't receive *any* echoes in that small spot in the center of the screen. Limit the use of STC when cruising in calm waters. This way, you won't miss small close-in targets.

8A158 When you examine the RADAR you notice that there is no target video in the center of the CRT. The blank spot gets smaller in diameter as you increase the range scale. If all of the front panel controls are properly adjusted, what would be the most probable faulty circuit?
A. The local oscillator is misadjusted
B. Video amplifier circuit
C. The IF amplifier circuit
D. The TR (TRL) Cell

ANSWER D: If you still have a blank spot in the middle of the tube, you might suspect misadjustment of the internal TR cell timing circuitry. If not properly timed, the TR cell desensitizes the receiver; this affects close-in echoes most.

8A159 While examining the shipboard radar sets, you notice on a particular indicator that the video representing the pier is distorted closest to the center of the PPI. (The video appears to bend in a concave fashion.) This is a primary indication of what?
A. The deflection coils need adjusting.
B. The centering magnets at the CRT neck need adjusting.
C. The waveguide compensation delay line needs adjusting.
D. The CRT filaments are weakening.

ANSWER C: This is a common problem when radar cables have been lengthened or shortened since the last timing adjustment. In the old days of waveguides, they called this "compensation delay line" adjusting. Now it's simply the "timing" circuit to compensate for different lengths of cable between the indicator unit and the antenna/TR unit. Aboard small boats where different cable lengths may be ordered, it is essential that you time the radar properly by observing a straight pier or breakwater, and adjust the timing control so the echo indeed looks straight—not concave or convex.

8A160 How do you eliminate stationary objects such as trees, buildings, bridges, etc., from the PPI presentation?
 A. Remove the discriminator from the unit.
 B. Use a discriminator as a second detector.
 C. Calibrate the IF circuit.
 D. Calibrate the local oscillator.
ANSWER B: For radar installations used on land for the vessel traffic system (VTS), using a discriminator as a second detector is one way to minimize return echoes from stationary objects.

8A161 The radar service person should take the following precautions to ensure that the magnet of the magnetron is not weakened:
 A. Keep metal tools away from the magnet.
 B. Do not subject it to excessive heat.
 C. Do not subject it to shocks or blows.
 D. All of these.
ANSWER D: The magnetron is a delicate network and can be damaged if a nearby wrench pulled into it by the strong magnetic field from its permanent magnet. Keep metal tools away from the magnetron, never subject the magnetron to excessive heat, and protect it from receiving shocks or blows.

8A162 While making repairs or adjustments to radar units:
 A. Wear fire-retardant clothing.
 B. Discharge all high-voltage capacitors to ground.
 C. Maintain the filament voltage.
 D. Reduce the magnetron voltage.
ANSWER B: Before working on any radar, be sure that all high-voltage capacitors are safely discharged. Also be sure that no one can accidentally turn on the equipment while you are working on it.

8A163 While removing a CRT from its operating casing, it is a good idea to:
 A. Discharge the first anode.
 B. Test the second anode with your fingertip.
 C. Wear gloves and goggles.
 D. Set it down on a hard surface.
ANSWER C: If ever you are required to pull out a defective cathode-ray tube and install a new one, wear protective gloves and goggles for safety.

8A164 In a marine radar set, a high VSWR is indicated at the magnetron output. The waveguide and rotary joint appear to be functioning properly. What component may be malfunctioning?
 A. The magnetron
 B. The waveform generator
 C. The STC circuit
 D. The waveguide array termination

ANSWER D: High VSWR at the magnetron output indicates a problem beyond the output to the antenna system. The incorrect answers are all components within the radar itself, not at the antenna. You would check the waveguide array termination for corrosion.

8A165 The VSWR of a microwave transmission line device might be measured using:
A. A dual directional coupler and a power meter.
B. A Network Analyzer.
C. A Spectrum Analyzer.
D. A dual directional coupler, a power meter, and a Network Analyzer.
ANSWER D: Instruments used to determine VSWR of a microwave transmission line or antenna system include a dual-directional coupler, a power meter, and a network analyzer that would also allow you to determine the cause of high VSWR. Now that most antenna units are integral to the T/R units, waveguide runs are either relatively short or non-existent.

8A166 Prior to removing, servicing or making measurements on any solid state circuit boards from the radar set, the operator should ensure that:
A. The waveguide is detached from the antenna to prevent radiation.
B. The magnetic field is present to prevent over-current damage or overheating from occurring in the magnetron.
C. The proper work surfaces and ESD grounding straps are in place to prevent damage to the boards from electrostatic discharge.
D. Only non-conductive tools and devices are used.
ANSWER C: Before working on any radar equipment, be sure the system is turned off and cannot be accidentally turned on while you are working on it. Also doublecheck that all work surfaces are well ESD grounded to prevent damage to circuit boards and components from electrostatic discharge.

8A167 Before removing either a primary or secondary airport surveillance radar from service for maintenance or repair, the operator must:
A. Notify the pilots of all planes in the area.
B. Ensure that the area is relatively clear of air traffic.
C. Notify Air Traffic Control.
D. Record the operating conditions of the radar in the proper maintenance log.
ANSWER C: If you plan to work on aeronautical airport radar equipment, you must notify Air Traffic Control.

8A168 Before ground testing of an aircraft radar, the operator should:
A. Ensure that the area in front of the antenna is clear of other maintenance personnel to avoid radiation hazards.
B. Be sure that the receiver has been properly shielded and grounded.
C. First test the transmitter connected to a matched load.
D. Measure power supply voltages to prevent circuit damage.
ANSWER A: Radar energy is at dangerous levels nearby the transmitting antenna. Before testing an aircraft radar on the ground, be sure that the area in front of the antenna is clear of other maintenance personnel so you don't bombard them with radiation from the radar.

8A169 The azimuth encoder of an airport surveillance radar appears to be malfunctioning. What must the operator do before performing maintenance on the unit?

A. Inform Air Traffic Control that the azimuth information is unreliable until maintenance is performed. Continue operation until scheduled maintenance can be accomplished.
B. Notify Air Traffic Control and cease operation immediately.
C. Notify Air Traffic Control that the PSR will be removed from service to avoid radiation hazard to maintenance personnel. Then, and only then, you may remove and replace the malfunctioning unit.
D. The redundant unit is automatically switched into service and operates independently. You may remove and replace the malfunctioning unit without affecting normal operation.

ANSWER D: If you are called to work on airport radar surveillance equipment, be sure the secondary redundant radar is switched into service before you remove and replace the malfunctioning unit. Sometimes the switch to turn on the redundant unit is not automatic, and you must switch it on manually.

8A170 It is reported that the radar is not receiving small targets. The most likely causes are:

A. Magnetron, IF amplifier, or receiver tuning.
B. PFN, crystals, or processor memory.
C. Crystals, Local oscillator tuning, or power supply.
D. Fuse blown, IF amp, or video processor.

ANSWER A: The magnetron is the main transmitting tube used for transmitting radar microwaves. If the tube gets weak, power output is reduced, and so are the echoes. Large targets still show up, but small targets with weak echoes won't have enough return signal strength because of the weak tube. A weak IF amp in the receiver or improperly tuned *outside* receiver can also cause small targets not to return a strong enough echo to show up on the display. All of the other answers are incorrect because any failure of those other components would lead to absolutely no reception of any targets.

8A171 The radar display has sectors of solid video (spoking). What would be the first thing to check?

A. Antenna information circuits failure
B. Frequency of raster scan
C. For interference from nearby ships
D. Constant velocity of antenna rotation

ANSWER C: Spoking is common in big harbors where other radars are turned on and operating in the vicinity. Spoking is interference from nearby ships operating radar within the same band that your radar is tuned to. Turning on the interference rejection circuits helps minimize spoking, but the IR circuitry might cause you to miss on-purpose "interference" echoes from a nearby RACON.

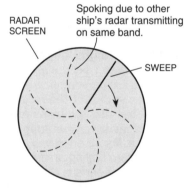

Spoking on Radar Screen

8A172 When replacing components of a radar transmitter, what safety precautions should be taken?
  A. All of these.
  B. De-energize equipment and turn off associated ship's power circuit breakers whenever possible.
  C. Adhere to safety notices appearing in the equipment's service manual.
  D. Ensure someone else is nearby.
ANSWER A: When working on a radar transmitter, de-energize the equipment and turn off the associated ship's power circuit, adhere to safety notices, and insure someone else is nearby in case of an emergency. All three answers are correct, making answer A correct.

8A173 Targets displayed on the radar display are not on the same bearing as their visual bearing. What should you first suspect?
  A. A bad reed relay in the antenna pedestal
  B. A sweep length misadjustment
  C. One phase of the yoke assembly is open
  D. Incorrect antenna position information
ANSWER D: Inside the antenna unit is a microswitch that triggers at *dead ahead.* If the microswitch is improperly calibrated, you receive incorrect antenna position information. Sometimes these microswitches get loose, and might be off as much as 10 or 15 degrees.

8A174 Range rings on the PPI indicator are oval in shape. Which circuit would you suspect is faulty?
  A. Timing circuit
  B. Video amplifier circuit
  C. Range marker circuit
  D. Sweep generation circuit
ANSWER D: If the range *rings* on the scope are more oval than round, you are going to need service of the sweep generation circuit for a slight adjustment to bring the rings back into a round shape. This is a yearly calibration process that should be checked every time you are inside the radar lower unit.

8A175 On a vessel with two radars, one has a different range indication on a specific target than the other. How would you determine which radar is incorrect?
  A. Triangulate target using the GPS and visual bearings.
  B. Check the sweep and timing circuits of both indicators for correct readings.
  C. Check antenna parallax.
  D. Use the average of the two indications and adjust both for that amount.
ANSWER B: The timing circuit, or sychronizer, performs a number of functions including control of range markers and pulse repetition frequency. If a radar cable between the antenna unit and the indicator unit is replaced, lengthened, or shortened, you need to recalibrate the sweep and timing circuits for correct readings. You know you have a timing circuit problem if a straight breakwater looks concave or convex on the screen. Consult the technical manual on how to recalibrate and synchronize the timing circuits.

8A176 Sea clutter on the radar scope cannot be effectively reduced using front panel controls. What circuit would you suspect is faulty?

A. The Sensitivity Time Control (STC) circuit
B. The False Target Eliminator (FTE) circuit
C. The Fast Time Constant (FTC) circuit
D. The Intermediate Frequency (IF) circuit

ANSWER A: Sea clutter is actually a positive indication that a marine radar is operating properly in the short-range setting. However, in heavy weather, sea clutter may sometimes cover up close-in targets such as small boats and buoys. Switching in the sensitivity time control (STC) circuit attenuates close-in echoes so you can better identify real close-in targets. Think of "STC" as *Short* timing control.

8A177 When monitoring the gate voltage of a power MOSFET in the switching power supply of a modern radar, you would expect to see the gate voltage change from "low" to "high" by how much?

A. 1 volt to 2 volts
B. 300 microVolts to 700 microvolts
C. Greater than 2 volts
D. 1.0 volt to 20.0 volts

ANSWER C: The MOSFET is the preferred transistor in switching power supplies, and the MOSFET gate voltage change is always greater than 2.0 volts.

8A178 Prior to making "power-on" measurements on a switching power supply, you should be familiar with the supply because of the following:

A. You need to know where the filter capacitors are so they can be discharged.
B. If it does not use a line isolation transformer you may destroy the supply with grounded test equipment.
C. It is not possible to cause a component failure by using ungrounded test equipment.
D. So that measurements can be made without referring to the schematic.

ANSWER B: When working on a switching power supply that is turned on, keep in mind that it may not use an isolation transformer and the internal resistance of your test equipment could destroy the transistors unless your test probe is specifically designed for use with switching power supplies.

8A179 In a radar using digital video processing, a bright, wide ring appears at a fixed distance from the center of the display on all digital ranges. The transmitter is operating normally. What receiver circuit would you suspect is causing the problem?

A. VRM circuit
B. Range ring generator
C. Video storage RAM or shift register
D. EBL circuit

ANSWER C: In diagnosing this symptom of a wide ring that appears on all digital ranges at the same distance from the center of the display, you should suspect a problem in the video random access memory (VRAM). It should generate proper range rings that change in proportion to the radar range you select.

8A180 The raster scan radar display has missing video in a rectangular block on the screen. Where is the most likely problem area?

A. Horizontal sweep circuit

B. Power supply

C. Memory area failure

D. Vertical blanking pulse

ANSWER C: When a modern raster scan radar all of a sudden develops a blank video block on the screen, part of the digitally stored signal is missing. Because it is only a small rectangular block on the screen, the most likely failure area is in the RAM or memory area. The other problems create failures of the entire video display, not a small block.

8A181 The ship's speed indication on the ARPA display can be set manually, but does not change with changes in the vessel's speed. What other indication would point to a related equipment failure?

A. "GYRO OUT" is displayed on the ARPA indicator.

B. "LOG OUT" is displayed on the ARPA indicator.

C. "TARGET LOST" is displayed on the ARPA indicator.

D. "NORTH UP" is displayed on the ARPA indicator.

ANSWER B: On an ARPA radar display, a "Log Out" error message indicates the unit is not receiving information from an external speed source such as GPS, digital paddlewheel information, or Loran. "Gyro Out" indicates a failure of heading information, not speed.

8A182 The ship's heading flash on a gyro-stabilized indicator tracks in the opposite direction of the ship's turn. What would be the most likely cause?

A. No gyro information to the indicator.

B. Two of the phases of the gyrocompass signal to the indicator are reversed.

C. The reed switch in the antenna pedestal is broken.

D. The compass drive circuits in the indicator are faulty.

ANSWER B: When the gyro wiring was connected to a terminal strip, it must have been miswired because the gyro is reading opposite of the true direction of the ship's turn. To minimize wiring errors, new radars are going exclusively to plugs and getting away from individual terminal strips.

8A183 While troubleshooting a memory problem in a raster scan radar, you discover that the "REFRESH" cycle is not operating correctly. What type of memory circuit are you working on?

A. DRAM

B. SRAM

C. ROM

D. PROM

ANSWER A: Dynamic random access memory (DRAM) requires a refresh cycle. SRAM does not because SRAM memory is changed when another bit replaces a stored bit. ROM and PROM memory are permanently stored.

8A184 Small targets in an area of very heavy swell are frequently being lost and reacquired by a collision avoidance radar. Each time a target is acquired, the antenna rotates several times before the target's course and speed can be displayed. What is the most likely cause of the delay?
   A. An intermittent connection in the CAS computer.
   B. The antenna sweep rate is affected by the sea conditions.
   C. The CAS computations require several sweeps of the antenna.
   D. The targets are moving between sweeps.
ANSWER C: The collision avoidance computations require several sweeps to confirm true targets from those of sea return. Due to the scattering of signals in heavy swells, it takes several sweeps to complete computations, and several consecutive sweeps are required from the same target to trigger the collision avoidance alarm.

8A185 An increase in the deflection on the magnetron current meter could likely be caused by:
   A. Insufficient pulse amplitude from the modulator.
   B. Too high a B+ level on the magnetron.
   C. A decrease of the magnetic field strength.
   D. A lower duty cycle, as from 0.0003 to 0.0002.
ANSWER C: In checking current at the magnetron, an increase in the deflection of the magnetron current could be caused by an older magnetron losing its permanent magnetic field.

8A186 An increase in magnetron current which coincides with a decrease in power output is an indication of what?
   A. The pulse length decreasing
   B. A high SWR
   C. A high magnetron heater voltage
   D. The external magnet weakening
ANSWER D: When a magnetron begins to grow weak, the problem is usually weakening of the permanent magnet, which results in an increase in magnetron current, a slight change in transmit frequency, and reduced power output (or no power at all).

8A187 Low or no mixer current could be caused by:
   A. A local oscillator frequency misadjustment.
   B. All of these.
   C. A TR cell failure.
   D. Mixer diode degradation.
ANSWER B: Low mixer current could be caused by the inputs of the local oscillator/TR cell failing, or by the mixer diode itself. Local oscillator frequency misadjustment can also cause low or no mixer current.

8A188 Silicon crystals are used in radar mixer and detector stages. Using an ohmmeter, how might a crystal be checked to determine if it is functional?
   A. Its resistance should be the same in both directions.
   B. Its resistance should be low in one direction and high in the opposite direction.
   C. Its resistance cannot be checked with a dc ohmmeter because the crystal acts as a rectifier.

D. It would be more appropriate to use a VTVM and measure the voltage drop across the crystal.

ANSWER B: Sometimes you are out there with nothing more than a VOM to run a basic check of a radar system that may not be operating properly. A silicon crystal looks similar to a diode if the crystal is functional; that is, resistance is low in one direction and high in the other.

8A189 In a radar unit, if the crystal mixer becomes defective, replace the:
  A. Crystal only.
  B. The crystal and the ATR tube.
  C. The crystal and the TR tube.
  D. The crystal and the klystron.

ANSWER C: The point-contact crystals in the mixer can be damaged by any current exceeding 1 mA. The TR switch (tube) helps to protect the crystal from damage. If the crystal goes out, it may have gone bad because of a defective TR tube. Check out the tube before replacing the crystal.

8A190 Radar interference on a communications receiver appears as:
  A. A varying tone.
  B. Static.
  C. A steady tone.
  D. A hissing tone.

ANSWER C: Interference from an energized radar heard over a communications receiver is a steady tone at the pulse repetition rate frequency. It sounds like a musical tone.

8A191 Radar interference to a communications receiver is eliminated by:
  A. Not operating other devices when radar is in use.
  B. Properly grounding, bonding, and shielding all units.
  C. Using a high pass filter on the power line.
  D. Using a link coupling.

ANSWER B: Grounding, bonding, and shielding all radar units keeps noise from being radiated out of the chassis.

8A192 A defective crystal in the AFC section will cause:
  A. No serious problems.
  B. Bright flashing pie sections on the PPI.
  C. Spiking on the PPI.
  D. Vertical spikes that constantly move across the screen.

ANSWER B: When a crystal goes out in the AFC section of a phosphorous PPI scope, bright flashing pie sections appear on the screen.

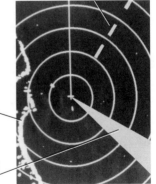

RACON
REFLECTOR
SIGNAL

LOS ANGELES
BREAKWATER

BRIGHT PIE SECTION
DUE TO DEFECTIVE
CRYSTAL

Effect of Defective Crystal
on Radar Screen

8A193 Before testing a radar transmitter, it would be a good idea to:
A. Make sure no one is on the deck.
B. Make sure the magnetron's magnetic field is far away from the magnetron.
C. Make sure there are no explosives or flammable cargo being loaded.
D. Make sure the Coast Guard has been notified.

ANSWER C: Never run a radar at the dock when the vessel is loading explosive or flammable cargo. Never energize a radar with anyone standing next to the rotating antenna unit.

8A194 How can a CRT be damaged?
A. By operating it at a high intensity
B. By operating it at a lower-than-normal intensity
C. By leaving a stationary high-intensity image on the screen
D. By operating it beyond its frequency limit

ANSWER C: You can damage both phosphorous and raster CRT's by turning up the range ring intensity too bright and allowing them to constantly run for days on end. You can almost always spot a little black burn mark in the center of radar CRT's.

8A195 If a CRT is dropped:
A. Most likely nothing will happen because they are built with durability in mind.
B. It might go out of calibration.
C. It might implode, causing damage to workers and equipment.
D. The phosphor might break loose.

ANSWER C: If a CRT is dropped, it could implode, causing glass to fly in all directions. Always handle a new or used radar CRT as if it were an expensive television picture tube. It is!

8A196 A high magnetron current indicates:
A. A defective AFC crystal.
B. A defective external magnetic field.
C. An increase in duty cycle.
D. A high standing wave ratio (SWR).

ANSWER B: If there is high magnetron current, chances are the permanent magnet inside the magnetron is getting weak.

8A197 If the TR box malfunctions:
A. The transmitter might be damaged.
B. The receiver might be damaged.
C. The klystron might be damaged.
D. Magnetron current will increase.

ANSWER B: If the TR tube malfunctions in the radar, it could cause radar energy to leak into the receiver section, permanently damaging the receiver.

8A198 How would you prolong the life of a spark gap?
A. By applying only low voltages to the unit.
B. By periodically reversing the polarity.
C. By applying extremely high voltages to it from time to time.
D. It does not matter because spark gaps are not used anymore.

ANSWER B: Periodically reversing the polarity of the spark gap circuit helps prolong the life of this tube.

8A199 If long-length transmission lines are not properly shielded and terminated:
- A. The silicon crystals can be damaged.
- B. Communications receiver interference might result.
- C. Overmodulation might result.
- D. Excessive RF loss can result.

ANSWER B: If long radar cables are not properly shielded or terminated, they radiate interference that can interfere with onboard communication receivers. This could also effect Loran reception, and quite possibly GPS reception.

8A200 You are troubleshooting a component on a printed circuit board in a RADAR system while referencing the following Truth Table in Diagram EL8-B. What kind of integrated circuit is the component?
- A. D-Type Flip-Flop, 3-State, Inverting
- B. Q-type Flip-Flop, Non-Inverting
- C. Q-type Directional Shift Register, Dual
- D. D to Q Convertor, 2-State

TRUTH TABLE

INPUTS			OUTPUT
$\overline{OE}$	CP	Dn	$\overline{Qn}$
L	⌐	H	L
L	⌐	L	H
L	L	X	No Change
H	X	X	Z

Note:
X = Don't care
Z = High impedance state
⌐ = Low-to-High transition

Diagram EL8-B

ANSWER A: The Dn column indicates this is a D-type flip-flop. OE is enable.

8A201 A circuit card in a radar system has just been replaced with a spare card. You notice the voltage level at point E in Diagram EL8-C is negative 4.75 volts when the inputs are all at 5 volts. The problem is:
- A. The 25 K resistor is open.
- B. The 100 K resistor has been mistakenly replaced with a 50 K resistor.
- C. The op amp is at the rail voltage.
- D. The 50 K resistor has been mistakenly replaced with a 25 K resistor.

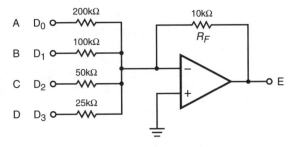

Diagram EL8-C

ANSWER D: The circuit is a summing amplifier that uses an inverting op amp operating from a power supply of $+5V$ and $-5V$. In the summing amplifiers of EL8-C, the output voltage is:

$$V_O = -R_F \left(\frac{V_A}{200 \text{ k}\Omega} + \frac{V_B}{100 \text{ k}\Omega} + \frac{V_C}{50 \text{ k}\Omega} + \frac{V_D}{25 \text{ k}\Omega} \right)$$

when $V_A = V_B = V_C = V_D = 5$ V

$$V_O = -10 \text{ k}\Omega \left(\frac{5}{200} \times 10^{-3} + \frac{5}{100} \times 10^{-3} + \frac{5}{50} \times 10^{-3} + \frac{5}{25} \times 10^{-3} \right)$$

$$V_O = -10 \times 10^3 (0.025 + 0.05 + 0.1 + 0.2) \times 10^{-3}$$
$$= -10 \times 0.375 = -3.75 \text{ V}$$

If the output $V_O = 4.75$, the current has increased by 0.1 mA, caused by the mistaken replacement of a 25-kΩ resistor for a 50-kΩ input R.

8A202 In the circuit contained in Diagram EL8-C, there are 5 volts present at points B and C, and there are zero volts present at points A and D. What is the voltage at point E?
    A. 3.75 Volts
    B. −3.75 Volts
    C. −1.5 Volts
    D. 4.5 Volts

ANSWER C: Using the formula developed for question 8A201, the output voltage is:

$$V_O = -10 \text{ k}\Omega \,(0 + 0.05 + 0.1 + 0) \times 10^{-3} = -10 \times 0.15 = -1.5 \text{ V}$$

Only inputs B and C generate input current, not A and D.

8A203 What does the schematic in Diagram EL8-D represent?
    A. A magnetron circuit
    B. A klystron oscillator
    C. An STC circuit
    D. An audio oscillator

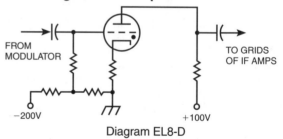

FROM MODULATOR

TO GRIDS OF IF AMPS

−200V   +100V

Diagram EL8-D

ANSWER C: This circuit in Diagram EL8-D is the sensitivity time control (STC). The purpose of this circuit is to reduce the sensitivity of the receiver to nearby targets, and yet allow the receiver to operate at full sensitivity for distant targets. This prevents sea return from nearby waves. It usually uses a thyratron tube that is filled with hydrogen—thus the black dot in the tube.

8A204 The circuit shown in Diagram EL8-E is being used as a time base generator for a CRT. How can this circuit be changed to produce the desired waveform shown?
    A. Increase the resistance of R4
    B. Increase the resistance of R2
    C. Increase the capacitance of C3
    D. Decrease the resistance of R2

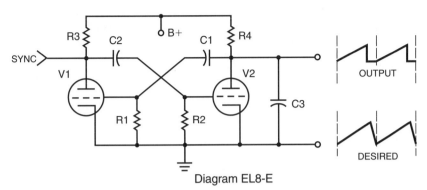

Diagram EL8-E

ANSWER B: Capacitors and resistors in series are often used in circuits with non-sinusoidal wave forms, such as the pulse-producing circuit in Diagram EL8-E. If you keep in mind that the time it takes for a capacitor to charge is proportional to resistance times capacitance, it's easy to see how changing the value of the resistor affects the waveform. If you increase R2's value, it takes longer for the capacitor C2 to discharge, so the waveform's fall time is not as steep. Because R3 and R4 are on the input side of the circuit, changing their values affects the rise time of the waveform instead of the fall time. Increasing the capacitance affects only the peak position in the waveform.

8A205 It is desired to modify the circuit in Diagram EL8-E to produce trapezoidal sweep voltages for an electromagnetic CRT. This can be accomplished by:
- A. Increasing the resistance of R4.
- B. Increasing the resistance of R3.
- C. Connecting a resistor between C3 and ground.
- D. Connecting another capacitor in series with C3.

ANSWER C: Placing a resistor between C3 and ground increases the RC time to charge C3 and causes the capacitor to charge or discharge slower. A trapezoidial waveform results because the sweep is slower and fall time is longer.

8A206 The two sinewaves in the Diagram EL8-F are being applied to the vertical and horizontal CRT deflection plates as shown. The display on the CRT screen will be:
- A. A straight line, inclined to the right.
- B. A straight line, inclined to the left.
- C. An ellipse.
- D. A circle.

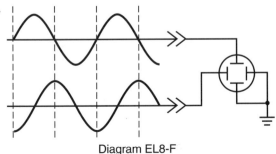

Diagram EL8-F

ANSWER D: Lissajous figures are produced by placing sinusoidal ac voltages on the deflection plates. If both are the same frequency ratio, and are 90 degrees out of phase, the display output is a circle.

8A207 The diagram in Diagram EL8-G shows a simplified radar mixer circuit using a crystal diode as the first detector. What is the output of the circuit when no echoes are being received?

    A. 60 MHz CW
    B. 4095 MHz CW
    C. 4155 MHz CW
    D. No output is developed

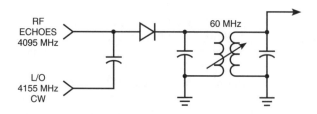

Diagram EL8-G

ANSWER D: When no echoes are received, the output from the circuit is zero because there is no output from the first detector.

8A208 In the circuit shown in Diagram EL8-H, the diode current is limited by:

    A. Power supply.
    B. R38.
    C. R37.
    D. The transistor.

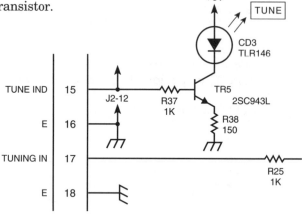

Diagram EL8-H

ANSWER B: The current path is from the +5 volt dc source to ground through the transistor and resistor R38. When the transistor is on, current is limited by that resistor.

8A209 The circuit shown in Diagram EL8-I is the output of a switching power supply. Measuring from the junction of CR6 and L1 to A GND with an oscilloscope, what waveform would you expect to see?
> A. Filtered DC
> B. Pulsating DC at line frequency
> C. AC at line frequency
> D. Pulsating DC much higher than line frequency

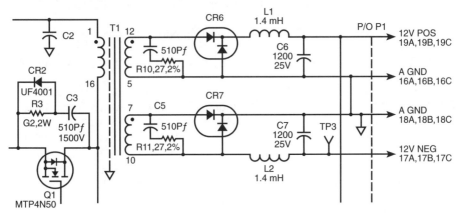

Diagram EL8-I

ANSWER D: Switching power supplies have eliminated the need for bulky transformers in radar equipment. Transistor "chopper" circuits create the pulsating dc, and what you see with an oscilloscope at the junction of CR6 and L1 to A GND is this pulsating dc at a much higher frequency than the line input. The down side of switching power supply circuits is the broad-band noise that is sometimes radiated to other onboard marine electronic navigation receivers.

8A210 With regard to the comparator shown in Diagram EL8-J, the input is a sinusoid. Nominal high level output of the comparator is 4.5 volts. Choose the most correct statement regarding the input and output.
> A. The rising edge of the output waveform trails the positive zero crossing of the input waveform by 45 degrees.
> B. The rising edge of the output waveform trails the negative zero crossing of the input waveform by 45 degrees.
> C. The rising edge of the output waveform trails the positive peak of the input waveform by 45 degrees.
> D. The leading edge of the output waveform occurs 180 degrees after positive zero crossing of the input waveform.

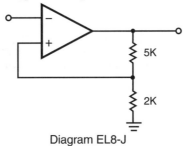

Diagram EL8-J

ANSWER D: The sinusoidal input is connected to the negative terminal. When the input sinewave goes positive, the output goes negative and is a square wave. In comparison, the two waveforms are 180 degrees out of phase.

8A211 The nominal output high of the comparator shown in Diagram EL8-J is 4.5 volts. Choose the most correct statement which describes the trip points.

    A. Upper trip point is 4.5 volts. Lower trip point is approximately 0 volts.

    B. Upper trip point is 2.5 volts. Lower trip point is approximately 2.0 volts.

    C. Upper trip point is 900 microvolts. Lower trip point is approximately 0 volts.

    D. Upper trip point is +1.285 volts. Lower trip point is −1.285 volts.

ANSWER D: In a comparator shown in Diagram EL8-J, we have assigned R_1 to the upper resistor and R_2 to the lower resistor. The comparator operates from a +5V and −5V supply. The output swings from +4.5V to −4.5V. The trip points are:

$$V_{UT} = \frac{R_2}{R_1 + R_2} V_{O(MAX)} \qquad V_{LT} = \frac{R_2}{R_1 + R_2}(-V_{O(MAX)})$$

$$V_{UT} = \frac{2{,}000}{5000 + 2000} \times 4.5 = 1.285V \qquad V_{LT} = \frac{2{,}000}{5000 + 2000} \times (-4.5) = -1.285V$$

8A212 In the circuit shown in Diagram EL8-K, what will be the output of the circuit?

    A. $\overline{ABC}$

    B. $\overline{AB} + CB$

    C. $AB + C$

    D. $ABC + \overline{BC}$

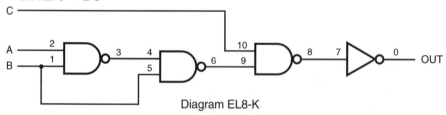

Diagram EL8-K

ANSWER D: DeMorgan's laws say: $\overline{A+B} = \overline{A} \cdot \overline{B}$ and $\overline{A \cdot B} = \overline{A} + \overline{B}$. Therefore, using these laws, the input at pin 9 will be $A \cdot B + \overline{B}$. Since pin 10 input = C, output 8 will be $\overline{(A \cdot B + \overline{B}) \cdot C}$. Output 0, since it is output 8 inverted, is $(A \cdot B + \overline{B}) \cdot C$ or $ABC + \overline{B}C$.

8A213 In the circuit shown in Diagram EL8-L, U5 pins 1 and 4 are high and both are in the reset state. Assume one clock cycle occurs of Clk A followed by one cycle of Clk B. What are the output states of the two 'D' type flip flops?

    A. Pin 5 low, Pin 9 low

    B. Pin 5 high, Pin 9 low

    C. Pin 5 low, Pin 9 high

    D. Pin 5 high, Pin 9 high

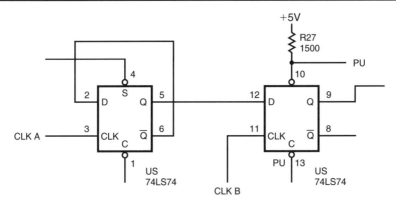

Diagram EL8-L

ANSWER D: Pins 1 and 4 do not have an effect on the logic condition. At the start, pins 5 and 9 are low, and pins 6 and 8 are high. One clock cycle of Clk A causes pins 5 and 12 to take the state of the D input on Pin 2, a high. Pins 6 and 2 go low with no effect. On the clock cycle at Clk B, pin 9 takes the state of 12, which is high.

8A214 Choose the selection from the truth table which is correct for the circuit shown in Diagram EL8-M.

    P Q R S
A. 0 1 0 1
B. 0 1 1 1
C. 1 0 0 1
D. 1 1 1 1

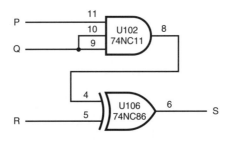

Diagram EL8-M

ANSWER B: For S = 1 out of a two-input XOR gate, both inputs must be different. For the three-input AND gate, the inputs must all be 1 for an output of 1. So, with P = 0, Q = 1, the output of the AND gate is zero. For S = 1, then R = 1.

8A215 In the circuit shown in Diagram EL8-N, which of the following is true?

A. With A and B high, Q1 is saturated and Q2 is off.
B. With either A or B low, Q1 is saturated and Q2 is off.
C. With A and B low, Q2 is on and Q4 is off.
D. With either A or B low, Q1 is off and Q2 is on.

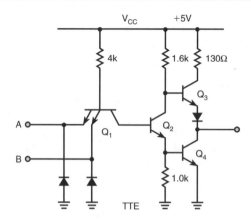

Diagram EL8-N

ANSWER B: When either A or B is low, Q_1 is turned on and saturates, which pulls the base of Q_2 low, turning off Q_2.

8A216 The block diagram of a typical radar system microprocessor is shown in Diagram EL8-O. Choose the most correct statement regarding this system.
   A. The ALU is used for address decoding.
   B. General registers are used for arithmetic manipulations.
   C. The control unit executes arithmetic manipulations.
   D. Address pointers are contained in the general registers.

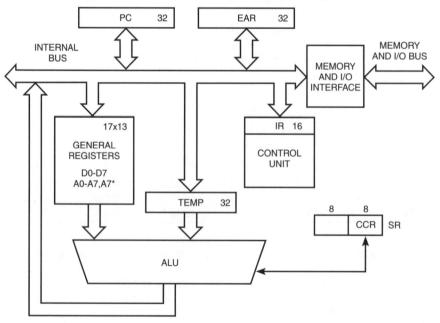

Diagram EL8-O

ANSWER D: Radar system microprocessors are very similar to home personal computers, or PCs. A microprocessor, by definition, is an integrated circuit, or a circuit that contains all the computational and control circuitry for a small computer. It can both input and output information on data pins which can be "bussed" to other circuitry or devices. Moving through a series of steps, information is sent through the data pins and then either inputted or bussed to other areas to perform certain tasks. The information is encoded to a unique pattern of "1's" and "0's and are arranged in either 4, 8, 16, or 32-bit groups depending on whether it is a 4-bit, 8-bit, 16-bit, or 32-bit microprocessor. The encoded information is then sent along these grouped data lines. The signal lines are grouped into 3 categories: data bus lines, address bus lines, and control bus lines. The address lines are controlled by a program counter in the general registers which "point" or sequence the counting of the "1's" and "0's" and provide the instructions to the radar unit. Addressing then, is program controlled and it is in the general registers where the radar's instructions are generated. It should be noted, however, that this sequence is normally altered by information brought into the microprocessor, so that the radar can be "programmed" to meet specific needs.

8A217 The block diagram of a typical radar system microprocessor is shown in Diagram EL8-O. Choose the most correct statement regarding this system.
   A. The ALU is used for address decoding.
   B. General registers are used for arithmetic manipulations.
   C. The ALU executes arithmetic manipulations.
   D. Address pointers are contained in the control unit.

ANSWER C: The ALU or Arithmetic Logic Unit is the computer's "brain", capable of adding, subtracting, multiplying and dividing binary numbers which a microprocessor executes according to its program. The ALU can be thought of as the "thinking" part of a computer system.

8A218 The block diagram of a typical radar system microprocessor is shown in Diagram EL8-O. Choose the most correct statement regarding this system.
   A. The ALU is used for address decoding.
   B. The Memory and I/O communicate with peripherals.
   C. The control unit executes arithmetic manipulations.
   D. The internal bus is used simultaneously by all units.

ANSWER B: The two memory systems, along with the CPU (Central Processing Unit), provide the operating instructions to the radar system. The I/O (Input/Output) device is primarily responsible for data transfers between the CPU and all other supporting circuits outside the CPU. It can be seen from the diagram that the memory and I/O buss will receive input from GPS, LORAN, GYN and boat speed in a marine system. Refer to the diagram in question 8A216.

8A219 In this Line-Driver/Coax/Line-receiver circuit shown in Diagram EL8-P, what component is represented by the blank box?
   A. 25-ohm resistor
   B. 51-ohm resistor
   C. 10-microhm inductor
   D. 20-microhm inductor

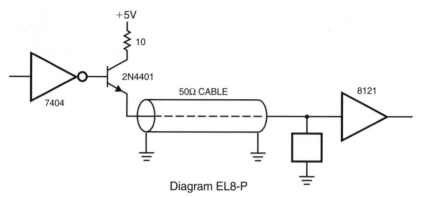

Diagram EL8-P

ANSWER B: A 51-ohm resistor is required to match the impedance of the cable. This allows efficient signal transfer and maintains a clean waveform.

8A220 With reference to the schematic shown in Diagram EL8-Q, choose the most correct statement.
   A. This is a low voltage circuit.
   B. The anode of V201 carries high voltage.
   C. The filament of V201 carries dangerous voltages.
   D. This is a video amplifier.

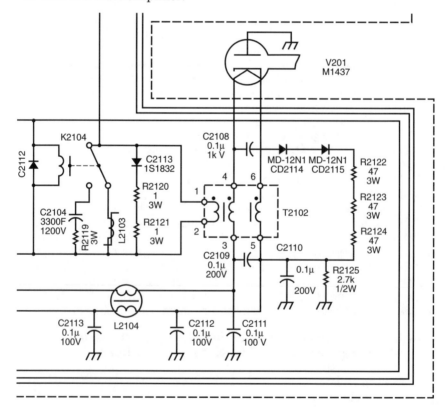

Diagram EL8-Q

ANSWER C: The 1-kilovolt input to the transformer and the isolation of the rectifier should tell you this is a high-voltage power supply. And the anode of the rectifier is grounded, so its voltage must be 0. This leave only answer C as a possibility.

8A221 If more light strikes the photodiode in Diagram EL8-R, there will be:
- A. More diode current.
- B. Less diode current.
- C. No change in diode current.
- D. There is wrong polarity on the diode.

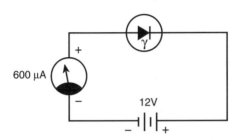

Diagram EL8-R

ANSWER A: More diode current passes when photons (light) strike the photo-diode.

8A222 What is the correct value of Rs in Diagram EL8-S, if the voltage across the LED is 1.9 Volts with 5 Volts applied and I_f max equals 40 milliamps?
- A. 75 ohms
- B. 4700 ohms
- C. 155 ohms
- D. 10000 ohms

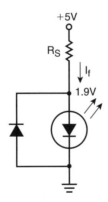

Diagram EL8-S

ANSWER A: If 1.9 volts is applied across the diode, then 3.1 volts is applied across R_s in Diagram EL8-S. Even though the forward bias current is typically between 10 and 20 mA, they expect you to use the stated maximum current of 40 milliamps. So 3.1 ÷ 0.04 = 77.5 ohms. A is the closest answer.

8A223 Visible light is measured between:
A. 100 and 1000 nanometers.
B. 10 and 100 nanometers.
C. 0.1 and 10 nanometers.
D. 1 and 10 nanometers.

ANSWER A: Visible light is measured between 100 and 1,000 nanometers.

8A224 What is a typical forward bias voltage across an LED when it is emitting light?
A. 0.3 volts
B. 0.7 volts
C. 1.2 volts
D. 1.7 volts

ANSWER D: A forward bias of 1.7 volts develops a little more than 10 milliamps of current through the light-emitting diode. This results in a visible glow. Twenty (20) milliamps causes an LED to glow very brightly, and 50 to 75 milliamps is the maximum you should ever run through an LED to illuminate it.

8A225 Laser light is produced by:
A. Stimulated emission.
B. Spontaneous emission.
C. Black magic.
D. Electricity.

ANSWER A: The type of light produced by a laser is called "stimulated emission." Look at the incorrect answers, and you won't forget "stimulated emission."

8A226 What is the primary purpose of an optical coupler?
A. To equalize voltage levels
B. To reduce power consumption
C. To provide voltage isolation
D. To couple optical signals

ANSWER C: An optical coupler provides excellent voltage isolation where the typical resistance is in the megohm area and capacitance is less than a pico-farad. Optical couplers are used in variable frequency oscillator circuits where spinning a radio dial causes other circuits to sense changes in frequency.

8A227 The voltage drop across a LED rated at 1300 nm is:
A. 1 volt.
B. 0 volts.
C. 1.5 volts.
D. 3 volts.

ANSWER A: An LED emits light when it is forward biased at a voltage greater than its band-gap voltage—1.5 volts for GaAlAs LEDs emitting at 800 to 900 nm and 1 volt for InGaAs LEDs emitting at 1300 nm. As with other forward-biased semiconductors, the voltage drop across the diode is considered to be a constant regardless of the applied current. To modulate the light output, the drive current is modulated, not the voltage. Modulation of voltage fed to a transistor causes the current passing through the LED, and the LEDs light output, to vary. Thus, for a 1300 nm LED, a voltage drop of 1 volt can be thought of as a constant.

8A228 Optical fibers used to carry radar signals are made of:
   A. Glass coated with plastic.
   B. Ultrapure glass.
   C. Plastic.
   D. All of these.

ANSWER D: Fiber optics are found in aeronautical and marine radio installations where a remote *head* in the cockpit controls the transmitting and receiving equipment located far away. The fiber optic cable has less loss than coax and is also immune to electrical noise. Fiber optics can be glass-coated with plastic, pure glass, plastic, or as Answer D describes, all of these.

8A229 Light emissions from an LED are modulated by:
   A. Voltages applied across the diode.
   B. Current passing through the diode.
   C. Illumination of the diode.
   D. An indirect voltage surge.

ANSWER B: Passing 5 milliamps or more of current through a light-emitting diode changes its intensity. Changing the intensity of an LED is a type of lightwave modulation.

8A230 Which wavelengths are standard for attenuation measurements in fiber optic cable?
   A. 850 nm
   B. 1300 nm
   C. 1550 nm
   D. All of these.

ANSWER D: LEDs (light-emitting diodes) that emit known wavelengths are used as test sources when measuring attenuation in fiber optic cables. LEDs with a 660 nm wavelength are used for plastic fibers. Short wavelengths (820, 850, and 870 nm) and longer wavelengths (1300 and 1550 nm) are also used. Having a known wavelength as a test source is critical in determining the loss (attenuation) caused by cracks, splices, and/or connectors along the cable.

Because each LED emits only one wavelength, many different test sources are needed to do complete testing. The light from the source might be continuous or modulated, depending on the test.

Laser light can also be used for testing fiber optics, but LEDs are less expensive and require much less power to operate.

8A231 The response time of a fiber optic system, (like those used to carry radar signal information), is:
   A. The square root of the sum of the squares of the response times of its individual components.
   B. The sum of the response times of its individual components.
   C. The average response time of its individual components.
   D. Not directly dependent upon individual response times.

ANSWER A: Since we are dealing with a round optical cable, we will use the square root of the sum of the squares of the response times of each of the individual components in the chain to obtain the total response time.

$$t_r = \sqrt{t_1^2 + t_2^2 + t_3^2 + ...}$$

8A232 What is the usual modulation method for fiber optic transmitters used to carry radar signal information?
A. Intensity modulation
B. Frequency modulation
C. Wavelength modulation
D. Voltage modulation
ANSWER A: By changing the current to an illuminated LED in response to electrical signals from an opto coupler, you create intensity modulation.

8A233 If the rise time of a radar system's fiber optic signal transmitter is 1 nanosecond, what is its theoretical bandwidth?
A. 1 MHz
B. 100 MHz
C. 350 MHz
D. 350 kHz
ANSWER C: Rise time for fiber optic signal transmitters can be defined as the time it takes light output to rise from 10% to 90% of its steady-state level. Fall time is the inverse, or the time it takes light output to drop from 90% to 10%. If the rise time and the fall times are equal (which we are assuming in this question), the formula for approximating bandwidth is:

    Bandwidth = 0.35/RISE TIME   Where:
BW = Bandwidth in **MHz**
Rise time = t_r in
microseconds

Thus, when t_r = 1 nanosecond = 1×10^{-9} = 0.001 microseconds

BW = 0.35/.001
= 350 MHz   Answer C.

8A234 What circuit element receives the drive voltage in a radar system's fiber optic signal transmitter?
A. A filter capacitor
B. A load-limiting resistor
C. A temperature sensor
D. A transistor
ANSWER D: It is the role of a transistor to receive the drive voltage in a radar system's fiber optic signal transmitter.

8A235 What is the special advantage of using plenum fiber optic cables to carry radar signal information aboard maritime vessels?
A. They are small in diameter and easily fit through wire-ways.
B. They meet stringent fire codes for running through wire-ways.
C. They are crush-resistant and can withstand 10,000 pounds per square inch of pressure.
D. They have special armor coatings that keep rodents from damaging them.
ANSWER B: The benefit of using plenum fiber optic cables to carry radar signals is that plenum fiber optics meet stringent fire codes enforced aboard large commercial ships.

8A236 Which of the following is *not* a type of multiplexing used in radar receiver signal transmitters sent through optical fibers?
   A. Wavelength-division multiplexing, with pulses transmitted at different wavelengths
   B. Time-division multiplexing, with pulses transmitted at different times
   C. Space-division multiplexing, with pulses transmitted at different positions in the fiber
   D. Frequency-division multiplexing, with analog signals transmitted at different frequencies
ANSWER C: This question asks which is *not* a type of multiplexing in fiber optics. Wavelength-division multiplexing, time-division multiplexing, frequency-division multiplexing are familiar methods used with fiber optics, but "space-division" multiplexing with pulses at different positions in the fiber is not a viable way of sending information down the fiber optic cable.

8A237 A radio wave will travel a distance of three nautical miles in:
   A. 6.17 microseconds
   B. 18.51 microseconds
   C. 37.0 microseconds
   D. 22.76 microseconds
ANSWER B: Radar and radio waves travel at approximately the speed of light; it takes 6.17 µs for a radar wave to travel one nautical mile from the transmitter. To travel three nautical miles, it would take 18.51 (6.17 × 3) microseconds to travel this distance, and your correct answer is 18.51 microseconds.
    The way you can remember the number 6 in microseconds per nautical mile is never to get "seasix".

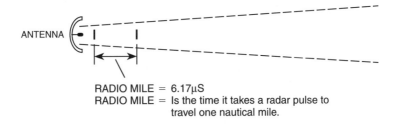

RADIO MILE = 6.17µS
RADIO MILE = Is the time it takes a radar pulse to travel one nautical mile.

Radio Mile

8A238 One radar mile is how many microseconds?
   A. 6.2
   B. 12.34
   C. 528.0
   D. .186
ANSWER B: It takes a radar pulse 6.17 microseconds to travel out one mile. One "radar mile" considers the echo return, so 2 × 6.17 = 12.34 microseconds. Be careful of answer A.

    Radar Mile = 2 × 6.17 µs   Where: Radar Mile = Time out and back to target at one nautical mile in µs.

8A239 If a target is 5 miles away, how long does it take for the radar echo to be received back at the antenna?
  A. 51.4 microseconds
  B. 61.7 microseconds
  C. 123 microseconds
  D. 30.75 microseconds

ANSWER B: If the target is 5 miles away, the radar wave must travel a total of 10 miles—5 miles to the target, and 5 miles return as an echo. Ten nautical miles times 6.17 microseconds per mile gives us about 61.7 microseconds. Watch out for Answer D; this is *not* a one-way trip!

8A240 How long would it take for a radar pulse to travel to a target 10 nautical miles away and return to the radar receiver?
  A. 123.4 microseconds
  B. 12.34 microseconds
  C. 1.234 microseconds
  D. 10 microseconds

ANSWER A: Be sure to read these types of questions carefully to see whether or not they want the range to the target, or the range to the target and back again as a time delay. It takes 6.17 microseconds for the radar wave to travel 1 nautical mile, or 10 nautical miles to the target in 61.7 microseconds, and the return echo would take another 61.7 microseconds, or a total time delay of 123.4 microseconds. Answer A is the best choice. Remember, up and back means times two of the target distance.

8A241 What is the distance in nautical miles to a target if it takes 308.5 microseconds for the radar pulse to travel from the radar antenna to the target and back.
  A. 12.5 nautical miles
  B. 25 nautical miles
  C. 50 nautical miles
  D. 2.5 nautical miles

ANSWER B: Divide 308.5 microseconds by 6.17 to find the total distance the signal travels (to the target and back to the radar receiver). Then, divide the answer by 2 to find the one-way distance to the target.

$$308.5 \div 6.17 = 50 \div 2 = 25$$

Answer B is the one!

8A242 If the PRF is 2500 Hz, what is the PRI?
  A. 40 microseconds
  B. 400 microseconds
  C. 250 microseconds
  D. 800 microseconds

ANSWER B: Pulse repetition interval (PRI) = 1 ÷ pulse repetition rate = 1/2,500 = 400 μ seconds.

PRF, pulse repetition frequency and pulse repetition rate (PRR) are the same thing — *the number of pulses per second.* Pulse repetition interval (PRI), sometimes called pulse repetition time (PRT), is 1/PRF, or the period of the pulse repetition frequency. PRI is the time from the leading edge of a pulse to the leading edge of the next pulse.

PRI = 1/PRF = 1 ÷ 2500 = 0.4 x 10^{-3} = 0.0004 = 400 × 10^{-6}

The period of the pulses is 400 microseconds

8A243 If the pulse repetition frequency (PRF) is 2000 Hz, what is the pulse repetition interval (PRI)?
  A. 0.05 seconds
  B. 0.005 seconds
  C. 0.0005 seconds
  D. 0.00005 seconds
ANSWER C: To calculate pulse repetition interval (PRI), divide 1 second by the pulse repetition frequency:

  1 ÷ 2000 cycles/second = 0.0005 seconds.

8A244 If the radar unit has a pulse repetition frequency (PRF) of 2000 Hz and a pulse width of 0.05 microseconds, what is the duty cycle?
  A. 0.0005
  B. 0.0001
  C. 0.05
  D. 0.001
ANSWER B: Here the radar has a pulse repetition rate of 2,000 Hz, and a pulse width of 0.05 microseconds. The duty cycle is calculated by dividing pulse width (0.05 × 10^{-6} seconds) by pulse repetition interval (1/2000):

  0.05 × 10^{-6} ÷ 1/2000 = 0.0001

8A245 A shipboard radar transmitter has a pulse repetition frequency (PRF) of 1,000 Hz, a pulse width of 0.5 microseconds, peak power of 150 KW, and a minimum range of 75 meters. Its duty cycle is:
  A. 0.05
  B. 0.005
  C. 0.0005
  D. 0.00005
ANSWER C: Duty cycle = pulse width ÷ pulse repetition interval. Because the pulse repetition interval equals 1/PRF. Then the duty cycle is equal to the pulse width times PRF.

  (0.5 × 10^{-6}) × (1000) = 0.5 × 10^{-3} = 0.0005

150 KW and 75 meters are not needed for the calculation.

8A246 A pulse radar has a pulse repetition frequency (PRF) of 400 Hz, a pulse width of 1 microsecond, and a peak power of 100 kilowatts. The average power of the radar transmitter is:
  A. 25 watts.
  B. 40 watts.
  C. 250 watts.
  D. 400 watts.
ANSWER B: Duty cycle = pulse width ÷ pulse repetition interval. Since the pulse repetition interval equals 1/PRF, then the duty cycle is equal to the pulse width times PRF duty cycle = PW × PRF, Then, 1 × 10^{-6} × 400 = 4 × 10^{-4} = 0.0004. Average power is equal to peak power times duty cycle, or 0.0004 × 100,000 = 40 watts.

8A247 A pulse radar transmits a 0.5 microsecond RF pulse with a peak power of 100 kilowatts every 1600 microseconds. This radar has:
  A. An average power of 31.25 watts.
  B. A PRF of 3200.
  C. A maximum range of 480 kilometers.
  D. A duty cycle of 3.125 percent.
ANSWER A: Duty cycle is equal ($0.5 \times 10^{-6} \div 1600 \times 10^{-6} = 0.5 \times 10^{-6} \div 1.6 \times 10^{-3} = 0.0003125$). Average power equals 100 KW $\times$ .0003125 = 31.125 watts.

8A248 If a radar transmitter has a pulse repetition frequency (PRF) of 900 Hz, a pulse width of 0.5 microseconds and a peak power of 15 kilowatts, what is its average power output?
  A. 15 kilowatts
  B. 13.5 watts
  C. 6.75 watts
  D. 166.67 watts
ANSWER C: Average power output = peak power $\times$ pulse repetition frequency $\times$ pulse width.

$$15,000 \times 900 \times (0.5 \times 10^{-6}) = 1.5 \times 10^4 \times 9 \times 10^2 \times 0.5 \times 10^{-6} = 6.75 \text{ watts}$$

Everything gets multiplied together, but remember 0.5 microseconds is 0.5×10^{-6}.

8A249 What is the average power if the radar set has a PRF of 1000 Hz, a pulse width of 1 microsecond, and a peak power rating of 100 kilowatts?
  A. 10 watts
  B. 100 watts
  C. 1000 watts
  D. None of these
ANSWER B: Multiply 100,000 watts $\times$ 1000 Hz $\times$ (1.0×10^{-6} seconds) = 100 watts. See the previous question for more details.

8A250 A search radar has a pulse width of 1.0 microsecond, a pulse repetition frequency (PRF) of 900 Hz, and an average power of 18 watts. The unit's peak power is:
  A. 200 kilowatts
  B. 180 kilowatts
  C. 20 kilowatts
  D. 2 kilowatts
ANSWER C: Peak power = average power $\div$ duty cycle. The duty cycle is 900 $\times$ (1×10^{-6}) = 0.0009 seconds. 18 watts $\div$ 0.0009 = 20,000 watts, or 20 KW as the correct answer.

8A251 For a range of 10 nautical miles, the radar pulse repetition frequency (PRF) should be:
  A. Approximately 8.1 kHz or less.
  B. 900 Hz.
  C. 18.1 kHz or more.
  D. 120.3 microseconds.
ANSWER A: Let's first calculate the amount of time for a radar wave to travel 10 miles out and 10 miles back:

$$10 \times 2 \times 6.17 = 123.4 \text{ microseconds}$$

The maximum pulse repetition frequency required to allow that signal to return is 1 divided by the time it takes the pulse to travel (in seconds), or

$$1 \div 0.0001234 \text{ seconds} = 8.104 \text{ kHz.}$$

8.1 kHz "or more" isn't right because a higher pulse repetition rate would not have enough time for the pulse to return to the receiver before another pulse was sent.

8A252 For a range of 30 nautical miles, the radar pulse repetition frequency should be:
    A. .27 kHz or less.
    B. 27 kHz or more.
    C. 2.7 kHz or less.
    D. 2.7 Hz or more.
ANSWER C: First calculate the amount of time for a radar wave to travel 30 nautical miles out, and 30 nautical miles back:

$$30 \times 2 \times 6.17 = 370.2 \text{ microseconds}$$

The maximum pulse repetition frequency required to allow that signal to return is 1 divided by the pulse time in seconds, or

$$1 \div 0.0003702 \text{ seconds} = 2701 \text{ Hz, which converted to kHz is 2.7 kHz,}$$
or less.

Watch out for answers that say, "or more." They would not work because the pulse repetition rate would not have enough time for the pulse to return to the receiver before another pulse was sent.

8A253 For a range of 5 nautical miles, the radar pulse repetition frequency should be:
    A. 16.2 Hz or more.
    B. 16.2 MHz or less.
    C. 1.62 kHz or more.
    D. 16.2 kHz or less.
ANSWER D: $5 \times 2 \times 6.17 = 61.7$ microseconds

$$1 \div 0.0000618 \text{ seconds} = 16181 \text{ Hz} = 16.2 \text{ kHz, } \textit{or less}$$

8A254 For a range of 100 nautical miles, the radar pulse repetition frequency should be:
    A. 8.1 kHz or less.
    B. 810 Hz or less.
    C. .81 MHz or more.
    D. 810 kHz or more.
ANSWER B: $100 \times 2 \times 6.17 = 1234$ microseconds

$$1 \div 0.001234 \text{ seconds} = 810 \text{ Hz, } \textit{or less}$$

8A255 U.S. Regulations limit exposure to microwave energy to a power density of 5 mW/centimeters squared. What is the average energy density transmitted by a radar across a one square foot surface area with the following pulse parameters: 1000 pulses per second, 55 kilowatts peak power, and a 3-microsecond pulse width? (Assume all the RF is focused on the 1 square-foot surface.)

    A. 55 watts/centimeters squared
    B. 178 miliwatts/centimeters squared
    C. The answer cannot be estimated
    D. 5 miliwatts/centimeters squared

ANSWER B: The American National Standards Institute (ANSI) limits the exposure to radar microwave energy to a power density of 5 milliwatts (mW) per square centimeter/per second. To calculate the average power density across a one square foot surface area, we first need to calculate the average power. The pulse width times the PRF gives the duty cycle in seconds ($0.000003 \times 1,000 = 0.003$ seconds.)

The average power is then calculated by multiplying the peak power by the duty cycle ($55,000 \times 0.003 = 165$ watts).

If all the power is focused on one square foot, you need to know how many square centimeters there are in a square foot ($(2.54 \times 12)^2 = 929$.)

Now, simply divide the average power by the centimeters in a square foot to find the amount of power per square centimeter (165 watts $\div$ 929 square cm $= 0.178$ W/cm^2 = 178 mW/cm^2. This is obviously more than the maximum exposure limit!

8A256 You are asked to determine if it is safe for maintenance personnel to work immediately in front of the antenna of an air traffic control SSR while it is operating. The radar transmits at a frequency of 1030 MHz with a peak power of 3 KW, a pulse width of 2.0 microseconds and a pulse repetition rate (PRR) of 250 Hz with an antenna whose dimensions are 1.5 meters high and 8.54 meters wide. What is the average power density immediately in front of the antenna?

    A. 2.05 milliwatts/centimeters squared
    B. .012 milliwatts/centimeters squared
    C. 2.05 watts/centimeters squared
    D. .73 milliwatts/centimeters squared

ANSWER B: The radar's duty cycle is $(2.0 \times 10^{-6}) \times 250$ Hz $= 0.0005$ seconds. The average power is 3,000 watts $\times$ 0.0005 seconds $= 1.5$ watts/sec $= 1500$ mW/sec.

The area of the antenna is 150cm $\times$ 854cm $= 128,100$ cm^2.

Now it is a simple matter to divide the power per seconds (in miliwatts) by number of square centimeters to get the answer (1,500 cm^2 mW/sec $\div$ 128,100 cm^2 = 0.0117 mW/sec per cm^2).

8A257 A radar operating at a frequency of 3.5 GHz has a wavelength of approximately:

    A. 0.86 centimeters.
    B. 8.6 centimeters.
    C. 0.8 meters.
    D. 0.86 meters.

ANSWER B: Wavelength in cm = 30,000/frequency in MHz ($30,000 \div 3,500 = 8.6$ cm). Remember for this formula the frequency is in MHz.

8A258 A radar transmitting at 6 GHz has a wavelength of:
  A. 0.5 cm.
  B. 5 cm.
  C. 50 cm.
  D. 500 cm.
ANSWER B: Use the formula at 8A257. 30,000 ÷ 6,000 = 5 cm.

8A259 How long is a half wavelength at 5400 MHz?
  A. 5.5 cm
  B. 2.7 cm
  C. 11 cm
  D. 55 cm
ANSWER B: Wavelength in centimeters is equal to 30,000 ÷ frequency in MHz.
30,000 ÷ 5,400 = 5.5 cm

  Half wavelength = 5.5 ÷ 2 = 2.7 cm

8A260 At the operating frequency of 3000 MHz, what is the distance between the waveguide and the receiver in Diagram EL8-T?
  A. 10 cm
  B. 5 cm
  C. 2.5 cm
  D. 1.25 cm

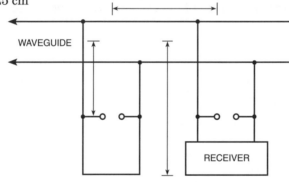

Diagram EL8-T

ANSWER B: In an S-band radar operating near 3,000 MHz, the distance between the waveguide and the receiver is 1/2 wavelength.

  (30,000 ÷ Frequency in MHz) × 0.5 = Distance
  (30,000 ÷ 3,000) × 0.5 = 5 cm.

8A261 At the operating frequency of 3000 MHz, what is the distance between the waveguide and the ATR spark gap in Diagram EL8-T?
  A. 10 cm
  B. 5 cm
  C. 2.5 cm
  D. 1.25 cm
ANSWER C: The distance to the spark gap in Diagram EL8-T is 1/4 wavelength from the waveguide.

  (30,000 ÷ Frequency in MHz) × 0.25 = Distance
  (30,000 ÷ 3,000) × 0.25 = 2.5 cm.

8A262 If the operating radar frequency is 3000 MHz, what is the distance between the waveguide and the spark gaps in older radar units?
  A. 10 cm
  B. 5 cm
  C. 2.5 cm
  D. 20 cm
ANSWER C: For an S-band radar operating at 3,000 MHz, the distance between the waveguide and the spark gap is 1/4 wavelength. Wavelength (cm) = 30,000 ÷ frequency in Mhz (30,000 ÷ 3,000 = 10 cm). One-quarter wavelength is 10 cm ÷ 4 = 2.5 cm.

8A263 The wide dimension of a rectangular waveguide for a given frequency band is:
  A. Approximately 1/4 wavelength.
  B. Approximately 1/2 wavelength.
  C. Dependent on pulse width.
  D. Not dependent upon the frequency band in use.
ANSWER B: The width of a waveguide is approximately 1/2 wavelength. In actual practice, the narrow dimension is about 1/2 the wide dimension. These dimensions provide only one dominant mode at the lowest operating frequency.

8A264 A CW radar is operating on 1.5 GHz. If this radar uses rectangular waveguide, the wide dimension of the waveguide will measure:
  A. 10 cm.
  B. 20 cm.
  C. 30 cm.
  D. 40 cm.
ANSWER A: The wide dimension is the wavelength divided by 2, which equals (30,000 ÷ 1500) ÷ 2 or 20 ÷ 2 = 10 cm.

8A265 Which of these conditions would require a waveguide with the smallest physical dimensions?
  A. A frequency of 5 GHz
  B. A frequency of 3500 MHz
  C. A wavelength of 10 cm
  D. A wavelength of 0.08 meters
ANSWER A: The highest frequency (and shortest wavelength) requires the smallest waveguide. In this question, the highest frequency is 5 GHz. The shortest wavelength is 10 cm, but 10 cm converts to 3 GHz.

8A266 A shipboard raster scan radar has a CRT with the following characteristics: 70 pixels per character, 80 characters per line, 25 lines per screen and it scans 100 screens per second. What is the minimum required bandwidth for the electron beam control signal?
  A. 210 MHz
  B. 0.21 MHz
  C. 2.1 MHz
  D. 21 MHz

ANSWER D: The frequency of the pulse train (F) equals (N/2)/T where N is the number of pixels on the screen and T is time to complete a full screen scan.

   $$F = \frac{N/2}{T}$$

Number of pixels (N) on the screen is 70 × 80 × 25 = 140,000 pixels. Scan time (T) is 1/100 = 0.01.

To get good scan definition, harmonics of the third order must be accommodated, so 3 F = BW. Since F = (140,000 ÷ 2)/0.01 = 7 MHz, the required bandwidth is 3 × 7 = 21 MHz.

8A267 A shipboard raster scan radar has a CRT with the following characteristics: 70 pixels per character, 80 characters per line, 25 lines per screen and it scans 60 screens per second. What is the minimum required bandwidth for the electron beam control signal?
  A. 4.2 MHz
  B. 6.8 MHz
  C. 8.8 MHz
  D. 12.6 MHZ

ANSWER D: Frequency of the pulse train (F) equals (N/2)/T where N = the number of pixels on the screen; T = the time to complete a full screen scan. Number of pixels on the screen = 70 × 80 × 25 = 140,000 total pixels. Scan time (T) is 1/60 = 0.017. In order to get good scan definition, harmonics of the third order must be accommodated, so F = 140,000 ÷ 2/0.017 = 4.2 MHz. Required bandwidth is 3 × 4.2 = 12.6 MHz.

8A268 A shipboard raster scan radar has a CRT with the following characteristics: 100 pixels per character, 100 characters per line, 30 lines per screen and it scans 60 screens per second. What is the minimum required bandwidth for the electron beam control signal?
  A. 26.5 MHz
  B. 8.8 MHz
  C. 12.6 MHz
  D. 30.2 MHz

ANSWER A: Same formula (N/2)/T. First multiply 100 pixels × 100 characters × 30 lines = 300,000 pixels. Scan time is 1/60 = 0.017. F = (300,000 ÷ 2) = 150,000 ÷ 0.017 = 8.8 MHz × 3 = 26.47 MHz, (26.5 MHz) the correct answer.

8A269 A shipboard raster scan radar has a CRT with the following characteristics: 100 pixels per character, 100 characters per line, 60 lines per screen and it scans 100 screens per second. What is the minimum required bandwidth for the electron beam control signal?
  A. 26.5 MHz
  B. 30 MHz
  C. 90 MHz
  D. 8.8 MHz

ANSWER C: Solve using the same formula (N/2)/T by multiplying 100 pixels × 100 characters × 60 lines = 600,000. Scan time is 1/100 = 0.01. (600,000 ÷ 2) = 300,000 ÷ 0.01 = 30 MHz × 3 = 90 MHz.

8A270 Radar range is measured by the constant:

A. 150 meters per microsecond.
B. 150 yards per microsecond.
C. 300 yards per microsecond.
D. 18.6 miles per microsecond.

ANSWER A: 150 meters is the round-trip distance that a microwave travels in one microsecond.

8A271 A Continuous-Wave radar is frequency modulated with a 50-Hz sinewave. At the output of the receiver phase detector, a phase delay of 18 degrees is measured. This indicates a target range of:

A. 15 kilometers.
B. 75 kilometers.
C. 150 kilometers.
D. 300 kilometers.

ANSWER C: Continuous-Wave radars are excellent long range radars because the lower frequencies used to produce the sinewaves cover an extremely long distance to go through their 360° signal cycle. The first thing we must do is calculate the wavelength of the 50-Hz sinewave:

▶ $$\lambda \text{ (in meters)} = \frac{300{,}000}{f(\text{in kHz})}$$ Where: λ = wavelength in **meters**
                                            f = frequency in **kilohertz**

Since $f = 50\text{Hz} = 0.05$ kHz

$$\lambda = \frac{3{,}000{,}000}{.05} = 6{,}000{,}000 \text{ meters} = 6{,}000 \text{ kilometers}$$

The 6,000 kilometers is the distance the 50-Hz sinewave must travel to complete one 360° cycle. The phase detector acts as a comparator between the transmitted signal and the returned echo. By comparing where in the 360° cycle the detected object is, distance to the object can be calculated. This comparison is called the phase-delay ratio, and is calculated:

$$\text{Phase-delay ratio} = \frac{\text{number of degrees of the received signal}}{360° \text{ (number of degrees in one complete signal cycle)}}$$

so, $$\frac{18}{360} = .05 \text{ (phase delay ratio)}$$

Next, we must multiply the phase delay ratio times the distance the 50-Hz sinewave would travel in one complete cycle: $0.05 \times 6{,}000$ kilometers = 300 kilometers. Since this the the the total distance the wave *traveled to the target and back*, we then divide by 2 to get the distance to the target: 300 kilometers/2 = 150 kilometers. Remember we used the λ formula where f is in kHz.

8A272 A Continuous Wave radar is frequency modulated with a 50-Hz sinewave. At the output of the receiver phase detector, a phase delay of 36 degrees is measured. This indicates a target range of:

A. 15 kilometers.
B. 75 kilometers.
C. 150 kilometers.
D. 300 kilometers.

ANSWER D: This is almost exactly like question 8A271, so refer back to it for more details. The first thing you must do is calculate the wavelength of the 50-Hz sinewave as you did in the previous question: 300,000 ÷ Frequency in kHz = Wavelength in Meters.

300,000 ÷ 0.05 = 6,000,000 meters (or 6,000 kilometers)

Next, find the phase delay ratio by dividing the phase delay by 360 degrees:

36 ÷ 360 = 0.1

Then, multiply the phase delay ratio by 6,000 and you have the out-and-back distance to the target (6,000 km × 0.1 = 600 km). Now, simply divide the out-and-back distance by 2 and you have the answer (600 km ÷ 2 = 300 km).

8A273 A Continuous Wave radar is frequency modulated with a 100-Hz sinewave. At the output of the receiver phase detector, a phase delay of 18 degrees is measured. This indicates a target range of:
  A. 15 kilometers.
  B. 75 kilometers.
  C. 150 kilometers.
  D. 300 kilometers.

ANSWER B: This question uses a different wave frequency than the previous two, but the process is the same. First, calculate the wavelength of the 100-Hz sinewave: 300,000 ÷ Frequency in kHz = Wavelength in Meters.

300,000 ÷ 0.1 = 3,000,000 meters (or 3,000 kilometers)

Next, find the phase delay ratio by dividing the phase delay by 360 degrees:

18 ÷ 360 = 0.05

Then, multiply the phase delay ratio by 3,000 and you have the out-and-back distance to the target (3,000 km × 0.05 = 150 km). Finally, divide the out-and-back distance by 2 and you have the distance to the target (150 km ÷ 2 = 75 km).

8A274 A Continuous Wave radar is frequency modulated with a 100-Hz sinewave. At the output of the receiver phase detector, a phase delay of 36 degrees is measured. This indicates a target range of:
  A. 15 kilometers.
  B. 75 kilometers.
  C. 150 kilometers.
  D. 300 kilometers.

ANSWER C: Again, refer to question 8A271 for the formulas and procedure.

Calculate the wavelength of the 100-Hz sinewave: 300,000 ÷ Frequency in kHz = Wavelength in Meters.

300,000 ÷ 0.1 = 3,000,000 meters (or 3,000 kilometers)

Find the phase delay ratio by dividing the phase delay by 360 degrees:

36 ÷ 360 = 0.1

Multiply the phase delay ratio by 3,000 and you have the out-and-back distance to the target (3,000 km × 0.1 = 300 km).

Divide the out-and-back distance by 2 and you have the distance to the target (300 km ÷ 2 = 150 km).

It's as easy as 1-2-3 (or 1-2-3-4 in this case)!

8A275 A Continuous Wave radar is frequency modulated with a 25-Hz sinewave. At the output of the receiver phase detector, a phase delay of 18 degrees is measured. This indicates a target range of:

    A. 15 kilometers.
    B. 75 kilometers.
    C. 150 kilometers.
    D. 300 kilometers.

ANSWER D: If you still don't have these continuous wave questions figured out, refer back to question 8A271. But you're probably becoming a expert at this by now.

Calculate the wavelength of the 25-Hz sinewave: 300,000 ÷ Frequency in kHz = Wavelength in Meters.

    300,000 ÷ 0.025 = 12,000,000 meters (or 12,000 kilometers)

Find the phase delay ratio by dividing the phase delay by 360 degrees:

    18 ÷ 360 = 0.05

Multiply the phase delay ratio by 12,000 and you have the out-and-back distance to the target (12,000 km × 0.05 = 600 km).

Divide the out-and-back distance by 2 and you have the distance to the target (600 km ÷ 2 = 300 km).

8A276 A Continuous Wave radar is frequency modulated with a 25-Hz sinewave. At the output of the receiver phase detector, a phase delay of 36 degrees is measured. This indicates a target range of:

    A. 15 kilometers.
    B. 150 kilometers.
    C. 300 kilometers.
    D. 600 kilometers.

ANSWER D: One more time...refer back to question 8A271 is you still need more information.

Calculate the wavelength of the 25-Hz sinewave: 300,000 ÷ Frequency in kHz = Wavelength in Meters.

    300,000 ÷ 0.025 = 12,000,000 meters (or 12,000 kilometers)

Find the phase delay ratio by dividing the phase delay by 360 degrees:

    36 ÷ 360 = 0.1

Multiply the phase delay ratio by 12,000 and you have the out-and-back distance to the target (12,000 km × 0.1 = 1200 km).

Divide the out-and-back distance by 2 and you have the distance to the target (1200 km ÷ 2 = 600 km).

8A277 A target pulse appears on the CRT 100 microseconds after the transmitted pulse. The target slant range is:
 A. 30 kilometers.
 B. 93 miles.
 C. 15 kilometers.
 D. 15,000 yards.
ANSWER C: Divide 100 microseconds by radar range 12.4 microseconds, and this gives you 8 nautical miles. Since1 nautical mile = 1.852 kilometers, multiply 8 times 1.852 kilometers for a slant range of 15 kilometers.

8A278 A gated LC oscillator, operating at 12.5 kHz, is being used to develop range markers. If each cycle is converted to a range mark, the range between markers will be:
 A. 120 kilometers.
 B. 12 kilometers.
 C. 1.2 kilometers.
 D. None of these.
ANSWER B: The time between pulses = 1/12.5kHz, which is 1/0.0125 MHz = 80 microseconds. 80 microseconds ÷ 12.4 microseconds = 6.45 nautical miles, or 6.45 × 1.852 kilometers = 12 kilometers.

8A279 On an "A" scope presentation, range markers are displayed at 200-microsecond intervals. A target pulse appears exactly half-way between the third and fourth range marks. The target slant range is:
 A. 75 kilometers.
 B. 105 kilometers.
 C. 150 kilometers.
 D. 210 kilometers.
ANSWER B: Range markers are intensified range rings on the radar monitor where the ship is in the center, and range rings surround it. On an "A" scope presentation, range markers are displayed at 200-microsecond intervals. Since one nautical mile = 1.852 km, a radio wave propagating one nautical mile takes 6.2 microseconds. An echo would be two times this, or 12.4 microseconds. Therefore, range markers are displayed at 200 ÷ 12.4 microseconds = 16.13 nautical miles, or 16.13 × 1.852 kilometers = 30 kilometers. Each range marker is then 30 kilometers. First to second would be 30-60 kilometers, and second to third would be 60-90 kilometers, and third to fourth would be 90- 120 kilometers. Since the echo is between the third and fourth "A" scope line, an answer of 105 kilometers is correct.

8A280 On an "A" scope presentation, range markers are displayed at 200-microsecond intervals. A target pulse appears exactly half-way between the third and fourth range marks. The range marks are being developed using a ringing oscillator. The operating frequency of the ringing oscillator must be:
 A. 500 Hz.
 B. 5,000 Hz.
 C. 10,000 Hz.
 D. 15,000 Hz.
ANSWER B: The ringing oscillator frequency is equal to 1/200 microseconds = $1/2 \times 10^{-4} = 0.5 \times 10^4 = 5,000$ Hz.

8A281 A radar transmitter is operating on 2.5 GHz and the reflex klystron local oscillator, operating at 2.56 GHz, develops a 60-MHz IF. If the magnetron drifts higher in frequency, the AFC system must cause the klystron repeller plate to become:
A. More positive.
B. More negative.
C. Less positive.
D. Less negative.
ANSWER B: The reflector is also called a repeller plate. It is negatively charged to repel the electrons. The more negatively charged the repeller plate, the lower the klystron frequency. The drift is high so the repeller must be more negative.

8A282 A radar magnetron develops a transmission frequency of 1250 MHz. To develop an IF of 60-MHz, the local oscillator must operate at:
A. 1.031 GHz.
B. 1.265 GHz.
C. 1.31 GHz.
D. 13.10 GHz.
ANSWER C: 1250 + 60 = 1310 MHz = 1.31 GHz. MHz to GHz, move the decimal point three places to the left.

8A283 The impedance total (Zo) of a transmission line can be calculated by Zo = $\sqrt{L/C}$ when inductance L and capacitance C are known. When a section of transmission line contains 250 microhenries of inductance and 1000 picofarads of capacitance, its impedance total (Zo) will be:
A. 50 ohms.
B. 250 ohms.
C. 500 ohms.
D. 1,000 ohms.
ANSWER C: Zo = $\sqrt{(250 \times 10^{-6}/1000 \times 10^{-12})}$ = $\sqrt{(0.25 \times 10^6)}$ = $\sqrt{250,000}$ = 500 ohms or Ω.

8A284 A certain length of transmission line has a characteristic impedance of 72 ohms. If the line is cut at its center, each half of the transmission line will have a Zo of:
A. 36 ohms.
B. 72 ohms.
C. 144 ohms.
D. The exact length must be known to determine Zo.
ANSWER B: Cutting a transmission line will not effect its characteristic impedance.

8A285 Energy travels down a section of transmission line at a rate of T=$\sqrt{LC}$. If a section of line has 100 microhenries of inductance and 1,000 picofarads of capacitance, how long will it take the leading edge of a pulse to travel the length of the line?
A. 100 microseconds
B. 0.01 microseconds
C. 316 nanoseconds
D. 31.6 nanoseconds

ANSWER C: All you do here is apply the formula T = √LC, but you must state the inductance is henries and the capacitance in farads.

T = $\sqrt{(100 \times 10^{-6})}$ henries × (1000×10^{-12}) farads = $\sqrt{100,000 \times 10^{-18}}$ seconds

$\sqrt{10 \times 10^{-14}}$ = 3.16 × 10^{-7} = 316 × 10^{-9} or 316 nanoseconds.

8A286 A basic constant frequency switching power supply regulator with an input voltage of 165 volts DC, and a switching frequency of 20 kHz, has an "ON" time of 27 microseconds when supplying 1 ampere to its load. What is the output voltage across the load?

    A. It cannot be determined with the information given.
    B. 305.55 volts DC
    C. 89.1 volts DC
    D. 165 volts DC

ANSWER C: The formula for the output voltage of a switching power supply is:

$$V_O = V_{IN} \times D \qquad \text{Where:} \quad \begin{aligned} V_O &= \text{Output voltage in \textbf{volts}} \\ V_{IN} &= \text{Input voltage in volts} \\ D &= \text{duty cycle} = t_{ON} \times f \end{aligned}$$

Since   D = 27 × 10^{-6} × 20 × 10^{+3} = 540 × 10^{-3}
then V_O = 165 × 0.54 = 89.1 volts

8A287 A directional coupler has an attenuation of -30 db. A measurement of 100 milliwatts at the coupler indicates the power of the line is:

    A. 10 watts.
    B. 100 watts.
    C. 1,000 watts.
    D. 10,000 watts.

ANSWER B: Minus 30 dB is 1,000 times down in power level. 1,000 times 0.1 watts is 100 watts.
    10 dB = 10 times
    20 dB = 100 times
    30 dB = 1,000 times

8A288 The high-gain IF amplifiers in a radar receiver may amplify a 2 microvolt input signal to an output level of 2 volts. This amount of amplification represents a gain of:

    A. 60 db.
    B. 100 db.
    C. 120 db.
    D. 1,000 db.

ANSWER C: Voltage ratio equals 2 volts ÷ 2 microvolts = 1 × 10^6. Voltage gain in dB is = 20 log 1,000,000. The $\log_{10} 10^6$ = 6. Thus db = 20 × 6 = 120 dB. Remember 2 microvolts is 2.0 × 10^{-6}.

8A289 Given a square wave with a frequency of 1,000 hertz to be converted to trigger pulses by an RC network, which combination of R and C will provide the sharpest pulses?

    A. R = 10 kilohms, C = 0.01 microfarad
    B. R = 20 kilohms, C = 0.005 microfarads
    C. R = 51 kilohms, C = 0.001 microfarad
    D. R = 100 kilohms, C= 0.002 microfarads

ANSWER C: We already know from our basic understanding of electronics that the time it takes for an RC circuit to charge or discharge a capacitor by 63% is T = R × C (Time Constant Formula). The first thing we must do is calculate the time it would take for each combination of R and C given as possible answer choices to produce a pulse:

A. $10 \times 10^3 \times 0.01 \times 10^{-6} = 0.000100$ or 100 microseconds
B. $20 \times 10^3 \times 0.005 \times 10^{-6} = 0.000100$ or 100 microseconds also
C. $51 \times 10^3 \times 0.001 \times 10^{-6} = 0.000051$ or 51 microseconds
D. $100 \times 10^3 \times 0.002 \times 10^{-6} = 0.000200$ or 200 microseconds

Here's a picture of what occurs in the circuit:

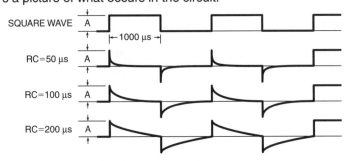

Forming Sharp Pulses from Square Wave

It really takes 5RC times to completely discharge or charge a capacitor; therefore, the PRT must be as long as 5RCs, other wise the pulse will have a pedestal. The PRT of 1 kHz is 1,000 microseconds so for answer C, 5 × 51 = 255 µs is much less than 1,000 µs.

8A290 What is the most common type of radar antenna used aboard commercial maritime vessels?
   A. Parabolic
   B. Truncated parabolic
   C. Slotted array
   D. Multi-element Yagi array
ANSWER C: The slotted array is commonly used on most X-band radars. It has a number of in-phase radiating slots which act as a stacked array, narrowing the horizontal beamwidth for increased resolution of individual targets.

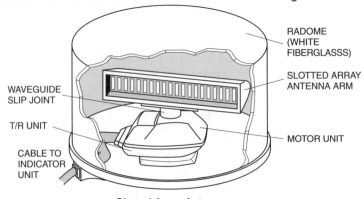

Slotted Array Antenna

8A291 A circulator provides what function in the RF section of a radar system?

A. It cools the magnetron by forcing a flow of circulating air.

B. It replaces the TR cell and functions as a duplexer.

C. It permits tests to be made to the thyristors while in use.

D. It transmits antenna position to the indicator during operation.

ANSWER B: A circulator eliminates the need of the older TR cell, and the circulator will then perform as a duplexer.

8A292 Why is coaxial cable often used for S-band installations instead of a waveguide:

A. Losses can be kept reasonable at S-band frequencies and the installation cost is lower.

B. A waveguide will not support the power density required for modern S-band radar transmitters.

C. S-band waveguide flanges show too much leakage and are unsafe for use near personnel.

D. Dimensions for S-band waveguide do not permit a rugged enough installation for use by ships-at-sea.

ANSWER A: Since S-band radars operate below 3,000 MHz, hard-line coax cable is sometimes employed instead of waveguide because installation is easier and cost is lower. Newer installations eliminate the need for long runs of waveguides because now the transmitter and receiver units are up inside the antenna—not down at the display.

8A293 Radar antenna direction must be sent to the display in all ARPA's or radar systems. How is this accomplished?

A. 3-phase synchros

B. 2-phase resolvers

C. Optical encoders

D. Any of these

ANSWER D: The modern ARPA radar system might employ synchros, resolvers, and optical encoders to reference the physical radar antenna direction to the scanned display. Any of these are found in ARPA systems.

8A294 A thick layer of rust and corrosion on the surface of the parabolic dish will have what effect?

A. No noticeable effect

B. Scatter and absorption of radar waves

C. Decrease in performance, especially for weak targets

D. Slightly out of focus PPI scope

ANSWER C: Older radars employ waveguide and a horn feed which concentrates the energy into a parabola. Rust and corrosion on the surface of the parabolic dish will decrease the performance to detect distant target echoes.

8A295 A shipboard acquisition or surface-search type of radar would have an antenna in the form of a:

A. Vertical orange peel.

B. Horizontal orange peel.

C. Parabolic dish.

D. Dipole with reflector and directors.

ANSWER B: A common type of search radar antenna is one that has a shape similar to the peel of an orange.

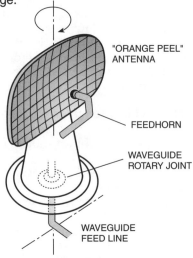

"ORANGE PEEL" ANTENNA

FEEDHORN

WAVEGUIDE ROTARY JOINT

WAVEGUIDE FEED LINE

Typical Surface Search Radar Antenna

8A296 To insert RF energy or extract RF energy from a waveguide, which of the following would not be used?
   A. A current loop
   B. A coupling capacitance
   C. An aperture window
   D. A voltage probe
ANSWER B: For this question, they want to know what would not be used to insert RF energy into a waveguide. All of the other answers would be appropriate for inserting RF energy, or extracting RF energy, from the waveguide.

8A297 Slotted waveguide arrays, when fed from one end exhibit:
   A. Frequency scan.
   B. High VSWR.
   C. Poor performance in rain.
   D. A narrow elevation beam.
ANSWER A: The slotted waveguide antenna is a popular one for small shipboard radar antenna systems. The slots are in-phase radiating openings, similar to a stacked array for narrow horizontal beamwidth. The slotted array also has low SWR, good performance in rain, and a regular vertical beamwidth, and the slotted waveguide array is said to be frequency scan.

8A298 A microwave transmission line constructed of a center conductor suspended between parallel conductive ground planes is called:
   A. Microstrip.
   B. Coax.
   C. Stripline.
   D. Waveguide.
ANSWER C: Stripline circuitry is used at microwave frequencies in both transmitting and receive circuits. In fact, small boat marine radars utilize stripline circuitry as the actual radiating antenna elements.

8A299 What device(s) might be used to allow electronic scanning in an array antenna?

A. PIN diode phase shifters
B. Solid state T/R switching
C. Ferrite circulators
D. A two-axis gimbal

ANSWER A: PIN diodes will allow phase shifting to different parts of a stripline antenna to allow electronic scanning of a fixed antenna array. Although big and small marine radars now employ a rotating antenna assembly, in a few more years phased array radars may eliminate the need for a rotating antenna system.

8A300 The radome of an aircraft radar must be kept clean in order to prevent:

A. Loss of range performance.
B. Reflected power from affecting the transmitter.
C. Changes to the beam shape of the antenna which might reduce accuracy.
D. All of these.

ANSWER D: It's important to keep the radome of aircraft or marine radars as clean as possible. This will increase range performance, decrease SWR, and will keep the beam shape of the antenna undistorted. A dirty or salt-encrusted radome can reduce range, create high SWR, and indeed could change the beam shape of the antenna output.

a. Mast Mounted

b. Stand Alone
Courtesy of ICOM America, Inc.

Marine Radar Radomes on Aircraft

8A301 A ferrite circulator is most commonly used in what portion of a radar system?

A. The antenna
B. The modulator
C. The duplexer
D. The receiver

ANSWER C: The duplexer action is similar to that of a T/R switch on a transceiver. There is enough isolation in the duplexer to keep transmit power from feeding back into, and destroying, the receiver. The duplexer circuit is accomplished by a ferrite circulator for simultaneous transmit and receive capabilities without the need of any mechanical switching.

8A302 Choose the most correct statement:
A. Radar antennas do not have side lobes.
B. X-Band radar gives a larger target return than S-Band radar.
C. X-Band radar requires more power for a given range than S-Band radar.
D. For a given antenna, as frequency increases, beamwidth decreases.

ANSWER D: The gain of a fixed diameter antenna is a function of the frequency. As the gain goes up, so does the frequency. This means that the capture of the signal is better due to a narrower beamwidth.

8A303 Proportionally, a very narrow beamwidth for a radar antenna indicates:
A. Proportionally higher frequencies are being transmitted and received.
B. Proportionally lower frequencies are being transmitted and received.
C. A narrow beamwidth does not coincide with frequency increases or decreases.
D. Proportionally narrow beamwidths indicate a pulse-type radar is being employed.

ANSWER A: Very narrow beamwidths are achieved at higher frequency microwave radar systems. The lower the frequency, the broader the beamwidth. A very narrow beamwidth allows a radar operator to see two individual targets at the same range as opposed to a broad beam yielding only one echo return.

8A304 Marine radar is dependent upon antenna beamwidth for:
A. Range resolution.
B. Bearing resolution.
C. Maximum receiver sensitivity.
D. Increased power output.

ANSWER B: Bearing resolution is a function of antenna arm length. The longer the rotating antenna element, the better the resolution with a narrower beamwidth. If the beamwidth is too wide, it is difficult to accurately determine the quantity of small targets being detected. Although the targets would show up, they would show up only as one big blob. A larger antenna system would discern each target as an individual echo.

8A305 The ferrite material in a circulator is used as a(an):
A. Electric switch.
B. Saturated reactor.
C. Loading element.
D. Phase shifter.

ANSWER D: A circulator is a 3- or 4-port device that passes energy from one port to the next port, but in only one direction. A ferrite material within the circulator has low-forward loss, and high-reverse loss. The effect is a phase shift from port to port.

8A306 A waveguide is used at radar microwave frequencies because:
A. It is easier to install than other feedline types.
B. It is more rugged than other feedline types.
C. It has lower transmission losses than other feedline types.
D. It is less expensive than other feedline types.

ANSWER C: Waveguide is used exclusively at radar microwave frequencies because losses are lower than coaxial feedline. Waveguide analysis is similar to open line; that is, two parallel lines. At microwave frequencies it is not practical to have open line because it must be insulated and high-quality insulators are hard to make at microwave frequencies.

8A307 Why should long horizontal runs of waveguide be avoided?
A. They must be insulated to prevent electric shock.
B. To prevent accumulation of condensation.
C. To prevent damage from shipboard personnel.
D. To minimize reception of horizontally polarized returns.

ANSWER B: Long horizontal runs of waveguide should be avoided because condensation may settle on the bottom of the waveguide, creating loss and the deterioration of the copper walls of the waveguide. After heavy weather, you notice that your radar is losing distant target response. Chances are moisture has leakeded into the waveguide. Check all joints to insure their gaskets are supple with no evidence of moisture inside the joint. Most of the problem occurs up in the antenna unit.

8A308 Loss of distant targets during and immediately after wet weather indicates:
A. High atmospheric absorption.
B. Dirt or soot on the rotary joint.
C. High humidity in the transmitter causing power supply loading.
D. A leak in waveguide or rotary joint.

ANSWER D: Moisture can accumulate and leak into the rotary waveguide joint, feeding the antenna. Moisture inside a waveguide can severely attenuate the radar signals. See figure of joint at question 8A295.

8A309 A right-angle bend in an X-band waveguide must have a radius greater than:
A. Three inches.
B. Six inches.
C. Two inches.
D. One inch.

ANSWER C: Sharp bends in an X-band waveguide run will cause signal transmission problems. A bend with a radius of 2 inches or greater will provide a smooth path with minimal "bumps" along the transmission line. Since most waveguide runs are short and confined to the antenna unit, the factory that has produced the equipment has already taken this into account.

8A310 A rotary joint is used to:
A. Couple two waveguides together at right angles.
B. Act as a switch between two waveguide runs.
C. Connect a stationary waveguide to the antenna array.
D. Maintain pressurization at the end of the waveguide.

ANSWER C: The rotary waveguide joint connects a piece of fixed waveguide from the transceiver to that on the antenna that is rotating. This joint should be inspected often to insure proper alignment and that there is no moisture that could leak in. See figure at question 8A295.

8A311 A typical shipboard radar antenna is a:

A. Rotary parabolic transducer.

B. Slotted waveguide array.

C. Phased planar array.

D. Dipole.

ANSWER B: The slotted waveguide array is the most popular type of marine radar antenna. Although we occasionally will see a parabolic reflector or a printed circuitboard antenna system, the slotted waveguide is preferred because of its low loss and predictable signal dispersion. See figure at 8A295.

8A312 At microwave frequencies, waveguides are used instead of conventional coaxial transmission lines because:

A. They are smaller and easier to handle.

B. They have considerably less loss.

C. They are lighter since they have hollow centers.

D. Moisture is never a problem with them.

ANSWER B: Waveguides at microwave frequencies are the preferred method of coupling microwave energy from the transmitter to the antenna because they have considerably less loss than coaxial cable.

8A313 Long horizontal sections of waveguides are not desirable because:

A. The waveguide can sag, causing loss of signal.

B. Moisture can accumulate in the waveguide.

C. Excessive standing waves can occur.

D. The polarization of the signal might shift.

ANSWER B: This question dates back to the old days of radar when the TR unit was down below. Back then, long horizontal runs of waveguide were avoided because they would accumulate moisture.

8A314 How is the signal removed from a waveguide or magnetron?

A. With a thin wire called a T-hook

B. With a thin wire called a J-Hook

C. With a coaxial connector

D. With a waveguide flange joint

ANSWER B: It is called the "J-Hook"—that section in the waveguide that actually removes the signal. See magnetrom cross section at question 2A98.

8A315 What should be done to the interior surface of a waveguide in order to minimize signal loss?

A. Fill it with nitrogen gas.

B. Paint it with nonconductive paint to prevent rust.

C. Keep it as clean as possible.

D. Fill it with a high-grade electrical oil.

ANSWER C: The inside surface of waveguide must be kept absolutely clean from dirt, moisture, or filings as the radar antenna unit is matched up to the TR section. Always cover the waveguide with a cap to insure it stays absolutely clean.

8A316 The following is true concerning waveguides:

A. Conduction is accomplished by the polarization of electromagnetic and electrostatic fields.

B. The magnetic field is strongest at the edges of the waveguide.

C. Ancillary deflection is employed.

D. The magnetic field is strongest at the center of the waveguide.

ANSWER B: The magnetic field is strongest along the copper edges of waveguide, and the electric field is strongest at the center of the waveguide. It's important that the insides of waveguide be as clean as possible. Since new ship installations don't require long waveguide runs, this waveguide question really applies to older (very old!) installations.

8A317 The polarization of the radiated electromagnetic energy through a waveguide is determined by:

A. Both E and H fields.

B. The vector sum of the E and H fields.

C. The H field alone.

D. The E field alone.

ANSWER D: The E field stands for "electric field", and this is what a vertical antenna radiates. Vertical lines of flux. The H field is the horizontal magnetic flux, and the electric and horizontal flux lines are 90 degrees apart. The plane that contains the E field is known as the plane of polarization. If someone says their antenna is vertically polarized, they are talking about the E field.

8A318 Which of these transmission line statements is *not* true?

A. Incident waves travel toward the load.

B. E and I are in phase on a flat line.

C. Standing waves travel down the line.

D. Reflected waves travel toward the source.

ANSWER C: Standing waves on a transmission line do not travel down the line. They are absorbed by the load. These standing waves have high voltage and current points periodically varying from the source to the load along the line.

8A319 Standing waves on a transmission line may be an indication that:

A. All energy is being delivered to the load.

B. Source and surge impedances are equal to Zo and ZL.

C. The line is terminated in impedance equal to Zo.

D. Some of the energy is not absorbed by the load.

ANSWER D: A transmission line with high SWR will sometimes feel warm to the touch. This is because some of the energy is not absorbed by the load (the antenna system) and points along the transmission line begin to get hot.

8A320 What is the purpose or function of the RADAR duplexer/circulator?

A. A coupling device that is used in the transition from a rectangular waveguide to a circular waveguide.

B. An electronic switch that allows the use of one antenna for both transmission and reception.

C. A modified length of waveguide that is used to sample a portion of the transmitted energy for testing purposes.

D. A dual section coupling device that allows the use of a magnetron as a transmitter.

ANSWER B: You can consider the duplexer and circulator as an electronic switch that allows the outgoing energy and incoming echo to switch between the antenna and the receiver without outgoing energy directly coupling into the receiver causing burnout.

8A321 Resistive losses in a waveguide are very small because:
A. The inner surface of the waveguide is large.
B. The inner surface of the waveguide is small.
C. The waveguide does not require a ground connection.
D. The heat remains in the waveguide and cannot dissipate.

ANSWER A: A waveguide provides a boundary which allows the wave to be propagated freely. Only the boundary's inner surface, which is relatively large in comparison with the wave, confines the energy. Waveguides with proper dimensions must be used for the particular band on which the radar is operating.

9

Taking the License Examinations

ABOUT THIS BOOK AND THIS CHAPTER

This book is intended to be a complete guide to the FCC's Commercial Radio Operator examination program — with emphasis on the *General Radiotelephone Operator License* (GROL) and the *Ship Radar Endorsement*. The GROL is, by far, the most popular of all commercial radio operator licenses. The GROL requires examinations taken from the Element 1 question pool and the Element 3 question pool. Element 1 question pool is found in Chapter 5, and Element 3 question pool is found in Chapter 6. Passing examinations on Element 1 and Element 3 question pools are also prerequisites for many of the other commercial licenses (See *Tables 3-1* and *3-2.*). Element 1 (radio law) is a requirement for the Marine Radio Operator Permit, all radiotelegraph certificates and both of the new Global Maritime Distress and Safety System (GMDSS) licenses. Element 3 (the technical questions) is also needed for the GMDSS Maintainer's license.

This book covers everything you need to know about the requirements to obtain any commercial license; however, only the Element 1, Element 3 and Element 8 question pools are included. Therefore, if you are striving to pass an examination for a license other than GROL, MROP or the Ship Radar Endorsement, additional question pools will be required. These are available from National Radio Examiners at toll-free 1-800-669-9594.

This chapter tells who is qualified to give examinations, how to find an examination location, how the examination will be given, and what happens after you complete an examination.

WHICH LICENSE DO I NEED?

The U.S. government in 1984 discontinued the requirement that an electronic technician hold a commercial radio operator license when installing, maintaining and repairing radio transmitting equipment in some of the domestic radio services. In its place, they substituted a program of industry technical certification. Chapter 2 details the requirements for the various types of licenses, permits and endorsements. Refer to it for the specific license requirements to make sure you are qualified and prepared for the examination.

WHO GIVES EXAMINATIONS

The Commercial Operator License Examination (COLE) System is the name of the Federal Communications Commission's privatized commercial radio operator testing program. The objective of this program is to meet the need of prospective licensees for frequent examination opportunities at convenient locations and to meet the need of industry for examinations that reflect state-of-the-art technology and contemporary operating conditions, at minimal expense to the government.

A Commercial Operator License Examination Manager (COLEM) is an organization that has been approved by the FCC to prepare and administer the various examination elements that must be passed by persons in order to be eligible to apply to the FCC for commercial radio operator licenses. They are independent contractors and each has an extensive nationwide commercial radio operator testing network.

EXAMINATION LOCATIONS

Now that commercial radio operator testing has been privatized, there are hundreds of locations across the United States where you can be administered commercial radio operator examinations. All approved examiners are required to be affiliated with one of the approved COLEMs. The examination location and time is determined by the examiners. The best way to find a nearby examination location is to simply telephone the COLEM's main office. There is a detailed list of the COLEMs in the Appendix giving telephone numbers, contact person, and examination availability. Most have toll-free "800" telephone numbers and their operators can give you the local telephone number of the nearest test center. Most COLEMs sponsor monthly or quarterly examinations, but some have them available daily on a walk-in basis. Many technical schools and colleges are also affiliated with a COLEM and many will administer the required tests as part of your tuition.

There is probably an examination site within a hour's travel time of your home! We strongly recommend that you telephone ahead to the examiners and let them know you are coming, since space at some test sessions may be limited. This will assure that a space will be reserved for you.

EXAMINATION FEES

You can expect to pay only one fee for your new FCC Commercial Radio Operator license. The *Examination Fee* is paid directly to the COLEM examiners. Another fee, payable only when you renew or replace your license is called an *Application Fee*. It is important to note that all fees are subject to change without notice. These fees are detailed in Chapter 2.

THE QUESTION POOLS AND YOUR EXAMINATION

Until the fall of 1993, the questions contained in the commercial radio operator tests were not readily known. Many publishers printed study manuals with review or sample questions, but they were not the word-for-word questions that you would be asked in an actual examination. This has now changed. The process (called the *Question Pool System)* now is similar to taking the written portion of a state automobile driver's license test. You obtain a license preparation manual such as this book, you study the questions, and you pass the test.

There is one, and only one, question pool for each written element. There no longer will be any "secret" questions appearing on any radio operator examinations. Everything — including the schematic diagrams — has been released by the FCC and no question, nor any of the multiple-choice answers, may be used in a question set administered to an examinee unless it appears in the question pool of the most recent release. The questions and answers in Chapters 5 (Element 1), Chapter 6 (Element 3) and Chapter 8 (Element 8) in this book are the same questions and answers that will appear on your examination. An explanation of the correct answer is included to help you understand and remember the answer. About fifteen percent of the questions in a pool are asked in any one question set.

The FCC will periodically update the questions that appear in every question pool and it is important that you study the current version. It is anticipated that the Element 1, 3, and 8 questions in this book will remain in effect indefinitely. Since every COLEM must use exactly the same question pool to construct their examinations, the examinations of one COLEM are no easier or harder than those of another.

Most of the remaining question pools have also been revised and released to the public during 1995, and will become effective by February 1, 1996. Although they usually contain more, FCC regulations (§13.215) require that each question pool must contain at least five times the number of questions required for a single examination.

TAKING THE EXAMINATION

You will be asked for positive identification before you are administered any commercial radio examination. One of these identification documents should be a state driver's license with your photograph on it. This is to prevent anyone from fraudulently taking the examination for another person. You will probably be asked to pay the examination fee before the tests are administered. The examination fee is due whether you pass or fail.

Examiners for each COLEM will administer the examinations in basically the same way. Some administer tests at a personal computer while others simply hand out a prepared written examination and you check the appropriate multiple-choice answers. You will answer each question by selecting one of four possible answers. All commercial radio examiners are required by the FCC to use exactly the same question pools, multiple choices, schematic diagrams and answers. Thus, if you study Chapter 5, 6 and 8 of this book, there will be no surprises when you are administered the Element 1 and 3 examinations for the Marine Radio Operator Permit (MROP), the General Radiotelephone Operator License (GROL) or Element 8 for the Ship

Radar Endorsement.

The Marine Radio Operator Permit requires passing only Element 1. You must correctly answer at least 75% (18) of the 24 questions selected from the Element 1 question pool of 170 questions.

The 100-question examination for the General Radiotelephone Operator License consists of two written multiple-choice examinations; one from Element 1 and one from Element 3. You must correctly answer at least 75% (18) of the 24 questions selected from the Element 1 question pool of 170 questions, and you must correctly answer at least 75% (57) of the 76 questions selected from the Element 3 question pool of 916 questions. There are 321 questions in Element 8, Ship Radar Techniques. You must correctly answer 38 of the 50 questions to pass.

You must abide by the instructions of the examiners. The amount of time that you will be allowed to answer the questions is at the discretion of the examiners administering the tests. As a general guideline, you will be given about one hour for the Element 1 examination and about two hours for the Element 3 examination.

HINTS ON PASSING THE TEST

- Be sure to get a good night's sleep. Don't stay up half the night studying!
- Be at the exam room early.
- Review the explanations of some of the most difficult answers. Try remembering the hints on the answers.
- Examinees are generally allowed to bring a calculator with them into the exam room, but the examiners may ask to clear its memory.
- You won't be allowed to leave the exam room once the examination starts — so make your bathroom stop beforehand.
- Read every question completely before trying to answer it. If you don't know the answer right away, go on to the next question. After the first pass, go back to the ones you skipped over. Don't leave any question unanswered; mark your best guess.
- On non-math questions, try eliminating obviously wrong choices. You have a better chance of guessing right if you can reduce the number of choices.
- Try working the math problems a number of ways until you get one of the answers.
- An interesting and very effective way to study for the GROL license is with a personal computer. Some COLEMs; e.g., National Radio Examiners, have low-price interactive computer software that allows you to study for the GROL and take sample examinations right at your IBM or IBM-compatible PC keyboard.

GRADING THE EXAMINATION

Your examiners will grade your examination(s) as soon as you turn in your answer sheet(s). They will immediately tell you your score and whether you passed the examination. They usually will not tell you which questions you missed. If you took your examinations at a computer keyboard, the PC will print out a scored answer sheet.

At the option of the examiner, you may retake a failed test after paying another examination fee. The rules (§13.209(c)) require that a different question set be re-administered to the examinee. Due to time constraints, however, examiners may ask you to appear at their next scheduled testing session.

Some applicants who pass Elements 1 and 3 will want to take the

Element 8 examination for the radar endorsement. This will take another hour and consists of 50 questions on the proper operation and servicing of ship radar equipment. If you pass, this endorsement will be placed on your GROL. Check with your COLEM on the availability of the Element 8 examination.

PROOF-OF-PASSING CERTIFICATE (PPC)

Once an applicant has passed any of the examination elements, the examiners will complete a *Proof-of-Passing Certificate* — or PPC. A PPC is documented evidence that an applicant has passed a test. An examinee is eligible to submit an application to the COLEM or FCC for a commercial radio operator license once he/she has accumulated the necessary original PPCs — or other credit document (either a commercial or amateur radio operator license) — indicating that he/she has completed the necessary examination requirements for a license. For example, a photocopy of a Marine Radio Operator Permit (MROP) will grant Element 1 credit toward the General Radiotelephone Operator License (GROL), or any or the radiotelegraph or GMDSS licenses. Likewise, a copy of a GROL confers Element 1 and 3 credit toward the GMDSS Maintainer's License. An Amateur Extra Class operator license grants Telegraphy Element 1 and 2 credit toward the Second and Third Class Radiotelegraph Operator's Certificate. (See § 13.9(c)(2).)

Just because you qualified for a lower class permit or license than you wanted does not mean you have to submit an application for the lower class permit or license. For example, let's say you wanted a GROL. You took the examinations for Element 1 and Element 3. You passed Element 1 but failed Element 3. You do not have to submit an application for a MROP. You will receive a PPC for Element 1 which you may hold until you pass Element 3, then you can use the Element 1 PPC as credit towards the GROL.

FILLING OUT THE FCC FORM 605

At present, either the applicant — or his/her agent (i.e. COLEM) — may submit FCC Form 605 and Schedule E for a new FCC Commercial Radio Operator and/or Maintainer License. Nearly all applications for new FCC Commercial Radio Operator and/or Maintainer licenses are now filed electronically by the COLEM that administers your examination. The examination team will have the current, correct version of FCC Form 605, and Schedules E and F for you to fill out at your exam session. You should complete the forms using black or blue ink (not pencil) using block, capital letters so that it is as legible as possible. This will help assure that you receive your license promptly.

You can obtain copies of FCC Form 605 and Schedules E and F without charge from the FCC forms supply house by calling (800) 481-FORM, or by downloading the forms from the FCC website at: http://wireless.fcc.gov/feesforms/obtaining.html.

RECEIVING YOUR LICENSE

In the past it has taken six to eight weeks to receive your commercial radio operator license from the Federal Communications Commission in Gettysburg, Pennsylvania. The FCC, however, now permits the electronic filing of commercial radio operator license applications. This is an optional program and at present no COLEM is required to file your application directly into the FCC's data processing system. Electronic filing dramatically reduces (to less than a week in some cases) the amount of time examinees must wait to receive their license. Your author's National Radio Examiners is one of the COLEMs that files all applications electronically.

Regardless of how long it takes your new license to arrive, all applicants may now begin their commercial radio operation once they have received the Proof-of-Passing Certificate (PPC) from their examiner. See §13.9(d) for further information on this conditional operating authority. Contact the FCC's Consumer Assistance hotline at 888-CALL FCC (888-325-5322) if your license has not arrived and your application has been outstanding for more than ninety days.

TESTING THE HANDICAPPED

Applicants who are afflicted with physical handicaps which would impair their ability to perform as a radio operator under emergency conditions may still be examined and licensed. Applicants are not denied a license solely on the basis of being disabled. The radio examiners are required to "accommodate" the disabled examinee by using special testing methods. Any required special equipment must be supplied by the disabled examinee.

A handicapped person who qualifies for a license employing special accommodative procedures and is unable to accomplish emergency duties involving the safety of life or property will have a special restrictive endorsement added to his/her license.

Appendix

COLE Managers – COMMERCIAL OPERATOR LICENSE EXAMINER MANAGERS

A COLE Manager is a non-government organization that has entered into a written agreement with the Federal Communications Commission to provide Commercial Radio Operator license testing to the public. There are nine such organizations and each charge a different examination fee. Applicants for the various commercial radio operator licenses, permits and endorsements should apply directly to one of these organizations.

COLEM	Examination Availability
National Radio Examiners P.O. Box 565206 Dallas, TX 75356-5206 Voice: (800) 669-9594 or (817) 860-3800, FAX: (817) 548-9594	All examination elements are available at more that 250 test centers across the U.S. and in some foreign countries. Fast electronic filing to the FCC results in almost immediate receipt of license. Fee: $35.00 per license. UPS or Fed Ex address: 2000 E. Randol Mill Rd., Suite 608-A, Arlington, TX 76011. **Contact: Larry Pollock** E-mail: colem@radioexaminers.com Web: www.radioexaminers.com
BFT Training Unlimited, Inc. P.O. Box 2677 Santa Rosa, CA 95405 Voice: (800) 821-0906 or (707) 792-5678 FAX: (707) 792-5677	All elements are available at regularly scheduled times or by appointment from 150 examiners throughout most of the United States and U.S. Territories, as well as other locations by prior arrangement. Fee: $75.00 per 1-2 exam elements, with a minimum $75.00 per exam session fee. **Contact: J. David Byrd,** E-mail: info@elkinstraining.com Web: www.elkinstraining.com
Electronic Technicians Association International, Inc. 5 Depot St. Greencastle, IN 46135 Voice: (800) 288-3824 or (765) 653-4301 FAX: (765) 653-4287	All elements are available at test sites located throughout all states, including all U.S. military installations (DANTES). Call for schedule information. Examinations are also available by appointment. Fee: $50.00 per 1 or 2 exam elements. **Contact: Richard Glass,** E-mail: eta@tds.net Web: www.eta-sda.com
Elkins International, Inc. P.O. Box 797666 Dallas, TX 75379 Voice: (888) 621-8876	All elements are available at regular scheduled times or by appointment at test centers throughout the U.S., American Territories and installations overseas. Fee: $25.00 per element with a $50.00 minimum per exam session. **Contact: Laura Elkins,** E-mail: elkins@eitn.com
International Society of Certified Electronics Technicians (ISCET) 3608 Pershing Ave. Fort Worth, TX 76107-4527 Voice: (817) 921-9101 FAX: (817) 921-3741	All elements are available by appointment from 360 examiners in 47 states, Guam, and selected foreign countries. (Alaska, Vermont, and Wyoming not presently available.) Fee: $50.00 - 90.00. **Contact: Alice Brown - Dept. 19** E-mail: alice@iscet.org Web: iscet.org
LaserGrade Computer Testing, Inc. 16821 SE McGillivray, Suite 201 Vancouver, WA 98683 Voice: (800) 211-2753 FAX: (360) 891-0958	All written elements are available at more than 500 sites throughout North America and internationally. Fee: $60.00 for first element and $30.00 for up to 2 additional elements per day. Computer based testing is offered on a daily basis throughout the year. **Contact: Linda Small,** E-mail: lsmall@lasergrade.com Web: www.lasergrade.com
Marine Technical Institute 1915 S. Andrews Avenue Fort Lauderdale, FL 33316 Voice: (954) 525-1014 FAX: (954) 764-0431	All elements are available daily by appointment at test sites in fifteen states. Fee: $50.00 for the first 2 elements and $25.00 for each additional element per sitting. Minimum sitting fee is $50.00. **Contact: David Greenspan,** E-mail: dgmti@aol.com

COLEM (Cont.)

Examination Availability

The National Association of Radio Telecommunications Engineers, Inc. (NARTE)
P.O. Box 678
167 Village Street,
Medway, MA 02053
Voice: (800) 89-NARTE or
FAX: (508) 533-3815

All elements are available at regularly scheduled times or by appointment at over 270 testing locations throughout the U.S. and at U.S. military installations (DANTES) worldwide. Fast electronic filing available for FCC License. Fee: $40.00 for up to 3 exam elements.
Contact: Sandy Felone - FCC Examination Registration and testing site information
E-mail: felone@narte.org
Web: www.narte.org

Sea School
8440 4th Street N.
St. Petersburg, FL 33702
Voice: (800) 237-8663
FAX: (727) 522-3155

All elements are available by appointment in 83 coastal cities and at 200 airport cities around the country. Fee: $25.00 - $55.00.
Contact: Mary Foster, E-mail: hqstaff@seaschool.com
Web: www.seaschool.com

FEDERAL COMMUNICATIONS COMMISSION

The Commercial Operator License Examination (COLE) program is regulated by the FCC's Wireless Telecommunications Bureau (WTB.)

 Federal Communications Commission
 WTB, Private Wireless Division
 445 12th Street, SW
 Washington, DC 20554

Contacts:
 J. Joy Alford, Administration,
 Room 8010, Tel. (202) 418-0680
 William T. Cross, Rulemaking,
 Room 5332, Tel. (202) 418-0691

All Commercial Radio Operator licenses are issued by the FCC's licensing facility located at:
 Federal Communications Commission
 1270 Fairfield Road
 Gettysburg, PA 17325-7245

Contact:
 Darlene Reeder
 Tel. (717) 338-2673

For information on the status of your Commercial Radio Operator license, permit or endorsement, call:
 FCC Consumer Assistance Hot Line
 Toll Free: (888) CALL FCC (225-5322)

FCC FIELD ORGANIZATION

The various FCC Field Offices may be contacted by mail or through the FCC's Consumer Assistance Hotline at (toll free) 888-CALL-FCC (322-5322).

FCC-Atlanta Field Office
3575 Koger Blvd., Suite #320
Duluth, Georgia 30096-4958

FCC-Boston Field Office
1 Batterymarch Park
Quincy, Massachusetts 02169-7495

FCC-Chicago Field Office
Park Ridge Office Center #306
1550 Northwest Highway
Park Ridge, Illinois 60068-1460

FCC-Columbia Field Office
9300 E. Hampton Dr.
Capitol Heights, MD 20743

FCC-Dallas Field Office
9330 LBJ Freeway, Room #1170
Dallas, Texas 75243-3429

FCC-Denver Field Office
215 S. Wadsworth Blvd., Suite #303
Lakewood, Colorado 80226-1544

FCC-Detroit Field Office
24897 Hathaway Street
Farmington Hills, Michigan 48335-1952

FCC-Kansas City Field Office
520 NE Colbern Rd., 2nd Floor
Lees Summit, MO 64086

FCC-Los Angeles Field Office
Cerritos Corporate Tower
1800 Studebaker Road, Room #660
Cerritos, California 90701-3684

FCC- New Orleans Field Office
2424 Edenborn Avenue, Suite 460
Metarie, LA 70001

FCC-New York City Field Office
201 Varick Street, Suite 1151
New York, New York 10014-4870

FCC-Philadelphia Field Office
One Oxford Valley Office Bldg. #404
2300 E. Lincoln Highway
Langhorne, Pennsylvania 19047-1859

FCC-San Diego Field Office
4542 Ruffner Street, Room #370
San Diego, California 92111-2216

FCC-San Francisco Field Office
5653 Stoneridge Dr., #105
Pleasanton, California 94588-8543

FCC-Seattle Field Office
11410 N.E. 122nd Way, Suite #312
Kirkland, Washington 98034-6927

FCC-Tampa Field Office
2203 N. Lois Avenue, Room #1215
Tampa, Florida 33607-2356

SUMMARY OF MOST USED QUESTION POOL FORMULAS

Question	Formula	Where:
3B22-25	**Radio Horizon Distance** $$d = \sqrt{17h}$$	D = Distance to radio horizon in **kilometers** h = Height of antenna in **meters**
3C5-8	**Counter Readout Error** Readout Error (in hertz) = f × a	f = Frequency in **MHz** being measured a = Counter accuracy in **parts per million**
3D4-8	**Resonance** $$X_L = X_C$$ $$2\pi f_r L = \frac{1}{2\pi f_r C}$$ $$f_r = \frac{1}{2\pi\sqrt{LC}}$$	X_L = Inductive reactance in **ohms** X_C = Capacitive reactance in **ohms** f_r = Resonant frequency in **hertz** L = Inductance in **henrys** C = Capacitance in **farads** π = 3.14
3D25-30,120	**True Power** For AC: $P_T = P_A \times P_F$ $P_A = E \times I$ $PF = \cos\phi$ For DC: $P_T = E \times I$	P_T = True power in **watts** P_A = Apparent power in **watts** ϕ = Phase angle in **degrees** PF = Power factor E = Applied voltage in **volts** I = Circuit current in **amps**
3D49-76	**Time Constants** $$\tau = RC$$	τ = Time Constant in **seconds** R = Resistance in **ohms** C = Capacitance in **farads**
	$$\tau = \frac{L}{R}$$	τ = Time Constant in **seconds** R = Resistance in **ohms** L = Inductance in **henries**
3D59-66, 129,130	**Capacitors** **Resistors** **Inductance** Series: $C_T = \dfrac{C_1 \times C_2}{C_1 + C_2}$ $R_T = R_1 + R_2$ $L_T = L_1 + L_2$ Parallel: $C_T = C_1 + C_2$ $R_T = \dfrac{R_1 \times R_2}{R_1 + R_2}$ $L_T = \dfrac{L_1 \times L_2}{L_1 + L_2}$	C = Capacitance in **farads** R = Resistance in **ohms** L = Inductance in **henries**
3D77-79,87	**Inductive Reactance** $$X_L = 2\pi fL$$	X_L = Inductive reactance in **ohms** L = Inductance in **henries** f = Frequency in **hertz** π = 3.14

CUT HERE

SUMMARY OF MOST USED QUESTION POOL FORMULAS (Continued)

Question	Formula	Where:
3D77-88,132	**Rectangular Coordinates** $Z = R \pm jX$ **Polar Coordinates** $Z = Z\,\underline{/\pm\Theta}$	Z = Impedance in **ohms** R = Resistance in **ohms** $+jX$ = Inductive reactance X_L in **ohms** $-jX$ = Capacitive reactance X_C in **ohms** Θ = Phase angle in **degrees** arctan = angle whose tangent is

Conversion:

$Z = R \pm jX$ to $Z\,\underline{/\pm\Theta}$ $\quad Z = \sqrt{R^2 + X^2}\,\underline{/\,\text{arctan}\,\frac{X}{R}}$

$Z\,\underline{/\pm\Theta}$ to $Z = R \pm jX$ $\quad R = Z\cos\Theta$

$\pm jX = Z\sin\Theta$

Question	Formula	
3D77-88	**Do Addition and Subtraction in Rectangular Coordinates** $Z_1 = R_1 + jX_1 \qquad Z_2 = R_2 - jX_2$ **Addition** $Z_T = (R_1 + jX_1) + (R_2 - jX_2)$ $Z_T = (R_1 + R_2) + j(X_1 - X_2)$ **Subtraction** $Z_T = (R_1 + jX_1) - (R_2 - jX_2)$ $Z_T = (R_1 - R_2) + j(X_1 + X_2)$	**Do Multiplication and Division in Polars Coordinates** $Z_1 = Z_1\,\underline{/\,\Theta_1} \qquad Z_2 = Z_2\,\underline{/\,\Theta_2}$ **Multiply** $Z_1 Z_2 = Z_1 \times Z_2\,\underline{/\,\Theta_1 + \Theta_2}$ **Divide** $\dfrac{Z_1}{Z_2} = \dfrac{Z_1\,\underline{/\,\Theta_1}}{Z_2\,\underline{/\,\Theta_2}} = \dfrac{Z_1}{Z_2}\,\underline{/\,\Theta_1 - \Theta_2}$

Question	Formula	Where:
3D81,86,88	**Capacitive Reactance** $X_C = \dfrac{1}{2\pi f C}$	X_C = Capacitive reactance in **ohms** C = Capacitance in **farads** f = frequency in **hertz** π = 3.14
	$X_C = \dfrac{10^6}{2\pi f C}$ **Impedance in Series** $Z_T = Z_1 + Z_2$	f = frequency in **MHz** C = Capacitance in **picofarads** Z = Impedance in **ohms** π = 3.14
	Impedance in Parallel $Z_T = \dfrac{Z_1 \times Z_2}{Z_1 + Z_2}$	**Special Case When $Z_1 = R$ and $Z_2 = \pm X$** $Z_T = \dfrac{RX\,\underline{/\pm 90°}}{\sqrt{R^2 + X^2}\,\underline{/\,\text{arctan}\,\frac{\pm X}{R}}}$
3D93-95,99	**Voltage Regulation** $\%\,\text{Reg} = \dfrac{E_{nl} - E_{fl}}{E_{fl}}$	E_{nl} = No-load voltage in **volts** E_{fl} = Full-load voltage in **volts** $\%\,\text{Reg}$ = Regulation in **percent**
3D98,103,105, 107-109,114, 143,168	**Power and Ampere Hours** $P = E \times I$ $I \times t$ = ampere hours	P = Power in **watts** E = DC Voltage in **volts** I = DC Current in **amperes** t = Time in **hours**
3D113,144, 162-165,166	**Power — E, I and R** $P = I^2 \times R \qquad P = \dfrac{E^2}{R}$	E = Voltage in **volts** I = Current in **amps** R = Resistance in **ohms** P = Power in **watts**

CUT HERE

SUMMARY OF MOST USED QUESTION POOL FORMULAS (Continued)

Question	Formula	Where:
3D101,102, 106,110,112, 115,137	**Turns Ratio** $$\frac{N_S}{N_P} = \frac{E_S}{E_P} \qquad \frac{N_P}{N_S} = \sqrt{\frac{Z_P}{Z_S}}$$	Z_P = Primary Z in **ohms** Z_S = Secondary Z in **ohms** N_P = Transformer Primary **turns** N_S = Transformer Secondary **turns** E_S = Transformer Secondary **volts** E_P = Transformer primary **volts**
3D148-149	**Motor Efficiency** $$P_{IN} \times Eff = P_{OUT}$$	P_{IN} = Input power in **watts** P_{OUT} = Output power in **watts** Eff = Motor efficiency in **percent**
3D162,169, 170,96,127,128 135,138,148, 142	**Ohm's Law** $$E = IR \qquad I = \frac{E}{R}$$ $$R = \frac{E}{I}$$	E = Voltage in **volts** I = Current in **amps** R = Resistance in **ohms**
3D186-189	**Sine Wave** $$V = A \, Sin \, \Theta$$	V = Instantaneous voltage in **volts** A = Amplitude of rotating vector in **volts** Θ = Angle of rotation in **degrees**
3F53	**Transistor Amplifier Load Resistor** $$R_L = \frac{(V_{CC})^2}{2P_O}$$	R_L = Load resistance in **ohms** V_{CC} = Collector voltage in **volts** P_O = Amplifier's power output in **watts**
3G12-15	**Deviation Ratio** $$Deviation\ Ratio = \frac{Maximum\ Carrier\ Frequency\ Deviation\ (in\ kHz)}{Maximum\ Modulation\ Frequency\ (in\ kHz)}$$	
3G16-20	**Modulation Index** $$Modulation\ Index = \frac{Deviation\ of\ FM\ Signal\ (in\ Hz)}{Modulating\ Frequency\ (in\ Hz)}$$	
3G44-51, 88-92,160	**RMS, Peak and Peak-to-Peak Voltage** $$V_{RMS} = 0.707\ V_{PK}$$ $$V_{PK} = 1.414\ V_{RMS}$$ $$V_{PP} = 2\ V_{PK}$$ $$V_{PK} = \frac{V_{PP}}{2}$$	V_{PP} = Peak-to-peak voltage in **volts** V_{PK} = Peak voltage in **volts** V_{RMS} = Root-mean-square voltage in **volts** The first two equations are good for sinewaves; the last two for symmetrical waveforms.
3G52-56	**Peak Envelope Power** $$PEP = P_{DC} \times Efficiency$$	PEP = Peak envelope power in **watts** P_{DC} = Input dc power in **watts** Eff = Efficiency in **decimals**

SUMMARY OF MOST USED QUESTION POOL FORMULAS (Continued)

Question	Formula	Where:
3G69-79	**Bandwidth for Digital** BW = baud rate + $(1.2 \times f_S)$	BW = Necessary bandwidth in **hertz** f_S = frequency shift in **hertz** Baud rate = Digital signal rate in **bauds**
3G69-79	**Bandwidth for CW** BW_{CW} = baud rate $\times$ wpm $\times$ fading factor	BW_{CW} = Necessary bandwidth in **hertz** wpm = Morse code signal rate in **words per minute** Fading factor = Constant of 5 for **CW**
3H31-34	**Physical Length vs Electric Length** $L = \dfrac{984 \, \lambda V}{f}$	L = Physical length in **feet** λ = Electrical length in **wavelength** V = Velocity factor of feedline f = Frequency in **MHz**
3H60-63	**Antenna Beamwidth** $\text{Beamwidth} = \dfrac{203}{(\sqrt{10})^x}$ where $X = \dfrac{A_G}{10}$	Beamwidth = Antenna beamwidth in **degrees** A_G = Antenna gain in **dB**
8A238-241, 251	**Nautical Radar Mile** Nautical Radar Mile = 2×6.17 μs	Radar mile = Time out and back to target in **microseconds** but calibrated on screen in **miles**. **Note:** All questions in the pools assume nautical miles. A nautical mile is 1.151 statute miles, or 6077.3 feet. A statute mile is 5280 feet.
8A242-248	**PRF, PRR, PRT** $PRT = \dfrac{1}{PRF}$ PRR = PRF	PRF = Pulse repetition **frequency** PRR = Pulse repetition **rate** PRT = Pulse repetition **time**
8A244-256	**Duty Cycle, Average Power** $\text{Duty Cycle} = \dfrac{PW}{1/PRF} = PW \times PRF$ Average Power = Peak Power $\times$ Duty Cycle	PW = Pulse width in **microseconds** Average Power in **watts** Peak Power in **watts**
8A257-262, 266-276	**Wavelength** $\lambda = \dfrac{300}{f \text{ (in MHz)}}$ $\lambda = \dfrac{300,000}{f \text{ (in kHz)}}$	λ = Wavelength in **cm** f = Wavelength in **MHz** λ = Wavelength in **meters** f = frequency in **kHz**
8A266-270	**Frequency of Raster Scan** $F = \dfrac{N/2}{T}$	N = Number of **pixels** T = Time to complete full screen scan in **seconds**

CUT HERE

Federal Communications Commission - Rules and Regulations

47 C.F.R. – PART 1 – PRACTICE AND PROCEDURE

PART 1 – APPLICATION REQUIREMENTS AND PROCEDURES

§ 1.913 Application forms; electronic and manual filing.

(a) *Application Forms.* Applicants and licensees in the Wireless Radio Services shall use the following forms and associated schedules for all applications:
*****

(4) *FCC Form 605, Quick-form Application for Authorization for Wireless Radio Services.* FCC Form 605 is used to apply for Amateur, Ship, Aircraft, and General Mobile Radio Service (GMRS) authorizations, as well as Commercial Radio Operator Licenses.

(b) *Electronic filing.* Except as specified in subparagraph (d) or elsewhere in this chapter, all applications and other filings using FCC Forms 601 through 605 or associated schedules must be filed electronically in accordance with the electronic filing instructions provided by ULS. For each Wireless Radio Service that is subject to mandatory electronic filing, this subparagraph is effective on (1) July 1, 1999, or (2) six months after the Commission begins use of ULS to process applications in the service, whichever is later. The Commission will announce by public notice the deployment date of each service in ULS.
*****

(d) *Manual filing.*

(1) ULS Forms 601, 603 and 605 may be filed manually or electronically by applicants and licensees in the following services:
*****

(2) Manually filed applications must be submitted to the Commission at the appropriate address with the appropriate filing fee.
*****

(4) Manually filed applications that do not require fees must be addressed and sent to Federal Communications Commission, 1270 Fairfield Road, Gettysburg, Pennsylvania 17325-7245

(5) Standard forms may be reproduced and the copies used in accordance with the provisions of Sec. 0.409 of this chapter.
*****

§ 1.923 Content of Applications

(a) *General.* Applications must contain all information requested on the applicable form and any additional information required by the rules in this chapter and any rules pertaining to the specific service for which the application is filed.

(h) *Taxpayer Identification Number (TIN).* Wireless applicants and licensees... are required to provide their Taxpayer Identification Numbers (TIN) (as defined in 26 U.S.C. §§ 6109) to the Commission, pursuant to the *Debt Collection Improvement Act of 1996 (DCIA).* Under the DCIA, the FCC may use an applicant or licensee's TIN for purposes of collecting and reporting to the Department of the Treasury any delinquent amounts arising out of such person's relationship with the Government. The Commission will not publicly disclose applicant or licensee TINs unless authorized by law, but will assign a "public identification number" to each applicant or licensee registering a TIN. This public identification number will be used for agency purposes other than debt collection.

(i) Unless an exception is set forth elsewhere in this chapter, each applicant must specify an address where the applicant can receive mail delivery by the United States Postal Service. This address will be used by the Commission to serve documents or direct correspondence to the applicant.
*****

§1.8001 FCC Registration Number (FRN).

(a) The FCC Registration Number (FRN) is a 10-digit unique identifying number that is assigned to entities doing business with the Commission.

(b) The FRN is obtained through the Commission Registration System (CORES) over the Internet at the CORES link or by filing FCC Form 160.

§1.8002 Obtaining an FRN.

(a) The FRN must be obtained by anyone doing business with the Commission.
*****

(b)(1) When registering for an FRN through the CORES, an entity's name, entity type, contact name and title, address, and taxpayer identifying number (TIN) must be provided. For individuals, the TIN is the social security number (SSN).

(2) Information provided when registering for an FRN must be kept current by registrants either by updating the information on-line at the CORES link at or by filing FCC Form 161 (CORES Update/Change Form).

(c) A business may obtain as many FRNs as it deems appropriate for its business operations. Each subsidiary with a different TIN must obtain a separate FRN. Multiple FRNs shall not be obtained to evade payment of fees or other regulatory responsibilities.

(d) An FRN may be assigned by the Commission, which will promptly notify the entity of the assigned FRN.

Federal Communications Commission - Rules and Regulations – Complete

47 C.F.R. – PART 13 – COMMERCIAL RADIO OPERATOR RULES

GENERAL

§ 13.1   Basis and purpose.
§ 13.3   Definitions.
§ 13.5   Licensed commercial radio operators required.
§ 13.7   Classification of operator licenses and endorsements.
§ 13.8   Authority conveyed.
§ 13.9   Eligibility and application for new license or endorsement.
§ 13.10   Licensee address.
§ 13.11   Holding more than one commercial radio operator license.
§ 13.13   Application for renewed or modified license.
§ 13.15   License term.
§ 13.17   Replacement license.
§ 13.19   Operator's responsibility.

EXAMINATION SYSTEM

§ 13.201 Qualifying for a commercial operator license or endorsement.
§ 13.203 Examination elements.
§ 13.207 Preparing an examination.
§ 13.209 Examination procedures.
§ 13.211 Commercial radio operator license examination.
§ 13.213 COLEM qualifications.
§ 13.215 Question pools.
§ 13.217 Records.

GENERAL PROVISIONS

§ 13.1 Basis and purpose.

(a) Basis. The basis for the rules contained in this Part is the Communications Act of 1934, as amended, and applicable treaties and agreements to which the United States is a party.

(b) Purpose. The purpose of the rules in this Part is to prescribe the manner and conditions under which commercial radio operators are licensed by the Commission.

§ 13.3 Definitions.

The definitions of terms used in Part 13 are:

(a) *COLEM.* Commercial operator license examination manager.

(b) *Commercial radio operator.* A person holding a license or licenses specified in § 13.7(b).

(c) *GMDSS.* Global Maritime Distress and Safety System.

(d) *FCC.* Federal Communications Commission.

(e) *International Morse code.* A dot-dash code as defined in International Telegraph and Telephone Consultative Committee (CCITT) Recommendation F.1 (1984), Division B, I. Morse code.

(f) *ITU.* International Telecommunication Union.

(g) *PPC.* Proof-of-Passing Certificate.

(h) *Question pool.* All current examination questions for a designated written examination element.

(i) *Question set.* A series of examination questions on a given examination selected from the current question pool.

(j) *Radio Regulations.* The latest ITU Radio Regulations to which the United States is a party.

§ 13.5 Licensed commercial radio operator required.

Rules that require FCC station licensees to have certain transmitter operation, maintenance, and repair duties performed by a commercial radio operator are contained in Parts 23, 80, and 87 of this chapter.

§ 13.7 Classification of operator licenses and endorsements.

(a) Commercial radio operator licenses issued by the FCC are classified in accordance with the Radio Regulations of the ITU.

(b) There are ten types of commercial radio operator licenses, certificates and permits (licenses). The license's ITU classification, if different from its name, is given in parenthesis.

(1) First Class Radiotelegraph Operator's Certificate.

(2) Second Class Radiotelegraph Operator's Certificate.

(3) Third Class Radiotelegraph Operator's Certificate (radiotelegraph operator's special certificate).

(4) General Radiotelephone Operator License (radiotelephone operator's general certificate).

(5) Marine Radio Operator Permit (radiotelephone operator's restricted certificate).

(6) Restricted Radiotelephone Operator Permit (radiotelephone operator's restricted certificate).

(7) Restricted Radiotelephone Operator Permit-Limited Use (radiotelephone operator's restricted certificate).

(8) GMDSS Radio Operator's License (general operator's certificate).

(9) Restricted GMDSS Radio Operator's License (restricted operator's certificate).

(10) GMDSS Radio Maintainer's License (technical portion of the first-class radio electronic certificate).

(c) There are six license endorsements affixed by the FCC to provide special authorizations or restrictions. Endorsements may be affixed to the license (s) indicated in parenthesis.

(1) Ship Radar Endorsement (First and Second Class Radiotelegraph Operator's Certificates, General Radiotelephone Operator License, GMDSS Radio Maintainer's License).

(2) Six Months Service Endorsement (First and Second Class Radiotelegraph

Operator's License).

(3) Restrictive endorsements relating to physical handicaps, English language or literacy waivers, or other matters (all licenses).

(4) Marine Radio Operator Permits shall bear the following endorsement: This Permit does not authorize the operation of AM, FM or TV broadcast stations.

(5) General Radiotelephone Operator Licenses issued after December 31, 1985, shall bear the following endorsement: This license confers authority to operate licensed radio stations in the Aviation, Marine and International Fixed Public Radio Service only. This authority is subject to: any endorsement placed upon this license; FCC orders, rules, and regulation; United States statutes; and the provisions of any treaties to which the United States is a party. This license does not confer any authority to operate broadcast stations. It is not assignable or transferable.

(6) (i) If a person is afflicted with an uncorrected physical handicap which would clearly prevent the performance of all or any part of the duties of a radio operator, under the license for which application is made, at a station under emergency conditions involving the safety of life or property, that person still may be issued the license if found qualified. Such a license shall bear a restrictive endorsement as follows:

This license is not valid for the performance of any operating duties, other than installation, service and maintenance duties, at any station licensed by the FCC which is required, directly or indirectly, by any treaty, statute or rule or regulation pursuant to statute, to be provided for safety purposes.

(ii) In the case of a license that does not require an examination in technical radio matters, the endorsement specified in (i) above will be modified by deleting the reference therein to installation, service, and maintenance duties.

(iii) In any case where an applicant who normally would receive or has received a commercial radio operator license bearing the endorsement prescribed by paragraph (i) above, indicates a desire to operate a station falling within the prohibited terms of the endorsement, the applicant may request in writing that such endorsement not be placed upon, or be removed from his or her license, and may submit written comments or statements from other parties in support thereof.

(iv) An applicant who shows that he has performed satisfactorily the duties of a radio operator at a station required to be provided for safety purposes during a period when he or she was afflicted by uncorrected physical handicaps of the same kind and to the same degree as the physical handicaps shown by his or her current application shall not be deemed to be within the provisions of paragraph (i) above.

(d) A Restricted Radiotelephone Operator Permit-Limited Use issued by the FCC to an aircraft pilot who is not legally eligible for employment in the United States is valid only for operating radio stations on aircraft.

(e) A Restricted Radiotelephone Operator Permit-Limited Use issued by the FCC to a person under the provision of § 303 (1)(2) of the Communications Act of 1934, as amended, is valid only for the operation of radio stations for which that person is the station licensee.

§ 13.8 Authority conveyed.

Licenses, certificates and permits issued under this part convey authority for the operating privileges of other licenses, certificates, and permits issued under this part as specified below:

(a) First Class Radiotelegraph Operator's Certificate conveys all of the operating authority of the Second Class Radiotelegraph Operator's Certificate, the Third Class Radiotelegraph Operator's Certificate, the Restricted Radiotelephone Operator Permit, and the Marine Radio Operator Permit.

(b) A Second Class Radiotelegraph Operator's Certificate conveys all of the operating authority of the Third Class Radiotelegraph Operator's Certificate, the Restricted Radiotelephone Operator Permit, and the Marine Radio Operator Permit.

(c) A Third Class Radiotelegraph Operator's Certificate conveys all of the operating authority of the Restricted Radiotelephone Operator Permit and the Marine Radio Operator Permit.

(d) A General Radiotelephone Operator License conveys all of the operating authority of the Marine Radio Operator Permit.

(e) A GMDSS Radio Operator's License conveys all of the operating authority of the Marine Radio Operator Permit.

(f) A GMDSS Radio Maintainer's License conveys all of the operating authority of the General Radiotelephone Operator License and the Marine Radio Operator Permit.

§ 13.9 Eligibility and application for new license or endorsement.

(a) If found qualified, the following persons are eligible to apply for commercial radio operator licenses:

(1) Any person legally eligible for employment in the United States.

(2) Any person, for the purpose of operating aircraft radio stations, who holds:

(i) United States pilot certificates; or

(ii) Foreign aircraft pilot certificates which are valid in the United States, if the foreign government involved has entered into a reciprocal agreement under which such foreign government does not impose any similar requirement relating to eligibility for employment upon United States citizens.

(3) Any person who holds a FCC radio station license, for the purpose of operating that station.

(4) Notwithstanding any other provisions of the FCC's rules, no person shall be eligible

to be issued a commercial radio operator license when

(i) The person's commercial radio operator license is suspended, or

(ii) The person's commercial radio operator license is the subject of an ongoing suspension proceeding, or

(iii) The person is afflicted with complete deafness or complete muteness or complete inability for any other reason to transmit correctly and to receive correctly by telephone spoken messages in English.

(b)(1) Each application for a new General Radiotelephone Operator License, Marine Radio Operator Permit, First Class Radiotelegraph Operator's Certificate, Second Class Radiotelegraph Operator's Certificate, Third Class Radiotelegraph Operator's Certificate, Ship Radar Endorsement, Six Months Service Endorsement, GMDSS Radio Operator's License, Restricted GMDSS Radio Operator's License, GMDSS Radio Maintainer's License and GMDSS Radio Operator/Maintainer must be filed on FCC Form 605 in accordance with § 1.913 of this chapter.

(2) Each application for a Restricted Radiotelephone Operator Permit or a Restricted Radiotelephone Operator Permit-Limited Use must be filed on FCC Form 605 in accordance with Sec. 1.913 of this chapter.

(c) Each application for a new General Radiotelephone Operator License, Marine Radio Operator Permit, First Class Radiotelegraph Operator's Certificate, Second Class Radiotelegraph Operator's Certificate, Third Class Radiotelegraph Operator's Certificate, Ship Radar Endorsement, GMDSS Radio Operator's License, Restricted GMDSS Radio Operator's License, GMDSS Radio Maintainer's License, or GMDSS Radio Operator/Maintainer License must be accompanied by the required fee, if any, and submitted in accordance with § 1.913 of this chapter. The application must include an original PPC(s) from a COLEM(s) showing that the applicant has passed the necessary examinations element(s) within the previous 365 days when the applicant files the application. If a COLEM files the application electronically on behalf of the applicant an original PPC(s) is not required. However, the COLEM must keep the PPC(s) on file for a period of 1 year.

(d) An applicant will be given credit for an examination element as specified below:

(1) An unexpired (or within the grace period) FCC-issued commercial radio operator license: The written examination and telegraphy Element(s) required to obtain the license held.

(2) An unexpired (or within the grace period) FCC-issued Amateur Extra Class operator license: Telegraphy Elements 1 and 2.

(e) Provided that a person's commercial radio operator license was not revoked, or suspended, and is not the subject of an ongo-

ing suspension proceeding, a person whose application for a commercial radio operator license has been received by the FCC but which has not yet been acted upon and who holds a PPC(s) indicating that he or she passed the necessary examination(s) within the previous 365 days, is authorized to exercise the rights and privileges of the operator license for which the application was received. This authority is valid for a period of 90 days from the date the application was received. The FCC, in its discretion, may cancel this temporary conditional operating authority without a hearing.

(f) Each application for a new Six Months Service Endorsement must be submitted in accordance with Sec. 1.913 of this chapter. The application must include documentation showing that:

(1) The applicant was employed as a radio operator on board a ship or ships of the United States for a period totaling at least six months;

(2) The ships were equipped with a radio station complying with the provisions of part II of title III of the Communications Act, or the ships were owned and operated by the U.S. Government and equipped with radio stations;

(3) The ships were in service during the applicable six month period and no portion of any single in-port period included in the qualifying six months period exceeded seven days;

(4) The applicant held a FCC-issued First or Second Class Radiotelegraph Operator's Certificate during this entire six month qualifying period; and

(5) The applicant holds a radio officer's license issued by the U.S. Coast Guard at the time the six month endorsement is requested.

(g) No person shall alter, duplicate for fraudulent purposes, or fraudulently obtain or attempt to obtain an operator license. No person shall use a license issued to another or a license that he or she knows to be altered, duplicated for fraudulent purposes, or fraudulently obtained. No person shall obtain or attempt to obtain, or assist another person to obtain or attempt to obtain, an operator license by fraudulent means.

§ 13.10 Licensee address.

In accordance with Section 1.923 of this chapter all applicants must specify an address where the applicant can receive mail delivery by the United States Postal Service except as specified below:

(a) Applicants for a Restricted Radiotelephone Operator Permit;

(b) Applicants for a Restricted Radiotelephone Operator Permit - Limited Use.

§ 13.11 Holding more than one commercial radio operator license.

(a) An eligible person may hold more than one commercial operator license except as follows:

(1) No person may hold two or more unexpired radiotelegraph operator's certificates at the same time;

(2) No person may hold any class of radiotelegraph operator's certificate and a Marine Radio Operator Permit;

(3) No person may hold any class of radiotelegraph operator's certificate and a Restricted Radiotelephone Operator Permit.

(b) Each person who is not legally eligible for employment in the United States, and certain other persons who were issued permits prior to September 13, 1982, may hold two Restricted Radiotelephone Operator Permits simultaneously when each permit authorizes the operation of a particular station or class of stations.

§ 13.13 Application for a renewed or modified license.

(a) Each application to renew a First Class Radiotelegraph Operator's Certificate, Second Class Radiotelegraph Operator's Certificate, Third Class Radiotelegraph Operator's Certificate, Marine Radio Operator Permit, GMDSS Radio Operator's License, Restricted GMDSS Radio Operator's License, GMDSS Radio Maintainer's License, or GMDSS Radio Operator/Maintainer License must be made on FCC Form 605. The application must be accompanied by the appropriate fee and submitted in accordance with §1.913 of this chapter.

(b) If a license expires, application for renewal may be made during a grace period of five years after the expiration date without having to retake the required examinations. The application must be accompanied by the required fee and submitted in accordance with Sec. 1.913 of this chapter. During the grace period, the expired license is not valid. A license renewed during the grace period will be effective as of the date of the renewal. Licensees who fail to renew their license within the grace period must apply for a new license and take the required examination(s).

(c) Each application involving a change in operator class must be filed on FCC Form 605. Each application for a commercial operator license involving a change in operator class must be accompanied by the required fee, if any, and submitted in accordance with Sec. 1.913 of this chapter. The application must include an original PPC(s) from a COLEM(s) showing that the applicant has passed the necessary examinations element(s) within the previous 365 days when the applicant files the application. If a COLEM files the application electronically on behalf of the applicant an original PPC(s) is not required. However, the COLEM must keep the PPC(s) on file for a period of 1 year.

(d) Provided that a person's commercial radio operator license was not revoked, or

suspended, and is not the subject of an ongoing suspension proceeding, a person holding a General Radiotelephone Operator License, Marine Radio Operator Permit, First Class Radiotelegraph Operator's Certificate, Second Class Radiotelegraph Operator's Certificate, Third Class Radiotelegraph Operator's Certificate, GMDSS Radio Operator's License, Restricted GMDSS Radio Operator's License, GMDSS Radio Maintainer's License, or GMDSS Radio Operator/Maintainer license, who has an application for another commercial radio operator license which has not yet been acted upon pending at the FCC and who holds a PPC(s) indicating that he or she passed the necessary examination(s) within the previous 365 days, is authorized to exercise the rights and privileges of the license for which the application is filed. This temporary conditional operating authority is valid for a period of 90 days from the date the application is received. This temporary conditional operating authority does not relieve the licensee of the obligation to comply with the certification requirements of the Standards of Training, Certification and Watchkeeping (STCW) Convention. The FCC, in its discretion, may cancel this temporary conditional operating authority without a hearing.

(e) An applicant will be given credit for an examination element as specified below:

(1) An unexpired (or within the grace period) FCC-issued commercial radio operator license: The written examination and telegraphy Element(s) required to obtain the license held; and

(2) An unexpired (or within the grace period) FCC-issued Amateur Extra-Class operator license: Telegraphy Elements 1 and 2.

§ 13.15 License Term.

(a) Commercial radio operator licenses are normally valid for a term of five years from the date of issuance, except as provided in paragraph (b) of this section.

(b) General Radiotelephone Operator Licenses, Restricted Radiotelephone Operator Permits, and Restricted Radiotelephone Operator Permits-Limited Use are normally valid for the lifetime of the holder. The terms of all Restricted Radiotelephone Operator Permits issued prior to November 15, 1953, and valid on that date, are extended to the lifetime of the operator.

§ 13.17 Replacement license.

(a) Each licensee or permittee whose original document is lost, mutilated, or destroyed must request a replacement. The application must be accompanied by the required fee and submitted to the address specified in part 1 of the rules.

(b) Each application for a replacement General Radiotelephone Operator License, Marine Radio Operator Permit, First Class Radiotelegraph Operator's Certificate, Second Class Radiotelegraph Operator's Certificate, Third Class Radiotelegraph Operator's Certificate, GMDSS Radio Operator's License, Restricted GMDSS Radio Operator License, GMDSS Radio Maintainer's License, or GMDSS Radio Operator/Maintainer license must be made on FCC Form 605 and must include a written explanation as to the circumstances involved in the loss, mutilation, or destruction of the original document.

(c) Each application for a replacement Restricted Radiotelephone Operator Permit must be on FCC Form 605.

(d) Each application for a replacement Restricted Radiotelephone Operator Permit-Limited Use must be on FCC Form 605.

(e) A licensee who has made application for a replacement license may exhibit a copy of the application submitted to the FCC or a photocopy of the license in lieu of the original document.

§ 13.19 Operator's responsibility.

(a) The operator responsible for maintenance of a transmitter may permit other persons to adjust that transmitter in the operator's presence for the purpose of carrying out tests or making adjustments requiring specialized knowledge or skill, provided that he or she shall not be relieved thereby from responsibility for the proper operation of the equipment.

(b) In every case where a station operating log or service and maintenance log is required, the operator responsible for the station operation or maintenance shall make the required entries in the station log. If no station log is required, the operator responsible for the service or maintenance duties which may affect the proper operation of the station shall sign and date an entry in the station maintenance records giving:

(1) Pertinent details of all service and maintenance work performed by the operator or conducted under his or her supervision;

(2) His or her name and address; and

(3) The class, serial number and expiration date of the license when the FCC has issued the operator a license, or the PPC serial number(s) and date(s) of issue when the operator is awaiting FCC action on an application;

(c) When the operator is on duty and in charge of transmitting systems, or performing service, maintenance or inspection functions, the license or permit document, or a photocopy thereof, or a copy of the application and PPC(s) received by the FCC, must be posted or in the operator's personal possession, and available for inspection upon request by a FCC representative.

(d) The operator on duty and in charge of transmitting systems, or performing service, maintenance or inspection functions, shall not be subject to the requirements of paragraph (b) of this section at a station, or stations of one licensee at a single location, at which the operator is regularly employed

and at which his or her license, or a photocopy, is posted.

EXAMINATION SYSTEM

§ 13.201 Qualifying for a commercial operator license or endorsement.

(a) To be qualified to hold any commercial radio operator license, an applicant must have a satisfactory knowledge of FCC rules and must have the ability to send correctly and receive correctly spoken messages in the English language.

(b) An applicant must pass an examination for the issuance of a new commercial radio operator license, other than the Restricted Radiotelephone Operator Permit and the Restricted Radiotelephone Operator Permit-Limited Use, and for each change in operator class. An applicant must pass an examination for the issuance of a new Ship Radar Endorsement. Each application for the class of license or endorsement specified below must pass, or otherwise receive credit for, the corresponding examination elements:

(1) First Class Radiotelegraph Operator's Certificate.

(i) Telegraphy Elements 3 and 4;

(ii) Written Elements 1, 5, and 6;

(iii) Applicant must be at least 21 years old;

(iv) Applicant must have one year of experience in sending and receiving public correspondence by radiotelegraph at a public coast station, a ship station, or both.

(2) Second Class Radiotelegraph Operator's Certificate.

(i) Telegraphy Elements 1 and 2;

(ii) Written Elements 1, 5, and 6.

(3) Third Class Radiotelegraph Operator's Certificate.

(i) Telegraphy Elements 1 and 2;

(ii) Written Elements 1 and 5.

(4) General Radiotelephone Operator License: Written Elements 1 and 3.

(5) Marine Radio Operator Permit: Written Element 1.

(6) GMDSS Radio Operator's License: Written Elements 1 and 7, or a Proof of Passing Certificate (PPC) issued by the United States Coast Guard or its designee representing a certificate of competency from a Coast Guard-approved training course for a GMDSS endorsement.

(7) Restricted GMDSS Radio Operator License: Written Elements 1 and 7R, or a Proof of Passing Certificate (PPC) issued by the United States Coast Guard or its designee representing a certificate of competency from a Coast Guard-approved training course for a GMDSS endorsement.

(8) GMDSS Radio Maintainer's License: Written Elements 1, 3, and 9.

(9) Ship Radar Endorsement: Written Element 8.

§ 13.203 Examination elements.

(a) A written examination (written Element) must prove that the examinee possesses the operational and technical qualifications to perform the duties required by a person holding that class of commercial radio operator license. Each written examination must be comprised of a question set as follows:

(1) Element 1 (formerly Elements 1 and 2): Basic radio law and operating practice with which every maritime radio operator should be familiar. 24 questions concerning provisions of laws, treaties, regulations, and operating procedures and practices generally followed or required in communicating by means of radiotelephone stations. The minimum passing score is 18 questions answered correctly.

(2) Element 3: General Radio telephone. 76 questions concerning electronic fundamentals and techniques required to adjust, repair, and maintain radio transmitters and receivers at stations licensed by the FCC in the aviation, maritime, and international fixed public radio services. The minimum passing score is 57 questions answered correctly.

(3) Element 5: Radiotelegraph operating practice. 50 questions concerning radio operating procedures and practices generally followed or required in communicating by means of radiotelegraph stations primarily other than in the maritime mobile services of public correspondence. The minimum passing score is 38 questions answered correctly.

(4) Element 6: Advanced radiotelegraph. 100 questions concerning technical, legal and other matters applicable to the operation of all classes of radiotelegraph stations, including operating procedures and practices in the maritime mobile services of public correspondence, and associated matters such as radio navigational aids, message traffic routing and accounting, etc. The minimum passing score is 75 questions answered correctly.

(5) Element 7: GMDSS radio operating practices. 76 questions concerning GMDSS radio operating procedures and practice sufficient to show detailed practical knowledge of the operation of all GMDSS subsystems and equipment; ability to send and receive correctly by radio telephone and narrow-band direct-printing telegraphy; detailed knowledge of the regulations applying to direct-printing telegraphy; detailed knowledge of the regulations applying to direct-printing telegraphy; detailed knowledge of the regulations applying to radio communications, knowledge of the documents relating to charges for radio communications and knowledge of those provisions of the International Convention for the Safety of Life at Sea which relate to radio; sufficient knowledge of English to be able to express oneself satisfactorily both orally and in writing; knowledge of and ability to perform each function listed in Section 80.1081; and

knowledge covering the requirements set forth in IMO Assembly Resolution on Training for Radio Personnel (GMDSS), Annex 3. The minimum passing score is 57 questions answered correctly.

(6) Element 8: Ship radar techniques. 50 questions concerning specialized theory and practice applicable to the proper installation, servicing and maintenance of ship radar equipment in general use for marine navigational purposes. The minimum passing score is 38 questions answered correctly.

(7) Element 9: GMDSS radio maintenance practices and procedures. 50 questions concerning the requirements set forth in IMO Assembly on Training for Radio Personnel (GMDSS), Annex 5 and IMO Assembly on Radio Maintenance Guidelines for the Global Maritime Distress and Safety System related to Sea Areas A3 and A4. The minimum passing score is 38 questions answered correctly.

(b) A telegraphy examination (telegraphy Elements) must prove that the examinee has the ability to send correctly by hand and to receive correctly by ear texts in the international Morse code at not less than the prescribed speed, using all the letters of the alphabet, numerals 0-9, period, comma, question mark, slant mark, and prosigns AR, BT and SK.

(1) Telegraphy Element 1: 16 code groups (CG) per minute.

(2) Telegraphy Element 2: 20 Plain Language (PL) words per minute.

(3) Telegraphy Element 3: 20 code groups (CG) per minute.

(4) Telegraphy Element 4: 25 Plain Language (PL) words per minute.

§ 13.207 Preparing an examination.

(a) Each telegraphy message and each written question set administered to an examinee for a commercial radio operator license must be provided by a COLEM.

(b) Each question set administered to an examinee must utilize questions taken from the applicable Element question pool. The COLEM may obtain the written question sets from a supplier or other COLEM.

(c) A telegraphy examination must consist of a plain language text or code group message sent in the international Morse code at no less than the prescribed speed for a minimum of five minutes. The message must contain each required telegraphy character at least once. No message known to the examinee may be administered in a telegraphy examination. Each five letters of the alphabet must be counted as one word or one code group. Each numeral, punctuation mark, and prosign must be counted as two letters of the alphabet. The COLEM may obtain the telegraphy message from a supplier or other COLEM.

§ 13.209 Examination procedures.

(a) Each examination for a commercial radio operator license must be administered at a location and a time specified by the COLEM. The COLEM is responsible for the proper conduct and necessary supervision of each examination. The COLEM must immediately terminate the examination upon failure of the examinee to comply with its instructions.

(b) Each examinee, when taking an examination for a commercial radio operator license, shall comply with the instructions of the COLEM.

(c) No examination that has been compromised shall be administered to any examinee. Neither the same telegraphy message nor the same question set may be re-administered to the same examinee.

(d) Passing a telegraphy examination.

(1) To pass a receiving telegraphy examination, an examinee is required to receive correctly the message by ear, for a period of 1 minute without error at the rate of speed specified in Section 13.203 for the class of license sought.

(2) To pass a sending telegraphy examination, an examinee is required to send correctly for a period of 1 minute at the rate of speed prescribed in Section 13.203 (b) for the class of license sought.

(e) Passing a telegraphy receiving examination is adequate proof of an examinee's ability to both send and receive telegraphy. The COLEM, however, may also include a sending segment in a telegraphy examination.

(f) The COLEM is responsible for determining the correctness of the examinee's answers. When the examinee does not score a passing grade on an examination element, the COLEM must inform the examinee of the grade.

(g) When the examinee is credited for all examination elements required for the commercial operator license sought, the examinee may apply to the FCC for the license.

(h) No applicant who is eligible to apply for any commercial radio operator license shall, by reason of any physical handicap, be denied the privilege of applying and being permitted to attempt to prove his or her qualification (by examination if examination is required) for such commercial radio operator license in accordance with procedures established by the COLEM.

(i) The COLEM must accommodate an examinee whose physical disabilities require a special examination procedure. The COLEM may require a physician's certification indicating the nature of the disability before determining which, if any, special procedures are appropriate to use. In the case of a blind examinee, the examination questions may be read aloud and the examinee may answer orally. A blind examinee wishing to use this procedure must make arrangements with the COLEM prior to the date the examination is desired.

(j) The FCC may:

(1) Administer any examination element itself.

(2) Readminister any examination element previously administered by a COLEM, either itself or by designating another COLEM to readminister the examination element.

(3) Cancel the commercial operator license(s) of any licensee who fails to appear for re-administration of an examination when directed by the FCC, or who fails any required element that is re-administered. In case of such cancellation, the person will be issued an operator license consistent with completed examination elements that have not been invalidated by not appearing for, or by failing, the examination upon readministration.

§ 13.211 Commercial radio operator license examination.

(a) Each session where an examination for a commercial radio operator license is administered must be managed by a COLEM or the FCC.

(b) Each examination for a commercial radio operator license must be administered as determined by the COLEM.

(c) The COLEM may limit the number of candidates at any examination.

(d) The COLEM may prohibit from the examination area items the COLEM determines could compromise the integrity of an examination or distract examinees.

(e) Within 10 days of completion of the examination element (s), the COLEM must provide the results of the examination to the examinee and the COLEM must issue a PPC to an examinee who scores a passing grade on an examination element.

(f) A PPC is valid for 365 days from the date it is issued.

§ 13.213 COLEM qualifications.

No entity may serve as a COLEM unless it has entered into a written agreement with the FCC. In order to be eligible to be a COLEM, then entity must:

(a) Agree to abide by the terms of the agreement;

(b) Be capable to serving as a COLEM;

(c) Agree to coordinate examinations for one or more types of commercial radio operator licenses and/or endorsements;

(d) Agree to assure that, for any examination, every examinee eligible under these rules is registered without regard to race, sex, religion, national origin or membership (or lack thereof) in any organization;

(e) Agree to make any examination records available to the FCC, upon request.

(f) Agree not to administer an examination to an employee, relative, or relative of an employee.

§ 13.215 Question pools.

The question pool for each written examination element will be composed of questions acceptable to the FCC. Each question pool must contain at least 5 times the number of questions required for a single examination. The FCC will issue public announcements detailing the questions in the pool for each element. COLEMs must use only the most recent question pool made available to the public when preparing a question set for a written examination element.

§ 13.217 Records.

Each COLEM recovering fees from exam-inees must maintain records of expenses and revenues, frequency of examinations administered, and examination pass rates. Records must cover the period from January 1 to December 31 of the preceding year and must be submitted as directed by the Commission. Each COLEM must retain records for 1 year and the records must be made available to the FCC upon request.

Federal Communications Commission - Rules and Regulations – Excerpts from

47 C.F.R. – PART 23 – INTERNATIONAL FIXED PUBLIC RADIOCOMMUNICATION SERVICES

§ 23.1 Definitions.

Assigned frequency. The frequency coinciding with the center of an authorized bandwidth of emission.

Authorized bandwidth. The maximum bandwidth authorized to be used by a station as specified in the station license. This shall be occupied bandwidth or necessary bandwidth, whichever is greater.

Authorized reference frequency. A frequency having a fixed and specific position with respect to the assigned frequency.

Authorized service. The transmission of public correspondence to a point of communication subject to such special provisions as may be contained in the license of the station.

Fixed public service. A radiocommunication service carried on between fixed stations open to public correspondence.

Fixed public press service. A limited radiocommunication service carried on between point-to-point telegraph stations, consisting of transmissions by fixed stations open to limited public correspondence, of news items, or other material related to or intended for publication by press agencies, newspapers, or for public dissemination.

Fixed station. Includes all apparatus used in rendering the authorized service at a particular location under a single instrument of authorization.

Frequency tolerance. The maximum permissible departure by the center frequency of the frequency band occupied by an emission from the assigned frequency or by the carrier, or suppressed carrier, from the reference frequency.

International fixed public radiocommunication service. A fixed service, the stations of which are open to public correspondence and which, in general, is intended to provide radiocommunication between any one of the 50 states or any other U.S. possession and any other point. Communications solely between Alaska, or any one of the contiguous 48 states (including the District of Columbia), and either Canada or Mexico are not deemed to be in the international fixed public radiocommunication service when such radiocommunications are transmitted on frequencies above 72 MHz.

International fixed public control service. A fixed service carried on for the purpose of communicating between transmitting stations, receiving stations, message centers or control points in the international fixed public radiocommunication service.

Occupied bandwidth. The frequency bandwidth such, that, below and above its upper frequency limits, the mean powers radiated are each equal to 0.5 percent of the total mean power radiated by a given emission.

Point-to-point telegraph station. A fixed station authorized for radiotelegraph communication.

Point-to-point telephone station. A fixed station authorized for radiotelephone communication.

Point of communication. A specific location designated in the license to which a station is authorized to communicate for the transmission of public correspondence.

Radiotelegraph. Includes types NØN, A1A, A2A, A3C, F1B, F2B, and F3C emission.

Radiotelephone. With respect to operation on frequencies below 30 MHz, means a system of radiocommunication for the transmission of speech or, in some cases, other sounds by means of amplitude modulation including double sideband (A3E), single sideband (R3E, H3E, J3E) or independent sideband (B3E) transmission.

§ 23.13 Types of emission.

Stations in the international fixed public radiocommunication services may be authorized to use any of the types of emission or combinations thereof - as well as new types which may be developed: *Provided,* That harmful interference to adjacent operations in not caused, *And provided further,* That the intelligence to be transmitted will use the bandwidth requested to a degree of efficiency compatible with the current state of the art.

§ 23.18 Authorization of power.

(a) *Authorized power.* ...is peak-envelope-power for transmitters having full, unkeyed carrier, single sideband or independent sideband emissions and mean power for transmitters having other emissions.

(b) *Use of minimum power.* In the interest of avoiding interference to other operations, all stations shall radiate only as much power as is necessary to ensure a satisfactory service.

§ 23.19 Use of directional antennas

Insofar as practicable, directional antennas, of type consistent with the current state of art, shall be used on all circuits for both transmitting and receiving.

§ 23.20 Assignment of frequencies

(a) Only those frequencies which are in accordance with the international and United States Table of Allocations (§2.106) may be authorized for use by stations in the Fixed Public and Fixed Public Press Services. Selection of specific frequencies within such bands shall be made by the applicant. Its availability for assignment as requested will be determined by a study of the probabilities of interference to and from existing services.

§23.24 Correspondents and points of communication.

Each instrument of authorization issued for fixed public or fixed public press service shall authorize communication to the points of communication and to the organizations, agencies, or persons specified therein only.

§ 23.29 License period and expiration time.

Licenses for stations operating in the fixed public radiocommunications services will be issued for a period of 10 years unless otherwise stated in the instrument of authorization.

§ 23.37 Station identification.

Every radiotelegraph or radiotelephone station in the International Fixed Public or Fixed Public Press Service shall transmit the identifying call sign or other approved identification signal on each of its assigned frequencies below 30 MHz on which energy is being transmitted. The call sign shall be transmitted at the beginning and end of each period of use of the frequency.

§ 23.46 Operators, class required and general duties.

(a) The operation and control of all transmitting apparatus licensed at a station in the international fixed public radiocommunication services shall be carried on only by a person holding a valid operator license issued by the Commission:

(b) Classes of operator licenses required:

(1) Radiotelegraph stations: Radiotelegraph first or second class license except;

(i) A Third Class Radiotelegraph Operator's Certificate or higher is required if manual Morse code keying is used for the purposes of identification or for sending service messages and no public correspondence is transmitted.

(2) Radiotelephone stations: General Radiotelephone Operator License.

§ 23.47 Station records.

(a) Station records shall be kept in an orderly manner, and in such detail that the data required is readily available.

§ 23.54 Use of double sideband radiotelephone

Use of double sideband radiotelephone transmissions, on frequencies below 30 MHz, shall not be transmitted except in cases where the foreign correspondent is unable to receive single sideband.

Federal Communications Commission - Rules and Regulations – Excerpts from

47 C.F.R. – PART 73 – RADIO BROADCAST SERVICES (Operator license no longer required)

Subpart H – Rules Applicable to All Broadcast Stations

§ 73.1350 Transmission system operation

(a) Each licensee is responsible for maintaining and operating its broadcast station in a manner which complies with the tech-nical rules set forth elsewhere in this part, and in accordance with the terms of the station authorization.

(b) Each licensee must designate a chief operator in accordance with § 73.1870. The licensee may designate one or more technically competent persons to adjust the transmitter operating parameters for compliance with the technical rules and the station authorization.

§ 73.1820 Station log

(a) Entries must be made in the station log either manually by a person designated by the licensee who is in actual charge or the transmitting apparatus, or by automatic devices...Indications of operating parameters that are required to be logged must be logged prior to any adjustment of the equipment.

§ 73.1870 Chief Operators

(a) The licensee of each AM, FM, or TV broadcast station must designate a person to serve as the station's chief operator. At times when the chief operator is unavailable or unable to act, (e.g. vacations, sickness), the licensee shall designate another person as the acting chief operator on a temporary basis.

(b)(3) The designation of the chief operator must be in writing with a copy of the designation posted with the station license.

(c) The chief operator is responsible for (1) inspections and calibrations of the trans-mission system...(2) periodic field monitoring, equipment performance measurements and other tests [and] (3) reviewing of the station records at least once a week.

Federal Communications Commission - Rules and Regulations – Excerpts from

47 C.F.R. – PART 80 – STATIONS IN THE MARITIME SERVICES

Subpart A - General Information

§80.5 Definitions.

Alaska-public fixed station. A fixed station in Alaska which is open to public correspondence and is licensed by the Commission for radio communication with Alaska-private fixed stations on paired channels.

Alaska-private fixed station. A fixed station in Alaska which is licensed by the Commission for radio communication within Alaska and with associated ship stations, on single frequency channels. Alaska-private fixed stations are also eligible to communicate with Alaska-public fixed stations on paired channels.

Associated ship unit. A portable VHF transmitter for use in the vicinity of the ship station with which it is associated.

Automated maritime telecommunications system (AMTS). An automatic, integrated and interconnected maritime communications system.

Automated mutual-assistance vessel rescue system (AMVER). An international system, operated by the U.S. Coast Guard, which provides aid to the development and coordination of search and rescue (SAR) efforts. Data is made available to recognized SAR agencies or vessels of any nation for reasons related to marine safety.

Bridge-to-bridge station. A radio station located on a ship's navigational bridge or main control station operating on a specified frequency which is used only for navigational communications, in the 156-162 MHz band.

Cargo ship safety radiotelegraphy certificate. A certificate issued after an inspection of a cargo ship radiotelegraph station which complies with the applicable Safety Convention radio requirements.

Cargo ship safety radiotelephony certificate. A certificate issued after inspection of a cargo ship radiotelephone station which complies with the applicable Safety Convention radio requirements.

Categories of ships. (1) When referenced in Part II of Title III of the Communications Act or the radio provisions of the Safety Convention, a ship is a *passenger ship* if it carries or is licensed or certificated to carry more than twelve passengers. A *cargo ship* is any ship not a passenger ship.

(2) A *commercial transport vessel* is any ship which is used primarily in commerce (i) for transporting persons or goods to or from any harbor(s) or port(s) or between places within a harbor or port area, or (ii) in connection with the construction, change in construction, servicing, maintenance, repair, loading, unloading, movement, piloting, or salvaging of any other ship or vessel.

(3) The term *passenger carrying vessel,* when used in reference to Part III, Title III of the Communications Act of the Great Lakes Radio Agreement, means any ship transporting more than six passengers for hire.

(4) *Power-driven vessel.* Any ship propelled by machinery.

(5) *Towing vessel.* Any commercial ship engaged in towing another ship astern, alongside or by pushing ahead.

(6) *Compulsory ship.* Any ship which is required to be equipped with radiotelecommunication equipment in order to comply with the radio or radio-navigation provisions of a treaty or statute to which the vessel is subject.

(7) *Voluntary ship.* Any ship which is not required by treaty or statute to be equipped with radiotelecommunications equipment.

Coast station. A land station in the maritime mobile service.

Commercial communications. Communications between coast stations and ship stations aboard commercial transport vessels, or between ship stations aboard commercial transport vessels, which relate directly to the purposes for which the ship is used including the piloting of vessels, movements of vessels, obtaining vessel supplies, and scheduling of repairs.

Day. (1) Where the word day is applied to the use of a specific frequency assignment or to a specific authorized transmitter power, its use means transmission on the frequency assignment or with the authorized transmitter power during that period of time included between one hour after local sunrise and one hour before local sunset.

(2) Where the word day occurs in reference to watch requirements, or to equipment testing, its use means the calendar day, from midnight to midnight, local time.

Digital selective calling (DSC). A synchronous system developed by the International Telecommunication Union Radiocommunication (ITU-R) Sector, used to establish contact with a station or group of stations automatically by means of radio. The operational and technical characteristics of this system are contained in Recommendations ITU-R M.493-10 and ITU-R M.541-8. (See subpart W of this part.)

Direction finder (radio compass). Apparatus capable of receiving radio signals and taking bearings on these signals from which the true bearing and direction of the point of origin may be determined.

Distress signal. The distress signal is a digital selective call using an internationally recognized distress call format in the bands used for terrestrial communication or an internationally recognized distress message format, in which case it is relayed through space stations, which indicates that a person, ship, aircraft, or other vehicle is threatened by grave and imminent danger and requests immediate assistance.

(1) In radiotelephony, the international distress signal consists of the enunciation of

the word "Mayday", pronounced as the French expression "m'aider". In case of distress, transmission of this particular signal is intended to ensure recognition of a radiotelephone distress call by stations of any nationality.

(2) For GMDSS, distress alerts result in an audible alarm and visual indication that a ship or person is threatened by grave and imminent danger and requests immediate assistance. These automatic systems contain sufficient information in the distress alert message to identify the vessel, prepare to assist and begin a search. However, except when transmitted via satellite EPIRB, the distress alert is just the initial call for help. Communication between the vessel or person in distress and the Rescue Coordination Center (RCC) or ship assisting should always follow.

Distress traffic. Distress traffic consists of all messages relating to the immediate assistance required by a person, ship, aircraft, or other vehicle in distress, including search and rescue communications and on-scene communications.

Environmental communications. Broadcasts of information about the environmental conditions in which vessels operate, i.e., weather, sea conditions, time signals adequate for practical navigation, notices to mariners, and hazards to navigation.

Fleet radio station license. An authorization issued by the Commission for two or more ships having a common owner or operator.

Global maritime distress and safety system (GMDSS). An International Maritime Organization (IMO) worldwide coordinated maritime distress system designed to provide the rapid transfer of distress messages from vessels in distress to units best suited for giving or coordinating assistance. The system includes standardized equipment and operational procedures, unique identifiers for each station, and the integrated use of frequency bands and radio systems to ensure the transmission and reception of distress and safety calls and messages at short, medium and long ranges.

Great Lakes. This term, used in this part in reference to the Great Lakes Radio Agreement, means all of Lakes Ontario, Erie, Huron (including Georgian Bay), Michigan, Superior, their connecting and tributary waters and the St. Lawrence River as far east as the lower exit of the St. Lambert Lock as Montreal in the Province of Quebec, Canada, but does not include any connecting and tributary waters other than: the St. Mary's River, the St. Clair River, Lake St. Clair, the Detroit River and the Welland Canal.

Harbor or port. Any place to which ships may resort for shelter, or to load or unload passengers or goods, or to obtain fuel, water, or supplies. This term applies to such places whether proclaimed public or not and whether natural or artificial.

Inland Waters. This term, as used in reference to waters of the United States, its territories and possessions, means waters that lie landward of the boundary lines of inland waters as contained in 33 CFR 80.01, as well as waters within its land territory, such as rivers and lakes, over which the United States exercises sovereignty.

INMARSAT. INMARSAT Ltd. is a private commercial company licensed in the United Kingdom.

Marine utility station. A station in the maritime mobile service consisting of one or more handheld radiotelephone units licensed under a single authorization. Each unit is capable of operation while being hand-carried by an individual. The station operates under the rules applicable to ship stations when the unit is aboard a vessel, and under the rules applicable to private coast stations when the unit is on land.

Maritime control communications. Communications between private coast and ship stations or between ship stations licensed to a state or local governmental entity, which relate directly to the control of boating activities or assistance to ships.

Maritime mobile repeater station. A land station at a fixed location established for the automatic retransmission of signals to extend the range of communication of ship and coast stations.

Maritime mobile-satellite service. A mobile-satellite service in which mobile earth stations are located on board ships. Survival craft stations and EPIRB stations may also participate in this service.

Maritime mobile service. A mobile service between coast stations and ship stations, or between ship stations, or between associated on-board communication stations. Survival craft stations and EPIRB stations also participate in this service.

Maritime mobile service identities (MMSI). An international system for the identification of radio stations in the maritime mobile service. The system is comprised of a series of nine digits which are transmitted over the radio path to uniquely identify ship stations, ship earth stations, coast stations, coast earth stations and groups of stations.

Maritime radiodetermination service. A maritime radiocommunication service for determining the position, velocity, and/or other characteristics of an object, or the obtaining of information relating to these parameters, by the propagation properties of radio waves.

Maritime support station. A station on land used in support of the maritime services to train personnel and to demonstrate, test and maintain equipment.

Navigable waters. This term, as used in reference to waters of the United States, its territories and possessions, means the waters shoreward of the baseline of its territorial sea and internal waters.

Navigational communications. Safety communications pertaining to the maneuvering of vessels or the directing of vessel movements. Such communications are primarily for the exchange of information between ship stations and secondary between ship stations and coast stations.

Noncommercial communications. Communication between coast stations and ship stations other than commercial transport ships, or between ship stations aboard other than commercial transport ships which pertain to the needs of the ship.

Non-selectable transponder. A transponder whose coded response is displayed on any conventional radar operating in the appropriate band.

On-board communication station. A low-powered mobile station in the maritime mobile service intended for use for internal communications on board a ship, or between a ship and its lifeboat and liferafts during lifeboat drills or operations, or for communication within a group of vessels being towed or pushed, as well as for line handling and mooring instructions.

On-board repeater. A radio station that receives and automatically retransmits signals between on-board communication stations.

Open sea. The water area of the open coast seaward of the ordinary low-water mark, or seaward of inland waters.

Operational fixed station. A fixed station, not open to public correspondence, operated by entities that provide their own radiocommunication facilities in the private land mobile, maritime or aviation services.

Passenger ship safety certificate. A certificate issued by the Commandant of the Coast Guard after inspection of a passenger ship which complies with the requirements of the Safety Convention.

Pilot. Pilot means a Federal pilot required by 46 U.S.C. 764, a state pilot required under the authority of 46 U.S.C. 211, or a registered pilot required by 46 U.S.C. 216.

Port operations communications. Communications in or near a port, in locks or in waterways between coast stations and ship stations or between ship stations, which relate to the operational handling, movement and safety of ships and in emergency to the safety of persons.

Portable ship station. A ship station which includes a single transmitter intended for use upon two or more ships.

Private coast station. A coast station, not open to public correspondence, which serves the operational, maritime control and business needs of ships.

Public coast station. A coast station that offers radio communication common carrier services to ship radio stations.

Public correspondence. Any telecommunication which the offices and stations must, by reason of their being at the disposal of the public, accept for transmission.

Radar beacon (RACON). A receiver-transmitter which, when triggered by a radar, automatically returns a distinctive signal which can appear on the display of the triggering radar, providing range, bearing and identification information.

Radioprinter operations. Communications by means of a direct printing radiotelegraphy system using any alphanumeric code, within specified bandwidth limitations, which is authorized for use between private coast stations and their associated ship stations on vessels of less than 1600 gross tons.

Safety communication. The transmission or reception of distress, alarm, urgency, or safety signals, or any communication preceded by one of these signals, or any form of radiocommunication which, if delayed in transmission or reception, may adversely affect the safety of life or property.

Safety signal. (1) The safety signal is the international radiotelegraph or radiotelephone signal which indicates that the station sending this signal is preparing to transmit a message concerning the safety of navigation or giving important meteorological warnings.

(2) In radiotelegraphy, the international safety signals consists of three repetitions of the group "TTT", sent before the call, with the letters of each group and the successive groups clearly separated from each other.

(3) In radiotelephony, the international safety signal consists of three oral repetitions of "Security", pronounced as the French word "Securite", sent before the call.

(4) For GMDSS, safety calls result in an audible alarm and visual indication that the station sending this signal has a very urgent message to transmit concerning the safety of navigation or giving important meteorological warnings.

Selectable transponder. A transponder whose coded response may be inhibited or displayed on a radar on demand by the operator of that radar.

Selective calling. A means of calling in which signals are transmitted in accordance with a prearranged code to operate a particular automatic attention device at the station whose attention is sought.

Ship earth station. A mobile earth station in the maritime mobile-satellite service located on board ship.

Ship or vessel. Ship or vessel includes every description of watercraft or other artificial contrivance, except aircraft, capable of being used as a means of transportation on water whether or not it is actually afloat.

Ship radio station license. An authorization issued by the Commission to operate a radio station onboard a vessel.

Ship station. A mobile station in the maritime mobile service located on-board a vessel which is not permanently moored, other than a survival craft station.

Station. One or more transmitters or a combination of transmitters and receivers, including the accessory equipment, necessary at one location for carrying on radiocommunication services.

Survival craft station. A mobile station in the maritime or aeronautical mobile service intended solely for survival purposes and located on any lifeboat, liferaft or other survival equipment.

Underway. A vessel is underway when it is not at anchor, made fast to the shore, or aground.

Urgency signal. (1) The urgency signal is the international radiotelegraph or radiotelephone signal which indicates that the calling station has a very urgent message to transmit concerning the safety of a ship, aircraft, or other vehicle, or of some person on board or within sight.

(2) In radiotelegraphy, the international urgency signal consists of three repetitions of the group "XXX", sent before the call, with the letters of each group and the successive groups clearly separated from each other.

(3) In radiotelephony, the international urgency signal consists of three oral repetitions of the group of words "PAN PAN", each word of the group pronounced as the French word "PANNE" and sent before the call.

(4) For GMDSS, urgency calls result in an audible alarm and visual indication that the station sending this signal has a very urgent message to transmit concerning the safety of a ship, aircraft, or other vehicle, or of some person on board or within sight.

Vessel traffic service (VTS). A U.S. Coast Guard traffic control service for ships in designated water areas to prevent collisions, groundings and environmental harm.

Watch. The act of listening on a designated frequency.

Subpart B - Applications and Licenses

§80.13 Station license required

(a) Except as provided in paragraph (c) of this section, stations in the maritime service must be licensed by the FCC either individually or by fleet.

(b) One ship station license will be granted for all maritime transmitting equipment on board a vessel.

(c) A ship station is licensed by rule and does not need an individual license issued by the FCC if the ship station is not subject to the radio equipment carriage requirements of any statute, treaty or agreement to which the United States is signatory, the ship station does not travel to foreign ports, and the ship station does not make international communications. A ship station licensed by rule is authorized to transmit radio signals using a marine radio operating in the 156-162 MHz band, any type of EPIRB, and any type of radar installation.

§80.15 Eligibility for station license.

(a) *General.* A station license cannot be granted to or held by a foreign government or its representative.

(b) *Public coast stations and Alaska public fixed stations.* A station license for a public coast station or an Alaska public fixed station cannot be granted to or held by:

(1) Any alien or the representative of any alien;

(c) *Private coast and marine utility stations.* The supplemental eligibility requirements for private coast and marine utility stations are contained in §80.501(a).

(d) *Ship stations.* A ship station license may only be granted to:

(1) The owner or operator of the vessel;

(2) A subsidiary communications corporation of the owner or operator of to vessel;

(3) A State or local government subdivision; or

(4) Any agency of the U.S. Government subject to section 301 of the Communications Act.

(e) *EPIRB stations.* (1) Class A or Class B EPIRB stations will be authorized for use on board the following types of vessels until December 31, 2006:(i) Vessels authorized to carry survival craft; or

(ii) Vessels expected to travel in waters beyond the range of marine VHF distress coverage which is generally considered to be more than 32 kilometers (approximately 20 miles) offshore; or

(iii) Vessels required to be fitted with EPIRBs to comply with U.S. Coast Guard regulations.

(2) A 406.025 MHz EPIRBs may be used by any ship required by U.S. Coast Guard regulations to carry an EPIRB or by any ship that is equipped with a VHF ship radio station.

§80.17 Administrative classes of stations.

(a) Stations in the Maritime Mobile Service are licensed according to class of station as follows:

(1) Public coast stations.

(2) Private coast stations.

(3) Maritime support stations.

(4) Ship stations. The ship station license may include authority to operate other radio station classes aboard ship such as; radionavigation, onboard, satellite, EPIRB, radiotelephone, radiotelegraph and survival craft.

(5) Marine utility stations.

(b) Stations on land in the Maritime Radiodetermination Service are licensed according to class of station as follows:

(1) Shore radiolocation stations.

(2) Shore radionavigation stations.

(c) Fixed stations in the Fixed Service associated with the maritime services are licensed as follows:

(1) Operational fixed stations.

(2) Alaska-public fixed stations.

(3) Alaska-private fixed stations.

§80.23 Filing of applications.

Rules about the filing of applications for radio station licenses are contained in this section.

(a) Each application must specify an ad-

APPENDIX – PART 80

dress in the United States to be used by the Commission in serving documents or directing correspondence to the licensee.

(b) An original of each application must be filed.

(c) Each application must be filed with the Federal Communications Commission, Gettysburg, PA 17326 unless otherwise noted on the application form. Applications requiring fees as set forth at part 1, subpart G of this chapter must be filed in accordance with §0.401(b) of the rules.

(d) One application for two or more new maritime utility stations may be submitted when the applicant and proposed area of operation for each station is the same.

(e) One application for transfer of control may be submitted for two or more stations subject to this part when the individual stations are clearly identified and the following elements are the same for all existing or requested station authorizations involved;

(1) Applicant;

(2) Specific details of basic request.

§80.25 License term.

(a) Licenses for ship stations in the maritime services will normally be issued for a term of ten years from the date of original issuance, major modification, or renewal.

(b) Licenses other than ship stations in the maritime services will normally be issued for a term of five years from the date of original issuance, major modifications, or renewal.

§80.29 Changes during license term.

(a) The following table indicates the required action for changes made during the license term:

Type of change	Required action
Mailing address	Written notice to the Commission
Name of licensee (without change in ownership, control or corporate structure).	Written notice to the Commission
Name of the vessel	Written notice to the Commission.

(b) Written notices must be sent to the Federal Communications Commission, Gettysburg, PA 17325.

§80.31 Cancellation of license.

When a station subject to this part which is not a communication common carrier permanently discontinues operation, the licensee must return the station license to the Commission's office at 1270 Fairfield Rd., Gettysburg, PA 17325, for cancellation.

§80.43 Equipment acceptable for licensing.

Transmitters listed in §80.203 must be type accepted for a particular use by the Commission based upon technical requirements contained in subparts E and F of this part.

§80.45 Frequencies.

When an application is submitted on FCC Form 503, the applicant must propose frequencies to be used by the station. The applicant must ensure that frequencies requested are consistent with the applicant's eligibility, the proposed class of station operation and the frequencies available for assignment as contained in subpart H of this part.

§80.47 Operation during emergency.

A station may be used for emergency communications when normal communication facilities are disrupted. The Commission may order the discontinuance of any such emergency communication service.

§80.53 Application for a portable ship station license.

(a) The Commission may grant a license permitting operation of a portable ship station aboard different vessels of the United States. Each application for a portable ship station must include a showing that:

(1) The station will be operated as an established class of station on board ship, and

(2) A station license for portable equipment is necessary to eliminate frequent application to operate a ship station on board different vessels.

§80.54 Automated Maritime Telecommunications System (AMTS) System Licensing.

AMTS licensees will be issued blanket authority for a system of coast stations and mobile units (subscribers). AMTS applicants will specify the maximum number of mobile units to be placed in operation during the license period.

§80.55 Application for a fleet station license.

(a) An applicant may apply for licenses for two or more radiotelephone stations aboard different vessels on the same application. Under these circumstances a fleet station license may be issued for operation of all radio stations aboard the vessels in the fleet.

(b) The fleet station license is issued on the following conditions:

(1) The licensee must keep a current list of vessel names and registration numbers authorized by the fleet license;

(2) The vessels do not engage in voyages to any foreign country;

(3) The vessels are not subject to the radio requirements of the Communications Act or the Safety Convention.

§80.56 Transfer of ship station license prohibited.

A ship station license may not be assigned. Whenever the vessel ownership is transferred, the previous authorization must be forwarded to the Commission for cancellation. The new owner must file for a new authorization.

§80.59 Compulsory ship inspections.

(a) *Inspection of ships subject to the Communications Act or the Safety Convention.* (1)

The FCC will not normally conduct the required inspections of ships subject to the inspection requirements of the Communications Act or the Safety Convention. Note: Nothing in this section prohibits Commission inspectors from inspecting ships. The mandatory inspection of U. S. vessels must be conducted by an FCC-licensed technician holding an FCC General Radiotelephone Operator License, GMDSS Radio Maintainer's License, Second Class Radiotelegraph Operator's Certificate, or First Class Radiotelegraph Operator's Certificate in accordance with the following table:

Minimum class of FCC license required by private sector technician to conduct inspection - only one license required

Category of vessel	GROL	GMDSS Maintainer	2nd Class Radio Telegr.	1st Class Radio Telegr.
Radio-telephone equipped vessels	X	X	X	X
Radio-telegraph equipped vessels	—	—	X	X
GMDSS equipped vessels	—	X	—	—

(2) A certification that the ship has passed an inspection must be entered into the ship's log by the inspecting technician. The technician conducting the inspection and providing the certification must not be the vessel's owner, operator, master, or employee or their affiliates. Additionally, the vessel owner, operator, or ship's master must certify in the station log that the inspection was satisfactory. There are no FCC prior notice requirements for any inspection pursuant to paragraph (a)(1) of this section. An inspection of the bridge-to-bridge radio stations on board vessels subject to the Vessel Bridge-to-Bridge Radiotelephone Act must be conducted by the same FCC-licensed technician.

(3) Additionally, for passenger vessels operated on an international voyage the inspecting technician must send a completed FCC Form 806 to the Officer in Charge, Marine Safety Office, United States Coast Guard in the Marine Inspection Zone in which the ship is inspected.

(4) In the event that a ship fails to pass an inspection the inspecting technician must make a log entry detailing the reason that the ship did not pass the inspection. Additionally, the technician must notify the vessel owner, operator, or ship's master that the vessel has failed the inspection.

(5) Because such inspections are intended to ensure the availability of communications capability during a distress the Commission will vigorously investigate reports of fraudulent inspections, or violations of the Communications Act or the Commission's Rules related to ship inspections. FCC-licensed technicians, ship owners or operators should report such violations to the Commission through its National Call Center at 1-888-CALL FCC (1-888-225-5322).

(b) *Inspection and certification of a ship subject to the Great Lakes Agreement.* The FCC will not inspect Great Lakes Agreement vessels. An inspection and certification of a ship subject to the Great Lakes Agreement must be made by a technician holding one of the following: an FCC General Radiotelephone Operator License, a GMDSS Radio Maintainer's License, a Second Class Radiotelegraph Operator's Certificate, or a First Class Radiotelegraph Operator's Certificate. The certification required by Sec. 80.953 must be entered into the ship's log. The technician conducting the inspection and providing the certification must not be the vessel's owner, operator, master, or an employee of any of them. Additionally, the vessel owner, operator, or ship's master must certify that the inspection was satisfactory. There are no FCC prior notice requirements for any inspection pursuant to Sec. 80.59(b)

(c) *Application for exemption.*

(1) Applications for exemption from the radio provisions of part II or III of title III of the Communications Act, the Safety Convention, or the Great Lakes Radio Agreement, or for modification or renewal of an exemption previously granted must be filed as a waiver request using FCC Form 605. Waiver requests must include the following information:

(i) Name of ship;

(ii) Call sign of ship;

(iii) Official number of ship;

(iv) Gross tonnage of ship;

(v) The radio station requirements from which the exemption is requested:

(A) Radiotelephone (VHF/MF);

(B) Radiotelegraph; and/or

(C) Radio direction finding apparatus;

(vi) File number of any previously granted exemption;

(vii) Detailed description of the voyages for which the exemption is requested, including:

(A) Maximum distance from nearest land in nautical miles;

(B) Maximum distance between two consecutive ports in nautical miles; and

(C) Names of all ports of call and an indication of whether travel will include a foreign port;

(viii) Reasons for the exemption:

(A) Size of vessel;

(B) Variety of radio equipment on board;

(C) Limited routes; and/or

(D) Conditions of voyages;

(ix) A copy of the U.S. Coast Guard Certificate of Inspection an indication of whether the vessel is certified as a Passenger or Cargo ship (for passenger ships, list the number of passengers the ship is licensed to carry); and

(x) Type and quantity of radio equipment on board, including:

(A) VHF Radio Installation (indicate if GMDSS approved);

(B) Single Side-Band (SSB) (indicate the band of operation, MF or HF and indicate if GMDSS approved);

(C) Category 1, 406 MHz EPIRB (GMDSS approved);

(D) NAVTEX Receiver (GMDSS approved);

(E) Survival Craft VHF (GMDSS approved);

(F) 9 GHz Radar Transponder (GMDSS approved);

(G) Ship Earth Station;

(H) 500 kHz Distress Frequency Watch Receiver;

(I) 2182 Radiotelephone Auto Alarm;

(J) Reserve Power Supply (capability); and

(K) Any other equipment.

(2) Feeable applications for exemption must be filed with Mellon Bank, Pittsburgh, Pennsylvania at the address set forth in Sec. 1.1102. Waiver requests that do not require a fee should be submitted via the Universal Licensing System or to: Federal Communications Commission, 1270 Fairfield Road, Gettysburg, Pennsylvania 17325-7245. Emergency requests must be filed with the Federal Communications Commission, Office of the Secretary, 445 Twelfth Street, SW., TW-B204, Washington, DC 20554.

Note: With emergency requests, do not send the fee, you will be billed.

(d) *Waiver of annual inspection.*

(1) The Commission may, upon a finding that the public interest would be served, grant a waiver of the annual inspection required by Section 362(b) of the Communications Act, 47 U.S.C. 360(b), for a period of not more than 90 days for the sole purpose of enabling a United States vessel to complete its voyage and proceed to a port in the United States where an inspection can be held. An informal application must be submitted by the ship's owner, operator or authorized agent. The application must be submitted to the Commission's District Director or Resident Agent in charge of the FCC office nearest the port of arrival at least three days before the ship's arrival. The application must include:

(i) The ship's name and radio callsign;

(ii) The name of the first United States port of arrival directly from a foreign port;

(iii) The date of arrival;

(iv) The date and port at which annual inspection will be formally requested to be conducted;

(v) The reason why an FCC-licensed technician could not perform the inspection; and

(vi) A statement that the ship's compulsory radio equipment is operable.

(2) Vessels that are navigated on voyages outside of the United States for more than 12 months in succession are exempted from annual inspection required by section 362(b) of the Communications Act, provided that the vessels comply with all applicable requirements of the Safety Convention, including the annual inspection required by Regulation 9, Chapter I, and the vessel is inspected by an FCC-licensed technician in accordance with this section within 30 days of arriving in the United States.

Subpart C - Operating Requirements and Procedures

STATION REQUIREMENTS-GENERAL
§80.61 Commission inspection of stations.

All stations and required station records must be made available for inspection by authorized representatives of the Commission.

§80.63 Maintenance of transmitter power.

(a) The power of each radio transmitter must not be more than that necessary to carry on the service for which the station is licensed.

(b) Except for transmitters using single sideband and independent sideband emissions, each radio transmitter rated by the manufacturer for carrier power in excess of 100 watts must contain the instruments necessary to determine the transmitter power during its operation.

STATION REQUIREMENTS-LAND STATIONS
§80.67 General facilities requirements for coast stations.

(a) All coast stations licensed to transmit in the band 156-162 MHz must be able to transmit and receive on 156.800 MHz and at least one working frequency in the band.

(b) All coast stations that operate telephony on frequencies in the 1605-3500 kHz band must be able to transmit and receive using J3E emission on the frequency 2182 kHz and at least one working frequency in the band.

§80.69 Facilities requirement for public coast stations using telephony.

Public coast stations using telephony must be provided with the following facilities.

(a) When the station is authorized to use frequencies in the 1605-3500 kHz band, equipment meeting the requirements of §80.67(b) must be installed at each transmitting location.

(b) The transmitter power on the frequency 2182 kHz must not exceed 50 watts carrier power for normal operation. During distress, urgency and safety traffic, operation at maximum power is permitted.

§80.70 Special provisions relative to coast station VHF facilities.

(a) Coast stations which transmit on the same radio channel above 150 MHz must minimize interference by reducing radiated power, by decreasing antenna height or by installing directional antennas. Coast stations at locations separated by less than 241 kilometers (150 miles) which transmit on the same radio channel above 150 MHz must also consider a time-sharing arrangement. The Commission may order station changes if agreement cannot be reached between the involved licensees.

(b) Coast stations which transmit on a radio channel above 150 MHz and are located within interference range of any station within Canada or Mexico must minimize interference to the involved foreign station(s), and must notify the Commission of any station changes.

§80.72 Antenna requirements for coast stations.

All emissions of a coast station a marine-utility station operated on shore using telephony within the frequency band 30-200 MHz must be vertically polarized.

STATION REQUIREMENTS - SHIP STATIONS
§80.79 Inspection of ship station by a foreign government.

The governments or appropriate administrations of countries which a ship visits may require the license of the ship station or ship earth station to be produced for examination. When the license cannot be produced without delay or when irregularities are observed, governments or administrations may inspect the radio installations to satisfy themselves that the installation conforms to the conditions imposed by the Radio Regulations.

§80.80 Operating controls for ship stations.

(a) Each control point must be capable of:
(1) Starting and discontinuing operation of the station;
(2) Changing frequencies within the same sub-band;
(3) Changing from transmission to reception and vice versa.
(4) In the case of stations operating in the 156-162 MHz bands, reducing power output to one watt or less in accordance with §80.215(e).1

(b) Each ship station using telegraphy must be capable of changing from telegraph transmission to telegraph reception and vice versa without manual switching.

(c) Each ship station using telephony must be capable of changing from transmission to reception and vice versa within two seconds excluding a change in operating radio channel.

(d) During its hours of service, each ship station must be capable of:
(1) Commencing operation within one minute;
(2) Discontinuing all emission within five seconds after emission is no longer desired.

(e) Each ship station using a multichannel installation for telegraphy (except equipment intended for use only in emergencies on frequencies below 515 kHz) must be capable of changing from one radio channel to another within:

(1) Five seconds if the channels are within the same sub-band; or
(2) Fifteen seconds if the channels are not within the same sub-band.

(f) Each ship station and marine-utility station using a multi-channel installation for telephony must be capable of changing from one radio channel to another within:
(1) Five seconds within the band 1605-3500 kHz; or
(2) Three seconds within the band 156-162 MHz.

(g)(1) Any telegraphy transmitter constructed since January 1, 1952, that operates in the band 405-525 kHz with an output power in excess of 250 watts must be capable of reducing the output power to 150 watts or less.

(2) The requirement of paragraph (g)(1) of this section does not apply when there is available in the same station a transmitter capable of operation on the international calling frequency 500 kHz and at least one working frequency within the band 405-525 kHz, capable of being energized by a source of power other than an emergency power source and not capable of an output in excess of 100 watts when operated on such frequencies.

§80.81 Antenna requirements for ship stations.

All telephony emissions of a ship station or a marine utility station on board ship within the frequency band 30-200 MHz must be vertically polarized.

§80.83 Protection from potentially hazardous RF radiation.

Any license or renewal application for a ship earth station that will cause exposure to radiofrequency (RF) radiation in excess of the RF exposure guidelines specified in §1.1307(b) of the Commission's Rules must comply with the environmental processing rules set forth in §1.1301-1.1319 of this chapter.

OPERATING PROCEDURES - GENERAL
§80.86 International regulations applicable.

In addition to being regulated by these rules, the use and operation of stations subject to this part are governed by the Radio Regulations and the radio provisions of all other international agreements in force to which the United States is a party.

§80.87 Cooperative use of frequency assignments.

Each radio channel is available for use on a shared basis only and is not available for the exclusive use of any one station or station licensee. Station licensees must cooperate in the use of their respective frequency assignment in order to minimize interference and obtain the most effective use of the authorized radio channels.

§80.88 Secrecy of communication.

The station licensee, the master of the ship, the responsible radio operators and any person who may have knowledge of the radio communications transmitted or received by a fixed, land, or mobile station subject to this part, or of any radiocommunication service of such station, must observe the secrecy requirements of the Communications Act and the Radio Regulations. See sections 501, 502, and 705 of the Communications Act and Article 23 of the Radio Regulations.

§80.89 Unauthorized transmissions.

Stations must not:

(a) Engage in superfluous radiocommunication.

(b) Use telephony on 243 MHz.

(c) Use selective calling on 2182 kHz or 156.800 MHz.

(d) When using telephony, transmit signals or communications not addressed to a particular station or stations. This provision does not apply to the transmission of distress, alarm urgency, or safety signals or messages, or to test transmissions.

(e) Transmit while on board vessels located on land unless authorized under a public coast station license. Vessels in the following situations are not considered to be on land for the purposes of this paragraph:

(1) Vessels which are aground due to a distress situation;

(2) Vessels in drydock undergoing repairs; and

(3) State or local government vessels which are involved in search and rescue operations including related training exercises.

(f) Transmit on frequencies or frequency bands not authorized on the current station license.

§80.90 Suspension of transmission.

Transmission must be suspended immediately upon detection of a transmitter malfunction and must remain suspended until the malfunction is corrected, except for transmission concerning the immediate safety of life or property, in which case transmission must be suspended as soon as the emergency is terminated.

§80.91 Order of priority of communications.

(a) All stations in the maritime mobile service and the maritime mobile-satellite service shall be capable of offering four levels of priority in the following order:

(1) Distress calls, distress messages, and distress traffic.

(2) Urgency communications.

(3) Safety communications.

(4) Other communications.

(b) In a fully automated system, where it is impracticable to offer all four levels of priority, category 1 shall receive priority until such time as intergovernmental agreements remove exemptions granted for such systems from offering the complete order of priority.

§80.92 Prevention of interference.

(a) The station operator must determine that the frequency is not in use by monitoring the frequency before transmitting, except for transmission of signals of distress.

(b) When a radiocommunication causes interference to a communication which is already in progress, the interfering station must cease transmitting at the request of either party to the existing communication. As between nondistress traffic seeking to commence use of a frequency, the priority is established under §80.91.

(c) Except in cases of distress, communications between ship stations or between ship and aircraft stations must not interfere with public coast stations. The ship or aircraft stations which cause interference must stop transmitting or change frequency upon the first request of the affected coast station.

§80.93 Hours of service.

(a) *All stations.* All stations whose hours of service are not continuous must not suspend operation before having concluded all communication required in connection with a distress call or distress traffic.

(b) *Public coast stations.* (1) Each public coast station whose hours of service are not continuous must not suspend operation before having concluded all communication involving messages or calls originating in or destined to mobile stations within range and mobile stations which have indicated their presence.

(2) Unless otherwise authorized by the Commission upon adequate showing of need, each public coast station authorized to operate on frequencies in the 3000-23,000 kHz band must maintain continuous hours of service.

(c) *Compulsory ship stations.* (1) Compulsory ship stations whose service is not continuous may not suspend operation before concluding all traffic originating in or destined for public coast stations situated within their range and mobile stations which have indicated their presence.

(2) For GMDSS ships, radios shall be turned on and set to proper watch channels while ships are underway. If a ship has duplicate GMDSS installations for DSC or INMARSAT, only one of each must be turned on and keeping watch.

(d) *Ships Voluntarily Fitting GMDSS Subsystems.* For ships voluntarily fitting GMDSS subsystems, radios shall be turned on and set to proper watch channels while ships are underway. If ship has duplicate GMDSS installations for DSC or INMARSAT, only one of each must be turned on and keeping watch.

(e) *Other than public coast or compulsory ship stations.* The hours of service of stations other than those described in paragraphs (b), (c), and (d) of this section are determined by the station licensee.

§80.95 Message charges.

(a) Charges must not be made for service of:

(1) Any public coast station unless tariffs for the service are on file with the Commission;

(2) Any station other than a public coast

station or an Alaska-public fixed station, except cooperatively shared stations covered by §80.503;

(3) Distress calls and related traffic; and

(4) Navigation hazard warnings preceded by the SAFETY signal.

(b) The licensee of each ship station is responsible for the payment of all charges accruing to any other station(s) or facilities for the handling or forwarding of messages or communications transmitted by that station.

§80.101 Radiotelephone testing procedures.

This section is applicable to all stations using telephony except where otherwise specified.

(a) Station licensees must not cause harmful interference. When radiation is necessary or unavoidable, the testing procedure described below must be followed:

(1) The operator must not interfere with transmissions in progress.

(2) The testing station's call sign, followed by the word "test", must be announced on the radiochannel being used for the test.

(3) If any station responds "wait", the test must be suspended for a minimum of 30 seconds, then repeat the call sign followed by the word "test" and listen again for a response. To continue the test, the operator must use counts or phrases which do not conflict with normal operating signals, and must end with the station's call sign. Test signals must not exceed ten seconds, and must not be repeated until at least one minute has elapsed. On the frequency 2182 kHz or 156.800 MHz, the time between tests must be a minimum of five minutes.

(b) Testing of transmitters must be confined to single frequency channels on working frequencies. However, 2182 kHz and 156.800 MHz may be used to contact ship or coast stations as appropriate when signal reports are necessary. Short tests on 4215 MHz are permitted by vessels equipped with MF/HF radios to evaluate the compatibility of the equipment for distress and safety purposes. U.S. Coast Guard stations may be contacted on 2182 kHz or 156.800 MHz for test purposes only when tests are being conducted by Commission employees, when FCC-licensed technicians are conducting inspections on behalf of the Commission, when qualified technicians are installing or repairing radiotelephone equipment, or when qualified ship's personnel conduct an operational check requested by the U.S. Coast Guard. In these cases the test must be identified as "FCC" or "technical."

(c) Survival craft transmitter tests must not be made within actuating range of automatic alarm receivers.

§80.102 Radiotelephone station identification.

This section applies to all stations using telephony which are subject to this part.

(a) Except as provided in paragraphs (d)

and (e) of this section, stations must give the call sign in English. Identification must be made:

(1) At the beginning and end of each communication with any other station.

(2) At 15 minute intervals when transmission is sustained for more than 15 minutes. When public correspondence is being exchanged with a ship or aircraft station, the identification may be deferred until the completion of the communications.

(b) Private coast stations located at drawbridges and transmitting on the navigation frequency 156.650 MHz may identify by use of the name of the bridge in lieu of the call sign.

(c) Ship stations transmitting on any authorized VHF bridge-to-bridge channel may be identified by the name of the ship in lieu of the call sign.

(d) Ship stations operating in a vessel traffic service system or on a waterway under the control of a U.S. Government agency or a foreign authority, when communicating with such an agency or authority may be identified by the name of the ship in lieu of the call sign, or as directed by the agency or foreign authority.

(e) Voice traffic in the INMARSAT system is closed to other parties except the two stations involved and the identification is done automatically with the establishment of the call. Therefore, it is not necessary for these stations to identify themselves periodically during the communication. For terrestrial systems using DSC to establish radiotelephone communications, the identification is made at the beginning of the call. In these cases, both parties must identify themselves by ship name, call sign or MMSI at least once every 15 minutes during radiotelephone communications.

(f) VHF public coast station may identify by means of the approximate geographic location of the station or the area it serves when it is the only VHF public coast station serving the location or there will be no conflict with the identification of any other station.

§80.103 Digital selective calling (DSC) operating procedures.

(a) Operating procedures for the use of DSC equipment in the maritime mobile service are as contained in ITU-R Recommendation M.541-8 and subpart W of this part.

(b) When using DSC techniques, coast stations and ship stations must use maritime mobile service identities (MMSI) assigned by the Commission or its designees.

(c) DSC acknowledgement of DSC distress and safety calls must be made by designated coast stations and such acknowledgement must be in accordance with procedures contained in ITU-R Recommendation M.541-8. Nondesignated public and private coast stations must follow the guidance provided for ship stations in ITU-R Recommendation

M.541-8 with respect to DSC "Acknowledgement of distress calls" and "Distress relays." (See subpart W of this part.)

(d) Group calls to vessels under the common control of a single entity are authorized. A group call identity may be created from an MMSI ending in a zero, assigned to this single entity, by deleting the trailing zero and adding a leading zero to the identity.

§80.104 Identification of radar transmissions not authorized.

This section applies to all maritime radar transmitters except radar beacon stations.

(a) Radar transmitters must not transmit station identification.

OPERATING PROCEDURES - LAND STATIONS
§80.105 General obligations of coast stations.

Each coast station or marine-utility station must acknowledge and receive all calls directed to it by ship or aircraft stations. Such stations are permitted to transmit safety communication to any ship or aircraft station.

§80.106 Intercommunication in the mobile service.

(a) Each public coast station must exchange radio communications with any ship or aircraft station at sea; and each station on shipboard or aircraft at sea must exchange radio communications with any other station on shipboard or aircraft at sea or with any public coast station.

(b) Each public coast station must acknowledge and receive all communications from mobile stations directed to it, transmit all communications delivered to it which are directed to mobile stations within range in accordance with their tariffs. Discrimination in service is prohibited.

§80.108 Transmission of traffic lists by coast stations.

(a) Each coast station is authorized to transmit lists of call signs in alphabetical order of all mobile stations for which they have traffic on hand. These traffic lists will be transmitted on the station's normal working frequencies at intervals of:

(1) In the case of telegraphy, at least two hours and not more than four hours during the working hours of the coast station.

(2) In the case of radiotelephony, at least one hour and not more than four hours during the working hours of the coast station.

(b) The announcement must be as brief as possible and must not be repeated more than twice. Coast stations may announce on a calling frequency that they are about to transmit call lists on a specific working frequency.

§80.110 Inspection and maintenance of tower markings and associated control equipment.

The owner of each antenna structure required to be painted and/or illuminated under the provisions of section 303(q) of the Communications Act of 1934, as amended, shall operate and maintain the antenna structure painting and lighting in accordance with part 17 of this chapter.

§80.111 Radiotelephone operating procedures for coast stations.

This section applies to all coast stations using telephony which are subject to this part.

(a) *Limitations on calling.* (1) Except when transmitting a general call to all stations for announcing or preceding the transmission of distress, urgency, or safety messages, a coast station must call the particular station(s) with which it intends to communicate.

(2) Coast stations must call ship stations by voice unless it is known that the particular ship station may be contacted by other means such as automatic actuation of a selective ringing or calling device.

(3) Coast stations may be authorized emission for selective calling on each working frequency.

(4) Calling a particular station must not continue for more than one minute in each instance. If the called station does not reply, that station must not again be called for two minutes. When a called station does not reply to a call sent intervals of two minutes, the calling must cease for fifteen minutes. However, if harmful interference will not be caused to other communications in progress, the call may be repeated after three minutes.

(5) A coast station must not attempt to communicate with a ship station that has specifically called another coast station until it becomes evident that the called station does not answer, or that communication between the ship station and the called station cannot be carried on because of unsatisfactory operating conditions.

(6) Calls to establish communication must be initiated on an available common working frequency when such a frequency exists and it is known that the called ship maintains a simultaneous watch on the common working frequency and the appropriate calling frequency(ies).

(b) *Time limitation on calling frequency.* Transmissions by coast stations on 2182 kHz or 156.800 MHz must be minimized and any one exchange of communications must not exceed one minute in duration.

(c) *Change to working frequency.* After establishing communications with another station by call and reply on 2182 kHz or 156.800 MHz coast stations must change to an authorized working channel for the transmission of messages.

(d) *Use of busy signal.* A coast station, when communicating with a ship station which transmits to the coast station on a radio channel which is a different channel from that used by the coast station for transmission, may transmit a "busy" signal whenever transmission from tile ship station is being received.

OPERATING PROCEDURES - SHIP STATIONS
§80.114 Authority of the master.

(a) The service of each ship station must at all times be under the ultimate control of the master, who must require that each operator or such station comply with the Radio Regulations in force and that the ship station is used in accordance with those regulations.

(b) These rules are waived when the vessel is under the control of the U.S. Government.

§80.115 Operational conditions for use of associated ship units.

(a) Associated ship units may be operated under a ship station authorization. Use of an associated ship unit is restricted as follows;

(1) It must only be operated on the safety and calling frequency 156.800 MHz or on commercial or noncommercial VHF intership frequencies appropriate to the class of ship station with which it is associated.

(2) Except for safety purposes, it must only be used to communicate with the ship station with which it is associated or with associated ship units of the same ship station. Such associated ship units may not be used from shore.

(3) It must be equipped to transmit on the frequency 156.800 MHz and at least one appropriate intership frequency.

(4) Calling must occur on the frequency 156.800 MHz unless calling and working on an intership frequency has been prearranged.

(5) Power is limited to one watt.

(6) The station must be identified by the call sign of the ship station with which it is associated and an appropriate unit designator.

§80.116 Radiotelephone operating procedures for ship stations.

(a) *Calling coast stations.* (1) Use by ship stations of the frequency 2182 kHz for calling coast stations and for replying to calls from coast stations is authorized. However, such calls and replies should be on the appropriate ship-shore working frequency.

(2) Use by ship stations and marine utility stations of the frequency 156.800 MHz for calling coast stations and marine utility stations on shore, and for replying to calls from such stations, is authorized. However, such calls and replies should be made on the appropriate ship-shore working frequency.

(b) *Calling ship stations.* (1) Except when other operating procedure is used to expedite safety communication, ship stations, before transmitting on the intership working frequencies 2003, 2142, 2638, 2738, or 2830 kHz, must first establish communications with other ship stations by call and reply on 2182 kHz. Calls may be initiated on an intership working frequency when it is known that the called vessel maintains a simultaneous watch on the working frequency and on 2182 kHz.

(2) Except when other operating procedures are used to expedite safety communications, the frequency 156.800 MHz must be used for call and reply by ship stations and marine utility stations before establishing communication on one of the intership working frequencies. Calls may be initiated on an intership working frequency when it is known that the called vessel maintains a simultaneous watch on the working frequency and on 156.800 MHz.

(c) *Change to working frequency.* After establishing communication with another station by call and reply on 2182 kHz or 156.800 MHz stations on board ship must change to an authorized working frequency for the transmission of messages.

(d) *Limitations on calling.* Calling a particular station must not continue for more than 30 seconds in each instance. If the called station does not reply, the station must not again be called until after an interval of 2 minutes. When a called station called does not reply to a call sent three times at intervals of 2 minutes, the calling must cease and must not be renewed until after an interval of 15 minutes; however, if there is no reason to believe that harmful interference will be caused to other communications in progress, the call sent three times at intervals of 2 minutes may be repeated after a pause of not less than 3 minutes. In event of an emergency involving safety, the provisions of this paragraph do not apply.

(e) *Limitations on working.* Any one exchange of communications between any two ship stations on 2003, 2142, 2638, 2738, or 2830 kHz or between a ship station and a private coast station on 2738 or 2830 kHz must not exceed 3 minutes after the stations have established contact. Subsequent to such exchange of communications, the same two stations must not again use 2003, 2142, 2638, 2738, or 2830 kHz for communication with each other until 10 minutes have elapsed.

(f) *Transmission limitation on 2182 kHz and 156.800 MHz.* To facilitate the reception of distress calls, all transmissions on 2182 kHz and 156.800 MHz (channel 16) must be minimized and transmissions on 156.800 MHz must not exceed 1 minute.

(g) *Limitations on commercial communication.* On frequencies in the band 156-162 MHz, the exchange of commercial communication must be limited to the minimum practicable transmission time. In the conduct of ship-shore communication other than distress, stations on board ship must comply with instructions given by the private coast station or marine utility station on shore with which they are communicating.

SPECIAL PROCEDURES-SHIP STATIONS
§80.141 General provisions for ship stations.

(a) *Points of communication.* Ship stations and marine utility stations on board ships are authorized to communicate with any station in the maritime mobile service.

(b) *Service requirements for all ship stations.* (1) Each ship station must receive and acknowledge all communications which are addressed to the ship or to any person on board.

(2) Every ship, on meeting with any direct danger to the navigation of other ships such as ice, a derelict vessel, a tropical storm, subfreezing air temperatures associated with gale force winds causing severe icing on superstructures, or winds of force 10 or above on the Beaufort scale for which no storm warning has been received, must transmit related information to ships in the vicinity and to the authorities on land unless such action has already been taken by another station. All such radio messages must be preceded by the safety signal.

(3) A ship station may accept communications for retransmission to any other station in the maritime mobile service. Whenever such messages or communications have been received and acknowledged by a ship station for this purpose, that station must retransmit the message as soon as possible.

(c) *Service requirements for vessels.* Each ship station provided for compliance with Part II of Title III of the Communications Act must provide a public correspondence service on voyages of more than 24 hours for any person who requests the service. Compulsory radiotelephone ships must provide this service for at least four hours daily. The hours must be prominently posted at the principal operating location of the station.

§80.143 Required frequencies for radiotelephony.

(a) Except for compulsory vessels, each ship radiotelephone station licensed to operate in the band 1605-3500 kHz must be able to receive and transmit J3E emission on the frequency 2182 kHz. Ship stations are additionally authorized to receive and transmit H3E emission for communications with foreign coast stations and with vessels of foreign registry. If the station is used for other than safety communications, it must be capable also of receiving and transmitting the J3E emission on at least two other frequencies in that band. However, ship stations which operate exclusively on the Mississippi River and its connecting waterways, and on high frequency bands above 3500 kHz, need be equipped with 2182 kHz and one other frequency within the band 1605-3500 kHz.

(b) Except as provided in paragraph (c) of this section, at least one VHF radiotelephone transmitter/receiver must be able to transmit and receive on the following frequencies:

(1) The distress, safety and calling frequency 156.800 MHz;

(2) The primary intership safety frequency 156.300 MHz;

(3) One or more working frequencies; and

(4) All other frequencies necessary for its service.

(c) Where a ship ordinarily has no requirement for VHF communications, handheld VHF equipment may be used solely to comply with the bridge-to-bridge navigational communication.

SHIPBOARD GENERAL PURPOSE WATCHES
§80.147 Watch on 2182 kHz.

Ship stations must maintain a watch on 2182 kHz as prescribed by §80.304(b).

§80.148 Watch on 156.8 MHz (Channel 16).

Until February 1, 2005, each compulsory vessel, while underway, must maintain a watch for radiotelephone distress calls on 156.800 MHz whenever such station is not being used for exchanging communications. For GMDSS ships, 156.525 MHz is the calling frequency for distress, safety, and general communications using digital selective calling and the watch on 156.8 MHz is provided so that ships not fitted with DSC will be able to call GMDSS ships, thus providing a link between GMDSS and non-GMDSS compliant ships. The watch on 156.800 MHz is not required:

(a) Where a ship station is operating only with handheld bridge-to-bridge VHF radio equipment under §80.143(c) of this part;

(b) For vessels subject to the Bridge-to-Bridge Act and participating in a Vessel Traffic Service (VTS) system when the watch is maintained on both the bridge-to-bridge frequency and a separately assigned VTS frequency.

VIOLATIONS
§80.149 Answer to notice of violation.

(a) Any person receiving official notice of violation of the terms of the Communications Act, any legislative act, executive order, treaty to which the United States is a party, terms of a station or operator license, or the rules and regulations of the Federal Communications Commission must within 10 days from such receipt, send a written answer, in duplicate, to the office of the Commission originating the official notice. If an answer cannot be sent or an acknowledgment made within such 10-day period by reason of illness or other unavoidable circumstances, acknowledgment and answer must be made at the earliest practicable date with a satisfactory explanation of the delay. The answer to each notice must be complete in itself and must not be abbreviated by references to other communications or answers to other notices. The answer must contain a full explanation of the incident involved and must set forth the action taken to prevent a continuation or recurrence. If the notice relates to lack of attention to or improper operation of the station or to log or watch discrepancies, the answer must give the name and license number of the licensed operator on duty.

Subpart D - Operator Requirements

§80.151 Classification of operator licenses and endorsements.

(a) Commercial radio operator licenses issued by the Commission are classified in accordance with the Radio Regulations of the International Telecommunication Union.

(b) The following licenses are issued by the Commission. International classification, if different from the license name, is given in parentheses. The licenses and their alphanumeric designator are listed in descending order.

(1) T-1. First Class Radiotelegraph Operator's Certificate.

(2) T-2. Second Class Radiotelegraph Operator's Certificate.

(3) G. General Radiotelephone Operator License (radiotelephone operator's general certificate).

(4) T-3. Third Class Radiotelegraph Operator's Certificate (radiotelegraph operator's special certificate).

(5) MP. Marine Radio Operator Permit (radiotelephone operator's restricted certificate).

(6) RP. Restricted Radiotelephone Operator Permit (radiotelephone operator's restricted certificate).

(7) GOL. GMDSS Radio Operator license (General Operator's Certificate).

(8) ROL. Restricted GMDSS Radio Operator license (Restricted Operator's Certificate).

(c) The following license endorsements are affixed by the Commission to provide special authorizations or restrictions. Applicable licenses are given in parentheses.

(1) Ship Radar endorsement (First and Second Class Radiotelegraph Operator's Certificate, General Radiotelephone Operator License).

(2) Six Months Service endorsement (First and Second Class Radiotelegraph Operator's Certificate).

(3) Restrictive endorsements; relating to physical handicaps, English language or literacy waivers, or other matters (all licenses).

SHIP STATION OPERATOR REQUIREMENTS
§80.155 Ship station operator requirements.

Except as provided in §80.177 and 80.179, operation of transmitters of any ship station must be performed by a person holding a commercial radio operator license or permit of the class required below. The operator is responsible for the proper operation of the station.

§80.156 Control by operator.

The operator on board ships required to have a holder of a commercial operator license or permit on board may, if authorized by the station licensee or master, permit an unlicensed person to modulate the transmitting apparatus for all modes of communication except Morse code radiotelegraphy.

§80.157 Radio officer defined.

A radio officer means a person holding a first or second class radiotelegraph operator's certificate Issued by the Commission who is employed to operate a ship radio station in compliance with Part II of Title III of the Communications Act. Such a person is also required to be licensed as a radio officer by the U.S. Coast Guard when employed to operate a ship radiotelegraph station.

§80.159 Operator requirements of Title III of the Communications Act and the Safety Convention.

(a) Each telegraphy passenger ship equipped with a radiotelegraph station in accordance with Part II of Title III of the Communications Act must carry one radio officer holding a first or second class radiotelegraph operator's certificate and a second radio officer holding either a first or second class radiotelegraph operator's certificate. The holder of a second class radiotelegraph operator's certificate may not act as the chief radio officer.

(b) Each cargo ship equipped with a radiotelegraph station in accordance with Part II of Title III of the Communications Act and which has a radiotelegraph auto alarm must carry a radio officer holding a first or second class radiotelegraph operator's certificate who has had at least six months service as a radio officer on board U.S. ships. If the radiotelegraph station does not have an auto alarm, a second radio officer who holds a first or second class radiotelegraph operator's certificate must be carried.

(c) Each cargo ship equipped with a radiotelephone station in accordance with Part II of Title III of the Communications Act must carry a radio operator who meets the following requirements:

(1) Where the station power does not exceed 1500 watts peak envelope power, the operator must hold a marine radio operator permit or higher class license.

(2) Where the station power exceeds 1500 watts peak envelope power, the operator must hold a general radiotelephone radio operator license or higher class license.

(d) Each passenger ship equipped with a GMDSS installation in accordance with subpart W of this part shall carry at least two persons holding an appropriate GMDSS Radio Operator License or, if the passenger ship operates exclusively within twenty nautical miles of shore, at least two persons holding either a GMDSS Radio Operator License or a Restricted GMDSS Radio Operator License, as specified in §13.7 of this chapter.

(e) Each ship transporting more than six passengers for hire equipped with a radiotelephone station in accordance with Part III of Title III of the Communications Act must carry a radio operator who meets the following requirements:

(1) Where the station power does not exceed 250 watts carrier power or 1500 watts peak envelope power, the radio operator must hold a marine radio operator permit or higher class license.

(2) Where the station power exceeds 250 watts carrier power or 1500 watts peak envelope power, the radio operator must hold a general radiotelephone operator license or higher class license.

§80.161 Operator requirements of the Great Lakes Radio Agreement.

Each ship subject to the Great Lakes Radio Agreement must have on board an officer or member of the crew who holds a marine radio operator permit or higher class license.

§80.163 Operator requirements of the Bridge-to-Bridge Act.

Each ship subject to the Bridge-to-Bridge Act must have on board a radio operator who holds a restricted radiotelephone operator permit or higher class license.

§80.165 Operator requirements for voluntary stations.

Minimum operator license

Ship Morse telegraph	T-2
Ship direct-printing telegraph	MP
Ship telephone, with or without DSC, more than 250 watts carrier power or 1,000 watts peak envelope power.	G
Ship telephone, with or without DSC, not more than 250 watts carrier power or 1,000 watts peak envelope power.	MP
Ship telephone, with or without DSC, notmore than 100 watts carrier power or 400 watts peak envelope power:	
Above 30 MHz	None[1]
Below 30 MHz	RP
Ship earth station	RP

[1]RP required for compulsory ships and international voyage.

GENERAL OPERATOR REQUIREMENTS
§80.167 Limitations on operators.

The operator of maritime radio equipment other than T-1, T-2, or G licensees, must not:

(a) Make equipment adjustments which may affect transmitter operation;

(b) Operate any transmitter which requires more than the use of simple external switches or manual frequency selection or transmitters whose frequency stability is not maintained by the transmitter itself.

§80.169 Operators required to adjust transmitters or radar.

(a) All adjustments of radio transmitters in any radiotelephone station or coincident with the installation, servicing, or maintenance of such equipment which may affect the proper operation of the station, must be performed by or under the immediate supervision and responsibility of a person holding a first or second class radiotelegraph operator's certificate or a general radiotelephone operator license.

(b) Only persons holding a first or second class radiotelegraph operator certificate must perform such functions at radiotelegraph stations transmitting Morse code.

(c) Only persons holding an operator certificate containing a ship radar endorsement

must perform such functions on radar equipment.

§80.175 Availability of operator licenses.

All operator licenses required by this subpart must be readily available for inspection.

§80.177 When operator license is not required.

(a) No radio operator authorization is required to operate:

(1) A shore radar, a shore radiolocation, maritime support or shore radionavigation station;

(2) A survival craft station or an emergency position indicating radio beacon;

(3) A ship radar station if:

(i) The radar frequency is determined by a nontunable, pulse type magnetron or other fixed tuned device, and

(ii) The radar is capable of being operated exclusively by external controls;

(4) An on board station; or

(5) A ship station operating in the VHF band on board a ship voluntarily equipped with radio and sailing on a domestic voyage.

(b) No radio operator license is required to install a VHF transmitter in a ship station if the installation is made by, or under the supervision of, the licensee of the ship station and if modifications to the transmitter other than front panel controls are not made.

(c) No operator license is required to operate coast telephone stations or marine utility stations.

(d) No radio operator license is required to install a radar station on a voluntarily equipped ship when a manual is included with the equipment that provides step-by-step instructions for the installation, calibration, and operation of the radar. The installation must be made by, or under the supervision of, the licensee of that ship station and no modifications or adjustments other than to the front panel controls are to be made to the equipment.

Subpart E - General Technical Standards

§80.201 Scope.

This subpart gives the general technical requirements for the use of frequencies and equipment in the maritime services. These requirements include standards for equipment authorization, frequency tolerance, modulation, emission, power and bandwidth.

§80.203 Authorization of transmitters for licensing.

(a) Each transmitter authorized in a station in the maritime services after September 30, 1986, ...must be type accepted by the Commission for Part 80 operations. The procedures for type acceptance are contained in Part 2. Transmitters of a model type accepted or type approved before October 1, 1986 will be considered type accepted for use in ship or coast stations as appropriate.

(3) Programming of authorized channels must be performed only by a person holding a first or second class radiotelegraph operator's

certificate or a general radiotelephone operator's license.

§80.213 Modulation requirements.
(a) Transmitters must meet the following modulation requirements:

(1) When double-sideband emission is used the peak modulation must be maintained between 75 and 100 percent;

(2) When phase or frequency modulation is used in the 156-162 MHz and 216-220 MHz bands the peak modulation must be maintained between 75 and 100 percent. A frequency deviation of ±5 kHz is defined as 100 percent peak modulation; and

(3) In single-sideband operation the upper sideband must be transmitted.

§80.215 Transmitter power.
(g) The carrier power of ship station radiotelephone transmitters, except portable transmitters, operating in the 156-162 MHz band must be at least 8 but not more than 25 watts. Transmitters that use 12-volt lead acid storage batteries as a primary power source must be measured with a primary voltage between 12.2 and 13.7 volts DC.

§80.227 Special requirements for protection from RF radiation.
As part of the information provided with transmitters for ship earth stations, manufacturers of each unit must include installation and operating instructions to help prevent human exposure to radiofrequency (RF) radiation in excess of the RF ANSI (1982) exposure guidelines.

Subpart F - Equipment Authorization for Compulsory Ships

§80.251 Scope.
(a) This subpart gives the general technical requirements for certification of equipment used on compulsory ships. Such equipment includes automatic-alarm-signal keying devices, survival craft radio equipment, watch receivers, and radar.

(b) The equipment described in this subpart must be type accepted.

§80.268 Requirements of radiotelephone installation.
All radiotelephone installations in radiotelegraph equipped vessels must meet the following conditions.

(a) The radiotelephone transmitter must be capable of transmission of A3E or H3E emission on 2182 kHz and must be capable of transmitting clearly perceptible signals from ship to ship during daytime, under normal conditions over a range of 150 nautical miles when used with an antenna system in accordance with paragraph (c) of this section. The transmitter must:

(1) Have a duty cycle which allows for transmission of the radiotelephone alarm signal described in Sec. 80.221.

(2) Provide 25 watts carrier power for A3E

emission or 60 watts peak power on H3E emission into an artificial antenna consisting of 10 ohms resistance and 200 picofarads capacitance or 50 ohms nominal impedance to demonstrate compliance with the 150 nautical mile range requirement.

(3) Have a visual indication whenever the transmitter is supplying power to the antenna.

(4) Have a two-tone alarm signal generator that meets Sec. 80.221.

(5) This transmitter may be contained in the same enclosure as the receiver required by paragraph (b) of this section. These transmitters may have the capability to transmit J2D or J3E transmissions.

(b)(1) The radiotelephone receiver must receive A3E and H3E emissions when connected to the antenna system specified in paragraph (c) this section and must be preset to 2182 kHz. The receiver must additionally:

(i) Provide an audio output of 50 milliwatts to a loudspeaker when the RF input is 50 microvolts. The 50 microvolt input signal must be modulated 30 percent at 400 Hertz and provide at least a 6 dB signal-to-noise ratio when measured in the rated audio bandwidth.

(ii) Be equipped with one or more loudspeakers capable of being used to maintain a watch on 2182 kHz at the principal operating position or in the room from which the vessel is normally steered.

(2) The receiver required by Sec. 80.805 may be used instead of this receiver. If the watch is stood at the place from which the ship is normally steered, a radiotelephone distress frequency watch receiver must be used for this purpose.

(3) This receiver may be contained in the same enclosure as the transmitter required by paragraph (a) of this section. These receivers may have the capability to receive J2D or J3E transmissions.

(c) The antenna system must be as nondirectional and efficient as is practicable for the transmission and reception of radio ground waves over seawater. The installation and construction of the required antenna must ensure, insofar as is practicable, proper operation in time of emergency. If the required antenna is suspended between masts or other supports subject to whipping, a safety link must be installed which under heavy stress will reduce breakage of the antenna, the halyards, or any other supporting elements.

(d) The radiotelephone installation must be provided with a device for permitting changeover from transmission to reception and vice versa without manual switching.

(e) An artificial antenna must be provided to permit weekly checks, without causing interference, of the automatic device for generating the radiotelephone alarm signal on frequencies other than the radiotelephone distress frequency.

(f) The radiotelephone installation must be located in the radiotelegraph operating room or in the room from which the ship is normally steered.

(g) Demonstration of the radiotelephone installation may be required by Commission representatives to show compliance with applicable regulations.

(h) The radiotelephone installation must be protected from excessive currents and voltages.

(i) The radiotelephone installation must be maintained in an efficient condition.

§80.273 Technical requirements for radar equipment.

(a) Radar installations on board ships that are required by the Safety Convention or the U.S. Coast Guard to be equipped with radar must comply with either the document referenced in subparagraph (1) of this paragraph or the applicable document referenced in subparagraphs (2) through (4) of this paragraph. These documents are incorporated by reference in accordance with 5 USC 552(a). These documents contain specifications, standards and general requirements applicable to shipboard radar equipment and shipboard radar installations. For purposes of this part the specifications, standards and general requirements stated in these documents are mandatory irrespective of discretionary language. Radar documents are available for inspection at the Commission Headquarters in Washington, D.C. or may be obtained from the Radio Technical Commission for Maritime Services (RTCM), Suite 600, 1800 Diagonal Road, Alexandria, Virginia 22314-2480.

Subpart G - Safety Watch Requirements and Procedures

COAST STATION SAFETY WATCHES
§80.301 Watch requirements.

(a) Each public coast station operating on telegraphy frequencies in the band 405-535 kHz must maintain a watch for classes A1A, A2B and H2B emissions by a licensed radiotelegraph operator on the frequency 500 kHz for three minutes twice each hour, beginning at x h.15 and x h.45 Coordinated Universal Time (UTC).

(b) Each public coast station licensed to operate in the band 1605-3500 kHz must monitor such frequency(s) as are used for working or, at the licensee's discretion, maintain a watch on 2182 kHz.

(c) Except for distress, urgency or safety messages, coast stations must not transmit on 2182 kHz during the silence periods for three minutes twice each hour beginning at x h.00 and x h.30 Coordinated Universal Time (UTC).

(d) Each public coast station must provide assistance for distress communications when requested by the Coast Guard.

§80.303 Watch on 156.800 MHz (Channel 16).

(a) During its hours of operation, each coast station operating in the 156-162 MHz band and serving rivers, bays and inland lakes except the Great Lakes, must maintain a safety watch on the frequency 156.800 MHz except when transmitting on 156.800 MHz.

SHIP STATION SAFETY WATCHES
§80.304 Watch requirement during silence periods.

Each ship station operating on telephony on frequencies in the band 1605-3500 kHz must maintain a watch on the frequency 2182 kHz. This watch must be maintained at least twice each hour for 3 minutes commencing at x h.00 and x h.30 Coordinated Universal Time (UTC) using either a loudspeaker or headphone. Except for distress, urgency or safety messages, ship stations must not transmit during the silence periods on 2182 kHz.

§80.305 Watch requirements of the Communications Act and the Safety Convention.

(a) Each ship of the United States which is equipped with a radiotelegraph station for compliance with Part II of Title III of the Communications Act or chapter IV of the Safety Convention must:

(1) Keep a continuous and efficient watch on 500 kHz by means of radio officers while being navigated in the open sea outside a harbor or port. In lieu thereof, on a cargo ship equipped with a radiotelegraph auto alarm in proper operating condition, an efficient watch on 500 kHz must be maintained by means of a radio officer for at least 8 hours per day in the aggregate, i.e., for at least one-third of each day or portion of each day that the vessel is navigated in the open sea outside of a harbor or port.

(2) Keep a continuous and efficient watch on the radiotelephone distress frequency 2182 kHz from the principal radio operating position or the room from which the vessel is normally steered while being navigated in the open sea outside a harbor or port. A radiotelephone distress frequency watch receiver having a loudspeaker and a radiotelephone auto alarm facility must be used to keep the continuous watch on 2182 kHz if such watch is kept from the room from which the vessel is normally steered. After a determination by the master that conditions are such that maintenance of the listening watch would interfere with the safe navigation of the ship, the watch may be maintained by the use of the radiotelephone auto alarm facility alone.

(3) Until February 1, 2005, keep a continuous and efficient watch on the VHF distress frequency 156.800 MHz from the room from which the vessel is normally steered while in the open sea outside a harbor or port. The watch must be maintained by a designated member of the crew who may perform other duties, relating to the operation or navigation of the vessel, provided such other duties do not interfere with the effectiveness of the

watch. Use of a properly adjusted squelch or brief interruptions due to other nearby VHF transmissions are not considered to adversely affect the continuity or efficiency of the required watch on the VHF distress frequency. This watch need not be maintained by vessels subject to the Bridge-to-Bridge Act and participating in a Vessel Traffic Services (VTS) system as required or recommended by the U.S. Coast Guard, when an efficient listening watch is maintained on both the bridge-to-bridge frequency and a separate assigned VTS frequency.

§80.310 Watch required by voluntary vessels.

Voluntary vessels not equipped with DSC must maintain a watch on 156.800 MHz (Channel 16) whenever the vessel is underway and the radio is not being used to communicate. Noncommercial vessels, such as recreational boats, may alternatively maintain a watch on 156.450 MHz (Channel 9) for call and reply purposes. Voluntary vessels equipped with VHF-DSC equipment must maintain a watch on either 156.525 MHz (Channel 70) or VHF Channel 16 aurally whenever the vessel is underway and the radio is not being used to communicate. Voluntary vessels equipped with MF-HF DSC equipment must have the radio turned on and set to an appropriate DSC distress calling channel or one of the radiotelephone distress channels whenever the vessel is underway and the radio is not being use to communicate. Voluntary vessels equipped with Inmarsat A, B, or C systems must have the unit turned on and set to receive calls whenever the vessel is underway and the radio is not being used to communicate.

DISTRESS, ALARM, URGENCY AND SAFETY PROCEDURES
§80.311 Authority for distress transmission.

A mobile station in distress may use any means at its disposal to attract attention, make known its position and obtain help. A distress call and message, however, must be transmitted only on the authority of the master or person responsible for the mobile station. No person shall knowingly transmit, or cause to be transmitted, any false or fraudulent signal of distress or related communication.

§80.312 Priority of distress transmissions.

The distress call has absolute priority over all other transmissions. All stations which hear it must immediately cease any transmission capable of interfering with the distress traffic and must continue to listen on the frequency used for the emission of the distress call. This call must not be addressed to a particular station. Acknowledgement of receipt must not be given before the distress message which follows it is sent.

§80.313 Frequencies for use in distress.

The frequencies specified in the bands below are for use by mobile stations in distress. The conventional emission is shown. When a ship station cannot transmit on the designated frequency or the conventional emission, it may use any available frequency or emission. Frequencies for distress and safety calling using digital selective calling techniques are listed in §80.359(b). Distress and safety NBDP frequencies are indicated by note 2 in §80.361(b).

Frequency band	Emission	Carrier frequency
1605 - 3500 kHz	J3E	2182 kHz.
118 - 136 MHz	A3E	121.500 MHz.
156 - 162 MHz	F3E, PON	156.800 MHz
		156.750 MHz.
243 MHz	A3N	243.000 MHz

The maximum transmitter power obtainable may be used.

§80.314 Distress signals.

(a) The international radiotelephone distress signal consists of the word MAYDAY, pronounced as the French expression "m'aider".

(b) These distress signals indicate that a mobile station is threatened by grave and imminent danger and requests immediate assistance.

§80.315 Distress calls.

(a) The radiotelephone distress call consists of:

(1) The distress signal MAYDAY spoken three times;

(2) The words THIS IS;

(3) The call sign (or name, if no call sign assigned) of the mobile station in distress, spoken three times.

(b) The procedures for canceling false distress alerts are contained in section 80.335 of this part.

§80.316 Distress messages.

(a) The radiotelephone distress message consists of:

(1) The distress signal MAYDAY;

(2) The name of the mobile station in distress;

(3) Particulars of its position;

(4) The nature of the distress;

(5) The kind of assistance desired;

(6) Any other information which might facilitate rescue, for example, the length, color, and type of vessel, number of persons on board.

(b) As a general rule, a ship must signal its position in latitude and longitude, using figures for the degrees and minutes, together with one of the words NORTH or SOUTH and one of the words EAST or WEST. In radiotelegraphy, the signal .-.-. must be used to separate the degrees from the minutes. When practicable, the true bearing and distance in nautical miles from a known geographical position may be given.

(c) The procedures for canceling false distress alerts are contained in section 80.335 of this part.

§80.327 Urgency signals.

(a) The urgency signal indicates that the calling station has a very urgent message to transmit concerning the safety of a ship, aircraft, or other vehicle, or the safety of a person. The urgency signal must be sent only on the authority of the master or person responsible for the mobile station.

(b) In radiotelegraphy, the urgency signal consists of three repetitions of the group XXX, sent with the individual letters of each group, and the successive groups clearly separated from each other. It must be transmitted before the call.

(c) In radiotelephony, the urgency signal consists of three oral repetitions of the group of words PAN PAN PAN transmitted before the call.

(d) The urgency signal has priority over all other communications except distress. All mobile and land stations which hear it must not interfere with the transmission of the message which follows the urgency signal.

§80.329 Safety signals.

(a) The safety signal indicates that the station is about to transmit a message concerning the safety of navigation or giving important meteorological warnings.

(b) In radiotelegraphy, the safety signal consists of three repetitions of the group TTT, sent with the individual letters of each group, and the successive groups clearly separated from each other. It must be sent before the call.

(c) In radiotelephony, the safety signal consists of the word SECURITY, pronounced as in French, spoken three times and transmitted before the call.

(d) The safety signal and call must be sent on one of the international distress frequencies (500 kHz or 8364 kHz radiotelegraph; 2182 kHz or 156.8 MHz radiotelephone). Stations which cannot transmit on a distress frequency may use any other available frequency on which attention might be attracted.

§80.331 Bridge-to-bridge communication procedure.

(a) Vessels subject to the Bridge-to-Bridge Act transmitting on the designated navigational frequency must conduct communications in a format similar to those given below:

(1) This is the (name of vessel). My position is (give readily identifiable position, course and speed) about to (describe contemplated action). Out.

(2) Vessel off (give a readily identifiable position). This is (name of vessel) off (give a readily identifiable position). I plan to (give proposed course of action). Over.

(3) (Coast station), this is (vessel's name) off (give readily identifiable position). I plan to (give proposed course of action). Over.

(b) Vessels acknowledging receipt must answer (Name of vessel calling). This is (Name of vessel answering). Received your call, and follow with an indication of their in-

tentions. Communications must terminate when each ship is satisfied that the other no longer poses a threat to its safety and is ended with "Out".

(c) Use of power greater than 1 watt in a bridge-to-bridge station shall be limited to the following three situations:

(1) Emergency.

(2) Failure of the vessel being called to respond to a second call at low power.

(3) A broadcast call as in paragraph (a)(1) of this section in a blind situation, e.g., rounding a bend in a river.

§80.334 False distress alerts.

A distress alert is false if it was transmitted without any indication that a mobile unit or person was in distress and required immediate assistance. Transmitting a false distress alert is prohibited and may be subject to the provisions of part 1, subpart A of this chapter if that alert:

(a) was transmitted intentionally;

(b) was not cancelled in accordance with §80.335;

(c) could not be verified as a result of either the ship's failure to keep watch on appropriate frequencies in accordance with §80.1123 or subpart G of this part, or its failure to respond to calls from the U.S. Coast Guard;

(d) was repeated; or

(e) was transmitted using a false identity.

§80.335 Procedures for canceling false distress alerts.

If a distress alert is inadvertently transmitted, the following steps shall be taken to cancel the distress alert.

(a) VHF Digital Selective Calling.

(i) Reset the equipment immediately;

(ii) Transmit a DSC distress alert cancellation (i.e., own ship's acknowledgment), if that feature is available;

(iii) Set to Channel 16; and

(iv) Transmit a broadcast message to "All stations" giving the ship's name, call sign or registration number, and MMSI, and cancel the false distress alert.

(b) MF Digital Selective Calling.

(i) Reset the equipment immediately;

(ii) Transmit a DSC distress alert cancellation (i.e., own ship's acknowledgment), if that feature is available;

(iii) Tune for radiotelephony transmission on 2182 kHz; and

(iv) Transmit a broadcast message to "All stations" giving the ship's name, call sign or registration number, and MMSI, and cancel the false distress alert.

(c) HF Digital Selective Calling.

(i) Reset the equipment immediately;

(ii) Transmit a DSC distress alert cancellation (i.e., own ship's acknowledgment), if that feature is available, on each frequency on which the distress alert was transmitted;

(iii) Tune for radiotelephony on the distress and safety frequency in each band in which a false distress alert was transmitted; and

(iv) Transmit a broadcast message to "All stations" giving the ship's name, call sign or registration number, and MMSI, and cancel the false distress alert frequency in each band in which a false distress alert was transmitted.

(d) INMARSAT ship earth station. Immediately notify the appropriate rescue coordination center that the alert is cancelled by sending a distress priority message by way of the same land earth station through which the false distress alert was sent. Provide ship name, call sign or registration number, and INMARSAT identity with the cancelled alert message.

(e) EPIRB. If for any reason an EPIRB is activated inadvertently, immediately contact the nearest U.S. Coast Guard unit or appropriate rescue coordination center by telephone, radio or ship earth station and cancel the distress alert.

(f) General and other distress alerting systems. Notwithstanding the above, ships may use additional appropriate means available to them to inform the nearest appropriate U.S. Coast Guard rescue coordination center that a false distress alert has been transmitted and should be cancelled.

Subpart H - Frequencies

RADIOTELEGRAPHY
§80.351 Scope.

[This section] describes the carrier frequencies and general uses of radiotelegraphy with respect to the following:

♦ Distress, urgency, safety, call and reply.
♦ Working.
♦ Digital selective calling (DSC).
♦ Narrow-band direct-printing (NB-DP).
♦ Facsimile.

[This section also contains all of the Radiotelephony, Radiodetermination, Ship Earth Station, Maritime frequencies assigned to Aircraft, Operational Fixed Station, Vessel Traffic Service (VTS), Automated System, Alaska Fixed Station, Maritime Support Station and Developmental Station frequencies.]

§80.375 Radiodetermination frequencies.

This section describes the carrier frequencies assignable to radiodetermination stations. Only direction-finding stations will be authorized on land.

Subpart I - Station Documents

§80.401 Station documents requirement.

Licensees of radio stations are required to have current station documents such as station licenses, operator authorizations, station logs, safety certificates, Part 80 (Rules and Regulations), maritime mobile call sign and coast/ship station lists ...and other station equipment records.

§80.403 Availability of documents.

Station documents must be readily available to the licensed operator(s) on duty during the hours of service of the station and to authorized Commission employees upon request.

§80.405 Station license.

(a) *Requirement.* Except as provided in section 80.13(c) of this part, stations must have an authorization granted by the Federal Communications Commission.

(b) *Application.* Application for authorizations in the maritime services must be submitted on the prescribed forms in accordance with subpart B of this part.

(c) *Posting.* The current station authorization or a clearly legible copy must be posted at the principal control point of each station. If a copy is posted, it must indicate the location of the original. When the station license cannot be posted, as in the case of a marine utility station operating at temporary unspecified locations, it must be kept where it will be readily available for inspection. The licensee of a station on board a ship subject to Part II or III of Title III of the Communications Act or the Safety Convention must retain the most recently expired ship station license in the station records until the first Commission inspection after the expiration date.

§80.409 Station logs.

(a) General requirements. Logs must be established and properly maintained as follows:

(1) The log must be kept in an orderly manner. The required information for the particular class or category of station must be readily available. Key letters or abbreviations may be used if their proper meaning or explanation is contained elsewhere in the same log.

(2) Erasures, obliterations or willful destruction within the retention period are prohibited. Corrections may be made only by the person originating the entry by striking out the error, initialing the correction and indicating the date of correction.

(3) Ship station logs must identify the vessel name, country of registry, and official number of the vessel.

(4) The station licensee and the radio operator in charge of the station are responsible for the maintenance of station logs.

(b) Availability and retention. Station logs must be made available to authorized Commission employees upon request and retained as follows:

(1) Logs must be retained by the licensee for a period of one year from the date of entry, and when applicable for such additional periods as required by the following paragraphs:

(i) Logs relating to a distress situation or disaster must be retained for three years from the date of entry.

(ii) If the Commission has notified the licensee of an investigation, the related logs must be retained until the licensee is specifically authorized in writing to destroy them.

(iii) Logs relating to any claim or complaint of which the station licensee has notice must be retained until the claim or complaint has been satisfied or barred by statute limiting the time for filing suits upon such claims.

(2) Logs containing entries required by paragraphs (e) and (f) of this section must be kept at the principal radiotelephone operating location while the vessel is being navigated. All entries in their original form must be retained on board the vessel for at least 30 days from the date of entry.

(3) Ship radiotelegraph logs must be kept in the principal radiotelegraph operating room during the voyage.

Subpart J - Public Coast Stations

STATIONS ON LAND

§80.451 Supplemental eligibility requirements.

A public coast station license may be granted to any person meeting the citizenship provisions of §80.15(b).

§80.453 Scope of communications.

Public coast stations provide ship/shore radiotelephone and radiotelegraph services.

(a) Public coast stations are authorized to communicate:

(1) With any ship or aircraft station operating in the maritime mobile service, for the transmission or reception of safety communication;

(2) With any land station to exchange safety communications to or from a ship or aircraft station;

(3) With Government and non-Government ship and aircraft stations to exchange public correspondence.

Subpart K - Private Coast Stations and Marine Utility Stations

§80.501 Supplemental eligibility requirements.

(a) A private coast station or a marine utility station may be granted only to a person who is:

(1) Regularly engaged in the operation, docking, direction, construction, repair, servicing or management of one or more commercial transport vessels or United States, state or local government vessels; or is

(2) Responsible for the operation, control, maintenance or development of a harbor, port or waterway used by commercial transport vessels; or is

(3) Engaged in furnishing a ship arrival and departure service, and will employ the station only for the purpose of obtaining the information essential to that service; or is

(4) A corporation proposing to furnish a nonprofit radio communication service to its parent corporation, to another subsidiary of the same parent, or to its own subsidiary where the party to be served performs any of the eligibility activities described in this section; or is

(5) A nonprofit corporation or association, organized to furnish a maritime mobile service solely to persons who operate one or more commercial transport vessels; or is

(6) Responsible for the operation of bridges, structures or other installations that area part of, or directly related to, a harbor, port or waterway when the operation of such facilities requires radio communications with vessels for safety or navigation; or is

(7) A person controlling public moorage facilities; or is

(8) A person servicing or supplying vessels other than commercial transport vessels; or is

(9) An organized yacht club with moorage facilities; or is

(10) A nonprofit organization providing noncommercial communications to vessels other than commercial transport vessels.

(b) Each application for station authorization for a private coast station or a marine utility station must be accompanied by a statement indicating eligibility under paragraph (a) of this section.

Subpart L - Operational Fixed Stations

§80.555 Scope of communication.

An operational fixed station provides control, repeater or relay functions for its associated coast station.

Subpart M - Stations in the Radiodetermination Service

§80.601 Scope of communications.

Stations on land in the Maritime Radiodetermination Service provide a radionavigation or radiolocation service for ships.

§80.603 Assignment and use of frequencies.

The frequencies available for assignment to shore radionavigation/radiolocation stations are contained in subpart H of this part.

§80.605 U.S. Coast Guard coordination.

(a) Radionavigation coast stations operated to provide information to aid in the movement of any ship are private aids to navigation. Before submitting an application for a radionavigation station, an applicant must obtain written permission from the cognizant Coast Guard District Commander at the area in which the device will be located. Documentation of the Coast Guard approval must be submitted with the application.

Subpart N - Maritime Support Stations

§80.653 Scope of communications.

(a) Maritime support stations are land stations authorized to operate at permanent locations or temporary unspecified locations.

(b) Maritime support stations are authorized to conduct the following operations:

(1) Training of personnel in maritime telecommunications;

(2) Transmissions necessary for the test and maintenance of maritime radio equipment at repair shops; and

(3) Transmissions necessary to test the technical performance of the licensee's public coast station(s) radiotelephone receiver(s); and

(4) Transmissions necessary for radar/racon equipment demonstration.

Subpart O - Alaska Fixed Stations

§80.701 Scope of service.

There are two classes of Alaska Fixed stations. Alaska-public fixed stations are common carriers, open to public correspondence, which operate on the paired duplex channels listed in Subpart H of this Part. Alaska-private fixed stations may operate on simplex frequencies listed in Subpart H of this Part to communicate with other Alaska private fixed stations or with ship stations, and on duplex frequencies listed in Subpart H of this Part when communicating with the Alaska public fixed stations. Alaska-private fixed stations must not charge for service, although third party traffic may be transmitted. Only Alaska public fixed stations are authorized to charge for communication services.

Subpart P - Standards for Computing Public Coast Station VHF Coverage

§80.751 Scope.

This subpart specifies receiver antenna terminal requirements in terms of power, and relates the power available at the receiver antenna terminals to transmitter power and antenna height and gain.

Subpart Q - [Reserved]

Subpart R - Technical Equipment Requirements for Cargo Vesse ls Not Subject to Subpart W

§80.851 Applicability.

The radiotelephone requirements of this subpart are applicable to all compulsory ships which are not required to comply with Subpart W in total or in part or are temporarily exempted from some of the Subpart W provisions.

§80.853 Radiotelephone station.

(a) The radiotelephone station is a radiotelephone installation and other equipment necessary for the proper operation of the installation.

(b) The radiotelephone station must be installed to insure safe and effective operation of the equipment and to facilitate repair. Adequate protection must be provided against the effects of vibration, moisture, and temperature.

(c) The radiotelephone station and all necessary controls must be located at the level of the main wheelhouse or at least one deck above the ship's main deck.

(d) The principal operating position of the radiotelephone station must be in the room from which the ship is normally steered while at sea.

§80.855 Radiotelephone transmitter.

(a) The transmitter must be capable of transmission of H3E and J3E emission on 2182 kHz, and J3E emission on 2638 kHz and at least two other frequencies within the band 1605 to 3500 kHz available for ship-to-shore or ship-to-ship communication.

(b) The duty cycle of the transmitter must permit transmission of the international radiotelephone alarm signal.

(c) The transmitter must be capable of transmitting clearly perceptible signals from ship to ship during daytime under normal conditions over a range of 150 nautical miles.

(d) The transmitter complies with the range requirement specified in paragraph (c) of this section if:

(1) The transmitter is capable of being matched to actual ship station transmitting antenna meeting the requirements of §80.863; and

(2) The output power is not less than 60 watts peak envelope power for H3E and J3E emission on the frequency 2182 kHz and for J3E emission on the frequency 2638 kHz into either an artificial antenna consisting of a series network of 10 ohms resistance and 200 picofarads capacitance, or an artificial antenna of 50 ohms nominal impedance. An individual demonstration of the power output capability of the transmitter, with the radiotelephone installation normally installed on board ship, may be required.

(e) The transmitter must provide visual indication whenever the transmitter is supplying power to the antenna.

(f) The transmitter must be protected from excessive currents and voltages.

(g) A durable nameplate must be mounted on the transmitter or made an integral part of it showing clearly the name of the transmitter manufacturer and the type or model of the transmitter.

(h) An artificial antenna must be provided to permit weekly checks of the automatic device for generating the radiotelephone alarm signal on frequencies other than the radiotelephone distress frequency.

§80.873 VHF radiotelephone transmitter.

(a) The transmitter must be capable of transmission of G3E emission on 156.300 MHz and 156.800 MHz, and on frequencies which have been specified for use in a system established to promote safety of navigation.

(b) The transmitter must be adjusted so that the transmission of speech normally produces peak modulation within the limits of 75 percent and 100 percent.

(c) The transmitter must deliver a carrier power between 8 and 25 watts into a 50 ohm effective resistance. Provision must be made for reducing the carrier power to a value between 0.1 and 1.0 watt.

§80.880 Vessel radio equipment.

(a) Vessels operated solely within twenty nautical miles of shore must be equipped with a VHF radiotelephone installation as described in this subpart, and maintain a continuous watch on Channel 16.

(b) Vessels operated solely within one hundred nautical miles of shore must be equipped with a medium frequency transmitter capable of transmitting J3E emission and a receiver capable of reception of J3E emission within the band 1710 to 2850 kHz, in addition to the VHF radiotelephone installation required by paragraph (a) of this section, and must maintain a continuous watch on 2182 kHz. Additionally, such vessels must be equipped with either:

(1) a single sideband radiotelephone capable of operating on all distress and safety frequencies in the medium frequency and high frequency bands listed in §80.369(a) and (b), on all the ship-to-shore calling frequencies in the high frequency bands listed in 80.369(d), and on at least four of the automated mutual-assistance vessel rescue (AMVER) system HF duplex channels (this requirement may be met by the addition of such frequencies to the radiotelephone installation required by paragraph (b) of this section); or

(2) if operated in an area within the coverage of an INMARSAT maritime mobile geostationary satellite in which continuous alerting is available, an INMARSAT ship earth station meeting the equipment authorization rules of parts 2 and 80 of this chapter.

§80.881 Equipment Requirements for Ship Stations

Vessels subject to this Subpart must be equipped as follows:

(a) A category 1, 406.0-406.1 MHz EPIRB meeting the requirements of §80.1061;

(b) A NAVTEX receiver meeting the requirements of §80.1101(c)(1);

(c) A Search and Rescue Transponder meeting the requirements of §80.1101(c)(6);

(d) A two-way VHF radiotelephone meeting the requirements of §80.1101(c)(7).

Subpart S - Compulsory Radiotelephone Installations for Small Passenger Boats

§80.901 Applicability.

The provisions of Part III of Title III of the Communication Act require United States vessels which transport more than six passengers for hire while such vessels are being navigated on any tidewater within the jurisdiction of the United States adjacent or contiguous to the open sea, or in the open sea to carry a radiotelephone installation complying with this subpart. The provisions of Part III do not apply to vessels which are equipped with a radio installation for compliance with Part II of Title Ill of the Act, or for compliance with the Safety Convention, or to vessels navigating on the Great Lakes.

§80.903 Inspection of radiotelephone installation.

Every vessel subject to Part III of Title III of the Communications Act must have a detailed inspection by the Commission of the prescribed installation once every five years. If after inspection the Commission determines that all relevant provisions of Part III of Title III of the Communications Act, the rules of the Commission, and the station license are met a Communications Act Safety Radiotelephone Certificate will be issued. The effective date of this certificate is the date the installation is found to be in compliance, or not more than one business day later.

§80.905 Vessel radio equipment.

(a) Vessels subject to Part III of Title III of the Communications Act that operate in the waters described in §80.901 of this section must, at a minimum, be equipped as follows:

(1) Vessels operated solely within the communications range of a VHF public coast station or U.S. Coast Guard station that maintains a watch on 156.800 MHz while the vessel is navigated must be equipped with a VHF radiotelephone installation. Vessels in this category must not operate more than 20 nautical miles from land.

(2) Vessels operated beyond the 20 nautical mile limitation specified in paragraph (a)(1) of this section, but not more than 100 nautical miles from the nearest land, must be equipped with a MF transmitter capable of transmitting J3E emission and a receiver capable of reception of J3E emission within the band 1710 to 2850 kHz, in addition to the VHF radiotelephone installation required by paragraph (a)(1) of this section. The MF transmitter and receiver must be capable of operation on 2670 kHz.

(3) Vessels operated more than 100 nautical miles but not more than 200 nautical miles from the nearest land must:

(i) Be equipped with a VHF radiotelephone installation;

(ii) Be equipped with an MF radiotelephone transmitter and receiver meeting the requirements of paragraph (a)(2) of this section; and

(iii) Be equipped with either:

(A) a single sideband radiotelephone capable of operating on all distress and safety frequencies in the medium frequency and high frequency bands listed in Sec. 80.369 (a) and (b), on all the ship-to-shore calling frequencies in the high frequency bands listed in Sec. 80.369(d), and on at least four of the automated mutual-assistance vessel rescue (AMVER) system HF duplex channels (this requirement may be met by the addition of such frequencies to the radiotelephone installation required by paragraph (a)(2) of this section); or

(B) if operated in an area within the coverage of an INMARSAT maritime mobile geostationary satellite in which continuous alerting is available, an INMARSAT ship earth station meeting the equipment authorization rules of parts 2 and 80 of this chapter;

(iv) Be equipped with a reserve power supply meeting the requirements of Sec. 80.917(b), 80.919, and 80.921, and capable of powering the single sideband radiotelephone or the ship earth station (including associated peripheral equipment) required by paragraph (a)(3)(iii) of this section;

(v) Be equipped with a NAVTEX receiver conforming to the following performance standards: IMO Resolution A.525(13) and ITU-R Recommendation 540;

(vi) Be equipped with a Category I, 406 MHz satellite emergency position-indicating radiobeacon (EPIRB) meeting the requirements of Sec. 80.1061; and,

(vii) Participate in the AMVER system while engaged on any voyage where the vessel is navigated in the open sea for more than 24 hours. Copies of the AMVER Bulletin are available at: AMVER Maritime Relations (G-NRS-3/AMR), U.S. Coast Guard, Building 110, Box 26, Governor's Island, N.Y. 10004-5034, telephone number (212) 668-7764.

(4) Vessels operated more than 200 nautical miles from the nearest land must:

(i) Be equipped with two VHF radiotelephone installations;

(ii) Be equipped with an MF radiotelephone transmitter and receiver meeting the requirements of paragraph (a)(2) of this section;

(iii) Be equipped with either:

(A) an independent single sideband radiotelephone capable of operating on all distress and safety frequencies in the medium frequency and high frequency bands listed in Sec. 80.369(a) and (b), on all of the ship-to-shore calling frequencies in the high frequency bands listed in Sec. 80.369(d), and on at least four of the automated mutual-assistance vessel rescue (AMVER) system HF duplex channels; or

(B) If operated in an area within the coverage of an INMARSAT maritime mobile geostationary satellite in which continuous alerting is available, an INMARSAT ship earth station meeting the equipment authorization rules of parts 2 and 80 of this chapter;

(iv) Be equipped with a reserve power supply meeting the requirements of Sec. 80.917(b), 80.919, and 80.921, and capable of powering the single sideband radiotelephone or the ship earth station (including associated peripheral equipment) required by paragraph (a)(4)(iii) of this section;

(v) Be equipped with a NAVTEX receiver conforming to the following performance standards: IMO Resolution A.525(13) and CCIR Recommendation 540;

(vi) Be equipped with a Category I, 406 MHz satellite emergency position-indicating radiobeacon (EPIRB) meeting the requirements of Sec. 80.1061;

(vii) Be equipped with a radiotelephone distress frequency watch receiver meeting the requirements of Sec. 80.269;

(viii) Be equipped with an automatic radiotelephone alarm signal generator meeting the requirements of Sec. 80.221; and

(ix) Participate in the AMVER system while engaged on any voyage where the vessel is navigated in the open sea for more than 24 hours. Copies of the AMVER Bulletin are available at: AMVER Maritime Relations, Battery Park Building, New York, NY 10004. Phone 212-668-7764; Fax 212-668-7684.

(b) For a vessel that is navigated within the communication range of a VHF public coast station or U.S. Coast Guard station, but beyond the 20 nautical mile limitation specified in paragraph (a)(1) of this section, an exemption from the band 1605 to 2850 kHz installation requirements may be granted if the vessel is equipped with a VHF transmitter and receiver. An application for exemption must include a chart showing the route of the voyage or the area of operation of the vessel, and the receiving service area of the VHF public coast or U.S. Coast Guard station. The coverage area of the U.S. Coast Guard station must be based on written information from the District Commander, U.S. Coast Guard, a copy of which must be furnished with the application. The coverage area of a public coast station must be computed by the method specified in subpart P of this part.

(c) The radiotelephone installation must be installed to insure safe operation of the equipment and to facilitate repair. It must be protected against the vibration, moisture, temperature, and excessive currents and voltages.

(d) A VHF radiotelephone installation or a remote unit must be located at each steering station except those auxiliary steering stations which are used only during brief periods for docking or for close-in maneuvering. A single portable VHF radiotelephone set meets the requirements of this paragraph if adequate permanent mounting arrangements with suitable power provision and antenna feed are installed at each operator steering station. Additionally, for vessels of more than 100 gross tons, the radiotelephone installation must be located at the level of the main wheelhouse or at least one deck above the vessel's main deck.

§80.909 Radiotelephone transmitter.

(a) The medium frequency transmitter must have a peak envelope output power of at least 60 watts for J3E emission on 2182 kHz and at least one ship-to-shore working frequency within the band 1605 to 2850 kHz enabling communication with a public coast station if the region in which the vessel is navigated is served by a public coast station operating in this band.

(b) The single sideband radiotelephone must be capable of operating on maritime frequencies in the band 1710 to 27500 kHz with a peak envelope output power of at least 120 watts for J3E emission on 2182 kHz and J3E emission on the distress and safety frequencies listed in §80.369(b).

(c) The transmitter complies with the power output requirements specified in paragraphs (a) or (b) of this section when:

(1) The transmitter can be adjusted for efficient use with an actual ship station transmitting antenna meeting the requirements of Sec. 80.923 of this part; and

(2) The transmitter, with normal operating voltages applied, has been demonstrated to deliver its required output power on the frequencies specified in paragraphs (a) or (b) of this section into either an artificial antenna consisting of a series network of 10 ohms effective resistance and 200 picofarads capacitance or an artificial antenna of 50 ohms nominal impedance. An individual demonstration of power output capability of the transmitter, with the radiotelephone installation normally installed on board ship, may be required.

(d) The single sideband radiotelephone must be capable of transmitting clearly perceptible signals from ship to shore. The transmitter complies with this requirement if it is capable of enabling communication with a public coast station on working frequencies in the 4000 to 27500 kHz band specified in Sec. 80.371(b) of this part under normal daytime operating conditions.

Subpart T - Radiotelephone Installation Required for Vessels on the Great Lakes

§80.951 Applicability.

The Agreement Between the United States of America and Canada for Promotion of Safety on the Great Lakes by Means of Radio, 1973, applies to vessels of all countries when navigated on the Great Lakes.

Subpart U - Radiotelephone Installations Required by the Bridge-To-Bridge Act

§80.1001 Applicability.

The Bridge-to-Bridge Act and the regulations of this part apply to the following vessels in the navigable waters of the United States:

(a) Every power-driven vessel of 300 gross tons and upward while navigating;

(b) Every vessel of 100 gross tons and upward carrying one or more passengers for hire while navigating;

(c) Every towing vessel of 26 feet (7.8 meters) or over in length, measured from end to end over the deck excluding sheer, while navigating: and

(d) Every dredge and floating plant engaged, in or near a channel or fairway, in operations likely to restrict or affect navigation of other vessels. An unmanned or intermittently manned floating plant under the control of a dredge shall not be required to have a separate radiotelephone capability.

Subpart V - Emergency Position Indicating Radiobeacons (EPIRBs)

§80.1051 Scope.

This subpart describes the technical and performance requirements for Classes A, B, and S, and Categories 1, 2, and 3 EPIRB stations.

§80.1053 Special requirements for Class A EPIRB stations.

Class A EPIRBs shall not be manufactured, imported, or sold in the United States on or after February 1, 2003. Operation of Class A EPIRB stations shall be prohibited after December 31, 2006. New Class A EPIRBs will no longer be certified by the Commission. Existing Class A EPIRBs must be operated as certified.

§80.1055 Special requirements for Class B EPIRB stations.

Class B EPIRBs shall not be manufactured, imported, or sold in the United States on or after February 1, 2003. Operation of Class B EPIRB stations shall be prohibited after December 31, 2006. New Class B EPIRBs will no longer be certified by the Commission. Existing Class B EPIRBs must be operated as certified.

§80.1059 Special requirements for Class S EPIRB stations.

Class S EPIRBs shall not be manufactured, imported, or sold in the United States on or after February 1, 2003. Operation of Class S EPIRB stations shall be prohibited after December 31, 2006. New Class S EPIRBs will no longer be certified by the Commission. Existing Class S EPIRBs must be operated as certified.

§80.1061 Special requirements for 406.0-406.1 MHz EPIRB stations.

(a) Notwithstanding the provisions in paragraph (b) of this section, 406.0-406.1 MHz EPIRBs must meet all the technical and performance standards contained in the Radio Technical Commission for Maritime Services document titled "RTCM Recommended Standards for 406 MHz Satellite Emergency Position Indicating Radiobeacons (EPIRBs)" version 2.1, dated August 22, 2000.

Subpart W - Global Maritime Distress and Safety System (GMDSS)

This Subpart contains the rules applicable to the Global Maritime Distress and Safety System (GMDSS).

§80.1065 Applicability.

(b) The regulations contained within this subpart apply to all passenger ships regardless of size and cargo ships of 300 tons gross tonnage and upwards as follows:

(3) Ships constructed on or after February 1, 1995, must comply with all requirements of this subpart.

(4) Ships constructed before February 1, 1995, must comply with all requirements of this subpart as of February 1, 1999.

§80.1067 Inspection of station.

(a) Ships must have the required equipment inspected at least once every 12 months.

(b) Certificates issued in accordance with the Safety Convention must be posted in a prominent and accessible place on the ship.

§80.1069 Maritime sea areas.

(a) For the purpose of this subpart, a ship's area of operation is defined as follows:

(1) Sea area A1. An area within the radiotelephone coverage of at least one VHF coast station in which continuous DSC alerting is available as defined by the International Maritime Organization.

(2) Sea area A2. An area, excluding sea area Al, within the radiotelephone coverage of at least one MF coast station in which continuous DSC alerting is available as defined by the International Maritime Organization.

(3) Sea area A3. An area, excluding sea areas Al and A2, within the coverage of an INMARSAT geostationary satellite in which continuous alerting is available.

(4) Sea area A4. An area outside sea areas Al, A2 and A3.

§80.1073 Radio operator requirements for ship stations.

(a) Ships must carry at least two persons holding GMDSS Radio Operator's Licenses as specified in §13.2 of this chapter for distress and safety radiocommunications purposes. The GMDSS Radio Operator's License qualifies personnel as GMDSS radio operator for the purposes of operating GMDSS radio installation, including basic equipment adjustments as denoted in knowledge requirements specified in §13.21 of this chapter.

(1) A qualified GMDSS radio operator must be designated to have primary responsibility for radiocommunications during distress incidents, except if the vessel operates exclusively within twenty nautical miles of shore, in which case a qualified restricted radio operator may be so designated.

(2) A second qualified GMDSS radio operator must be designated as backup for distress and safety radiocommunications, except if the vessel operates exclusively within twenty nautical miles of shore, in which case a qualified restricted GMDSS radio operator may be so designated.

(b) A qualified GMDSS radio operator, and a qualified backup, as specified in paragraph (a) of this section, must be:

(1) Available to act as the dedicated radio operator in cases of distress as described in §80.1109(a);

(2) Designated to perform as part of normal routine each of the applicable communications described in §80.1109(b);

(3) Responsible for selecting HF DSC guard channels and receiving scheduled maritime safety information broadcasts;

(4) Designated to perform communications described in §80.1109(c);

(5) Responsible for ensuring that the watches required by §80.1123 are properly maintained; and

(6) Responsible for ensuring that the ship's navigation position is entered into all installed DSC equipment, either automatically through a connected or integral navigation receiver, or manually at least every four hours when the ship is underway.

§80.1074 Radio maintenance personnel for at-sea maintenance.

(a) Ships that elect the at-sea option for maintenance of GMDSS equipment (see §80.1105) must carry at least one person who qualifies as a GMDSS radio maintainer, as specified in paragraph (b) of this section, for the maintenance and repair of equipment specified in this subpart. This person may be, but need not be, the person designated as GMDSS radio operator as specified in §80.1073.

(b) The following licenses qualify personnel as GMDSS radio maintainers to perform at-sea maintenance of equipment specified in this subpart. For the purposes of this subpart, no order is intended by this listing or the alphanumeric designator.

(1) GM: GMDSS Maintainer's License;

(2) GB: GMDSS Operator's/Maintainer's License.

(3) Until February 1, 1999:

(i) TB1: First Class Radiotelegraph Operator's Certificate;

(ii) TB2: Second Class Radiotelegraph Operator's Certificate; or,

(iii) G: General Radiotelephone Operator License.

(c) While at sea, all adjustments of radio installations, servicing, or maintenance of such installations that may affect the proper operation of the GMDSS station must be performed by, or under the immediate supervision and responsibility of, a qualified GMDSS radio maintainer as specified in paragraph (b) of this section.

(d) The GMDSS radio maintainer must possess the knowledge covering the requirements set forth in IMO Assembly on Training for Radio Personnel (GMDSS), Annex 5 and IMO Assembly on Radio Maintenance Guidelines for the Global Maritime Distress and Safety System related to Sea Areas A3 and A4.

§80.1075 Radio records.

A record must be kept, as required by the Radio Regulations and §80.409 (a), (b) and (e), of all incidents connected with the radiocommunication service which appear to be of importance to safety of life at sea.

§80.1105 Maintenance requirements.

(a) Equipment must be so designed that the main units can be replaced readily, without elaborate recalibration or readjustment.

Where applicable, equipment must be constructed and installed so that it is readily accessible for inspection and on-board maintenance purposes. Adequate information must be provided to enable the equipment to be properly operated and maintained.

(c) On ships engaged on voyages in sea areas Al and A2, the availability must be ensured by duplication of equipment, shore-based maintenance, or at-sea electronic maintenance capability, or a combination of these.

(d) On ships engaged on voyages in sea areas A3 and A4, the availability must be ensured by using a combination of at least two of the following methods: duplication of equipment, shore-based maintenance, or at-sea electronic maintenance capability.

(I) If the shore-based maintenance method is used, the following requirements apply:

(1) Maintenance services must be completed and performance verified and noted in the ship's record before departure from the first port of call entered after any failure occurs.

(j) If the at-sea maintenance method is used, the following requirements apply:

(1) Adequate additional technical documentation, tools, test equipment, and spare parts must be carried onboard ship to enable a qualified maintainer as specified in §80.1074 to perform tests and localize and repair faults in the radio equipment.

(2) Only persons that comply with the requirements of §80.1074 may perform at-sea maintenance on radio installations required by this subpart.

§80.1109 Distress, urgency, and safety communications.

(a) Distress traffic consists of all messages relating to the immediate assistance required by the ship in distress, including search and rescue communications and on-scene communications. Distress traffic must as far as possible be on [GMDSS] frequencies.

(b) Urgency and safety communications include: navigational and meteorological warnings and urgent information; ship-to-ship safety navigation communications; ship reporting communications; support communications for search and rescue operations; other urgency and safety messages and communications relating to the navigation, movements and needs of ships and weather observation messages destined for an official meteorological service.

(c) Intership navigation safety communications are those VHF radiotelephone communications conducted between ships for the purpose of contributing to the safe movement of ships. The frequency 156.650 MHz is used for intership navigation safety communications (see §80.1077).

§80.1111 Distress alerting.

(a) The transmission of a distress alert indicates that a mobile unit or person is in distress and requires immediate assistance. The distress alert is a digital selective call using a distress call format in bands used for terrestrial radiocommunication or a distress message format, which is relayed through space stations.

(b) The distress alert must be sent through a satellite either with absolute priority in general communication channels or on exclusive distress and safety frequencies or, alternatively, on the distress and safety frequencies in the MF, HF, and VHF bands using digital selective calling.

(c) The distress alert must be sent only on the authority of the person responsible for the ship, aircraft or other vehicle carrying the mobile station or the mobile earth station.

(d) All stations which receive a distress alert transmitted by digital selective calling must immediately cease any transmission capable of interfering with distress traffic and must continue watch on the digital selective call distress calling channel until the call has been acknowledged to determine if a coast station acknowledges the call using digital selective calling. Additionally, the station receiving the distress alert must set watch on the associated distress traffic frequency for five minutes to determine if distress traffic takes place. The ship can acknowledge the call using voice or narrowband direct printing as appropriate on this channel to the ship or to the rescue authority.

§80.1113 Transmission of a distress alert.

(a) The distress alert must identify the station in distress and its position. The distress alert may also contain information regarding the nature of the distress, the type of assistance required, the course and speed of the mobile unit, the time that this information was recorded and any other information which might facilitate rescue.

§80.1123 Watch requirements for ship stations.

(a) While at sea, all ships must maintain a continuous watch:

(1) On VHF DSC channel 70, if the ship is fitted with a VHF radio installation;

(2) On the distress and safety DSC frequency 2187.5 kHz, if the ship is fitted with an MF radio installation;

(3) On the distress and safety DSC frequencies 2187.5 kHz and 8414.5 kHz also on at least one of the distress and safety DSC frequencies 4207.5 kHz, 6312 kHz, 12577 kHz, or 16804.5 kHz appropriate to the time of day and the geographical position of the ship, if the ship is fitted with an MF/HF radio installation (this watch may be kept by means of a scanning receiver); and

(4) For satellite shore-to-ship distress alert, if the ship is fitted with an INMARSAT ship earth station.

(b) While at sea, all ships must maintain radio watches for broadcasts of maritime safety information on the appropriate frequency or frequencies on which such information is broadcast for the area in which the ship is navigating.

(c) Until February 1, 2005, every ship while at sea must maintain, when practicable, a continuous listening watch on VHF Channel 16. This watch must be kept at the position from which the ship is normally navigated or at a position which is continuously manned.

(d) Every ship required to carry a radiotelephone watch receiver must maintain, while at sea, a continuous watch on the radiotelephone distress frequency 2182 kHz. This watch must be kept at the position from which the ship is normally navigated or at a position which is continually manned.

(e) On receipt of a distress alert transmitted by use of digital selective calling techniques, ship stations must set watch on the radiotelephone distress and safety traffic frequency associated with the distress and safety calling frequency on which the distress alert was received.

(f) Ship stations with narrow-band direct printing equipment must set watch on the narrow-band direct-printing frequency associated with the distress alert signal If it indicates the narrow-band direct-printing is to used for subsequent distress communications. If practicable, they should additionally set watch on the radio telephone frequency associated with the distress alert frequency.

Subpart X - Voluntary Radio Installations

GENERAL

§80.1151 Voluntary radio operations

Voluntary ships must meet the rules applicable to the particular mode of operation as contained in the following subparts of this part and as modified by §80.1153:

§80.1153 Station log and radio watches.

(a) Licensees of voluntary ships are not required to operate the ship radio station or to maintain radio station logs.

(b) When a ship radio station of a voluntary ship is being operated, appropriate general purpose watches must be maintained in accordance with §80.146 (500 kHz), §80.147 (2182 kHz) and §80.148 (156.800 MHz - Channel 16).

VOLUNTARY TELEGRAPHY
§80.1155 Radioprinter.

Radioprinter operations provide a record of communications between authorized maritime mobile stations.

(a) Supplementary eligibility requirements. Ships must be less than 1600 gross tons.

(b) Scope of communication.

(1) Ship radioprinter communications may be conducted with an associated private coast station.

(2) Ships authorized to communicate by radioprinter with a common private coast station may also conduct intership radioprinter

operations.

(3) Only those communications which are associated with the business and operational needs of the ship are authorized.

§80.1165 Assignment and use of frequencies.

Frequencies for general radiotelephone purposes are available to ships in three radio frequency bands. Use of specific frequencies must meet the Commission's rules concerning the scope of service and the class of station with which communications are intended. The three frequency bands are:

(a) *156-158 MHz (VHF/FM Radiotelephone).* Certain frequencies within this band are public correspondence frequencies and they must be used as working channels when communicating with public coast stations. Other working frequencies within the band are categorized by type of communications for which use is authorized when communicating with a private coast station or between ships. Subpart H of this Part lists the frequencies and types of communications for which they are available.

(b) *1600-4000 kHz (SSB Radiotelephone).* Specific frequencies within this band are authorized for single sideband (SSB) communications with public and private coast stations or between ships. The specific frequencies are listed in Subpart H of this Part.

(c) *4000-23000 kHz (SSB Radiotelephone).* Specific frequencies within this band are authorized for SSB communications with public and private coast stations. The specific frequencies are listed in Subpart H of this Part.

RADIODETERMINATION
§80.1201 Special provisions for cable-repair ship stations.

(a) A ship station may be authorized to use radio channels in the 285-315 kHz band in Region 1 and 285-325 kHz in any other region for cable repair radiodetermination purposes under the following conditions:

(1) The radio transmitting equipment attached to the cable-marker buoy associated with the ship station must be described in the station application;

(2) The call sign used for the transmitter operating under the provisions of this section is the call sign of the ship station followed by the letters "BT" and the identifying number of the buoy.

(3) The buoy transmitter must be continuously monitored by a licensed radiotelegraph operator on board the cable repair ship station; and

(4) The transmitter must operate under the provisions in §80.375(b).

Federal Communications Commission - Rules and Regulations – Excerpts from

47 C.F.R. – PART 87 – STATIONS IN THE AVIATION SERVICES

Subpart A - General Information

§87.5 Definitions.

Aeronautical advisory station (unicom). An aeronautical station used for advisory and civil defense communications primarily with private aircraft stations.

Aeronautical enroute station. An aeronautical station which communicates with aircraft stations in flight status or with other aeronautical enroute stations.

Aeronautical fixed service. A radiocommunication service between specified fixed points provided primarily for the safety of air navigation and for the regular, efficient and economical operation of air transport. A station in this service is an aeronautical fixed station.

Aeronautical Mobile Off-Route (OR) Service. An aeronautical mobile service intended for communications, including those relating to flight coordination, primarily outside national or international civil air routes. (RR)

Aeronautical Mobile Route (R) Service. An aeronautical mobile service reserved for communications relating to safety and regularity of flight, primarily along national or international civil air routes. (RR)

Aeronautical Mobile-Satellite Off-Route (OR) Service. An aeronautical mobile-satellite service intended for communications, including those relating to flight coordination, primarily outside national and international civil air routes. (RR)

Aeronautical Mobile-Satellite Route (R) Service. An aeronautical mobile-satellite service reserved for communications relating to safety and regularity of flights, primarily along national or international civil air routes. (RR)

Aeronautical mobile service. A mobile service between aeronautical stations and aircraft stations, or between aircraft stations, in which survival craft stations may also participate; emergency position-indicating radiobeacon stations may also participate in this service on designated distress and emergency frequencies.

Aeronautical multicom station. An aeronautical station used to provide communications to conduct the activities being performed by, or directed from, private aircraft.

Aeronautical radionavigation service. A radionavigation service intended for the benefit and for the safe operation of aircraft.

Aeronautical search and rescue station. An aeronautical station for communication with aircraft and other aeronautical search and rescue stations pertaining to search and rescue activities with aircraft.

Aeronautical station. A land station in the aeronautical mobile service. In certain instances an aeronautical station may be located, for example, on board ship or on a platform at sea.

Aeronautical utility mobile station. A mobile station used on airports for communications relating to vehicular ground traffic.

Air carrier aircraft station. A mobile station on board an aircraft which is engaged in, or essential to, the transportation of passengers or cargo for hire.

Aircraft earth station (AES). A mobile earth station in the aeronautical mobile-satellite service located on board an aircraft.

Aircraft station. A mobile station in the aeronautical mobile service other than a survival craft station, located on board an aircraft.

Airport. An area of land or water that is used or intended to be used for the landing and takeoff of aircraft, and includes its buildings and facilities, if any.

Airport control tower (control tower) station. An aeronautical station providing communication between a control tower and aircraft.

Automatic weather observation station. A land station located at an airport and used to automatically transmit weather information to aircraft.

Aviation service organization. Any business firm which maintain facilities at an airport for the purposes of one or more of the following general aviation activities: (a) Aircraft fueling; (b) aircraft services (e.g. parking, storage, tie-downs); (c) aircraft maintenance or sales; (d) electronics equipment maintenance or sales; (e) aircraft rental, air taxi service or flight instructions; and (f) baggage and cargo handling, and other passenger or freight services.

Aviation services. Radio-communication services for the operation of aircraft. These services include aeronautical fixed service, aeronautical mobile service, aeronautical radiodetermination service, and secondarily, the handling of public correspondence on frequencies in the maritime mobile and maritime mobile satellite services to and from aircraft.

Aviation support station. An aeronautical station used to coordinate aviation services with aircraft and to communicate with aircraft engaged in unique or specialized activities. (See Subpart K)

Civil Air Patrol station. A station used exclusively for communications of the Civil Air Patrol.

Emergency locator transmitter(ELT). A transmitter of an aircraft or a survival craft actuated manually or automatically that is used as an alerting and locating aid for survival purposes.

Emergency locator transmitter (ELT) test station. A land station used for testing ELTs or for training in the use of ELTS.

Expendable Launch Vehicle (ELV). A booster rocket that can be used only once to launch a payload, such as a missile or space vehicle.

Flight test aircraft station. An aircraft station used in the testing of aircraft or their major components.

Flight test land station. An aeronautical station used in the testing of aircraft or their major components.

Glide path station. A radionavigation land station which provides vertical guidance to aircraft during approach to landing.

Instrument landing system (ILS). A radionavigation system which provides aircraft with horizontal and vertical guidance just before and during landing and, at certain fixed points, indicates the distance to the reference point of landing.

Instrument landing system glide path. A system of vertical guidance embodied in the instrument landing system which indicates the vertical deviation of the aircraft from its optimum path of descent.

Instrument landing system localizer. A system of horizontal guidance embodied in the instrument landing system which indicates the horizontal deviation of the aircraft from its optimum path of descent along the axis of the runway or along some other path when used as an offset.

Land station. A station in the mobile service not intended to be used while in motion.

Localizer station. A radionavigation land station which provides horizontal guidance to aircraft with respect to a runway center line.

Marker beacon station. A radionavigation land station in the aeronautical radio-navigation service which employs a marker beacon. A marker beacon is a transmitter which radiates vertically a distinctive pattern for providing position information to aircraft.

Mean power (of a radio transmitter). The average power supplied to the antenna transmission line by a transmitter during an interval of time sufficiently long compared with the lowest frequency encountered in the modulation taken under normal operating conditions.

Microwave landing system. An instrument landing system operating in the microwave spectrum that provides lateral and vertical guidance to aircraft having compatible avionics equipment.

Mobile service. A radiocommunication service between mobile and land stations, or between mobile stations. A mobile station is intended to be used while in motion or during halts at unspecified points.

Operational fixed station. A fixed station, not open to public correspondence, operated by and for the sole use of persons operating their own radiocommunication facilities in the public safety, industrial, land transportation, marine, or aviation services.

Peak envelope power (of a radio transmitter). The average power supplied to the antenna transmission line by a transmitter during one radio frequency cycle at the crest of the modulation envelope taken under normal operating conditions.

Private aircraft station. A mobile station on board an aircraft not operated as an air carrier. A station on board an air carrier aircraft weighing less than 12,500 pounds maximum certified takeoff gross weight may be licensed as a private aircraft station.

Racon station. A radionavigation land station which employs a racon. A racon (radar beacon) is a transmitter-receiver associated with a fixed navigational mark, which when triggered by a radar, automatically returns a distinctive signal which can appear on the display of the triggering radar, providing range, bearing and identification information.

Radar. A radiodetermination system based upon the comparison of reference signals with radio signals reflected, or re-transmitted, from the position to be determined.

Radio altimeter. Radionavigation equipment, on board an aircraft or spacecraft, used to determine the height of the aircraft or spacecraft above the Earth's surface or another surface.

Radiobeacon station. A station in the radionavigation service the emissions of which are intended to enable a mobile station to determine its bearing or direction in relation to the radiobeacon station.

Radiodetermination service. A radiocommunication service which uses radiodetermination. Radiodetermination is the determination of the position, velocity and/or other characteristics of an object, or the obtaining of information relating to these parameters, by means of the propagation of radio waves. A station in this service is called a radiodetermination station.

Radiolocation service. A radiodetermination service for the purpose of radiolocation. Radiolocation is the use of radiodetermination for purposes other than those of radionavigation.

Radionavigation land test stations. A radionavigation land station which is used to transmit information essential to the testing and calibration of aircraft navigational aids, receiving equipment, and interrogators at predetermined surface locations. The Maintenance Test Facility (MTF) is used primarily to permit maintenance testing by aircraft radio service personnel. The Operational Test Facility (OTF) is used primarily to permit the pilot to check a radionavigation system aboard the aircraft prior to takeoff.

Radionavigation service. A radiodetermination service for the purpose of radionavigation. Radionavigation is the use of radiodetermination for the purpose of navigation, including obstruction warning.

Re-usable launch vehicle (RLV). A booster rocket that can be recovered after launch, refurbished and relaunched.

Surveillance radar station. A radionavigation land station in the aeronautical radionavigation service employing radar to display the presence of aircraft within its range.

Survival craft station. A mobile station in the maritime or aeronautical mobile service intended solely for survival purposes and located on any lifeboat, life raft or other survival equipment.

VHF Omni directional range station (VOR). A radionavigation land station in the aeronautical radionavigation service providing direct indication of the bearing (omnibearing) of that station from an aircraft.

Subpart B - Applications and Licenses

§87.17 Scope.
This subpart contains the procedures and requirements for the filing of applications for radio station licenses in the aviation services.

§87.18 Station license required.
(a) Except as noted in paragraph (b) of this section, stations in the aviation service must be licensed by the FCC either individually or by fleet.

(b) An aircraft station is licensed by rule and does not need an individual license issued by the FCC if the aircraft station is not required by statute, treaty, or agreement to which the United States is signatory to carry a radio, and the aircraft station does not make international flights or communications. Even though an individual license is not required, an aircraft station licensed by rule must be operated in accordance with all applicable operating requirements, procedures, and technical specifications found in this part.

§87.19 Basic eligibility.
(a) General. Foreign governments or their representatives cannot hold station licenses.

(b) Aeronautical enroute and aeronautical fixed stations. The following persons cannot hold an aeronautical enroute or an aeronautical fixed station license.

(1) Any alien or the representative of any alien;

(2) Any corporation organized under the laws of any foreign government;

(3) Any corporation of which any officer or director is an alien;

(4) Any corporation of which more than one-fifth of the capital stock is owned of record or voted by aliens or their representatives or by a foreign government or its representative, or by a corporation organized under the laws of a foreign country; or

(5) Any corporation directly or indirectly controlled by any other corporation of which more than one-fourth of the capital stock is owned of record or voted by aliens, their representatives, or by a foreign government or its representatives, or by any corporation organized under the laws of a foreign country, if the Commission finds that the public interest will be served by the refusal or revocation of such license.

§87.21 Standard forms to be used.
Applications must be submitted on prescribed forms which may be obtained from the Commission in Washington, DC 20554 or from any of its field offices.

§87.27 License term.
(a) Licenses for aircraft stations will normally be issued for a term of ten years from the date of original issuance, major modification or renewal.

(b) Licenses other than aircraft stations in the aviation services will normally be issued for a term of five years from the data or original issuance, major modification or renewal.

§87.35 Cancellation of license.
When a station permanently discontinues operation, the license must be returned to the Commission, Gettysburg, PA 17326.

§87.39 Equipment acceptable for licensing.
Transmitters listed in this part must be type accepted for a particular use by the Commission based upon technical requirements contained in Subpart D of this part.

§87.41 Frequencies.
(a) *Applicant responsibilities.* The applicant must propose frequencies to be used by the station consistent with the applicant's eligibility, the proposed operation and the frequencies available for assignment. Applicants must cooperate in the selection and use of frequencies in order to minimize interference and obtain the most effective use of stations. See Subpart E and the appropriate Subpart applicable to the class of station being considered.

(b) *Licensing limitations.* Frequencies are available for assignment to stations on a shared basis only and will not be assigned for the exclusive use of any licensee. The use of any assigned frequency may be restricted to one or more geographical areas.

(c) *Government frequencies.* Frequencies allocated exclusively to federal government radio stations may be licensed. The applicant for a government frequency must provide a satisfactory showing that such assignment is required for inter-communication with government stations or required for coordination with activities of the federal government. The Commission will coordinate with the appropriate government agency before a government frequency is assigned.

(d) *Assigned frequency.* The frequency coinciding with the center of an authorized bandwidth of emission must be specified as the assigned frequency. For single sideband emission, the carrier frequency must also be specified.

§87.43 Operation during emergency.

A station may be used for emergency communications in a manner other than that specified in the station license or in the operating rules when normal communication facilities are disrupted. The Commission may order the discontinuance of any such emergency service.

Subpart C - Operating Requirements and Procedures

OPERATING REQUIREMENTS

§87.69 Maintenance tests.

The licensee may make routine maintenance tests on equipment other than emergency locator transmitters if there is no interference with the communications of any other station. Procedures for conducting tests on emergency locator transmitters are contained in Subpart F.

§87.71 Frequency measurements.

A licensed operator must measure the operating frequencies of all landbased transmitters at the following times:

(a) When the transmitter is originally installed;

(b) When any change or adjustment is made in the transmitter which may affect an operating frequency; or

(c) When an operating frequency has shifted beyond tolerance.

§87.73 Transmitter adjustments and tests.

A general radiotelephone operator must directly supervise and be responsible for all transmitter adjustments or tests during installation, servicing or maintenance of a radio station. A general radiotelephone operator must be responsible for the proper functioning of the station equipment.

§87.75 Maintenance of antenna structure marking and control equipment.

The owner of each antenna structure required to be painted and/or illuminated under the provisions of Section 303(q) of the Communications Act of 1934, as amended, shall operate and maintain the antenna structure painting and lighting in accordance with part 17 of this chapter.

§87.77 Availability for inspections.

The licensee must make the station and its records available for inspection upon request.

§87.87 Classification of operator licenses and endorsements.

(a) Commercial radio operator licenses issued by the Commission are classified in accordance with the Radio Regulations of the International Telecommunication Union.

(b) The following licenses are issued by the Commission. International classification, if different from the license name, is given in parentheses. The licenses and their alphanumeric designator are listed in descending order.

(1) T-1 First Class Radiotelegraph Operator's Certificate
(2) T-2 Second Class Radiotelegraph Operator's Certificate
(3) G General Radiotelephone Operator License (radiotelephone operator's general certificate)
(4) T-3 Third Class Radiotelegraph Operator's Certificate (radiotelegraph operator's special certificate)
(5) MP Marine Radio Operator Permit (radiotelephone operator's restricted certificate)
(6) RP Restricted Radiotelephone Operator Permit (radiotelephone operator's restricted certificate)

§87.89 Minimum operator requirements.

(a) A station operator must hold a commercial radio operator license or permit, except as listed in paragraph (d).

(b) The minimum operator license or permit required for operation of each specific classification is:

MINIMUM OPERATOR LICENSE OR PERMIT

Land stations, all classes
• All frequencies except VHF telephony transmitters providing domestic service .... RP

Aircraft stations, all classes
• Frequencies below 30 MHz allocated exclusively to aeronautical mobile services .... RP
• Frequencies below 30 MHz not allocated exclusively to aeronautical mobile services .... MP or higher
• Frequencies above 30 MHz not allocated exclusively to aeronautical mobile services and assigned for international use .... MP or higher

• Frequencies above 30 MHz
not assigned for international use .... none
• Frequencies not used solely
for telephone or exceeding 250
watts carrier power or 1000
watts peak envelope power .... G or higher

(c) The operator of a telephony station must directly supervise and be responsible for any other person who transmits from the station, and must ensure that such communications are in accordance with the station license

(d) No operator license is required to:

(1) Operate an aircraft radar set, radio altimeter, transponder or other aircraft automatic radionavigation transmitter by flight personnel;

(2) Test an emergency locator transmitter or a survival craft station used solely for survival purposes;

(3) Operate an aeronautical enroute station which automatically transmits digital communications to aircraft stations;

(4) Operate a VHF telephony transmitter providing domestic service or used on domestic flights.

§87.91 Operation of transmitter controls.

The holder of a marine radio operator permit or a restricted radiotelephone operator permit must perform only transmitter operations which are controlled by external switches. These operators must not perform any internal adjustment of transmitter frequency determining elements. Further, the stability of the transmitter frequencies at a station operated by these operators must be maintained by the transmitter itself. When using an aircraft radio station on maritime mobile service frequencies the carrier power of the transmitter must not exceed 250 watts (emission A3E) or 1000 watts (emission R3E, H3E, or J3E).

OPERATING PROCEDURES

§87.103 Posting station license.

(a) Stations at fixed locations. The license or a photocopy must be posted or retained in the station's permanent records.

(b) Aircraft radio stations. The license must be either posted in the aircraft or kept with the aircraft registration certificate. If a single authorization covers a fleet of aircraft, a copy of the license must be either posted in each aircraft or kept with each aircraft registration certificate.

(c) Aeronautical mobile stations. The license must be retained as a permanent part of the station records.

§87.105 Availability of operator permit or license.

All operator permits or licenses must be readily available for inspection.

§87.107 Station identification.

(a) Aircraft station. Identify by one of the following means:

(1) Aircraft radio station call sign.

(2) Assigned FCC control number (assigned to ultralight aircraft).

(3) The type of aircraft followed by the characters of the registration marking ("N" number) of the aircraft, omitting the prefix letter "N". When communication is initiated by a ground station, an aircraft station may use the type of aircraft followed by the last three characters of the registration marking.

(4) The FAA assigned radiotelephony designator of the aircraft operating organization followed by the flight identification number.

(5) An aircraft identification approved by the FAA for use by aircraft stations participating in an organized flying activity of short duration.

(b) Land and fixed stations. Identify by means of radio station call sign, its location, its assigned FAA identifier, the name of the city area or airport which it serves, or any additional identification required. An aeronautical enroute station which is part of a multistation network may also be identified by the location of its control point.

(c) Survival craft station. Identify by transmitting a reference to its parent aircraft. No identification is required when distress signals are transmitted automatically. Transmissions other than distress or emergency signals, such as equipment testing or adjustment, must be identified by the call sign or by the registration marking of the parent aircraft followed by a single digit other than 0 or 1.

(d) Exempted station. The following types of stations are exempted from the use of a call sign: Airborne weather radar, radio altimeter, air traffic control transponder, distance measuring equipment, collision avoidance equipment, racon, radio relay, radionavigation land test station (MTF), and automatically controlled aeronautical enroute stations.

§87.109 Station logs.

A station at a fixed location in the international aeronautical mobile service must maintain a written or automatic log in accordance with Paragraph 3.5, Volume II, Annex 10 of the ICAO Convention.

§87.111 Suspension or discontinuance of operation.

The licensee of any airport control tower station or radionavigation land station must notify the nearest FAA regional office upon the temporary suspension or permanent discontinuance of the station. The FAA center must be notified again when service resumes.

Subpart D - Technical Requirements.

This subpart contains the power, frequency stability, bandwidth, emission, modulation and acceptability of transmitters for licensing requirements.

Subpart E - Frequencies.

This subpart contains class of station symbols and a frequency table which lists assignable frequencies.

Subpart F - Aircraft Stations.

This subpart covers communications limitations of domestic and foreign aircraft stations and requirements for public correspondence. Aircraft stations must limit their communications to the necessities of safe, efficient, and economic operation of aircraft and the protection of life and property.

Subpart G - Aeronautical Advisory Stations (Unicoms)

This subpart covers the guidelines for unicom communications which must be limited to the necessities of safe and expeditious operation of aircraft such as condition of runways, types of fuel available, wind conditions, weather information, dispatching, or other necessary information.

Subpart H - Aeronautical Multicom Stations

The communications of an aeronautical multicom station (multicom) must pertain to activities of a temporary, seasonal or emergency nature involving aircraft in flight. Communications are limited to directing or coordinating ground activities from the air or aerial activities from the ground.

Subpart I - Aeronautical Enroute and Aeronautical Fixed Stations

Aeronautical enroute stations provide operational control communications to aircraft along domestic or international air routes. Operational control communications include the safe, efficient and economical operation of aircraft, such as fuel, weather, position reports, aircraft performance, and essential services and supplies. Public correspondence is prohibited. Aeronautical fixed stations provide non-public point-to-point communications service pertaining to safety, regularity and economy of flight.

Subpart J - Flight Test Stations

The use of flight test stations is restricted to the transmission of necessary information or instructions relating directly to tests of aircraft or components thereof.

Subpart K - Aviation Support Stations

Aviation support stations are used for the following types of operations:

(a) Pilot training;

(b) Coordination of soaring activities between gliders, tow aircraft and land stations;

(c) Coordination of activities between free balloons or lighter-than-air aircraft and ground stations;

(d) Coordination between aircraft and aviation service organizations located on an airport concerning the safe and efficient portal-to-portal transit of the aircraft, such as the types of fuel and ground services available, and

(e) Promotion of safety of life and property.

Subpart L - Aeronautical Utility Mobile Stations

Aeronautical utility mobile stations provide communications for vehicles operating on an airport movement area. An airport movement area is defined as the runways, taxiways and other areas utilized for taxiing, takeoff and landing of aircraft, exclusive of loading ramp and parking areas.

Subpart M - Aeronautical Search and Rescue Stations

Aeronautical search and rescue land and mobile stations must be used only for communications with aircraft and other aeronautical search and rescue stations engaged in search and rescue activities.

Subpart N - Emergency Communications

This Subpart provides the rules governing operation of stations in the Aviation Services during any national or local emergency situation constituting a threat to national security or safety of life and property.

Subpart O - Airport Control Tower Stations

Airport control tower stations (control towers) and control tower remote communications outlet stations (RCOS) must limit their communications to the necessities of safe and expeditious operations of aircraft operating on or in the vicinity of the airport. Control towers and RCOs provide air traffic control services to aircraft landing, taking off and taxing on the airport as well as aircraft transmitting the airport traffic area.

Subpart P - Operational Fixed Stations

An operational fixed station provides control, repeater or relay functions for its associated aeronautical station.

Subpart Q - Stations in the Radiodetermination Service

Stations in the aeronautical radiodetermination service provide radionavigation and radiolocation services which must be limited to aeronautical navigation, including obstruction warning.

Subpart R - Civil Air Patrol Stations

Civil Air Patrol land and mobile stations must be used only for training, operational and emergency activities of the Civil Air Patrol. They may communicate with other land and, mobile stations of the Civil Air Patrol. When engaged in training or on actual missions in support of the U.S. Air Force, Civil Air Patrol stations may communicate with U.S. Air Force stations on the frequencies specified in Subpart E.

Subpart S - Automatic Weather Observation Stations

Automatic weather observation stations provide up-to-date weather information including the time of the latest weather sequence, altimeter setting, wind speed and direction, dewpoint, temperature, visibility and other pertinent data needed at airports having neither a full-time control tower nor a full-time FAA Flight Service Station.

Glossary

Alternator: A device for generating alternating current electricity by moving conductors across magnetic field lines.

AM – Amplitude Modulation: An audio signal such as a voice modulates a carrier wave. This creates a signal which, when transmitted, carries information. Intelligence is determined by varying the intensity of the radio wave.

Auto Alarm: Also called an automatic alarm. A device which monitors the distress frequencies and alerts personnel when traffic is received. Its fundamental purpose is to stand watch when the radio operator is not on duty.

Authorized Bandwidth: The maximum permitted band of frequencies as specified in the FCC authorization to be occupied by an emission. Includes a total of the frequency departure above and below the carrier frequency.

Authorized Frequency: The radio frequency assigned to a station by the FCC and specified in the station license or other instrument of authorization.

Authorized Power: Unless transmitting distress calls, the minimum amount of output power necessary to carry on the telecommunications for which the station is licensed. (See § 80.63(a)

Avionics: The electronics including the radio system aboard an aircraft which must be maintained by a General Radiotelephone Operator.

Azimuth: In marine radar, the bearing in degrees measured in a clockwise direction.

Bandwidth: The amount of frequency spectrum space taken up by 99% of a radiated signal.

Base Station: A fixed land station used to communicate with mobile radio stations installed in motor vehicles.

Break: Phrase spoken just before a brief pause in radiotelephone conversation to allow the other station to acknowledge your transmission.

Bridge-to-Bridge Station: A VHF radio station located on a ship's navigational bridge used for navigational purposes.

Calling Frequency: Agreed-upon frequency which stations use to initially call one another. Upon contact, both stations switch over to another "working" frequency so others may use the calling frequency.

Call Sign: A unique identifier issued to a radio station by the FCC as an aid to enforcement of radio regulations. A call sign identifies the country of origin and individual station.

Cargo Vessel: Any ship not licensed to carry more than 12 passengers.

Carrier: An alternating-current wave radiated from a transmitter of constant frequency with no modulation present. A carrier is a radio wave intended to "carry" the modulation or information.

Cathode-Ray Tube (CRT): An indicator device which consists of an electron gun which focuses an electron beam on a fluorescent screen. The screen is painted when the beam is diverted by four deflection plates or a magnetic field deflection system.

Clear (or Out): Word used in radiotelephony to indicate that a transmission has ended and no response is expected.

Clutter: Random echoes created by water waves, rain or snow that appear on a radarscope and which tend to obscure weak echoes from small objects.

Coast Station: Land-based radio station for the maritime services. Class-I stations provide long distance communications to ships at sea. Class-II coast stations provide regional service. A public coast station is open to public correspondence; a limited coast station may not transmit telecommunications for the public.

Coaxial Cable: A two-conductor transmission line consisting of a single center wire surrounded by a braided metal shield. It is called "hard line" when the outer conductor is solid metal. The insulator between the two conductors is called the dielectric. The outer non-conductive rubber sheath protects the cable from the weather.

Commercial Operator Licensing Examination Manager (COLEM): Entity approved by the FCC to provide commercial radio license exams.

Communications Act of 1934: The basic document for controlling telecommunications in the United States.

Communications Priority: The order of priority of communications in the mobile service is: (1) Distress messages, (2) Communications preceded by the urgency signal, (3) Communications preceded by the safety signal, (4) Radio direction finding communications, (5) Navigation and safe movement of aircraft, (6) Navigation, movement of ships, and weather observations, (7) Government radiotelegrams, (8) Government communications for which priority has been requested, (8) Communications relating to previously exchanged and; (9) All other communications.

Compass Bearing: The azimuth (compass direction) based on the north magnetic pole. True north is azimuth zero degrees, east is 90°, south 180° and west 270°.

Compulsory Ship: Ship required by international law to carry radio equipment, licensed radio operators and to keep logs.

COSPAS-SARSAT System: An international satellite-based search and rescue system established by Canada, France, the USA and Russia. Used to detect and locate land, sea and airborne radio beacons.

Crystal: Piezoelectric material (such as quartz) that mechanically vibrates at specific frequencies when an a-c current is applied. Crystals are very stable frequency oscillators.

Decibel: Abbreviated dB. A means of measuring levels of voltage, current or power. Assuming no transmission line losses, an antenna with a 3-dB gain radiates about twice the transmitter power. A 10-db gain represents a power gain of 10, 20-db: 100. dBd refers to antenna power gain over a half wave dipole; dBi is the power gain over an isotropic antenna.

Demodulation: The process of extracting information from a transmitted radio-frequency signal.

Deviation: In FM, the maximum amount that a frequency-modulated signal changes from the center frequency.

Digital Selective Calling (DSC): An automatic calling system which allows a specific station to be contacted. An "all ships" call can also be made for distress alerting and navigation safety communications.

Distress: Requiring immediate assistance. Distress traffic in radio communications receives the highest priority because distress calls indicate imminent disaster. MAYDAY is the radiotelephone distress signal; SOS in radiotelegraphy.

Distress Frequency: An internationally recognized frequency set aside for distress traffic such as 2182 kHz (single sideband), 156.8

MHz (FM) and 500 kHz (telegraph.) 121.5 MHz (AM double sideband-full carrier) is the aircraft distress channel.

Doppler Effect: In radar, the fact that a return echo changes in frequency when it encounters a moving object. The radar echo is shifted higher or lower as the target moves closer or further away from the radar installation. The magnitude of the shift determines the speed of the remote object. Stationary objects do not show up on Doppler radar.

Downlink: A one-way wide angle radio beam from a communications satellite in earth orbit to a station on the surface of the earth. See uplink, transponder

Dummy Antenna: A artificial antenna device used to prevent an antenna from radiating an interfering signal into space. Its resistance is similar to the impedance of an actual antenna.

Duplex: Two-way communications with both stations transmitting and receiving on different frequencies.

Duplexer: In radar, a high-speed switching device that permits an antenna to be used for both transmitting pulses and receiving return echoes.

Earth Station: A station in the earth-space service located on the surface of the earth or on a ship or aircraft.

Echo: In radar, a return radio-frequency pulse reflected from a remote object which is detected a few milliseconds or microseconds after transmission. The round trip time can accurately be converted to distance.

Effective Radiated Power: The amount of radiated power from a transmitter taking into consideration transmission line losses and the gain of the antenna over a half-wave dipole. The actual power that is radiated into space.

Emergency Locator Transmitter (ELT): An airborne distress alerting beacon operating on 121.5 MHz or 243 MHz that is detected by the COPAS-SARSAT polar-orbiting satellites. The ELT is the aviation counterpart of the maritime EPIRB. It is battery operated and transmits automatically when activated.

Emergency Position Indicating Radio Beacon (EPIRB): A battery-operated radio transmitter in the maritime mobile service that activates upon the sinking of a ship to facilitate search and rescue operations.

Emission Designator: A series of three alphanumeric characters used to identify radio signal properties. For example: A1A, manual radiotelegraphy. (A=Double-sideband ampli-tude modulation, full carrier; 1=One chan-nel of digital modulation; A=Morse code for manual reception.)

Enhanced Group Calling (EGC): A feature of the INMARSAT SafetyNET System that permits the addressing of messages to a group (or all vessels) in specific geographical areas.

Facsimile: The transmission and reception of fixed images by converting scanned lines to digital signals. The lines are redrawn on paper at the receiving site. An important maritime use is the transmission of weather maps for satellites.

FCC – Federal Communications Commission: The official telecommunications agency in the United States. Among its duties is the allo-cation and regulation of radio frequency assignments within a framework of interna-tional agreements.

Fiber Optics: The conveying of information by transmitting light waves down a thin thread of glass or plastic. A signal modulates a light emitting diode (LED) which is fed into an optical fiber. Fiber optic cables are inexpen-sive, light in weight, immune to electromag-netic interference and can carry more signals than coaxial cables.

Field Strength: An amount of received signal power intensity at a given distance mea-sured in volts per meter.

FM – Frequency Modulation: The process of varying a radio signal to convey intelligence by changing the transmitting frequency.

Frequency Allocation: A radio frequency or band of frequencies internationally or na-tionally assigned by an authorized body.

Frequency Deviation: VHF-FM transmitters in the maritime service are determined to be operating properly (100% modulation) when the maximum amount by which the carrier frequency changes either side of center frequency is plus-or-minus ($\pm$) 5 kHz.

Frequency Shift Keying (FSK): A digital method of transmission commonly used in radioteletype. Unlike on/off Morse keying, an FSK carrier is always present but the frequency shifts when the key is down. The mark and space frequencies are relatively close, between 60 and 850 Hz.

Full Duplex: Simultaneous two-way communi-cation in both directions on separate frequencies. Normal two-way conversation, similar to the telephone, is possible.

Gain: The difference between the input and output current, voltage or power. In antennas, the power difference between a reference antenna and its effective radiated power (ERP). Gain is usually expressed in decibels (dB).

General Radiotelephone Operator License (GROL): License issued by the FCC to individuals qualified to service, maintain, repair and operate radiotelephony communications equipment.

Geostationary Satellite: An orbiting satellite positioned 22,285 miles above the equator has exactly a 24-hour orbital period. Since this is the same as the rotation of the earth, the satellite appears to hang motionless in space.

GHz – Gigahertz: term for one billion cycles per second. Also 1,000 megahertz (MHz).

Global Maritime Distress and Safety System (GMDSS): An automated ship-to-shore dis-tress alerting system that uses satellites and advanced terrestrial communications sys-tems. It is coordinated throughout the world by the IMO, the International Maritime Organization. It picks up radio distress messages and relays them to the proper authorities.

Global Positioning System (GPS): Also called Navstar, the GPS uses multiple satellites to provide worldwide positional fix capability. This is accomplished by measuring the propagation time of satellite signals at the GPS receiver.

Ground: A connection with the earth to estab-lish ground potential. A common connection in an electrical or electronic circuit. The area directly below an antenna. With Marconi antennas, the ground becomes one-half of the antenna.

Great Lakes Radio Agreement: A rule applying to all ships in the Great Lakes region. Tech-nical requirements are stated for radio equipment and operators.

Half Duplex: Two-way communications over two separate channels or frequencies but not at the same time.

Harmful Interference: Any emission or radia-tion which interrupts or degrades a radio communications service operating in accor-dance with the rules. Operators must never deliberately interfere with any radio signal.

Harmonics: Spurious signals that show up at integer multiples of the main frequency.

Hertz: A measure of frequency equal to one cycle per second identified as hertz with lower case h.

Hertz Antenna: A one-half wave dipole antenna that is usually fed at its center and horizontally polarized.

HF – High Frequency: The radio frequency band that occupies 3 MHz to 30 MHz.

ILS – Instrument Landing System: An electronic aircraft guidance system using a radio beam to direct a pilot along a glide path.

INMARSAT: A four geostationary satellite network operated by the International Mobile Satellite Organization. It provides telex, telephone, data, fax and SafetyNET (maritime safety information, weather and navigational warning) service for ships at sea. INMARSAT is responsible for the space segment of GMDSS.

Interference: The presence of unwanted atmospheric noise or man-made signals that obstructs or inhibits the reception of radio communications.

International Fixed Public Radio Service: A point-to-point radio communications service open to public correspondence.

International Maritime Organization (IMO): A United Nations agency headquartered in London specializing in safety of shipping and preventing ships from polluting the seas.

International Phonetic Alphabet: Worldwide method of substituting words for individual letters to increase understanding.

International Radiotelegraph Alarm Signal: A signal consisting of a series of twelve dashes sent in one minute. This signal activates automatic devices to inform the operator that traffic is being received on a distress frequency.

International Radiotelephone Alarm Signal: A signal consisting of two sinusoidal audio tones transmitted alternately. This signal activates automatic devices to inform the operator that traffic is being received on a distress frequency.

International Telecommunication Union – ITU: The worldwide governing body controlling wire and radio telecommunications.

Ionosphere: An electrically charged frequency sensitive portion of the upper atmosphere that has the ability to reflect radio waves back to earth.

Isotropic Antenna: A theoretical antenna in free space that radiates an RF signal equally well in all directions. Usually used as a reference point in the measurement of antennas. The gain of a half-wave antenna over an isotropic radiator is 2.15 dB.

Kilohertz (kHz): One thousand cycles per second.

Klystron: A vacuum tube used to generate and amplify microwave alternating current signals. The most common are multicavity (amplifiers) and reflex (oscillators).

Land-Mobile Service: A mobile communications service between land-based movable and permanently located base stations - or between land-based movable stations.

LF – Low Frequency: This band occupies 30 kHz to 300 kHz.

Licensee: The entity to which a radio station is licensed by the Federal Communications Commission.

License Term: Ship and aircraft stations are licensed for five years. Most commercial radio operators are licensed for five years except Restricted Permits and General Radiotelephone Operator Licenses are issued for the life of the operator.

Line of Sight: The unobstructed distance to the horizon. Communication above the VHF level is usually by direct "line of sight" radio-wave propagation. Range depends upon antenna height and terrain.

Log: Diary of radio communications kept by the station operator. It must contain frequencies used, any technical problems encountered, what action has been taken to correct technical problems, and if any distress traffic has been intercepted. The log is the written report of the station's performance and activities.

Loran-C: Acronym for LOng RAnge Navigation. Loran-C is a system of radio transmitters broadcasting low-frequency pulses to allow ships to determine their positions. The difference in time it takes for pulses from different transmitters to reach a ship allows Loran-C to determine the ship's location very accurately.

Magnetron: An oscillator tube that produces high-power microwave pulses for radar.

Marconi Antenna: A one-quarter wavelength antenna fed at one end and operated against a good RF grounding system.

Marine Radio Operator Permit (MROP): A permit earned by passing a 24 question examination on regulations, operating techniques and practices in the maritime services. The MROP is granted by passing Element 1 – Radio Law. MROP holders may not make internal adjustments to radio transmitting equipment.

Maritime: Relating to navigation or commerce on the seas.

Maritime Mobile Radio Service: A two-way mobile communications service between ships, or ships and coast stations.

Maritime Mobile Repeater Station: A land station at a fixed location established for the automatic retransmission of signals to extend the range of communication of ship and coast stations.

Maritime Safety Information (MSI): Important navigational and meteorological information.

Master: A person licensed to command a merchant ship.

Mayday: A word spoken three times spoken during radiotelephone distress messages.

MF – Medium Frequency: This radio frequency band occupies 300 kHz to 3,000kHz (or 0.3 MHz to 3 MHz.)

Megahertz (MHz): One million cycles per second. Also 1,000 kilohertz (kHz).

Microwaves: Radio waves generally beginning at 1,000 MHz or 1 GHz. Most microwave activity is in the 1 to 50 GHz range.

Mission Control Center (MCC): Land-based rescue authorities who collect, store and exchange distress and alerting information.

Modem: A modulator/demodulator that converts digital signals to audio tones for transmission over wirelines or via radio wave. The process is reversed at the receiver.

Modulation: The process of modifying a radio wave so that information may be transmitted. The desired signal is superimposed onto a higher "carrier" radio frequency. A radio wave has three basic properties that can be varied: amplitude, frequency and phase.

Multihop – Multipath: Radio waves can bounce back and forth between the earth and the ionosphere or follow more than one route to a receiving point.

Multiplexing: The combining of two or more signals and transmitting them through a single cable or on a single radio-frequency channel. The signals are separated at the receiving end.

Nautical Mile: The fundamental unit of distance used in navigation. One nautical mile = 1.15 statute miles (or 6,080 feet). One knot is one nautical mile per hour.

Navigational Communications: Safety communications exchanged between ships and/or coast stations concerning the maneuvering of vessels.

NAVTEX: Navigational Telex, an international, automated system for instantly distributing maritime navigational warnings, weather forecasts, search and rescue notices and similar information to ships on 518 kHz worldwide.

Omega: A radio navigation system relying on eight land transmitters throughout the world. Ships carry special receivers to listen to the transmitters and determine from the information carried exactly where the ship is at all times. Omega relies on phase differences between received signals.

Omni-Directional: Performing equally well in all directions.

Over: Word spoken in radiotelephone conversation to indicate that it is the other station's turn to speak.

Over-the-Horizon (OTH) Radar: A method of tracking remote objects by refracting high-frequency (HF) radio signals off of the ionosphere down towards the earth and analyzing the return echoes.

Overmodulation: Driving a transmitter over its designed parameters causes adjacent frequency interference. Can be caused by shouting into a microphone. Peak modulation should not exceed 100%.

PAN: The internationally recognized radiotelephone urgency signal. The words "PAN PAN" are spoken three times in succession to indicate that an urgent message will follow.

Parabolic Antenna: A receiving or transmitting reflector (dish) antenna that focuses the RF energy into a narrow beam to insure maximum gain.

Parasitic Oscillations: Unwanted spurious signals at frequencies removed from the operating frequency. Parasitics can cause distortion, loss of power and possible interference to others. They are usually eliminated by shielding, RF chokes and neutralization.

Part 13: The rules issued by the FCC governing commercial radio operators.

Part 23: The group of rules issued by the FCC governing stations in the international fixed public radio communication services.

Part 73: The group of rules issued by the FCC governing radio broadcasting services.

Part 80: The group of rules issued by the FCC governing stations in the maritime services.

Part 87: The group of rules issued by the FCC governing stations in the aviation services.

Passenger Ship: Any ship carrying more than twelve paying passengers. (Six passengers when used in reference to the Great Lakes Radio Agreement.)

Peak Envelope Power (PEP): Method of measuring the output power of a single sideband signal since no carrier is transmitted.

Personal Locator Beacon (PLB): An land-based satellite beacon that is detected by the CO-PAS-SARSAT polar-orbiting satellites.

Phonetic Alphabet: A system of substituting easily understood words for corresponding letters.

Plan Position Indicator (PPI): A cathode ray tube display map used to highlight target objects illuminated by a scanning radar antenna.

Power-Driven Vessel: Any ship propelled by machinery.

Priority of Communications: Maritime service rules require that priority be given to: (1) distress calls concerning grave and imminent danger, (2) urgent messages concerning safety of life and property, and (3) traffic concerning navigational safety and weather warnings. (See §80.91)

Propagation, Radio Wave: The method of radio wave travel which may be along the earth's surface, directly through space or reflected from the upper atmosphere.

Public Correspondence: Any third-party telecommunication (message, image or voice traffic) that must be accepted for transmission.

Pulse: In radar, a repetitious burst of radio-frequency energy of very short duration which is directed at a remote object.

Pulse Repetition Rate (PPR): Also, the same as PRF, the pulse repetition frequency. The number of radar pulses transmitted in a second. PPRs usually vary between 800 and 2,000 per second.

Pulse Repetition Time (PRT): The time between the leading edge of one transmitted pulse and the leading edge of the next pulse. The period of the pulse repetition rate.

Radar: Acronym for RAdio Detection And Ranging. A method of tracking objects by analyzing reflected microwave radio signals or echoes. Doppler radar is used to measure speed.

Radar Mile: The time required for a radio frequency signal to travel from the transmitter to a target one nautical mile away and back. A radar mile is considered to be 12.3 microseconds.

Radio Direction Finding (RDF): The art of determining the direction of a radio signal using a radio receiver with a signal strength indicator and a rotatable directional antenna. The location of a radio signal requires bearings to be taken by two radio stations.

Radio-Frequency Wave: An electromagnetic wave that travels at the speed of light; 186,282 statute miles or 162,000 nautical miles per second. Its frequency ranges from 10 kHz to 3000 GHz.

Radio Operator: The FCC licensed operator in charge of the station who is responsible for its proper use and operation.

Radioprinter: A means of exchanging alphanumeric codes by direct printing.

Radio Services: Radio operations are classified into services according to the nature and purpose of the transmission.

Radiotelephony: Method of transmitting voice over radio waves.

Relative Bearing: The direction when the reference point is the ship's heading. The number of degrees port or starboard of the bow.

Repeater: A receiver/transmitter installation that receives signals on one channel and instantly retransmits them on another frequency usually at higher power from tall antennas. A duplexer keeps the transmitter from overloading the receiver. The advantage of a repeater is increased radio range. A repeater on a satellite is called a transponder.

Rescue Coordination Center (RCC): In GMDSS, the unit responsible for organizing and conducting search and rescue operations within a specific geographical area.

Resolution: In radar, the capability to separate two objects that are close to each other in range or bearing and to show them as two distinct echoes on the radarscope.

Restricted Radiotelephone Operator Permit: Permit allowing certain radio privileges in the aviation, broadcast and maritime services. No examination is required.

Roger: A word in radiotelephone conversation to indicate that you have received and understood all of the other station's transmission.

Safety Communications: A radio transmission indicating that a station is about to transmit an important navigation or weather warning.

Safety Convention: International agreement which spells out certain safety requirements for on-board radio equipment and operators.

Sea Area: Radio equipment on ships is considered in terms of its range and the areas in which the vessel will travel. There are four watchkeeping areas. Sea area A1 is within VHF communications range (20-30 miles); A2 within the coverage of a shore-based MF coast station (about 100 miles); A3 within coverage of an INMARSAT geostationary satellite and A4 is the remaining polar areas of the world.

Search and Rescue Radar Transponders (SARTs): These are portable devices which are taken into a survival craft when abandoning ship. When switched on, they transmit a series of dots on a rescuing ship's 9-GHz radar display. SARTs can also notify persons in distress that a rescue ship or aircraft is within range.

Security: The word "SECURITY" is spoken three times prior to the transmission of a safety message.

Secrecy of Communications: Other than broadcasts to the general public, persons may not divulge the content of telecommunications nor use the information obtained to benefit anyone other than the intended recipient.

Selective Calling: A coded transmission to a particular radio station. Other stations do not hear it.

Selectivity: The ability of a radio receiver to separate the desired signal from other signals.

Separation: A method of minimizing mutual interference by spacing stations using the same frequency at required distance intervals.

Sensitivity: The ability of a radio receiver to respond to weak input signals.

Ship Earth Station: A mobile satellite station located on board a vessel.

Silent Period: The three-minute duration of time during a continuous watch on a distress frequency when a maritime radio operator must not transmit.

Simplex: Two-way communications with both stations transmitting and receiving on the same frequency. Only one station may transmit at a time.

Single Sideband (SSB): Method of transmitting radiotelephony where one sideband is filtered out and the carrier suppressed or reduced. SSB is more efficient than double sideband signals since it takes up less radio spectrum. Emission: J3E

SITOR: Acronym for **SI**plex **T**eleprinter **O**n **R**adio. Error-free teleprinting of news and weather over the high-frequency radio bands.

SOS: The radiotelegraphy distress signal sent three times followed by DE ("this is") and the call sign of the station in danger.

Standing-Wave Ratio (SWR): The ratio of the current values at the maximum and minimum points on a transmission line. There are no standing waves when the load impedance matches the line impedance. SWR of 3:1 and less is generally considered satisfactory. Standing waves are measured with an SWR meter or reflectometer.

Station Authorization: Any construction permit, license or special temporary authorization issued by the FCC for activating a radio station.

Statute Mile: 5,280 feet. Unit of distance commonly used on land in the United States. One statute mile equals 0.8684 nautical miles.

Sunspot Cycle: The height, thickness and intensity of the ionosphere from which radio waves are reflected vary according to a cycle of approximately 11 years.

Survival Craft Station: A mobile station on a lifeboat, life raft or other survival equipment aboard a ship or aircraft intended for emergency purposes

Sweep: In radar, the line across the cathode-ray tube that is synchronized with the rotation of the scanning antenna.

Target: In radar, a term frequently used to denote a boat, buoy, island or other object that is radar conspicuous and produces an echo on the radarscope.

Telecommunication: The transfer of sound, images or other intelligence by electromagnetic means.

Telegraphy: The process of sending and receiving information through the use of Morse code.

Telephony: The process of exchanging information through the use of speech transmissions.

Teleprinter: A mechanical typewriter–like device that prints text sent over wire or radio circuits. In a radioteleprinter, a modem converts the audio output of a receiver into electrical impulses to drive the individual keys. See modem.

Traffic: Radio messages exchanged between stations.

Transducer: A device that converts one form of energy into another. In depth sounder applications, electronic impulses are converted into sound impulses and vice versa.

Translator, Broadcast: A relay station used to improve the reception of weak television and FM broadcast signals in remote locations. Translator equipment rebroadcast the input signal on another frequency or channel.

Transmission Line: The conduit by which radio frequency energy is transferred from the transmitter to the antenna.

Transponder: A device in an orbiting satellite that receives uplink (transmitted) signals from earth and downlinks (retransmits) them back to earth. A transponder is a wide coverage space repeater. See uplink, downlink.

Traveling-Wave Tube (TWT): A vacuum tube used to amplify UHF and microwave frequencies.

Type Acceptance: Radio equipment that has met FCC specifications. All transmitters in the Fixed, Aviation and Maritime Services must be "type accepted." Type acceptance is based on data submitted by the manufacturer. "Type Approval" is granted after tests are made by FCC technical personnel.

UHF – Ultra High Frequency: This radio frequency band occupies 300 MHz to 3,000 MHz (or 0.3 GHz to 3 GHz).

Universal Coordinated Time – (UTC): Sometimes called Greenwich Mean Time (GMT), the time appearing at the zero meridian near Greenwich, England. UTC is the standard for time throughout the world.

Universal Licensing System (ULS): New FCC licensing system using one integrated database to electronically issue and maintain radio operator licenses.

Uplink: The ground-based frequency on which an orbiting satellite receives its radio signals from earth. See downlink, transponder.

Upper Sideband: The information carrying portion of the signal just above the amplitude modulated (AM) carrier frequency which is reduced or eliminated before transmission.

Urgency Communication: Urgent message concerning the safety of a ship, aircraft, other vehicle or person. Urgency traffic has slightly lower priority than distress traffic.

Vertical Polarization: Standard method of orienting maritime antennas operating at frequencies above 30 MHz: perpendicular to the ground or water. Polarization is determined by the direction of the electric component of the electromagnetic field. Vertically oriented antennas produce vertically polarized waves.

Vessel Traffic Service (VTS): Traffic management service operated by the U.S. Coast Guard in certain water areas to prevent ship collisions, groundings and environmental harm.

VHF – Very High Frequency: This radio frequency band occupies 30 MHz to 300 MHz. (or 3,000 kHz to 30,000 kHz).

Violation Notices: Notification from the FCC of a rule infraction. A written response must be made within 10 days containing a full explanation and action taken to prevent reoccurrence.

VLF – Very Low Frequency: This band occupies 3 kHz to 30 kHz.

Voluntary Ship: A ship not required to carry radio equipment, licensed radio operators or keep logs. When radio equipment is carried, however, appropriate listening watches must be maintained on 2182 kHz.

WWV: The precise standard frequency and time transmissions of the National Bureau of Standards.

Watch: The act of keeping close observation on distress frequencies for any distress messages.

Waveguide: A low-loss circular or rectangular hollow pipe-like feedline used at UHF and microwave frequencies. The waves are carried inside the pipe.

Wilco: Phrase spoken in radiotelephony to acknowledge that a message has been received and that the receiving station will comply. "Wilco" is short for "will comply."

Working Frequency: A frequency establishing for conducting communications after first being established on a Calling Frequency.

Index